Geophysical Monograph Series

Including

IUGG Volumes
Maurice Ewing Volumes
Mineral Physics Volumes

Geophysical Monograph 148

Mid-Ocean Ridges: Hydrothermal Interactions Between the Lithosphere and Oceans

Christopher R. German
Jian Lin
Lindsay M. Parson
Editors

American Geophysical Union
Washington, DC

Published under the aegis of the AGU Books Board

Library of Congress Cataloging-in-Publication Data

Mid-ocean ridges : hydrothermal interactions between the lithosphere and oceans / Christopher R. German, Jian Lin, Lindsay M. Parson, editors.
      p. cm.- (Geophysical monograph ; 148)
Includes bibliographical references.
ISBN 0-87590-413-0 (alk. paper)
1. Hydrothermal circulation (Oceanography). 2. Mid-ocean ridges-Research. 3. Hydrothermal vents. 4. Sea-floor spreading. 5. Earth-Crust. I. German, Christopher R. II. Lin, Jian. III. Parson, Lindsay M.

GC171.M53 2004
55.1/36'2-dc22

                            2004062292

ISBN 0-87590-413-0
ISSN 0065-8448

Copyright 2004 by the American Geophysical Union
2000 Florida Avenue, N.W.
Washington, DC 20009

Cover: The black smoker chimney "Candelabra" located within the Logatchev hydrothermal field, Mid-Atlantic Ridge 15°N. Photo taken with the Breman Quest 4000 ROV during cruise M60/3 of the German Research vessel Meteor (Chief Scientist Thomas Kuhn). Copyright Universität Bremen 2004 (courtesy of Colin Devey, InterRidge).

CONTENTS

PREFACE

Mid-ocean ridges play an important role in the plate-tectonic cycle of our planet. Extending some 50–60,000 km across the ocean-floor, the global mid-ocean ridge system is the site of creation of the oceanic crust and lithosphere that covers more than two thirds of the Earth's exterior. Approximately 75% of Earth's total heat flux occurs through oceanic crust, much of it at mid-ocean ridges through complex processes associated with magma solidification, heat transfer, and cooling of young oceanic lithosphere. While the majority of this heat loss occurs through conduction, approximately one third of the total heat loss at mid-ocean ridges is influenced by a convective process: hydrothermal circulation.

Hydrothermal circulation facilitates the cycling of energy and mass between the solid Earth and the oceans, influencing the composition of the Earth's lithosphere, oceans and even, albeit largely indirectly, the atmosphere. Hydrothermal circulation arises when seawater percolates downward through fractured oceanic crust: the seawater is heated and undergoes chemical modification through reaction with the host rock, reaching maximum temperatures that can exceed 400°C. At these temperatures the fluids become extremely buoyant and rise rapidly back to the seafloor, where they are expelled into the overlying water column. Active sites of hydrothermal discharge can provide extreme ecological niches that can host unique chemosynthetic fauna, previously unknown to science.

The first identification of submarine hydrothermal venting and their accompanying chemosynthetically-based communities in the late 1970s remains one of the most exciting discoveries in modern science. A quarter of a century later, however, many of the processes that control magma supply and replenishment beneath the axial ridge-crest, the distributions of hydrothermal activity around the global ridge system, the timescales over which individual hydrothermal systems remain active or become renewed, and the detailed circulation pathways for hydrothermal fluids within young oceanic crust remain only incompletely understood.

Although such questions are complex, we are beginning to make important progress in two key fields of mid-ocean ridge study which, taken together, have allowed us to develop new understandings. A first key area concerns the structure and development of young ocean crust and the mechanisms by which heat is transferred from the Earth's interior to the

lithosphere. Geophysical techniques allow acoustic and geophysical mapping of the fine-scale tectonic and volcanic morphology of the ocean floor from both surface ships and deep-submergence vehicles. New seismic and electromagnetic technologies provide us with ever-improving images of the interior of the oceanic crust and upper mantle. In parallel with the above, direct observations can be used, on the modern seafloor, in materials recovered from ocean drilling and in ophiolites that are outcrops of fossilized mid-ocean ridge crest uplifted and preserved on land. Detailed petrological analyses allow researchers to derive information from the thermodynamic equilibrium between different mineral phases and reconstruct thermal cooling histories of crust and mantle rocks. Theoretical modeling and laboratory experiments further integrate different observations into coherent conceptual models of how melts are formed in the mantle and rise to the shallow depths at which they form oceanic crust and how the oceanic lithosphere is fractured and cracked, providing permeability conduits for hydrothermal circulation.

The second area of mid-ocean ridge research that allows us a better understanding of the coupling of the solid earth and oceans is submarine hydrothermal activity. Again, this area of research has undergone rapid progress in the past decade using a combination of techniques: remote detection using surface ships and deep-tow vehicles, direct submersible-based observations at the seafloor, instruments installed on the seafloor for substantial periods of time, numerical modeling and complementary experimentation in the laboratory. Systematic surveys, detecting and determining the magnitude of physical and chemical anomalies imparted to the water column above mid-ocean ridges, permit determination of the geographic locations of seafloor hydrothermal vent-sites co-registered with geological information about their volcanic/tectonic setting. Sampling and analysis of the chemically-enriched fluids discharging from these vent-sites at the seafloor can provide first interpretations of the detailed mineral-water interactions taking place at depth beneath the sea-bed at these sites and also, through time-series data and repeat sampling programmes, how those processes are progressing with time. By conducting complementary experimental procedures at high pressure and high temperature in the laboratory, we can test hypotheses developed from direct observations alone to refine our understanding of the processes active within the inaccessible seafloor. Numerical modeling has an important role to play here, providing valuable insight into the extent to which a limited number of direct "snap-shot" observations can be used to extrapolate to a longer-term understanding of how hydrothermal systems may evolve. Those models, in addition, can also help us predict what impact hydrothermal circulation may have, in return,

Mid-Ocean Ridges: Hydrothermal Interactions Between the Lithosphere and Oceans
Geophysical Monograph Series 148
Copyright 2004 by the American Geophysical Union
10.1029/148GM00

upon the thermal structure and, hence, geologic behaviour of its host oceanic crust.

The structure of the material within this volume reflects the two key areas in which most recent progress has been made: the 4-D structure (including temporal evolution) of mid-ocean ridges and submarine hydrothermal activity. To illustrate where our current understanding begins—and more importantly, ends—Chapter 1 investigates the extent to which coupling of hydrothermal activity and formation and development of young oceanic crust are currently understood. The key message here is not how much we know but how much there remains for us to know.

This is followed by a series of chapters (Chapters 2–6) that explain what we already know about mid-ocean ridge formation and thermal structure and how this information has been obtained—from geophysical remote sensing, from sea-floor imaging and mapping, from theoretical modeling, and from petrological sampling coupled with mineralogical and geochemical analysis. Chapter 7 describes an important laboratory study of thermal cracking of rocks and discusses its implications for rock fracture and thermal structure at mid-ocean ridges.

The second major section of the book (Chapters 8–13) details our current state-of-the-art concerning on-axis hydrothermal circulation along the global ridge-crest. Important chapters here include an understanding of the geologic controls of hydrothermal activity on a fast-spreading ridge and, a particularly new development, the temporal variability of hydrothermal activity at any given vent-site. A further contribution describes current knowledge of the distributions of hydrothermal activity along the global ridge-crest and how this may vary with spreading rate, an important, increasingly recognized theme, which is further elucidated in chapters on flow mechanisms within different hydrothermal systems and a specific consideration of the processes driving ultramafic-hosted hydrothermal systems. The latter are likely to be unique to slow- and ultraslow-spreading ridge crests.

The book builds directly on two earlier AGU volumes although we do not attempt to reconstruct the full breadth of scope of either of those volumes: Seafloor Hydrothermal Systems: Physical, Chemical, Biologic and Geological Interactions (Humpris, S.E., Zierenberg, R.A., Mullineaux, L.S. and R.E.Thomson, Editors, AGU, 1995) and Faulting and Magmatism at Mid-Ocean Ridges (Buck, W.R., Delaney, P.T., Karson, J.A. and Lagabrielle, Y., Editors, AGU, 1998). Rather, the work presented here pulls together just those most recent developments within each of two sub-fields of mid-ocean ridge research that are most relevant to understanding how magmatic heat supply, tectonic deformation and seafloor hydrothermal circulation might all interact, with an increasing emphasis on Earth's less well-understood slow spreading ridges.

The volume derives primarily from presentations and discussions at the first InterRidge Theoretical Institute that was held in Italy in September 2002. The majority of the chapters derive directly from a two-day short course held at the University of Pavia at the start of that meeting and establish the current international state-of-the-art. Additional chapters have been contributed by workshop participants based on discussions that ensued during the later workshop phase of the meeting, at Sestri Levante and provide examples of cutting-edge research in this active field that expand the breadth and diversity of the finished volume.

As editors, we are particularly grateful to InterRidge, the European Science Foundation, and the US NSF-sponsored RIDGE 2000 Program for their support of this endeavour and particularly to our co-organisers of the 1st InterRidge Theoretical Institute: Agnieszka Adamczewska and Kensaku Tamaki (Japan), Ricardo Tribuzzio (Italy), Mathilde Cannat (France) and Andy Fisher (USA). We are, of course, very grateful to all the authors for this volume, who worked diligently to meet our deadlines and did all required to bring this project to a successful outcome. We particularly appreciate the efficiency, understanding and expertise of the AGU staff who have worked so effectively with us to ensure the timely publication of this book. We are grateful, too, to our reviewers who devoted much time and effort to ensure the high quality of the final volume. Finally, we would like to pay particular tribute to Allan Graubard, our acquisitions editor, for his support and encouragement throughout the entire project.

Christopher R. German
Challenger Division for Seafloor Processes
Southampton Oceanography Centre, UK

Jian Lin
Department of Geology and Geophysics
Woods Hole Oceanographic Institution, USA

Lindsay M. Parson
Challenger Division for Seafloor Processes
Southampton Oceanography Centre, UK

The Thermal Structure of the Oceanic Crust, Ridge-Spreading and Hydrothermal Circulation: How Well Do We Understand Their Inter-Connections?

Christopher R. German

Southampton Oceanography Centre, Southampton, UK

Jian Lin

Woods Hole Oceanographic Institution, Woods Hole, Massachusetts, USA

Understanding the complex interplay between the geological processes active at mid-ocean ridges and the overlying ocean through submarine hydrothermal circulation remains a fundamental goal of international mid-ocean ridge research. Here we reflect on some aspects of the current state of the art (and limits thereto) in our understanding of the transfer of heat from the interior of the Earth to the ocean. Specifically, we focus upon the cooling of the upper oceanic crust and its possible relationship to heat transfer via high-temperature hydrothermal circulation close to the ridge axis. For fast- and intermediate-spreading ridges we propose a simple conceptual model in which ridge extension is achieved via episodic diking, with a repeat period at any given location of ca. 50 years, followed by heat removal in three stages: (a) near-instantaneous generation of "event plumes" (~5% of total heat available); (b) an "evolving" period (ca. 5 years) of relatively fast heat discharge (~20%); (c) a decadal "quiescent" period (~75%). On the slow-spreading Mid-Atlantic Ridge, our understanding of the mechanisms for the formation of long-lived, tectonically-hosted hydrothermal vent fields such as TAG and especially Rainbow are more problematic. We argue that both hydrothermal cooling which extends into the lower crust and heat release from serpentinisation could contribute to the required heat budget. Slow-spreading ridges exhibit much greater irregularity and episodic focussing of heat sources in space and time. This focussing may sustain the high heat flow required at the Rainbow and TAG sites, although the detailed processes are still poorly understood.

1. INTRODUCTION

Mid-ocean ridges play an important role in the plate-tectonic cycle of our planet. Extending some 50–60,000 km across the

Mid-Ocean Ridges: Hydrothermal Interactions Between the Lithosphere and Oceans
Geophysical Monograph Series 148
Copyright 2004 by the American Geophysical Union
10.1029/148GM01

ocean-floor, the global mid-ocean ridge system is the site of the creation of oceanic crust and lithosphere at a rate of ~3.3 km^2 yr^{-1} [*Parsons*, 1981; *White et al.*, 1992]. Approximately 75% of the total heat flux from the interior of the Earth (ca. 32 of 43 TW) occurs through this oceanic crust, which covers some two-thirds of the surface of the planet. Much of this heat is released along mid-ocean ridges through complex processes of magma solidification and rapid cooling of the young oceanic lithosphere. While the majority of this heat loss occurs through conduction, approximately one third of the total heat loss at mid-ocean ridges is effected by a convective process termed hydrothermal circulation [e.g., *Stein and Stein*, 1994].

Hydrothermal circulation plays a significant role in the cycling of energy and mass between the solid Earth and the oceans (see syntheses by *Humphris et al.* [1995] and *German and Von Damm* [2003] for detailed reference-lists). This circulation, which is driven by the heat loss at mid-ocean ridges, affects the composition of the Earth's oceans and, indeed, atmosphere. Hydrothermal circulation occurs when seawater percolates downward through fractured oceanic crust along the volcanic mid-ocean ridge system: the seawater is first heated and then undergoes chemical modification through reaction with the host rock as it continues downward, reaching maximum temperatures that can exceed 400°C . At these temperatures the fluids become extremely buoyant and rise rapidly back to the seafloor, where they are expelled into the overlying water column. Active sites of hydrothermal discharge provide an extreme ecological niche that is home to a variety of quite unique chemosynthetic fauna, previously unknown to science [*Van Dover*, 2000]. The first identification of submarine hydrothermal venting and their accompanying chemosynthetically-based communities in the late 1970s remains one of the most exciting discoveries in modern science. A quarter of a century later, however, many of the processes that control the distributions of hydrothermal activity around the global ridge crest, the detailed circulation path for hydrothermal fluids within young oceanic crust and how such processes vary with time, all remain only incompletely understood. The purpose of this volume, which arose from the first InterRidge Theoretical Institute (IRTI) held in Italy in September 2002, is to bring together expertise from two discrete communities, studying mid-ocean ridge geological processes and submarine hydrothermal systems, respectively; these communities had been progressing their research largely independently over the preceding decade.

The first of these groups had been tasked through an Inter-Ridge Working Group to investigate the 4-D (space and time) Architecture of Mid-Ocean Ridges. This research community focused on fundamental geological processes of the building of oceanic crust and lithosphere at ridges through a combination of acoustic and geophysical mapping of the ocean floor, seismic/electro-magnetic imaging, near- and on-bottom geological observations, rock sampling and geochemical/petrological analyses, ocean drilling, laboratory experiments, and theoretical and numerical modelling and syntheses. The highlights of the achievements of this community in the last decade include the direct measurements of the depth and dimension of magma lenses beneath fast-spreading, intermediate-rate, and hotspot-influenced ridges [*Sinha et al.*, 1998; *Detrick et al.*, 2002a; *Sinha and Evans*, this vol-

ume]; complete mapping of volcanic and tectonic fabrics of selected ridge spreading segments at all spreading rates, including the ultra-slow spreading Gakkel Ridge under the Arctic ocean [*Michael et al.*, 2003] and the Southwest Indian Ridge [*Sauter et al.*, 2002; *Lin et al.*, 2002; *Dick et al.*, 2003]; discovery of previously unrecognized forms of seafloor morphological features that indicate inherent and complex magmatic/tectonic cycles of seafloor spreading, including amagmatic ridges in ultra-slow spreading ridges and long-lived detachment faults in slow and other magma-starved ridges [*Tucholke and Lin*, 1994; *Cann et al.*, 1997; *Tucholke et al.*, 1998; *Searle et al.*, 2003]; and direct sampling of the mantle rocks along a segment of the Mid-Atlantic Ridge near the Fifteen–Twenty fracture zone [*Kelemen et al.*, 2003]. Land-based laboratory and theoretical studies have likewise produced valuable models and testable hypotheses on the focusing of 3-D mantle and melt flows beneath mid-ocean ridges [*Parmentier and Phipps Morgan*, 1990; *Lin and Phipps Morgan*, 1992; *Magde et al.*, 1997]; the thermal state of the oceanic crust as a function of magma supply and hydrothermal circulation at ridge crests [*Phipps Morgan and Chen*, 1993; *Shaw and Lin*, 1996; *Chen and Lin*, 2004, *Chen*, this volume]; integrated petrological models on the compositional variations among mid-ocean ridge crustal and mantle rocks [*Langmuir et al.* 1992; *Michael and Cornell*, 1998]; predictions of the varying styles of seafloor faulting and earthquakes and their dependence on the degree of tectonic extension and rock types [*Searle and Escartin*, this volume]; the causes of the dramatic differences in ridge crest topography between fast- and slow-spreading ridges; and the potentially important role of water and/or serpentinites in controlling the rheology and, thus, the tectonic deformation of the oceanic lithosphere [*Hirth et al.*, 1998; *Searle and Escartin*, this volume].

The second community involved in the development of both the IRTI workshop and the ideas presented in this volume were co-ordinated through the InterRidge Working Group studying Global Distributions of Seafloor Hydrothermal Venting. Previously, a state-of-the-art volume detailing many of the interacting subdisciplines of modern hydrothermal research had been compiled by *Humphris et al.* [1995] based on a US Ridge Theoretical Institute addressing physical, chemical, biological and geological interactions in seafloor hydrothermal venting. More recently, *German and Von Damm* [2003] have synthesised core information available at the time of publication of that earlier volume, together with new information that has only been obtained during the intervening decade. There have been two key discoveries in this time that have quite revolutionised our assumptions about seafloor venting to the oceans. The first of these has been the recognition that temporal evolution of any given seafloor hydrothermal system can vary profoundly [e.g., *Von Damm*, this volume] while the second is that hydrothermal activity may be far more wide-spread than had previously been recognised occurring along ridges of all spreading rates and, hence, being present in all ocean basins [e.g., *Baker and German*, this volume]. For the decade following discovery of high-temperature venting, what was remarkable was the constancy in temperature and com-position of erupting vent fluids even though individual sites might exhibit quite distinct compositions. The best example of this constancy was the first high-temperature system dis-covered at 21°N on the East Pacific Rise, which appears to have remained unperturbed for ~20 years [*Von Damm et al.*, 2002]. Another example is the TAG hydrothermal field at the Mid-Atlantic Ridge which has thus-far maintained a near-constant end-member fluid composition from 1986 until 2003, even after being subject to ODP drilling [*Edmonds et al.*, 1996]. What has become apparent in the past decade, however, is that such constancy may not be the norm. Both the tem-perature and composition of fluids exiting any given vent site can change significantly with time, apparently in direct response to volcanic extrusion and/or dike intrusion beneath the seafloor [e.g., *Butterfield and Massoth*, 1994; *Butterfield et al.*, 1997; *Von Damm et al.*, 1995, 1997]. Perhaps the best-studied such site, to date, is that at 9°50'N on the East Pacific Rise, which last underwent a volcanically eruptive episode in April 1991 [*Haymon et al.*, 1993]. A key question of active current debate is: what has controlled the continuing evolution of fluid temperatures and compositions at this site over the subsequent decade?

A major discovery in the past ten years of hydrothermal research has been the recognition that hydrothermal activity can occur in reasonable abundance along slow-spreading ridges [e.g., *German et al.*, 1996b] as well as along faster ridges that are supplied by significantly greater magmatic heat fluxes. Previously, *Baker et al.* [1996] had predicted that the abundance of venting along any section of ridge crest should scale directly with the available magmatic heat flux and, hence, with ridge spreading rate. The discovery of hydrothermal activity along some of the world's slowest spread-ing ridges – notably the SW Indian Ridge [*German et al.*, 1998; *Bach et al.* 2002] and the Gakkel Ridge in the Arctic Ocean [*Edmonds et al.*, 2003] – appears to contradict, or at least modify, this hypothesis. Where does this "additional" hydrothermal flux originate? Recent discoveries on the Mid-Atlantic Ridge have revealed a new class of tectonically-hosted vent-sites including the low-temperature Saldanha and Lost City sites [*Barriga et al.*, 1998; *Kelley et al.*, 2001] and the high-temperature Logatchev and Rainbow hydrothermal fields [e.g., *Charlou et al.*, 2002]. The latter, in particular, appears to impart exceptionally high fluxes of heat and chem-icals into the surrounding ocean [*Thurnherr & Richards*, 2001; *German et al.*, submitted].

Considering the above, the timeliness of this volume becomes self-evident. While the important role of hydrother-mal circulation in regulating ocean (hence, atmospheric) com-positions may be well established, what remains unclear is how that hydrothermal circulation is regulated itself. This is not something that can become understood just by studying hydrothermal flow alone. Instead, an integrated approach is required in which the transfer of heat throughout the ocean-ridge system is taken as the key master variable that drives all other aspects of ridge and hydrothermal interactions dur-ing its transfer from the interior of the Earth. That is the over-arching theme of the chapters that follow. In the remainder of this chapter, however, we use some key case studies to review our current understanding of the processes that sustain these hydrothermal systems and to highlight key problems where our level of understanding can be described as incom-plete, at best.

2. CASE STUDY I: THE FAST-SPREADING NORTHERN EAST PACIFIC RISE AND THE INTERMEDIATE-RATE JUAN DE FUCA RIDGE.

2.1. The East Pacific Rise at 9°–10°N

The East Pacific Rise (EPR) between 9° and 10°N has become one of the most intensively studied sections of ridge crest anywhere on Earth due primarily to the attention received during the US RIDGE programme over the past decade or more – a commitment renewed with the selection of this area as an integrated study site that will continue to be studied in particular detail during the new RIDGE 2000 initiative (http://r2k.bio.psu.edu). Of interest to the discussion we pres-ent below is that this region also lies within the northern most limits of a NOAA Acoustic Hydrophone Array (AHA), which continuously monitors seismic activity along the entire ca. 2,000 km of the EPR between 10°S and 10°N [*Fox et al.*, 2001; http://www.pmel.noaa.gov/vents/acoustics.html].

The ridge at this location has a full spreading rate of 110 mm/yr [*Carbotte and Macdonald*, 1992] and is underlain by a seismic low-velocity zone, interpreted to represent a melt lens at a depth of ca. 1.5 km below the seafloor [*Detrick et al.*, 1987]. The area attracted particular attention from hydrother-mal researchers in the early 1990's after it was recognised that an episode of dike emplacement and volcanic extrusion had occurred toward the northern end of the previously sur-veyed region at 9°09'–9°54'N in April 1991 [*Haymon et al.*, 1991, 1993]. Time series studies of vent-fluid compositions at this site (near 9°50'N) have subsequently continued to show evolving chemical compositions, in direct response to this volcanic episode, over more than a decade [e.g., *Von Damm et al.*, 1995, 1997; *Von Damm*, this volume].

2.2. Magmatic Heat Release Is Episodic

When considering the steady-state magmatic heat flux available to this section of ridge crest, one approach is to take that described by *Sinha and Evans* [this volume], which predicts a total magmatic thermal flux of 5.0×10^{10} Watts along the 600 km of global ridge crest that are spreading at a full rate of ~110 mm/yr. This yields a unit heat flux at the EPR 9–10°N of 8.3×10^7 W/km. However, the value calculated using that approach represents the total heat emplaced magmatically within the full 6 km thickness of fresh-formed oceanic crust. If we only consider the heat emplaced by diking above an EPR magma lens (at a depth of 1.5 km), the heat flux delivered immediately below the seafloor and right at the ridge axis reduces to ca. 25% of the total predicted by *Sinha and Evans* [this volume], i.e., ca. 2.1×10^7 W/km.

Let us now consider a situation in which all of this upper oceanic crust heat flux (conveniently expressed as 21 MW/km or 210 MW/10km) was dissipated by high-temperature hydrothermal circulation that is focussed close to the ridge axis. Reported heat fluxes from individual high-temperature focussed-flow vent-sites typically fall in the range 10–100MW [e.g., *Speer and Rona*, 1989; *Bemis et al.*, 1993; *Lupton*, 1995; *Rudnicki et al.*, 1994; *German et al.*, 1996a; *Rudnicki & German*, 2002]. While integrated heat-fluxes including diffuse flow may often exceed this value [e.g., *Lowell & Germanovich*, this volume] our focus, initially, is solely upon high-temperature "black smoker" flow. Here, therefore, we assume that a representative heat-flux from any given "black smoker" hydrothermal field is of the order 100 MW. In that case, the steady-state heat flux associated with dike emplacement would appear sufficient to sustain one such hydrothermal field approximately every 4–5 km along the ridge axis. While identification of individual vent sites is not straightforward at areas recently perturbed by volcanic eruptive processes [e.g., *Von Damm*, this volume], such a high frequency of venting appears greater than what is observed along more "quiescent" sections of the EPR where a mean spacing of ~8–15 km between adjacent vent sites appears to be closer to the "norm" [e.g., *Fornari et al.*, this volume]. Thus, what is observed at nominal "steady-state" heat flux on the northern EPR appears to be a factor of ~2–3 lower than the above prediction. It would appear, therefore, that the "quiescent" mode of black smoker hydrothermal flow—as perhaps best exemplified by that witnessed at the EPR 21°N vent sites throughout the past 23 years—may only account for some ca. 33–50% of the steady-state heat flux available to support hydrothermal circulation through the upper oceanic crust, close to the axis of fast-spreading ridges.

What, then, is the fate of the "missing" hydrothermal heat flux not accounted for in the above calculations? One possibility is that it is manifest, at steady state, in the form of diffuse hydrothermal flow that is not incorporated within the entrainment process close to black smoker hydrothermal plumes and, hence, remains unsampled by "plume-integration" heat-flux studies [e.g., *Converse et al.*, 1984]. What also merits consideration, however, is whether there is a significant component of hydrothermal flux that occurs in close association with periods of dike intrusion and/or eruptive activity at the ridge axis that is not well represented – indeed, perhaps completely overlooked – in estimates based on steady-state "quiescent" stages of hydrothermal discharge. A major barrier to evaluating the possible importance of this issue, to date, has been our lack of understanding of the episodicity with which dike emplacement and/or volcanic eruption should be expected to recur at any given location along the global ridge crest. While the time-averaged spreading rates of all mid-ocean ridges are reasonably well constrained [e.g., *DeMets et al.*, 1994], what remains elusive is any clear understanding of the size and frequency of the "quantum events" by which mid-ocean ridges achieve this time-averaged growth.

2.3. Dike Emplacement: The "Quantum Events" of Fast-Spreading Ridges?

An important new development in understanding mechanisms of plate-spreading at mid-ocean ridges has come from the use of the NOAA Acoustic Hydrophone Array (AHA), which has now provided data monitoring episodes of seismic activity along the EPR from 10°S to 10°N since 1996 [*Fox et al.*, 2001]. During the five years of 1996–2001, for example, "clusters" of seismic activity have been observed on five separate occasions on the EPR axis centred at latitudes of ~8°39'N, 7°17'N, 3°25'N, 2°46'S, and 5°46'S (marked by yellow ellipsoids in Plate 1). Each of these hydroacoustically determined event clusters was located toward the centre of a different second-order volcanic segment of the EPR rather than on the transform faults like most of the hydroacoustic events. The seismic swarm near the 8°39'N area occurred on March 2, 2001, where evidence for a hydrothermal plume was subsequently identified through a "rapid response" effort reported by *Bohnenstiehl et al.* [2003]. Because these seismic swarms occurred right at the axis of a fast-spreading ridge, where the brittle thickness of the lithosphere is expected to be thin, we hypothesize that these seismic swarms are associated with magmatic emplacement rather than tectonic faulting—such interpretations have yet to be confirmed from independent observations, except at 8°39'N.

If our interpretation is correct, however, it would imply that each of these seismic swarms is indicative of a fresh episode of dike intrusion. This dike intrusion may also be accompanied by volcanic extrusive activity at the seafloor, although the lat-

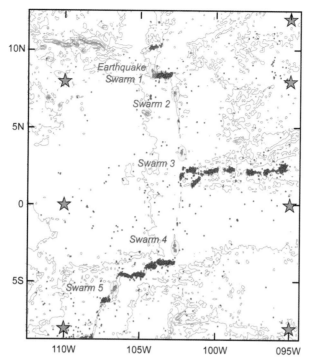

Plate 1. Map showing earthquakes on the East Pacific Rise during 1996–2001. Black thin lines show bathymetric contours. Blue stars mark the location of 7 NOAA hydroacoustic sensors. Small red dots show earthquakes determined by the NOAA hydroacoustic array during 1996–2001. Original earthquake data are from Fox et al. [2001] and the plot is modified from Gregg et al. [2003]. Yellow ellipsoids denote locations of five clusters of earthquake swarms on the EPR axis away from major transform faults. We interpret these earthquake swarms to be potentially associated with diking events on the EPR, although only the event cluster near 8°39'N has been confirmed to be associated with hydrothermal discharge [Bohnenstiehl et al., 2003].

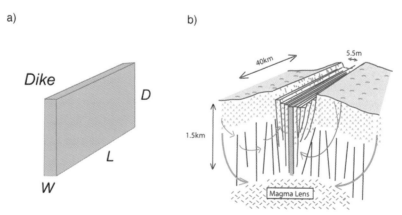

Plate 2. a) Simple schematics of a "model" dike intrusion measuring W = 5.5 m, D = 1.5 km and L = 40 km as used in our calculations; b) illustration [after Alt, 1985] showing possible routes of hydrothermal circulation (arrows) through a section of fast- or medium-spreading ridge crest and an idealised dike emplacement (vertical orange slab) which rises 1.5 km above an axial magma lens, measures 5.5m in width and extends 40km along-axis. The heat release from crystallization and cooling of a dike above an axial melt lens may be sufficient to generate "event plumes" (days–weeks), release heat rapidly through "evolving" hydrothermal systems (months–years) and yet still maintain more "quiescent" high temperature hydrothermal flow over timescales of decades.

ter is not required to have occurred. The interpretation is based on a comparison of this recent seismic activity along the EPR 10°S–10°N section with that reported from ground-truthed seafloor volcanic eruption events witnessed previously, both by submersible and using the SOSUS array, along the Juan de Fuca Ridge [e.g., *Dziak et al.*, 1995; *Fox and Dziak*, 1998; *Dziak and Fox*, 1999]. The fact that the ground-truthed seismic swarm at 8°39'N was, indeed, accompanied by hydrothermal discharge not previously reported [*Bohnenstiehl et al.*, 2003] encourages us in the validity of our approach. Another important calibration point for the NOAA hydroacoustic data set is obtained from the recognition that no significant seismic swarm activity has been reported from the segment at 9–10°N on the EPR throughout the past 5 years – an area where repeat Alvin dives can confirm that no further volcanic eruptions have, indeed, occurred.

Within the AHA survey area, an average occurrence has been recorded of one seismic swarm per year, with each such swarm—assumed here to be indicative of volcanic eruptions— appearing to extend over some ca. 30–50 km along axis. For the whole surveyed area (from 10°S to 10°N, ca. 2,200 km in total distance) to undergo one such "quantum event" of ridge spreading, a series of 45–75 episodes, each measuring 50–30 km in extent, would be required. This suggests that, on average, a repeat event for dike-intrusion/volcanic-extrusion at any one vent site on the fast-spreading EPR should only occur approximately once every 50 years. To a first-order approximation, this time scale does not appear inconsistent with much that is already known from the northern East Pacific Rise: first, that hydrothermal activity can continue unperturbed over decadal time-scales, e.g., at 21°N EPR over 20–25 years [*Von Damm et al.*, 2002]; second, that recently perturbed sites (e.g., 9°50'N) can continue to evolve/relax back toward relatively stabilised conditions over a period of approximately one decade following dike emplacement/volcanic eruption [*Von Damm*, this volume; *Von Damm et al.*, submitted].

If episodic spreading, as described above, does dominate the growth of oceanic crust at ridge axes and, further, if such events do only recur once every ~50 years (range 45–75 years) then, for a full spreading rate of 110 mm/yr [*DeMets et al.*, 1994], each such spreading event on the northern EPR should lead to an "instantaneous" extension, across the ridge axis, of ~5–8 m. This figure is also in reasonable agreement with what has previously been observed. For example, *Wright et al.* [1995] conducted a detailed study of seafloor images at the EPR 9°50'N site and showed that active hydrothermal venting occurred in close association with cracking at the fourth-order segment scale. These features were interpreted as "eruptive" cracks because they were sited directly within fresh lava-flows. Maximum crack widths in that study measured up to 7 m and the majority (even after partial infilling of indi-

vidual fissures by lava flows) fell into a range of ca. 2–4 m extension across strike. In the extreme, those authors argued for up to 20% extension across the axial zone of the ridge crest, effected through eruptive fissuring, during the 1991 episode of volcanic extrusion.

A final consideration is the volume of magma delivery that would be required to generate a single diking event, should this represent the "quantum event" of plate-tectonic spreading at fast-spreading ridge axes. *Gregg et al.* [1996] have estimated that the volume of extrusive flow alone at EPR 9°N associated with the 1991 eruption was ca. 1×10^6 m^3. Assuming a depth of 1.5 km to the top of the underlying magma lens along this section of the EPR [*Detrick et al.*, 1987], emplacement of a dike measuring 5–8 m wide, striking along the ridge axis would require a much greater volume of magma that measured 5–8 m x 1.5 km x 1 km = 7.5–12 $\times 10^6$ m^3 to be supplied for every 1 km of along-axis extent of that dike (Plate 2). When compared with the dimensions of an axial magma lens, which measures ca. 100 m thick and some 2 km wide across axis [*Kent et al.*, 1990; *Singh et al.*, 1999], however, we see that this dike volume could readily be supplied by the extraction of just ca. 5% of the volume of a magma lens in a single eruptive episode. This would be followed by some decades of magma chamber replenishment before any further magmatic extraction would be expected to recur. A final calculation shows that this volume of lava could be supplied from the magma chamber to the dike over a time frame of ca. 35 days assuming comparability with observed magma supply rates of 3–3.5 m^3/sec, as reported previously from Kilauea, Hawaii [*Tilling and Dvorak*, 1993] and Krafla, Iceland [*Einarsson*, 1978; *Trygvasson*, 1980]. It does not seem unreasonable, therefore, to conclude that the dike-emplacement mechanism should indeed serve as a reasonable approximation for mechanisms of ridge growth on fast- and intermediate-spreading ridges, wherever an axial magma lens has been identified to occur.

2.4. Heat Released by "Quantum" Diking Events

Having established the frequency with which dike-emplacement/volcanic eruptions might be expected to recur at any given ridge crest location, we can now return to the problem of whether a significant hydrothermal heat flux might be associated with such perturbations. In the previous section we estimated that emplacement of a single dike, extending 30–50 km along axis, should occur approximately once every 50 years at any given section of the fast-spreading northern EPR. Further, for the full-spreading rate of 110 mm/yr at the NEPR, each such diking event should lead to a broadening of the ridge axis by ca. 5.5 m (with a range of 5–8 m). For simplicity, we initially consider the case for an idealised dike that strikes 40 km along-axis, exhibits a constant across-axis width of 5.5 m, and extends 1.5 km down to the top of the underlying magma lens (Plate 2). The heat that would be released through cooling of this dike from its emplacement temperature to a reference "hydrothermal" temperature of 350°C [c.f., *Sinha and Evans*, this volume] would consist of two components: the latent heat released due to crystallization of the basaltic magma and the heat released during cooling of that solidified basaltic material from 1,250°C to the reference temperature of 350°C [e.g., *Sinha and Evans*, this volume].

2.4.1. Latent heat released due to crystallization of basaltic magma. The heat that would be released from crystallization of the lavas within our idealised dike at *in situ* emplacement temperatures, $E_{crystallization}$, can be calculated as the simple product of the dike volume V_{dike}, the density of the magmatic material undergoing crystallization ρ_{crust}, and the latent heat of fusion for basaltic material H_f. Thus:

$$E_{crystallization} = V_{dike} \times \rho_{crust} \times H_f$$

For the idealised dike described above, V_{dike} = 5.5 m x 1.5 km x 40 km = 3.3 x 10^8 m³. Then, using values of ρ_{crust} = 2,750 kg m⁻³ and H_f = 5 x 10^5 J kg⁻¹ [*Cannat et al.*, this volume; *Sinha and Evans*, this volume], we obtain a total heat release predicted to accompany crystallization, $E_{crystallization}$ = 4.54 x 10^{17} J, which equates to $E'_{crystallization}$ =1.14 x 10^{16} J/km along axis.

2.4.2. Heat released during cooling of solidified basaltic material, 1,250°C to 350°C. To determine the heat released during cooling of the solidified basaltic dike, a separate calculation is required in which the total energy available $E_{cooling}$ is a product of the dike volume V_{dike}, the density of the basaltic matter ρ_{crust}, the specific heat of basaltic crust $C_{p\,crust}$, and the temperature change ΔT for which the calculations are being conducted. Thus:

$$E_{cooling} = V_{dike} \times \rho_{crust} \times C_{p\,crust} \times \Delta T$$

Again, the assumed dike volume for these calculations is V_{dike} = 3.3 x 10^8 m³ and the appropriate basalt density ρ_{crust} = 2,750 kg m⁻³. Then, for a value of $C_{p\,crust}$ = 1,050 J kg⁻¹ K⁻¹ and taking ΔT = 900 K, which is the temperature change associated with cooling the magmatic body from 1250°C to the 350°C reference temperature [*Sinha and Evans*, this volume], we calculate a "cooling" heat release that totals $E_{cooling}$ = 8.58 x 10^{17} J, which equates to ca. $E'_{cooling}$ = 2.15 x 10^{16} J/km along axis. Thus, in sum, the total heat released from this idealised diking event should be $E_{dike} = E_{crystallization} + E_{cooling}$ =

1.31 x 10^{18} J, which is equivalent to E'_{dike} = 3.23 x 10^{16} J/km along the full 40 km strike of this dike.

2.5. The CoAxial Segment of the Juan de Fuca Ridge

The same calculations illustrated above can also be repeated for a differing, ~40 km idealised dike emplacement on the Juan de Fuca Ridge with a full-spreading rate of 70 mm/yr. There, the representative "quantum" dike width for the equivalent of 50 years' extension would be reduced to 3.5 m, consistent with reported estimates based on studies of the CoAxial segment of the Juan de Fuca Ridge [3–5 m: *Cherkouai et al.*, 1997; 2–4m: *Baker et al.*, 1998]. Another important difference is that for the Juan de Fuca Ridge calculations we increase the depth to the top of the axial magma chamber to 2.5 km. This assumption is based on a recent multi-channel seismic reflection experiment showing that the depth to the top of the magma lens varies in the range 2.3–2.8 km for various ridge segments of the Juan de Fuca Ridge [*Canales et al.*, 2002; *Carbotte et al.*, 2002; *Detrick et al.*, 2002b]. With these variations in our idealised dike volume, the equivalent total amount of heat available remains remarkably similar (presumably coincidentally so) with a value of E_{dike} = 1.39 x 10^{18} J. This value for the Juan de Fuca Ridge is of particular interest because *Baker et al.* [1998] carried out extensive hydrothermal heat-flow determinations during the 2–3 year period following the June 1993 seafloor dike-intrusion event in the CoAxial segment. We now have a framework within which to compare our model with independent studies of the hydrothermal flow apparently stimulated by that event.

2.6. Three Stages of Heat Release Associated with a Diking Event

2.6.1. The "Event-Plume" Period. In their study at the Juan de Fuca Ridge, *Baker et al.* [1998] were able to calculate the heat released into "event plumes", immediately after the dike emplacement event. They then compared those fluxes with the flow of heat (and chemicals) released during a phase of rapidly diminishing "chronic" (black-smoker type) high-temperature hydrothermal plume discharge, which continued over a further 2–3 years. Although *Baker et al.* [1998] calculated that the total heat content of the three event plumes detected within days to weeks of dike emplacement was high, $E_{event\,plume}$ = ca. 2 x 10^{16} J, they also concluded that this contribution to the total flux at the CoAxial segment was small when compared to the integrated heat flux released through the diminishing "chronic" hydrothermal discharge monitored from plume surveys over the subsequent 2–3 years. The estimated heat fluxes from this region dropped dramatically from values in the range 20–40 GW during the first few weeks

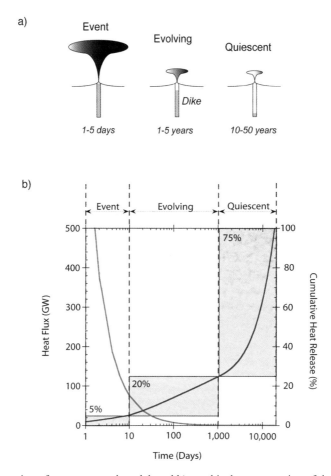

Plate 3. a) Cartoon illustration of our conceptual model, and b) graphical representation of the same model for the heat flux (red curve) and corresponding cumulative heat release (blue curve) as a function of time following dike emplacement at a fast- or medium-spreading ridge crest. Here we identify three time-periods. During the "event" period (measured in days–weeks) event plumes form releasing large quantities of heat into the water column. The "evolving" period (months–years) is characterized by continuing rapid discharge of heat via vigorous hydrothermal venting which evolves over a period of years, up to a decade. This is followed by the "quiescent" period (more than one decade), during which sustained, less-vigorous venting dominates the cumulative heat-flux from the system. Here we discuss only heat fluxes associated with such a progression from "event" to "evolving" to "quiescent" flow. A more detailed treatment of the changes in chemical compositions of the fluids during these different stages is presented by Von Damm [this volume] for the case of EPR 9°N. More general discussions are presented by German & Von Damm [2003] and references therein.

after the diking event (July –August 1993) to values significantly less than ~1 GW by June 1995. Integrating under the curve for measured decreasing heat flow over this time period, *Baker et al.* [1998] calculated a total heat flow from their 2–3 year study of decreasing hydrothermal plume discharge to represent a total energy release of ca. $E_{evolving} = 3.5 \times 10^{17}$ J. This is approximately one order of magnitude greater than was calculated for the initial event plume systems and led *Baker et al.* [1998] to conclude that event plume discharge, released at the instant that a dike is emplaced into the oceanic crust, could not account for more than ca. 5% of the total rapidly-released (≤2 years) heat flow associated with such a diking event.

2.6.2. The "Evolving" Period. Baker et al. [1998] found that the heat and fluid flow at their two study sites on the CoAxial segment not only diminished but ceased almost completely over 2–3 years. This is in contrast to the observations from the immediately adjacent Cleft segment of the Juan de Fuca Ridge, where sustained venting has been observed over more than 11 years [e.g., *Baker*, 1994]. We believe that this discrepancy can be resolved by the observation that the rapidly-diminishing "chronic" or "evolving" plume discharge monitored by *Baker et al.* [1998] at the CoAxial segment between 1993 and 1996 ($E_{evolving} = 3.5 \times 10^{17}$ J) was significantly less than the total heat available for cooling the entire dike to the top of the magma chamber ($E_{dike} = 1.39 \times 10^{18}$ J). The amount of heat reported by *Baker et al.* [1998] to have been released rapidly, in the 2–3 years following dike emplacement, equates to no more than ca. 25% of the total heat available from crystallizing and cooling of the entire volume of the dike to hydrothermal temperatures of 350°C. From one perspective, therefore, this rapid cooling phase following dike emplacement might be seen as equivalent to the heat released from cooling of the uppermost ~25% of the dike in volume, i.e., to a depth of ca. 600–650 m. Again, this is in good agreement with *Baker et al.* [1998]'s preferred explanation that they witnessed cooling of a dike to a depth of ~500 m. The remaining ca. 75% of the heat from the diking event ($E_{quiescent} = 1.04 \times 10^{18}$ J), then, must be released much more slowly, over subsequent decades.

2.6.3. The "Quiescent" Period. Comprehensive plume survey data for "event plumes" and "chronic" heat release are only available for the CoAxial segment of the Juan de Fuca Ridge. If the results of these studies are taken to be representative of the partitioning associated with all dike-emplacement events we can now calculate what further amount of heat should also be available that is not released through vigorous, but rapidly evolving venting in the years immediately following a dike event. Here we estimate the potential for longevity of such heat flow if it is extracted over a decadal time

scale in the form of less-vigorous, relatively "quiescent" blacksmoker hydrothermal flow. For the Juan de Fuca Ridge, our calculations estimate that ca. 25% of the total heat available from a dike emplaced above an axial magma lens ($E_{dike} = 1.39 \times 10^{18}$ J) was released by rapid heat removal (including event plumes) over a time scale of a few years ($E_{event\ plume\ +\ evolving} = 3.5 \times 10^{17}$ J). For the EPR, the same 25% proportion would represent a similar total heat content of $E_{event\ plume} + E_{evolving} = 0.25 \times 1.31 \times 10^{18}$ J $= 3.3 \times 10^{17}$ J. Subtracting this from the total heat emplaced by a 5.5 m wide, 40 km long dike at the EPR would yield a residual heat content, $E_{quiescent} = 9.8 \times 10^{17}$ J. If this residual heat is extracted in the form of prolonged, but less vigorous, high-temperature hydrothermal venting, it would be sufficient to sustain a total hydrothermal heat flux of 800 MW along a 40 km section of the EPR for approximately 50 years, which is the estimated time period between repeat diking events. We note that such "quiescent" high-temperature hydrothermal heat flow has been observed along the northern EPR for ≥20 years [e.g., 21°N: *Von Damm et al.*, 2002]. At a mean spacing of one hydrothermal vent site every 8–15 km along axis (see earlier), this predicted long-term high-temperature hydrothermal heat flux of 800 MW along a 40 km section of the ridge crest could easily be accommodated by, for example, 4 vent sites each with a power of 200 MW persisting over 50 years, or 5 such sites at 160 MW each persisting over the same time interval.

Thus we suggest a conceptual model (Plate 3) in which dike emplacement is accompanied by a minor proportion (ca. 5%) of the total available heat being released quasi-immediately (days) into "event plumes" followed by a more significant rate of energy release associated with rapidly-diminishing flow (order 10s of GW to < 1GW) through "chronic" or "evolving" hydrothermal plumes, in which some further ~ 20–30% of the total heat emplaced by diking might be released on the order of 5 years. This would then be followed by a much more "quiescent", extended period of heat flow (ca. 50 years) during which the remaining ca. 60–80% of the heat emplaced by diking was discharged through long-term high-temperature hydrothermal flow. Such hydrothermal flow could be emitted through relatively low-power hydrothermal vent systems spaced at regular intervals, ca. 10 km apart, along the ridge axis. Encouragingly, this conceptual model appears to be reasonably consistent with field data for both long-term and rapidly evolving high-temperature hydrothermal systems along the axes of the fast-spreading northern EPR and the intermediate-spreading Juan de Fuca Ridge. At sites where dike emplacement is known to have occurred (EPR 9°50'N and CoAxial segment of the Juan de Fuca Ridge), there has been rapid evolution of the hydrothermal system after volcanism has ceased, with flow diminishing rapidly over a matter of a few years and certainly less than a decade [*Baker et al.*, 1998; *Von*

Damm, this volume]. At sites where no such dike emplacement has been observed, however, relatively low-power hydrothermal vent sites (ca. 100 MW each) have apparently continued relatively unperturbed over periods extending over decadal time-scales [e.g., Cleft segment of the Juan de Fuca Ridge: *Baker*, 1994; EPR 21°N : *Von Damm et al.*, 2002].

2.7. Potential Deficiencies in This Simplified "Unifying" Model

Above, we have used a thermodynamic approach to generate a conceptual heat-balance model that can plausibly explain the inter-relationship between heat supply from magmatic emplacement at ridge crests and heat removal through high-temperature hydrothermal flow. In the next section we use the same approach to highlight where difficulties arise when one attempts to apply this same simplistic model to slower-spreading ridge crests. Before addressing that point, however, we must point out two key issues that the above, simplified model has overlooked.

First, to construct a simple heat-balance alone and yet not consider the mechanisms by which that heat might be extracted from crust to ocean, is to ignore a large body of work conducted into the mechanisms by which hydrothermal flow takes place through a porous/permeable medium. This work was commenced more than 20 years ago [e.g., *Strens and Cann*, 1982; *Lowell and Rona*, 1985] and is reviewed in detail by *Lowell and Germanovich* [this volume]. To briefly illustrate the point, the availability of the appropriate amount of heat in a system does not necessarily imply a mechanism by which fluid can then be extracted at a particular preferred temperature (e.g., the 350°C of a high-temperature "black smoker") over any designated amount of time. For example, while extracting heat through hydrothermal circulation from the top of a freshly-emplaced dike, the rest of that dike will also be cooling by conduction into the host surrounding rock. The standard conductive time scale for this process is $\tau \sim W^2/\kappa$ where W is dike width and thermal diffusivity $\kappa \sim 10^{-6} m^2/sec$. For the extreme case of conduction into cold rock (i.e., maximum heat removal not involving hydrothermal circulation) this yields $\tau \sim 1$ year for a 5.5m-wide dike (see above) and implies virtually complete cooling to background temperatures in just one decade – a time scale much less than the 50 years predicted by our calculations. [See *Lowell & Germanovich*, this volume, for more detailed discussions on this topic].

A second important issue is that the calculation in the previous paragraph, for dike-cooling by conduction, also appears to converge with the most recent field data from the northern East Pacific Rise. Although the EPR 21°N site has remained stable over more than a decade of measurements [cf., *Von Damm et al.* 1985, 2002] the most recent report indicates that

some change in composition may finally be occurring. Even more relevant, *Von Damm* [this volume] concludes from inter-comparison of fluid compositions and temperatures at the Bio9, Bio9' and P vents at EPR 9°50'N that fresh magmatic replenishment may have already recommenced at this location and fresh eruptions may be imminent, within little more than ten years of the prior eruptions in 1991. Clearly, such activity would be completely inconsistent with the model we have presented above: we have constructed an eminently testable hypothesis which the EPR 9°N Integrated Study Site of the Ridge 2000 programme should be uniquely well placed to test in the decade ahead. If *Von Damm* [this volume] is correct, then we may not have long to wait to see the results.

3. CASE STUDY II: THE SLOW-SPREADING NORTHERN MID-ATLANTIC RIDGE

If we set aside the mechanistic shortcomings of the "quantum event" model discussed above, it remains the case that our heat-budget approach, which could explain the links between crustal growth and hydrothermal cooling at fast- and intermediate-spreading ridges, appears to bear little or no relevance to slow-spreading ridges, particularly in the instances of large discrete hydrothermal fields such as those found at the TAG [e.g., *Rona et al.*, 1986] and Rainbow [*Charlou et al.*, 2002] sites on the northern Mid-Atlantic Ridge. In this section we examine processes that may be active along such slow-spreading ridges.

3.1. The TAG Hydrothermal Field: An Unusually Large High-Temperature Vent System

To date, a total of seven sites of high-temperature hydrothermal venting have been identified along the slow-spreading Mid-Atlantic Ridge (MAR) together with the low-temperature Lost City hydrothermal field [*Kelley et al.*, 2001]. Three of the high temperature fields have all been found in close proximity to one another close to the Azores Triple Junction: Menez Gwen (38°N), Lucky Strike (37°N) and Rainbow (36°N). To the south are the Broken Spur (29°N), TAG (26°N) and Snake Pit (23°N) sites, which are followed at 15°N by the Logatchev hydrothermal field [e.g., *Baker and German*, this volume]. The first site of high-temperature hydrothermal venting to be located on the MAR was the TAG hydrothermal field at 26°N [*Rona et al.*, 1986], which was long thought to be particularly anomalous. The next sites of venting to be discovered, the Snake Pit and Broken Spur vent fields, are both directly associated with axial summit fracturing along active volcanic ridges. These two vent fields reveal both hydrothermal deposits and volcano-tectonic settings [e.g., *Fouquet et al.*, 1993; *Murton et al.*, 1995] that are quite reminiscent of high-temperature hydrothermal vents along the fast-spreading EPR [e.g., *Wright et al.*, 1995]. The TAG hydrothermal field, in contrast, is much larger than the majority of seafloor hydrothermal systems and is hosted in tectonically-controlled terrain close to the base of the east wall of the rift valley, i.e., away from the active ridge axis [*Kleinrock and Humphris*, 1996]. The site of active venting at TAG is atop a 50 m tall, 150 m diameter mound of hydrothermal sulfide and sulfate deposits. This mound is much larger than most seafloor hydrothermal settings but comparable to numerous volcanic-hosted massive sulfide deposits found on land including some of the larger sulfide deposits from the Troodos ophiolite, Cyprus [e.g., *Hannington et al.*, 1998].

Humphris and Cann [2000] have calculated the total energy that would be required to construct the TAG hydrothermal mound based primarily upon the Fe budget of the TAG system. (A similar although simpler, estimate of this type is also presented by *Lowell & Germanovich* [this volume]). Those calculations indicate that formation of the TAG hydrothermal field requires a total input of $E_{TAG} = 1-2 \times 10^{19}$ J. Further, *Humphris and Cann* [2000] speculate that such deposition could be achieved through a brief period of high heat flow, for example, from the precipitation of Fe from high-temperature fluids flowing at a rate of 650 kg/s (with an associated power output of ca. 1,000 MW) over a period of 100–1,000 years. Because this rate of heat supply was significantly greater than can be provided from steady-state flux along a short section of slow-spreading ridge crest, the preferred explanation presented for these calculations was that hydrothermal circulation at TAG was driven by short periods of intense activity driven by episodic magmatic supply lasting tens of years, separated by thousands of years of inactivity [*Humphris and Cann*, 2000]. Such a timeline for TAG would be entirely consistent with the chronology of *Lalou et al.* [1998] and *You and Bickle* [1998].

3.2. The Rainbow Hydrothermal Field: An Even Larger High-Temperature Vent System

The Rainbow hydrothermal field, like TAG, is a fault-controlled hydrothermal system that is significantly larger than most typical volcanically-hosted vent sites. This system, which is associated with a non-transform discontinuity, occupies an area on the seafloor measuring ca. 300 m along axis and 100 m across axis. It hosts at least 10 discrete clusters of active black smoker (≤364°C) chimneys. The current output for the site, estimated from hydrothermal plume heat-flux studies, is 2.3 GW [*Thurnherr & Richards*, 2001; *Thurnherr et al.*, 2002], i.e., significantly stronger than TAG [*Humphris and Cann*, 2000]. What is also important to note is that these fluxes at Rainbow represent a direct like-for-like comparison with the high-temperature fluxes considered for the EPR (above). At

Rainbow, although fluxes are calculated from an integrated plume study, field observations reveal a distinct absence of diffuse flow at the Rainbow vent site [*Desbruyeres et al.*, 2001] and this is confirmed by the negligible concentrations of dissolved Rn-222 measurable in the Rainbow non-buoyant plume – even directly above the vent site [*Cooper,*1999]. In more typical hydrothermal settings, high Rn-222 concentrations are observed in hydrothermal plumes and these fluxes are dominated by contributions from extremely enriched diffuse flow "ground waters" which are entrained into the turbulent buoyant plumes rising above individual black smokers. At Rainbow, such diffuse flow is essentially absent and all the measured heat-flux, according to He:heat, CH_4:heat, Mn:heat and Fe:heat ratios appears to be dominated by high-temprature "black-smoker" type fluid flow [*German et al.*, submitted].

Furthermore, this level of output from the Rainbow hydrothermal field has been sustained throughout the past 8,000–12,000 years, based on a comparison of modern plume signals, 12-month sediment-trap fluxes, and long-term records preserved in sediment cores raised from beneath the dispersing non-buoyant plume [*Cave et al.*, 2002]. If all the above field observations are correct, the apparent energy release implied over the past 10,000 years would be ca. $E_{Rainbow} = 7.3 \times 10^{20}$ J, i.e., more than an order of magnitude greater than that estimated for the TAG hydrothermal mound (although 6 such mounds, the others inactive, have also been reported for the TAG ridge segment). As we will see later, supply of this amount of energy is extremely problematic. Is it valid, therefore, to use the values listed above which have been derived from oceanographic and biogeochemical, rather than geophysical observations? First, we have confidence in the heat-fluxes calculated from hydrothermal plume studies because of the close convergence between values calculated from "instantaneous" plume observations made during on-station operations in 1997 [*Thurnherr and Richards*, 2001] and those calculated from long-term current meter moorings deployed between 1997 and 1998 [*Thurnherr et al.*, 2002]. Second, we observe that geochemical fluxes from the present day plume (as collected in sediment traps) [*Cave*, 2002] agree directly with the average fluxes calculated for the surface mixed layer of underlying sediments [*Cave et al.*, 2002]. Reassuringly, the sediment trap data were collected during the same time interval as the long-term oceanographic data used to calculate heat flux [*Thurnherr et al.*, 2002] and so those two data sets are certainly directly related. Of course, the average flux into the sediments' surface mixed layer does not necessarily reflect a constant flux. Rather, during the time to accumulate this much sediment (7–10 cm, 3–4 ky) fluxes may have waxed and waned to values both higher and lower than those associated with the 1997–98 study. Even if such were the case, however, the time-averaged flux during this period would still have been indistinguishable from that of a constant flux identical to the modern day. On the assumption that chemical:heat fluxes at Rainbow have also remained unchanged throughout this period, the inference must be that the time-averaged heat flux during accumulation of the surface mixed layer (3–4ky) has also persisted at a value indistinguishable from the present. Finally, below the surface mixed layer, C-14 dating shows a steady accumulation of sediments at 2.5–3.5 cm/kyr beneath the Rainbow plume with no distinguishable change in hydrothermal plume fluxes (c.f. surface mixed layer, sediment traps) extending back to 8–12 kyr [*Cave et al.*, 2002]. On this basis we surmise that hydrothermal activity at the Rainbow vent-site has been sustained at, or close to, modern-day values over a time scale of ca.10,000 years. How might such large amounts of energy be supplied through hydrothermal circulation within the MAR rift valley?

One first approach is to apply the same considerations that we used previously for the fast- and intermediate-spreading ridges. At a full spreading rate of 26 mm/yr, the total crustal accretion within the past 10,000 years is ~260 m. The mean to the depth of the axial magma lenses on the EPR and Juan de Fuca Ridge is ~2 km, although no steady-state magma lens has so far been imaged along any section of the northern MAR that is not hotspot influenced [*Detrick et al.*, 1990]. If we apply our previous methodologies and initially assume that hydrothermal circulation can only access heat in the upper crust to a depth of ~2 km, we calculate that the heat available from the upper ocean crust sums to $E'_{upper\ crust} = E'_{crystallization}$ (7.15 x 10^{17} J/km, crystallization of basaltic magma) + $E'_{cooling}$ (1.35 x 10^{18} J/km, cooling of solidified basalt from 1,250 to 350°C) = 2.06 x 10^{18} J/km for each 1 km section along axis. Interestingly, this calculation suggests that it would be possible to provide all the heat required to form the TAG mound (E_{TAG} = 2 x 10^{19} J) by integration of the entire available heat release associated with formation of the upper 2 km of the oceanic crust along 10 km of the MAR axis, over a period of 10,000 years or, perhaps more relevantly, over just 2–2.5 km along axis over the 40,000–50,000 year lifetime of the TAG mound [*Humphris and Cann*, 2000]. In the case of the Rainbow hydrothermal field, by contrast, the situation is much more problematic. To provide all the heat calculated to have been released through the Rainbow hydrothermal system would appear to require an integration of all the available heat emplaced by formation of the uppermost ocean crust along ca. 350 km of the slow-spreading MAR axis. This seems clearly unrealistic.

3.3. Alternative Heat Sources at Slow-Spreading Ridges: Serpentinisation and Deeper Cooling?

From their distinctive end-member vent fluid compositions, it is already well-established that the Rainbow hydrothermal

fluids are fed not only by some magmatic heat source at depth but that they must also be interacting with serpentinising ultramafic rocks [*Charlou et al.,* 2002; *Douville et al.*, 2002]. Our next calculation, therefore, considers whether such serpentinisation reactions might play a significant role in balancing the heat budget calculated for Rainbow. For a latent heat $H_{serpentinisation}$ = 2.5 x 10^5 J/kg for the serpentinisation of peridotites [*Fyfe and Lonsdale*, 1981; *Lowell and Rona*, 2002] the heat budget for Rainbow could be entirely balanced by the serpentinisation of $M_{serpentinisation}$ = 2.92 x 10^{15} kg of ultramafic rocks or a volume of $V_{serpentinisation}$ = 8.85 x 10^{11} m^3 for mantle density of $\rho_{peridotite}$ = 3,300 kg m^{-3}. The thickness of the serpentinised mantle layer along the MAR must be quite variable but is not well known. For illustration purposes, however, if one assumes that mantle serpentinisation could proceed down into the top ca. 2 km of the upper mantle, the required volume of serpentinisation calculated above would be equivalent to extraction of all the heat available from serpentinisation beneath a 260-m wide (10,000 yrs of crustal growth) mantle block that extends ~1,700 km along axis. Even at such an extreme (~0.1 km^3/yr serpentinisation rate), the predicted average heat release associated with such serpentinisation would be ca. $E'_{serpentinisation}$ = 4.12 x10^{17} J/km over this 10,000 year period, i.e., only ~20% of the heat flux that could be sustained, over the same length scale, by cooling of the upper 2 km of the overlying oceanic crust. Thus we argue that while serpentinisation could contribute to the heat budget, it would be unlikely to provide the main source of heat required to drive the Rainbow hydrothermal system. (NB: By contrast, much lower rates of serpentinisation would be required to sustain the low-temperature ultramafic-hosted Lost City hydrothermal field: 5x10^{-3} to 1x10^{-4} km^3/yr [*Früh-Green et al.*, 2003].)

One possible explanation for Rainbow is that hydrothermal circulation at the ridge axis could penetrate deeper at slow-spreading ridges than at fast- and intermediate-spreading ridges. On the EPR and Juan de Fuca Ridge, it seems unlikely that fluid circulation could penetrate deeper than the upper 1.4–2.5 km of the oceanic crust because that is the depth to the axial magma lens. On slow-spreading ridges, by contrast, microearthquake studies have shown brittle failure to a depth of 5–8 km along the MAR [*Toomey et al.*, 1988; *Kong et al.*, 1992; *Wolfe et al.*, 1995]. If we repeat our earlier calculations for 260 m of extension across the MAR but allow heat extraction by hydrothermal circulation to extend to the base of a 6-km thick crust, the amount of heat potentially available through crystallization and cooling of the whole crust layer is increased three-fold to $E_{whole\ crust}$ = 6.18 x 10^{18} J/km over a 10,000 year period. To that can be added the amount of heat available from serpentinisation of the upper 2 km of the underlying upper-mantle lithosphere over the same period (1.18 x 10^{18} J/km), yielding a potential energy budget over the time frame of the Rainbow hydrothermal field's activity that totals $E_{whole\ crust\ +\ serpentinisation}$ = 7.36 x 10^{18} J/km. Of course, it remains debatable whether serpentinisation can indeed generate significant amounts of heat at Rainbow [*Allen & Seyfried*, 2004]. Nevertheless, because even the maximum contribution from such a process is small it does not affect our overall outcome, that at such a heat flow rate, ~100 km of the MAR axis would be required to match the total heat budget for the Rainbow site ($E_{Rainbow}$ = 7.3 x 10^{20} J) over the past 10,000 years.

We conclude this section, therefore, with the observation that hydrothermal circulation at the Rainbow vent field could have continued uninterrupted throughout the past 10,000 years IF it were supplied by all the heat available from cooling the entire 6 km section of the ocean crust emplaced within this time (± that released from serpentinisation of the underlying lithospheric mantle to a depth of 2 km), along ~100 km of the MAR ridge axis; this is essentially the full length of both ridge segments immediately north and south of the non-transform discontinuity that hosts the Rainbow vent field. While this continues to appear to be a rather extreme solution, we do already know that hydrothermal fluids can circulate laterally over some tens of kilometres at depth, as revealed from recent vent-fluid compositions along the Endeavour segment of the Juan de Fuca Ridge [*M.D.Lilley*, pers. comm., 2003]. Furthermore, while the preferred modern heat flux value for the Rainbow vent field is high (2.3 GW), the error margins associated with the calculations are rather broad [0.5–3.1 GW: *Thurnherr et al.*, 2002]. At the highest extreme, of course, the above calculations would require an increase of >33% in the length of ridge axis affected, to ~150 km. The lower bound for that same heat flux estimate, by contrast, would reduce our calculated values to <25% of the current requirements, indicating thorough cooling along no more than 24 km of axial ridge crest, i.e., to 12 km along axis, north and south of the vent site.

What remains problematic, however, is that—as for the EPR—all we have identified here is a plausible mechanism to reconcile total heat budgets at Rainbow and TAG, for crustal production and hydrothermal cooling. Once again, however, the detailed mechanisms by which such heat extraction could take place have not been explained. At Rainbow, in particular, vent-fluid compositions provide clear evidence for phase-separation at depth which, if circulation does indeed penetrate to 5–6 km, implies temperatures in excess of 500°C are likely at depth [*Douville et al.*, 2002]. Fluid circulation models cannot readily explain how such sustained high temperatures could be maintained both at depth and in exiting high-temperature fluids over the 10 ky timescales we have inferred [see *Lowell & Germanovich*, this volume].

3.4. Is There Missing Heat Flux Associated With Megamullion Detachment Fault Formation?

While the focused heat flow required to sustain Rainbow as an uninterrupted hydrothermal system with a power of 2.3 GW over 10,000 years at one single site appears large compared to steady-state heat flow available along axis [*Sinha and Evans*, this volume], it is relatively small considering the spatial and temporal focusing of magma accretion along slow-spreading ridges. The formation of "megamullion" detachment fault structures [*Cann et al.*, 1997; *Tucholke et al.*, 1998] is an extreme example of such focusing. During megamullion formation, magmatic crustal thickness is greatly reduced or is absent (Fig. 1) implying that the magma supply that would otherwise be delivered to the megamullion location may be diverted to other locations along the ridge axis. Megamullion features can be continuous along axis over distances of ca. 20 km and in time over periods of up to 2 million years [*Tucholke et al.*, 1998]. At steady state, a slow-spreading ridge such as the central northern MAR (full spreading rate of 26 mm/yr) should extend by 26 km/Myr. The potentially missing magmatic crust from the formation of a megamullion detachment fault that lasts 2 Myr and extends 20 km along axis, therefore, could have a maximum volume of $V_{megamullion}$ = 6 km x 52 km x 20 km = 6.24 x 10^{12} m^3. The minimum amount of heat required to be preferentially extracted from along this section of ridge crest, therefore, can be calculated from the same two components discussed earlier; crystallization of the basaltic/gabbroic oceanic crust and cooling of that material from emplacement temperatures of ca. 1,250°C to a standard reference temperature of 350°C. Repeating those calculations described previously but with the much larger volume of $V_{megamullion}$ = 6.24 x 10^{12} m^3, we calculate that the missing heat from the formation of such a megamullion is $E'_{megamullion}$ = 2.5 x 10^{22} J/km. For comparison, this much heat could only be extracted from the oceanic crust by a vent site of the 2.3 GW power observed at Rainbow if that vent site continued uninterrupted for a period of ~350 kyr, which is much longer than has been reported from the sedimentary record [*Cave et al.*, 2002]. This argues for the importance of considering non-steady state magma and heat supply at slow-spreading ridges.

4. SUMMARY AND CONCLUSIONS

In this opening chapter to this volume we have emphasised the importance of understanding the interplay between episodic volcanic/tectonic processes at mid-ocean ridges and their interactions with the overlying ocean through convective hydrothermal circulation.

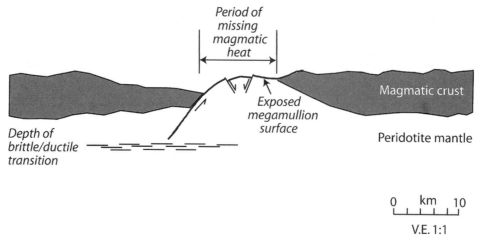

Figure 1. Schematic diagram of a megamullion detachment fault surface [modified from Tucholke et al., 1998]. During the formation of a megamullion detachment fault, the magmatic oceanic crust (shaded area) is significantly reduced or absent and, thus, much less heat release is expected from crystallization and cooling of this material. However, serpentinisation of the exposed peridotite mantle (white area) could contribute a small amount of heat flux.

For fast- and intermediate-spreading ridges we have proposed a simple conceptual model in which ridge extension in the upper crust is achieved via episodic diking events with a quantum extensional dimension. We argue that the heat available from emplacement of such a dike, which extends from the seafloor down to the roof of an axial magma chamber, could be matched by hydrothermal cooling in three stages: (a) instantaneous heat loss during a period in which "event plumes" are formed; (b) an "evolving" period of relatively fast discharge of heat through rapid cooling of the upper part of the dike in the form of diminishing but vigorous high-temperature hydrothermal flow; and (c) a decadal "quiescent" period, in which the residual heat available from cooling of the lower section of the dike is sufficient to sustain high-temperature flow for decades, albeit in the form of discrete and less-vigorous "black smoker" vent fields. While this approach matches thermal budgets and appears consistent with field data, to date, it cannot readily be reconciled with current fluid-circulation models. Further, our hypothesis may prove to be in need of significant refinement/re-evaluation if (for example) the EPR 9°N site suffers an eruption event comparable to 1991 within the next few years.

On the slow-spreading Mid-Atlantic Ridge, the apparent heat fluxes required to sustain the long-lived, tectonically-hosted TAG and Rainbow hydrothermal vent fields appear to be at least one order of magnitude greater than can readily be explained by steady-state cooling of the upper crust, alone. Hydrothermal cooling of the lower crust and heat release from serpentinisation could both contribute to the required heat budget. Furthermore, slow-spreading ridges exhibit much greater irregularity and episodic focussing of heat sources in space and time, as exemplified by the formation of megamullion detachment faults that can last ca.1–2 Ma. While focussing of heat sources, both spatially and temporally, along the MAR could be large enough to sustain high heat flow at the Rainbow and TAG sites, the mechanisms by which such heat could be extracted from the underlying lithosphere and delivered to such long-lived, highly-localised, vent-sites remains poorly understood.

The episodic nature of mid-ocean ridge and hydrothermal processes at ridges of different spreading rates is a theme that recurs throughout the chapters that follow. Much of the complexity arises from the inherent "quantum" nature of volcanic/tectonic events at mid-ocean ridges. What we have illustrated here is that tectono-magmatic and hydrothermal fluxes can become quite decoupled one from another – especially along slow-spreading ridges. This may explain how sections of even the slowest-spreading ridges can become magmatically robust and/or host vigorous hydrothermal systems, beyond what could be predicted based on long-term steady state models.

We urge that future investigations should strive to employ integrated geophysical and hydrothermal approaches to further our understanding of mid-ocean ridge systems. By recognising geophysical and hydrothermal approaches that cannot be reconciled at the present time, we identify those processes that most urgently need to be better understood.

Acknowledgments. This manuscript has benefited from many discussions with participants of the InterRidge Theoretical Institute (IRTI). Particular thanks are due to Karen Von Damm (UNH) and our reviewers, Bob Lowell (Georgia Tech) and Bill Seyfried (U.Minnesota)—not least for help in clarifying quite how much remains to be understood. We are grateful to Trish Gregg for creating a draft version of Plate 1. Significant funding for the IRTI was provided by the US RIDGE-2000 Program and the European Science Foundation. Additional support was provided by the InterRidge Office and the University of Pavia. CRG is funded through NERC's Core Strategic Research at SOC, JL is supported by NSF grant OCE-0129741 and a grant from the WHOI Deep Ocean Exploration Institute. This is Woods Hole Oceanographic Institution Contribution Number 11092.

REFERENCES

Allen, D. E. and W. E.Seyfried Jr., Serpentinisation and heat generation: constraints from Lost City and Rainbow hydrothermal systems, *Geochimica et Cosmochimica Acta 68,* 1347–1354, 2004.

Alt, J., Subseafloor processes in Mid-Ocean Ridge hydrothermal systems, *AGU Geophys. Monogr. 91,* 85–114, 1995.

Bach, W., N. R. Banerjee, H. J. B. Dick, and E. T. Baker, Discovery of ancient and active hydrothermal systems along the ultra-slow spreading Southwest Indian Ridge 10°–16°E, *Geochem. Geophys. Geosyst. 3,* 10.1029/2001GC000279, 2002.

Baker, E. T., A six-year time series of hydrothermal plumes over the Cleft segment of the Juan de Fuca Ridge. *J. Geophys. Res., 99,* 4889–4904, 1994.

Baker, E. T., Y. J. Chen, and J. Phipps Morgan, The relationship between near-axis hydrothermal cooling and the spreading rate of mid-ocean ridges, *Earth Planet. Sci. Lett., 142,* 137–145, 1996.

Baker, E. T., G. J. Massoth, R. A. Feely, G. A. Cannon, and R. E. Thomson, The rise and fall of the CoAxial hydrothermal field, 1993–1996, *J. Geophys. Res., 103,* 9791–9906, 1998.

Baker, E. T., and C. R. German, On the global distribution of hydrothermal vents, *this volume.*

Barriga, F. J. A. S., Y. Fouquet, A. Almeida, M. J. Biscoito, J. L. Charlou, R. L. P. Costa, A. Dias, A. M. S. F. Marques, J. M. A. Miranda, K. Olu, F. Porteiro and M. G. P. S.Queiroz. Discovery of the Saldanha Hydrothermal Field on the FAMOUS segment of the MAR (36° 30'N). *EOS, Trans. Am. Geophys. U.,* 79(45), F67 (abstr.), 1998.

Bemis, K. G., R. P. von Herzen and M. J. Mottl, Geothermal heat flux from hydrothermal plumes on the JDFR, *J. Geophys. Res.,* 98, 6351–6369, 1993.

Bohnenstiehl, D. R., M. Tolstoy, C. G. Fox, R. P. Dziak, E. Chapp, M. Fowler, J. Haxel, C. Fisher, C. Van Hilst, R. Laird, R. Collier, J. Cowen, M. Lilley, K. Simons, S.M. Carbotte, J.R. Reynolds, and C. H. Langmuir, Anomalous seismic activity at 8° 37–42'N on the East Pacific Rise: Hydroacoustic detection and site investigation, *Ridge 2000 Events, 1,* 18–20, 2003.

Butterfield, D. A., and G. J. Massoth, Geochemistry of North Cleft segment vent fluids: temporal changes in chlorinity and their possible relation to recent volcanism, *J. Geophys. Res., 99,* 4951–4968, 1994.

Butterfield, D. A., I. R. Jonasson, G. J. Massoth, R. A. Feely, K. K. Roe, R. E. Embley, J. F. Holden, R. E. McDuff, M. D. Lilley, and J. R. Delaney, Seafloor eruptions and evolution of hydrothermal fluid chemistry, *Phil. Trans. Roy. Soc. Lond. A, 355,* 369–387, 1997.

Canales, J. P., R. S. Detrick, S. M. Carbotte, G. Kent, A. Harding, J. Diebold, and M. Nedimovic, Multichannel seismic imaging along the Vance and Cleft segments of the southern Juan de Fuca Ridge, *EOS, Trans. AGU, 83,* Fall Meet. Suppl. Abstract T12B–1317, F1353, 2002.

Cann, J. R., D. K. Blackman, D. K. Smith, E. McAllister, B. Janssen, S. Mello, E. Avgerinos, A. Pascoe, and J. Escartin, Corrugated slip surfaces formed at North Atlantic ridge-transform intersections, *Nature, 385,* 329–332, 1997.

Cannat, M., J. Cann and J. McLennan, Some har drock constraints on the supply of heat to mid-ocean ridges. *This volume.*

Carbotte, S., and K. Macdonald, East Pacific Rise 8°–10°30'N: Evolution of ridge segments and discontinuities from SeaMARC II and three-dimensional magnetic studies, *J. Geophys. Res., 97,* 6959–6982, 1992.

Carbotte, S. M., R. S. Detrick, G. Kent, J. P. Canales, J. Diebold, A. Harding, M. Nedimovic, D. Epstein, I. Cochran, E. Van Ark, J. Dingler, and A. Jacobs, A multi-channel seismic investigation of ridge crest and ridge flank structure along the Juan de Fuca Ridge, *EOS, Trans. AGU, 83,* Fall Meet. Suppl. Abstract T72C-07, F1327, 2002.

Cave, R. R., A geochemical study of hydrothermal signals in marine sediments: the Rainbow hydrothermal area, 36°N on the Mid-Atlantic Ridge. *PhD Thesis, U.Southampton,* 154pp., 2002.

Cave, R. R., C. R. German, J. Thomson and R. W. Nesbitt, Fluxes to sediments from the Rainbow hydrothermal plume, 36°14'N on the MAR, *Geochim. Cosmochim. Acta, 66,* 1905–1923, 2002.

Charlou, J. L., J. P. Donval, Y. Fouquet, P. Jean-Baptiste, and N.Holm. Geochemistry of high H2 and CH4 vent fluids issuing from ultramafic rocks at the Rainbow hydrothermal field (36°14'N, MAR), *Chemical Geology, 191,* 345–359, 2002.

Chen, Y. J., Modeling the thermal state of the oceanic crust, *this volume.*

Chen, Y. J. and J. Lin, High sensitivity of ocean ridge thermal structure to changes in magma supply: the Galapagos Spreading center, *Earth Planet. Sci. Lett., 221,* 263–273, 2004.

Cherkaoui, A. S. M., W. S. D. Wilcock, and E. T. Baker, Thermal fluxes associated with the 1993 diking event on the CoAxial segment, Juan de Fuca Ridge: A model for the convective cooling of the dike, *J. Geophys. Res. 102,* 24,887–24,902, 1997.

Converse, D. R., H. D. Holland, and J. M. Edmond, Flow rates in the axial hot springs of the East Pacific Rise (21°N): Implications for the heat budget and the formation of massive sulfide deposits. *Earth Planet. Sci. Lett., 69,* 159–175, 1984.

Cooper, M. J., Geochemical investigations of hydrothermal fluid flow and mid-ocean ridges. *PhD Thesis, U.Cambridge,* 218pp., 1999.

DeMets, D., R. G. Gordon, D. F. Argus, and S. Stein, Effect of recent revisions to the geomagnetic reversal time scale on estimates of current plate motions, *Geophys. Res. Lett., 21,* 2191–2194, 1994.

Desbruyeres, D., M. Biscoito, J. C. Caprais, A. Colaco, T. Comtet, P. Crassous, Y. Fouquet, A. Khripounoff, N. le Bris, K. Olu, R. Riso, P. M. Sarradin, M. Segonzac and A. Vangriesheim. Variations on deep-sea hydrothermal vent-communities on the Mid-Atlantic Ridge near the Azores plateau, *Deep Sea Res., 48,* 1325–1346, 2001.

Detrick, R. S., P. Buhl, E. Vera, J. Mutter, J. Orcutt, J. Madsen, and T. Brocher, Multi-channel seismic imaging of a crustal magma chamber along the East Pacific Rise, *Nature, 326,* 35–42, 1987.

Detrick, R. S., J. C. Mutter, P. Buhl, and I. I. Kim, I. I., No evidence from multichannel reflection data for a crustal magma chamber in the MARK area on the Mid-Atlantic Ridge, *Nature, 347,* 61–64, 1990.

Detrick, R. S., J. M. Sinton, G. Ito, J. P. Canales, M. Behn, T. Blacic, B. Cushman, J. E. Dixon, D. W. Graham, and J. J. Mahoney, Correlated geophysical, geochemical, and volcanological manifestations of plume-ridge interaction along the Galápagos Spreading Center, *Geochem. Geophys. Geosyst., 3(10),* 8501 doi:10.1029/2002GC000350, 2002a.

Detrick, R. S., S. M. Carbotte, E. Van Ark, J. P. Canales, G. Kent, A. Harding, J. Diebold, and M. Nedimovic, New multichannel seismic constraints on the crustal structure of the Endeavour Segment, Juan de Fuca Ridge: Evidence for a crustal magma chamber, *EOS, Trans. AGU, 83,* Fall Meet. Suppl. Abstract T12B-1316, F1353, 2002b.

Dick, H. J. B., J. Lin, and H. Schouten, An ultraslow-spreading class of ocean ridge, *Nature, 426,* 405–412, 2003.

Douville E., J. L. Charlou, E. H. Oelkers, P. Bienvenu, C. F. J. Colon, J. P. Donval, Y. Fouquet, D. Prieur, and P. Appriou, The Rainbow vent fluids (36 degrees 14' N, MAR): the influence of ultramafic rocks and phase separation on trace metal content in Mid-Atlantic Ridge hydrothermal fluids, *Chem. Geol., 184,* 37–48, 2002.

Dziak, R. P., C. G. Fox, and A. E. Schreiner, The June–July 1993 seismo-acoustic event at CoAxial segment, Juan de Fuca Ridge: Evidence for a lateral dike injection, *Geophys. Res. Lett., 22,* 135–138, 1995.

Dziak, R. P. and C. G. Fox, The January 1998 earthquake swarm at Axial Volcano, Juan de Fuca Ridge: Hydroacoustic evidence of seafloor volcanic activity, *Geophys. Res. Lett., 26,* 3429–3432, 1999.

Edmonds, H. N., C. R. German, D. R. H. Green, Y. Huh, T. Gamo, and J. M. Edmond, Continuation of the hydrothermal fluid chemistry time series at TAG and the effects of ODP drilling, *Geophys. Res. Lett., 23,* 3487–3489, 1996.

Edmonds, H. N., P. J. Michael, E. T. Baker, D. P. Connelly, J. E. Snow, C. H. Langmuir, H. J. B. Dick, R. Mühe, C. R. German, and D. W. Graham, Discovery of abundant hydrothermal venting on the ultraslow-spreading Gakkel ridge in the Arctic Ocean, *Nature, 421,* 252–256, 2003.

Einarsson, P., S-wave shadows in the Krafla caldera in NE Iceland: Evidence for a magma chamber in the crust, *Bull. Volcanol., 41,* 1–9, 1978.

Fornari, D., M. A. Tivey, H. Schouten, M. Perfit, K. Von Damm, D. Yoerger, A. Bradley, M. Edwards, R. Haymon, T. Shank, D. Scheirer, and P. Johnson, Submarine lava flow emplacement processes at the East Pacific Rise 9° 50'N: Implications for hydrothermal fluid circulation in the upper ocean crust, *this volume.*

Fouquet, Y., A. Wafik, P. Cambon, C. Mevel, G. Meyer, and P. Gente, Tectonic setting and mineralogical and geochemical zonation in the Snake Pit sulfide deposit (Mid-Atlantic Ridge at 23°N), *Econ. Geol., 88,* 2018–2036, 1993.

Fox, C. G., W. E. Radford, R. P. Dziak, T.-K. Lau, H. Matsumoto, and A. E. Schreiner, Acoustic detection of a seafloor spreading episode on the Juan de Fuca Ridge using military hydrophone arrays, *Geophys. Res. Lett., 22,* 131–134, 1995.

Fox, C. G., and R. P. Dziak, Hydroacoustic detection of volcanic activity on the Gorda Ridge, February–March 1996, *Deep-Sea Res. II, 45,* 2513–2530, 1998.

Fox, C. G., H. Matsumoto, and T.-K.A. Lau, Monitoring Pacific Ocean seismicity from an autonomous hydrophone array, *J. Geophys. Res., 106,* 4183–4206, 2001.

Früh-Green, G. L., D. S.Kelley, S. M.Bernasconi, J. A. Karson, K. A. Ludwig, D. A. Butterfield, C.Boschi and G.Proskurowski, 30,000 years of hydrothermal activity at the Lost City vent field, *Science* 301, 495–498, 2003.

Fyfe, W. S., and P. Lonsdale, Ocean floor hydrothermal activity, in *The Sea,* vol. 7, *The Oceanic Lithosphere,* ed. Emiliano, Wiley, New York, pp. 589–638, 1981.

German, C. R., G. P. Klinkhammer and M. D. Rudnicki, The Rainbow hydrothermal plume, 36°15'N, MAR, *Geophys. Res. Lett.* 23, 2979–2982, 1996a.

German, C. R., L. M. Parson, and the HEAT Scientific Team, Hydrothermal exploration near the Azores Triple-Junction: tectonic control of venting at slow-spreading ridges? *Earth Planet. Sci. Lett., 138,* 93–104, 1996b.

German, C. R., E. T. Baker, C. Mevel, K. Tamaki, and the FUJI Scientific Team, Hydrothermal activity along the southwest Indian Ridge, *Nature, 395,* 490–493, 1998.

German, C. R. and K. L.Von Damm. Hydrothermal Processes, *in Treatise on Geochemistry* edited by K.K.Turekian & H.D.Holland, *Vol. 6: The Oceans and Marine Geochemistry* edited by H.Elderfield, Elsevier, Oxford, 2003.

German C. R., A. M. Thurnherr, J. Radford-Knöery J-L. Charlou, P. Jean-Baptiste, H. N. Edmonds, and the FLAME I & II Cruise Participants, Integrated geochemical and heat fluxes from a submarine hydrothermal field (Rainbow, Mid-Atlantic Ridge) and the importance of high-temperature venting for global ocean cycles. *Nature, submitted.*

Gregg, P. T., D. K. Smith, and J. Lin, Spatial and temporal patterns in the seismicity of the Equatorial Pacific and possible earthquake triggering at the East Pacific Rise (abstract), *EOS, Trans. AGU,* 2003.

Gregg, T. K. P., D. J. Fornari, M. R. Perfit, R. M. Haymon and J. H. Fink, Rapid emplacement of a mid-ocean ridge lava flow on the East Pacific Rise at 9 degrees 46'–51'N, *Earth Planet. Sci. Lett., 144,* E1–E7, 1996.

Hannington, M. D., A. D. Galley, P. M. Herzig and S. Petersen, Comparison of the TAG mound and stockwork complex with Cyprus-type massive sulfide deposits. *Proc. Ocean Drilling Program, Sci. Results, 158,* 389–415, 1998.

Haymon, R. M., D. J. Fornari, M. H. Edwards, S. Carbotte, D. Wright, and K. C. Macdonald, Hydrothermal vent distribution along the East Pacific Rise crest (9°09'–54'N) and its relationship to magmatic and tectonic processes on fast-spreading mid-ocean ridges, *Earth Planet. Sci. Lett., 104,* 513–534, 1991.

Haymon, R. M., D. J. Fornari, K. L. Von Damm, M. D. Lilley, M. R. Perfot, J. M. Edmond, W. C. Shanks III, R. A. Lutz,

J. M. Grebmeier, S. Carbotte, D. Wright, E. McLaughlin, M. Smith, N. Beedle, and E. Olson, Volcanic eruption of the mid-ocean ridge along the East Pacific Rise crest at 9°45–52'N: Direct submersible observations of seafloor phenomena associated with an eruption in April 1991, *Earth Planet. Sci. Lett., 119,* 85–101, 1993.

Hirth, G., J. Escartin, and J. Lin, The rheology of the lower oceanic crust: Implications for lithospheric deformation at mid-ocean ridges, *AGU Geophys. Monogr., 106,* 291–303, 1998.

Humphris, S. E., R. A. Zierenberg, L. S. Mullineaux, and R. E. Thompson, eds., Seafloor Hydrothermal Systems: Physical, Chemical, Biologic and Geological Interactions, *AGU Geophys. Monogr. 91,* 466 pp., 1995.

Humphris, S. E. and J. R. Cann, Constraints on the energy and chemical balances of the modern TAG and ancient Cyprus seafloor sulfide deposits, *J. Geophys. Res., 105,* 28477–28488, 2000.

Kelemen, P. B., E. Kikawa, D. J. Miller, et al., Ocean Drilling Program, Leg 209 Scientific Prospectus, Drilling Mantle Peridotite Along the Mid-Atlantic Ridge from 14° to 16°N (http://www-odp.tamu.edu/publications/prosp/209_prs/209toc.html), 2003

Kelley, D. S., J. A. Karson, D. K. Blackman, G. L. Früh-Green, D. A. Butterfield, M. D. Lilley, E. J. Olson, M. O. Schrenk, K. K. Roe, G. T. Lebon, P. Rivizzigno, and the AT3–60 shipboard party, An off-axis hydrothermal vent field near the Mid-Atlantic Ridge at 30°N, *Nature, 412,* 145–149, 2001.

Kent, G. M., A.J. Harding, and J. A. Orcutt, Evidence for a smaller magma chamber beneath the East Pacific Rise at 9–degrees–30'N, *Nature, 344,* 650–653, 1990.

Kleinrock, M. C., and S. E. Humphris, Structural controls on seafloor hydrothermal activity at the TAG active mound, *Nature, 382,* 149–153, 1996.

Kong, L. S., S. C. Solomon, and G. M. Purdy, Microearthquake characteristics of a mid-ocean ridge along-axis high, *J. Geophys. Res., 97,* 1,659–1,685, 1992.

Lalou, C., J. L.Reyss, and E. Brichet, Age of sub-bottom sulfide samples at the TAG active mound, *Proc. Ocean Drilling Progr., Sci. Results, 158,* 111–117, 1998.

Langmuir, C. H., E. M. Klein, and T. Plank, Petrological systematics of mid-ocean ridge basalts: Constraints on melt generation beneath ocean ridges, *AGU Geophys. Mongr. 71,* 1992.

Lin, J., and J. Phipps Morgan, The spreading rate dependence of three-dimensional midocean ridge gravity structure, *Geophys. Res. Lett., 19,*13–16, 1992.

Lin, J., H. J. B. Dick, and H. Schouten, Geophysical constraints on crustal accretion and hotspot-ridge interaction along the western SW Indian Ridge, *InterRidge SWIR Workshop,* Southampton, UK, April 2002.

Lowell, R. P., and L. N. Germanovich, Hydrothermal processes at mid-ocean ridges: Results from scale analysis and single-pass models, *this volume.*

Lowell, R. P., and P .A. Rona, Hydrothermal models for the generation of massive sulfide ore deposits, *J .Phys. Res., 90,* 8769–8783, 1985.

Lowell, R. P., and R. A. Rona, Seafloor hydrothermal systems driven by the serpentinisation of peridotite, *Geophys. Res. Lett., 29,* 10.1029/2001GL014411, 2002.

Lupton, J. E., Hydrothermal plumes, near and far field. *AGU Geophys. Monogr. 91,* 317–346, 1995.

Magde, L. S., D. W. Sparks, and R. S. Detrick, The relationship between buoyant mantle flow, melt migration, and gravity bull's eyes at the Mid-Atlantic Ridge between 33°N and 35°N, *Earth Planet. Sci. Lett. ,* 148, 59–67, 1997.

Michael, P. J., and W. C. Cornell, Influence of spreading rate and magma supply on crystallization and assimilation beneath mid-ocean ridges: Evidence from chlorine and major element chemistry of mid-ocean ridge basalts, *J. Geophys. Res., 103,* 18,325–18,356, 1998.

Michael, P. J., C. H. Langmuir, H. J. B. Dick, J. E. Snow, S. L. Goldstein, D. W. Graham, K. Lehnert, G. Kurras, W. Jokat, R. Mühe, and H. N. Edmonds. Magmatic and amagmatic seafloor generation at the ultraslow-spreading Gakkel ridge, Arctic Ocean. *Nature, 423,* 956–961, 2003.

Murton, B. J., C. Van Dover, and E. Southward, Geological setting and ecology of the Broken Spur hydrothermal vent-field: 29°10'N on the Mid Atlantic Ridge, *Geol. Soc. Spec. Publ. 87,* 33–42, 1995.

Parmentier, E. M., and J. Phipps Morgan, Spreading rate dependence of three-dimensional structure in oceanic spreading centers, *Nature, 348,* 325–328, 1990.

Parsons, B., The rates of plate creation and consumption, *Geophys. J. R. Astr. Soc., 67,* 437–448., 1981.

Phipps Morgan, J., and Y. J. Chen, The genesis of oceanic crust: Magma injection, hydrothermal circulation and crustal flow, *J. Geophys. Res., 98,* 6283–6297, 1993.

Rona, P. A., G. Klinkhammer, T. A. Nelsen, J. H. Trefry, and H. Elderfield, Black smokers, massive sulfides and vent biota at the Mid-Atlantic Ridge, *Nature, 321,* 33–37, 1986.

Rudnicki, M. D., R. H. James and H. Elderfield, Near-field variability of the TAG non-buoyant plume, 26°N, Mid-Atlantic Ridge, *Earth Planet. Sci. Lett.* 127, 1–10, 1994.

Rudnicki, M. D. and C. R. German, Temporal variability of the hydrothermal plume above the Kairei vent field, 25°S, Central Indian Ridge, *Geoghem., Geophys., Geosyst.* 3, doi: 10.1029/2001GC000240, 2002.

Sauter, D., L. Parson, V. Mendel, C. Rommevaux-Jestin, O. Gomez, A. Briais, C. Mével, K. Tamaki, and the FUJI Scientific Team, TOBI sidescan sonar imagery of the very slow-spreading Southwest Indian Ridge: Evidence for along-axis magma distribution, *Earth Planet. Sci. Lett., 199,* 81–95, 2002.

Searle, R. C., M. Cannat, K. Fujioka, C. Mevel, H. Fujimoto, A. Bralee, and L. Parson, The FUJI Dome: a large detachment fault near 64°E on the every slow-spreading southwest Indian Ridge, *Geochem. Geophys. Geosyst. 4,* doi:10.1029/2003GC000519, 2003.

Searle, R. C., and J. Escartin, The rheology of oceanic lithosphere and the morphology of mid-ocean ridges, *this volume.*

Shaw, W. J., and J. Lin, Models of ocean ridge lithospheric deformation: Dependence on crustal thickness, spreading rate, and segmentation, *J. Geophys. Res. ,101 ,* 17977–17993, 1996.

Sinha, M. C., S. C. Constable, C. Peirce, A. White, G. Heinson, L. M. MacGregor, and D. A. Navin, Magmatic processes at slow spreading ridges: Implications of the RAMESSES experiment at 57º 45' N on the Mid-Atlantic Ridge, *Geophs. J. Int.,* 135, 731–745, 1998.

Sinha, M., and R. Evans, Geophysical constraints upon the thermal regime of the ocean crust, *this volume*.

Singh, S. C., J. S. Collier, A. J. Harding, G. M. Kent, and J. A. Orcutt, Seismic evidence for a hydrothermal layer above the solid roof of the axial magma chamber at the southern East Pacific Rise, *Geology*, 27, 219–222, 1999.

Speer, K. G. and P. A. Rona, A model of an Atlantic and Pacific hydrothermal plume, *J. Geophys. Res.* 94, 6213–6220, 1989.

Stein, C. A., and S. Stein, Constraints on hydrothermal heat flux through the oceanic lithosphere from global heat flow, *J. Geophys. Res., 99*, 3081–3095, 1994.

Strens, M. R. and J. R. Cann, A model of hydrothermal circulation in fault zones at mid-ocean ridge crests, *Geophys. J. Roy. Astr. Soc., 71*, 225–240, 1982.

Thurnherr, A. M. and K. J. Richards. Hydrography and high-temperature heat flux of the Rainbow hydrothermal site (36°14'N, Mid-Atlantic Ridge). *J. Geophys. Res.* 106, 9411–9426.

Thurnherr, A. M., K. J. Richards, C. R. German, G. F. Lane-Serff, and K.G.Speer. Flow and mixing in the rift valley of the Mid-Atlantic Ridge, *J. Phys. Oceangr., 32*, 1763–1778, 2002.

Tilling, R. I., and J. J. Dvorak, Anatomy of a basaltic volcano, *Nature, 363*, 125–133, 1993.

Toomey, D. R., S. C. Solomon, G. M. Purdy, Microearthquakes beneath the median valley of the mid-Atlantic ridge near 23°N: Tomography and tectonics, *J. Geophys. Res., 93*, 9093–9112, 1988.

Trygvasson, E., Subsidence events in the Krafla area, north Iceland, 1975–1979, *J. Geophys., 47*, 141–153, 1980.

Tucholke, B. E., and J. Lin, A geological model for the structure of ridge segments in slow-spreading ocean crust, *J. Geophys. Res., 99*, 11,937–11,958, 1994.

Tucholke, B. E., J. Lin, and M. C. Kleinrock, Megamullions and mullion structure defining oceanic metamorphic core complexes on the Mid-Atlantic Ridge, *J. Geophys. Res., 103*, 9857–9866, 1998.

Van Dover, C. L., The Ecology of Deep-Sea Hydrothermal Vents. Princeton University Press, 424pp., 2000.

Von Damm K. L., J. M. Edmond, B. Grant, C. I. Measures, B. Walden and R.F. Weiss, Chemistry of submarine hydrothermal solutions at 21°N, East Pacific Rise. *Geochim. Cosmochim. Acta* 49, 2197–2220, 1985.

Von Damm, K. L., S. E. Oosting, R. Kozlowski, L. G. Buttermore, D. C. Colodner, H. N. Edmonds, J. M. Edmond, and J. M. Grebmeier, Evolution of East Pacific Rise hydrothermal vent fluids following a volcanic eruption, *Nature, 375,* 47–50, 1995.

Von Damm, K.vL., L. G. Buttermore, S. E. Oosting, A. M. Bray, D. J. Fornari, M. D. Lilley, and W. C.Shanks III, Direct observation of the evolution of a seafloor "black smoker" from vapor to brine, *Earth Planet. Sci. Lett., 149,* 101–112, 1997.

Von Damm, K. L., C. M. Parker, R. M. Gallant, J. P. Loveless, and The AdVenture 9 Science Party, Chemical evolution of hydrothermal fluids from EPR 21°N: 23 years later in a phase separating world, *Eos Trans. AGU, 83*, Abstract V61B-1365, F1421, 2002.

Von Damm, K. L., Evolution of the hydrothermal system at East Pacific Rise 9°50'N: Geochemical evidence for changes in the upper crust, *this volume*.

Von Damm, K. L., A. M. Bray, M. K. Brockington, L. G. Buttermore, R. M. Gallant, K. O'Grady, S. E. Oosting, and C. M. Parker, Reassessing mid-ocean ridge hydrothermal fluxes: implications of eruptive events and phase separation, *Earth Planet. Sci. Lett., submitted*.

White, R. S., D. McKenzie, and R. K. O'Nions, Oceanic crustal thickness from seismic measurements and rare earth element inversions, *J. Geophys. Res., 97,* 19683–19715, 1992.

Wolfe, C. J., G. M. Purdy, D. R. Toomey, and S. C. Solomon, Microearthquake characteristics and crustal velocity structure at 29°N on the Mid-Atlantic Ridge: The architecture of a slow-spreading segment, *J. Geophys. Res., 100*, 24,449–24,472, 1995.

Wright, D. J., R. M. Haymon, and D. J. Fornari, Crustal fissuring and its relationship to magmatic and hydrothermal processes on the East Pacific Rise crest (9° 12' – 54'N), *J. Geophys. Res.*, 100, 6097–6210, 1995.

You, C. F., and M. J. Bickle, Evolution of an active sea-floor massive sulfide deposit, *Nature, 394,* 668–671, 1998.

C. R. German, Southampton Oceanography Centre, European Way, Southampton SO14 3ZH, UK.

J. Lin, Department of Geology and Geophysics, Woods Hole Oceanographic Institution, Woods Hole, Massachusetts 02543.

Geophysical Constraints Upon the Thermal Regime of the Ocean Crust

Martin C. Sinha

Southampton Oceanography Centre, European Way, Southampton, United Kingdom

Rob L. Evans

Department of Geology and Geophysics, Woods Hole Oceanographic Institution, Woods Hole, Massachusetts

Geophysical measurements and models constrain the total rate of production of crustal material and the flux of thermal energy over the global ridge system. Flux estimates based on basin-scale compilations of heat-flow measurements or on 1-dimensional geodynamic models of melt generation and plate cooling provide a useful, but only a partial view of the crustal thermal regime. The rate of heat supply to the crust depends both on complex patterns of flow in the mantle, and on how this flow supplies magma (a major carrier of advective heat flux) to the crust. Much progress is still needed if we are to understand the thermal regime at and close to the crust-mantle boundary, and hence the extent to which segmentation and other variations in crustal structure may be inherited from the mantle. Within the lower and middle crust, the thermal regime is dominated by the presence (or otherwise) of crustal magma chambers. Over the last decade, geophysical data have provided a progressively more sophisticated understanding of these features, at all spreading rates. Correlations and quantitative links between new models of magma chamber structure and what is known from other disciplines about the overlying hydrothermal circulation system remain weak. Significant unknowns also still remain regarding the patterns and pathways of hydrothermal circulation within the crust. High resolution geophysical data are now beginning to provide quantitative constraints on the physical structure (overall porosity, and interconnectedness of pore spaces) of the permeable crust. The same observations and methods are also beginning to allow us to detect *in situ* variations in the properties of the fluids themselves, to depths equivalent to the base of layer 2, and on horizontal scales of several kilometres. In one case this has provided glimpses of what may be two-layer hydrothermal convection, related to phase separation. How flow patterns are influenced by key tectonic parameters such as spreading rate and ridge morphology remains an open issue. Also unknown is the extent to which shallow circulation may be driven by newly injected dikes, and the spatial and temporal scales of the resultant thermal perturbations. Lastly, we must consider the case where high and low temperature hydrothermal circulation is occurring in the absence of any significant crustal magma body. Are such systems related to the cooling of rocks that have recently crystallized from basaltic magmas? Do serpentinization reactions play a significant role? And how widespread are such circulation regimes?

Mid-Ocean Ridges: Hydrothermal Interactions Between the Lithosphere and Oceans
Geophysical Monograph Series 148
Copyright 2004 by the American Geophysical Union
10.1029/148GM02

1. INTRODUCTION

The existence of constructive plate boundaries in the form of mid-ocean ridges is a fundamental facet of plate tectonics. Over the last three decades the mid-ocean ridges have been studied in ever-increasing detail, confirming the initial hypothesis that they are sites of lithospheric generation, but also leading to many new insights into how this process operates, and about the chain of phenomena which stem from the consequent fluxes of heat and magma—affecting not only the face of the solid earth, but also the chemistry and biota of the oceans.

The mid-ocean ridge system stretches for almost 60,000 km around the globe, much of it forming a continuously connected chain of plate boundaries. The ridges account for 80% of the Earth's volcanism, and produce a flux of basaltic magma from the mantle which is sufficient to completely resurface more than 60% of the planet on a time scale of less than 100 Ma. The ridge axis heat flux is overwhelmingly transported from the newly-formed lithosphere into the hydrosphere by hydrothermal circulation, which takes place at a range of temperatures from just above that of ambient sea water up to temperatures commonly around 360° C and in some cases over 400° C. At mid-ocean ridges, the relatively stable stratifications which usefully characterise the outer parts of the Earth's structure in most locations (asthenosphere and lithosphere; mantle and crust; solid earth and water column) break down, and direct interactions and feedbacks occur between these components.

Recognition of the range and scale of the interactions occurring at ridges has led, over the last fifteen years, to the emergence of a self-consciously interdisciplinary international community of mid-ocean ridge scientists. Pursuit of interdisciplinary science does, however, bring its own difficulties. Not least of these is the problem that specialists from one discipline often have only a sketchy understanding of key background information from other disciplines. We have therefore consciously targeted this paper not towards our fellow geophysicists, but towards our mid-ocean ridge colleagues from other disciplines. In attempting such a review we are almost certain to get the balance wrong from time to time—most likely by over-generalising and over-simplification. We hope that, despite such shortcomings, this paper can make a useful contribution to the spirit and practice of interdisciplinary enquiry.

2. KEY PROCESSES AND GLOBAL FLUXES

2.1. Key Processes

Ridges occur where lithospheric plates diverge at a constructive plate boundary, and new seafloor is formed by vol-
canism. Ridges and the lithospheric plates which originate at them form an integral part of global mantle convection, and there has been considerable research devoted to developing unified models of mantle circulation that generate realistic plate geometries and spreading patterns [e.g., *Bercovici*, 1993, 1998; *Dumoulin et al.*, 1998 and references therein]. This unification is far from complete, however. Most mid-ocean ridge research tends to treat the ridges in isolation from the global mantle circulation, and regards the upwelling of asthenospheric mantle material beneath the ridge as a passive response to plate spreading. As we shall see, this process alone is sufficiently complex that significant unknowns remain. Whether we view sub-ridge mantle upwelling as a passive response to, or the cause of, plate spreading, this upwelling asthenosphere provides the source both of material for lithospheric growth and of the thermal energy flux associated with it.

Initially, near the base of the upwelling mantle column, the heat energy is carried along by advection with the moving, hot asthenosphere. Higher in the rising column, thermal conduction begins to contribute significantly to the net energy flux. One of the consequences of this is that the thermal structure of the newly forming lithosphere is strongly dependent on the ridge's spreading rate – faster spreading ridges are underlain by higher temperatures at any given depth than slower spreading ridges. This in turn influences lithospheric rheology, leading to the formation of thicker lithosphere and deep median valleys at some (slow spreading) mid-ocean ridges, but not at other (generally faster spreading) ones [e.g., *Chen and Morgan*, 1990].

An inevitable consequence of the upwelling of asthenospheric mantle beneath ridges is the initiation of partial melting of the mantle rocks, to form magmas of basaltic composition. The melting is due to the decreasing pressure experienced by the mantle as it rises closer to the seafloor—the pressure drop reduces its solidus temperature, causing the onset of melting and melt segregation. The liquid magmas have viscosities many orders of magnitude lower than those of the surrounding mantle rocks, and significantly lower densities—causing the melt, once it forms, to rise rapidly to the top of the mantle. In this way, once partial melting begins, a significant fraction of the heat flux, as well as material flux, is partitioned into the basaltic melt, and is then rapidly advected upwards.

Once the rising melt reaches crustal levels, it crystallises, releasing latent heat. Further heat is released by cooling of the newly formed crust to temperatures some hundreds of degrees Celsius below its crystallization temperature. Virtually all of the heat carried in the basaltic melt is transferred into seawater by hydrothermal circulation. The hydrothermal heat flux in turn drives chemical exchanges between the crust and the circulating seawater. The influences do not act in only one direction, however. The efficiency and depth of penetration of hydrothermal circulation have a major impact on the depth and dimensions of basaltic magma bodies in the crust. Given a constant rate of replenishment with new melt, a magma chamber will grow until the rate of heat loss from its margins balances the heat supplied by incoming melt. Broadly speaking, the rate of heat loss depends on the surface area of the magma body, and the rate of heat loss per unit area. If heat transfer across the margins is efficient, melt bodies remain small. This balance exerts one of the most important influences on the physical architecture of the newly formed crust. The same factors control the high- and low-temperature alteration of the upper crust, and the deposition of hydrothermal precipitates—further influencing the physical and chemical properties of the crust as it forms and matures.

2.2. Basin Scale Thermal Structure of the Oceanic Lithosphere

The most striking topographic signature in the oceans is the increase in depth of the seafloor away from the mid-ocean ridges. This results from the thickening, cooling and contraction of the newly created lithosphere as it moves away from the spreading axis, combined with isostasy. Both the gross topography and the plate thickening can be explained in terms of the evolving thermal structure of the uppermost mantle. Thermal models which provide excellent fits to global datasets are reviewed in *Stein and Stein* [1996].

The lithosphere can be defined either in terms of its rheological (mechanical) properties, or in terms of the mode of heat transport. Mechanically, the upper part of the lithosphere is cool enough to behave elastically, and corresponds to the rigid part of the oceanic plate. The maximum focal depth of earthquakes coincides approximately with the 700° C isotherm [*Wiens and Stein*, 1983]. This observation defines a 'mechanical boundary layer' which reaches a thickness, in mature oceanic lithosphere, of 40 to 50 km. An alternative rheological definition is in terms of the flexural thickness of oceanic plates, determined from their response over geologically long time scales to the imposition of gravitational loads. The flexural approach results in an 'elastic thickness' whose base coincides approximately with the 400°C isotherm, corresponding to a depth in mature oceanic lithosphere of 20 to 25 km [*Calmant et al.*, 1990]. Deformation in the lower, hotter part of the lithosphere is ductile, so this part contributes little to the strength of the lithosphere. None the less it forms part of the lithosphere, because it moves with the plate rather than convecting as part of the asthenosphere; and heat flux upwards through it is therefore dominated by conduction, rather than by convection. The 1450° C isotherm corresponds to the transition between these two dominant forms of heat transport in

the mantle [*Stein and Stein*, 1996]—and so corresponds also to a sharp change in geothermal gradient, from adiabatic (about 0.3° C km^{-1}) in the asthenosphere, to at least 10°C km^{-1} in the plate. This transition defines the lithosphere in terms of a 'thermal boundary layer'. The general thermal structure of an oceanic plate from *Stein and Stein's* [1996] GDH1 model is reproduced in Plate 1.

Cooling of the oceanic lithosphere is the major source of the heat flux from the planetary interior to the external environment. At ages greater than about 2 Ma, we can represent the thermal structure of oceanic lithosphere (including the lower half to two-thirds of the crust) to a good approximation simply by a linear vertical gradient in temperature from ~ 1450° C at the base of the plate to ~ 0° C at the seafloor. There are, though, some relatively minor small scale variations, both at the base of the plate due to variations in temperature of the underlying asthenosphere, and in the uppermost part of the crust due to continuing low-temperature hydrothermal circulation [*Stein and Stein*, 1994].

Measurements of heat flow through young (<65 Ma) ocean floor show a significant discrepancy between the observed values and those predicted by the thermal model that incorporates only conductive cooling. Observed values are consistently too low—or in other words, part of the heat flux appears to be 'missing'. Heat flow is generally measured on parts of the seafloor blanketed with sediment, and shows only the conductive component of the heat flux. The discrepancy is attributed to convective transport of a significant proportion of the total heat flux, through the action of hydrothermal vents close to the axis or porous, diffuse discharge at greater distances. The proportion of heat flux transported by hydrothermal flow depends primarily on crustal age, although sediment thickness has a secondary effect [*Stein and Stein*, 1994]. The 'missing' conductive heat flow indicates that at the ridge axis, convective transport by hydrothermal circulation dominates the heat flux from the crust into the water column; but that this diminishes progressively, and at ages above ~ 65 Ma virtually all of the remaining heat flux is by conduction.

2.3. Global Rates of Production and Associated Fluxes

In order to address the thermal regime of the ocean crust, the first prerequisite is a knowledge of the total (i.e., global) thermal and material fluxes that are involved in plate construction. The initial step is straightforward. We can calculate the areal rate **B** of generation of new lithosphere by integrating the full spreading rate **s** over the length of all spreading ridges:

$$B = \int s\,dl$$

Table 1. Length of ridge, magmatic flux and magmatically transported thermal flux vs spreading rate. Assumptions as for Table 2

Quantity	Unit	10	30	50	70	90	110	130	150	Iceland	Global Total / All Rates
Spreading Rate	mm a^{-1}	10	30	50	70	90	110	130	150	20	
Length of Ridge	km	14800	15200	6000	10700	6400	600	400	3900	500	58500
Rate of plate generation	km^2 a^{-1}	0.148	0.456	0.3	0.749	0.576	0.066	0.052	0.585	0.01	2.9
Global magma flux	km^3 a^{-1}	0.9	2.7	1.8	4.5	3.5	0.4	0.3	3.5	0.3	18
Global magma flux	kg s^{-1}	7.7 x 10^4	2.4 x 10^5	1.6 x 10^5	3.9 x 10^5	3.0 x 10^5	3.5 x 10^4	2.7 x 10^4	3.1 x 10^5	2.2 x 10^4	1.6 x 10^6
Global magmatic thermal flux	Watt	1.2 x 10^{11}	3.8 x 10^{11}	2.5 x 10^{11}	6.2 x 10^{11}	4.7 x 10^{11}	5.5 x 10^{10}	4.3 x 10^{10}	4.9 x 10^{11}	3.5 x 10^{10}	2.5 x 10^{12}

Plate geometries and spreading rates, determined from numerous sources—e.g., earthquake data, seafloor magnetic anomalies, transform fault orientations, geodetic measurements of present day plate motions, etc.—are well known [e.g., *de Mets et al.*, 1990; *Small and Danyushevsky*, 2003]. In Table 1, the first two rows show the total length of ridge, subdivided according to spreading rate in intervals of 20 mm yr^{-1} full spreading rate, and the area of new lithosphere created per year. The table shows that there are approximately 58,500 km of spreading center currently active, and that these generate approximately 3 km^2 of new lithosphere per year.

Using these data, we can next make a quantitative assessment of the material and heat fluxes associated with magma emplacement and crustal formation. To do this, we rely on the position of the Moho discontinuity, which can be readily located by seismic methods; and on the assumption that it represents a compositional boundary at the base of the crustal section, between ultramafic mantle rocks below, and basaltic crust formed by crystallization of magma above. If on this basis we assume a mean oceanic crustal thickness, then taking the product of this crustal thickness and the areal spreading rate provides an estimate of the total volumetric rate of crustal formation, and hence of the magmatic flux, for each interval of spreading rate. This estimate is also shown in Table 1. We have estimated the mean thickness of oceanic crust to be 6 km, based on compilations of regional to basin scale mean seismic crustal thickness estimates (e.g., *Solomon and Toomey* [1992], *Cannat et al.* [2004]). At slow spreading ridges crustal thickness varies significantly along-axis, depending on distance from segment centers. Although the crust tends to be thicker at segment centers, and much thinner towards segment ends [e.g., *Tolstoy et al.*, 1993], 6 km provides a good estimate of the mean. At spreading rates above about 70 mm yr^{-1}, 6 km is again a good mean estimate [e.g., *Vera et al.*, 1990; *Turner et al.*, 1999]. For very slow spreading rates (< 15 mm yr^{-1}), 6 km is an almost certainly an overestimate—the crust produced at such ridges tends to be thinner than this [e.g., *Muller et al.*, 1997; 1999; *Jokat et al.*, 2003]. However, the very slow spreading ridges contribute such a small share of the total magmatic flux that this discrepancy is unlikely to lead to a serious error in the global figure. Iceland—part of the ridge system that overlies a major mantle plume, and where crustal thickness is in excess of 20 km and may reach 35 km in places—has been included in the table as a special case, represented by a 500 km length of ridge producing on average 25 km thick crust—but this again has surprisingly little influence on the global figure.

A number of interesting features are highlighted by Table 1. Firstly, slow spreading ridges (0 to 40 mm yr^{-1}) make up by far the largest category in terms of ridge length (Figure 1(a)). The second largest category is at intermediate spreading rates, from 60 to 80 mm yr^{-1}. This category includes many back-

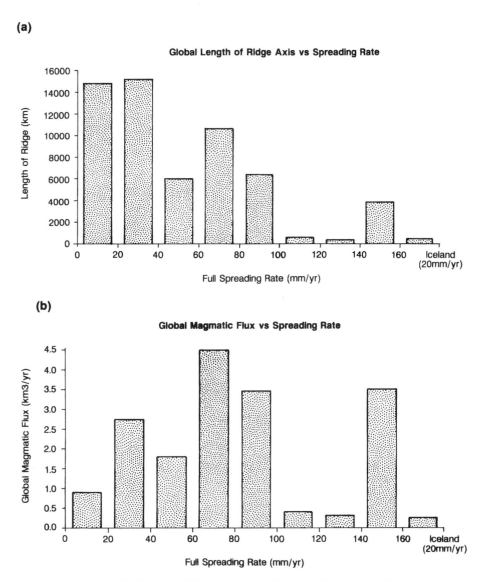

(a)

Global Length of Ridge Axis vs Spreading Rate

(b)

Global Magmatic Flux vs Spreading Rate

Figure 1. (a) Histogram of the global length of ridge axis against full spreading rate. (b) Histogram of the global length of ridge axis against rate of magma production.

arc spreading centers, as well as some spreading centers in the major ocean basins. Fast spreading ridges constitute a significant fraction of the total only because we tend to include in that category a very wide range of spreading rates, from < 90 mm yr^{-1} to > 150 mm yr^{-1}. The table also shows the resultant magmatic fluxes (Figure 1(b)), and in this case we see that melt production is dominated by the fast spreading ridges, especially at around 90 and 150 mm yr^{-1}—although the intermediate spreading rate (70 mm yr^{-1}) ridges are also a major contributor. Summing over all spreading rates we see that the ridges produce approximately 18 km^3 of new crust from basaltic melt per year (equivalent to 1.6 x 10^6 kg s^{-1} of melt production) globally.

Using published values for the latent heat of crystallization and the specific heat of basalts [*Cannat et al., this volume*] (Table 3), we can use the magmatic fluxes to estimate the thermal flux exported from the mantle by the melt (Table 2). The specific heat of MORB composition rocks varies significantly with temperature, ranging from less than 1000 Jkg^{-1}K^{-1} at 350°C to more than 1300 Jkg^{-1}K^{-1} at temperatures above 1200°C. We have estimated a value of 1200 Jkg^{-1}K^{-1} as being representative. Similarly, *Cannat et al.* [*this volume*] note that there remain some uncertainties in the value of latent heat of crystallization of MORB, but suggest a representative value of 0.5 MJ kg-1. In cooling the magma from a totally molten state (1250°C) to a temperature of 350°C (a temperature rep-

Table 2. Magmatic and thermal fluxes associated with magma emplacement and cooling, per km of ridge, as a function of spreading rate

Quantity	Unit									Iceland
Spreading Rate	mm a⁻¹	10	30	50	70	90	110	130	150	20
Magmatic flux per km of ridge	km³ a⁻¹	6×10^{-5}	1.8×10^{-4}	3.0×10^{-4}	4.2×10^{-4}	5.4×10^{-4}	6.6×10^{-4}	7.8×10^{-4}	9.0×10^{-4}	5.0×10^{-4}
Magmatic flux per km of ridge	kg s⁻¹	5.2	15.7	26.1	36.6	47.1	57.5	68.0	78.4	43.6
Latent heat of crystallization of magma per km of ridge (a)	Watt	2.6×10^{6}	7.8×10^{6}	1.3×10^{7}	1.8×10^{7}	2.4×10^{7}	2.9×10^{7}	3.4×10^{7}	3.9×10^{7}	2.2×10^{7}
Heat released during cooling of crust to 350 C per km of ridge (b)	Watt	5.6×10^{6}	1.7×10^{7}	2.9×10^{7}	4.0×10^{7}	5.1×10^{7}	6.2×10^{7}	7.3×10^{7}	8.5×10^{7}	4.7×10^{7}
Total magmatic thermal flux per km of ridge (a + b)	Watt	8.2×10^{6}	2.5×10^{7}	4.2×10^{7}	5.8×10^{7}	7.5×10^{7}	9.1×10^{7}	1.07×10^{8}	1.24×10^{8}	6.9×10^{7}

Table 3. Values of parameters used in estimating net global thermal and material fluxes

Quantity	Symbol	Value	Units	Reference
Specific heat of basaltic crust	$C_{p\,crust}$	1200	J kg⁻¹ K⁻¹	*Cannat et al., 2004*
Latent heat of crystallization of basaltic magma	$H_{f\,crust}$	5×10^{5}	J kg⁻¹	*Cannat et al., 2004*
Mean density of oceanic crust	ρ_{crust}	2750	kg m⁻³	*Solomon & Toomey* [1992]
Mean thickness of oceanic crust	h_{crust}	6	km	See text
Equilibrium thickness of oceanic plate	A	95	km	*Stein & Stein* [1996]
Temperature at base of oceanic plate	Tm	1450	C	*Stein & Stein* [1996]
Specific heat of lithosphere	$C_{p\,m}$	1.171×10^{3}	J kg⁻¹ K⁻¹	*Stein & Stein* [1996]
Density of mantle	ρ_m	3330	kg m⁻³	*Stein & Stein* [1996]
Thermal equilibration age of oceanic plate	t_{eq}	70	Ma	*Stein & Stein* [1996]
Temperature at top of oceanic plate	T	0	C	*Stein & Stein* [1996]

resentative of black-smoker hydrothermal fluids), only about 30% of the thermal energy released is due to latent heat, while almost 70% of it is due to cooling of solidified material. This suggests that the presence of magma may not be a prerequisite for high-temperature hydrothermal systems - the presence of hot but solidified material is likely to be the more important heat source. Summing over all spreading rates, we infer that the total thermal flux carried by basaltic melt at ridge axes is approximately 2.5 x 10^{12} Watts.

The 'missing heat flux' revealed by conductive heat flow measurements provides the most reliable means of estimating the total hydrothermal heat flux through mid-ocean ridges and their flanks. *Elderfield and Schultz* [1996] estimate this value to be 9±2 x 10^{12} W globally, of which 3±1 x 10^{12} W is released over an age range of 0 to 1 Ma, and the remainder is released over an age range of 1 to 65 Ma. This compares with a global oceanic heat flux (conductive plus hydrothermal) of 32 x 10^{12} W, and a global (oceanic plus continental) heat flux from earth's interior of 43 x 10^{12} W. Interestingly, the global magmatic thermal flux estimate from Table 1 of 2.5 x 10^{12} W is in good agreement with the 'missing heat flux' estimate by *Elderfield and Schultz* for crust up to 1 Ma old of 3±1 x 10^{12} W. The similarity of these numbers suggests that the magmatic heat flux into the newly formed lithosphere is approximately balanced by the axial (<1 Ma) hydrothermal heat flux out into the water column.

Another way of looking at these figures is in terms of the net flux of material and heat per square km of newly created sea floor (Table 4.) It is important to bear in mind that the numbers in Table 4 relate only to the part of the thermal flux carried advectively into the newly formed crust by the migrating magma, and associated with immediate cooling to a temperature of 350ºC. There is a further, ongoing thermal flux associated with cooling of the crust away from the ridge axes, to temperatures lower than 350ºC (e.g., we would expect that by ages of a few tens of Ma, the temperature even at the base of the crust would have dropped to less than 100ºC); and much more importantly, there is also the large thermal flux associated with progressive cooling of the solid residual mantle itself, as the plate thickens and ages as it moves away from the spreading center. This solid mantle thermal flux is expressed by *Stein and Stein* [1996] as

$$\Delta Q = \int_{z=0}^{a} A \big(T(z) - T_0 \big) C_{pm} \, \rho_m \, dz$$

where ΔQ is the heat lost over surface area A, $T(z)$ is the temperature of the plate as a function of depth beneath the sea floor, and other parameters are as listed in Table 3. *Stein and Stein* [1996] have estimated this from their thermal models using those values, and find that the heat released in cooling the lithospheric mantle until it reaches its equilibrium thermal

Table 4. Typical fluxes per km^2 of new sea floor

Description	Quantity
Volume of melt	6 x 10^9 m^3
Mass of melt	1.65 x 10^{13} kg
Latent heat released	0.8 x10^{19} J
Heat released during cooling of solid from 1250°C to 350°C	1.8 x 10^{19} J
Total magmatic heat	2.6 x 10^{19} J

structure at an age of about 70 Ma is approximately 27 x 10^{19} J per square km. This compares with the magmatic (axial) thermal flux of only 2.6 x 10^{19} J per km^2 (Table 4). Thus, by a plate age of 70 Ma, the process of seafloor spreading will have released for each square km of plate nearly 30 x 10^{19} J of thermal energy into the ocean; of which about 90% (27 x 10^{19} J) will have resulted from gradual thickening and cooling of the lithosphere, and only about 10% (<3 x 10^{19} J) will have resulted from crystallization of magma and rapid cooling of the crust to a temperature of 350°C, close to the ridge axis.

3. THERMAL REGIME OF THE MANTLE BENEATH MID-OCEAN RIDGES

3.1. Mantle Flow Patterns

As the mantle upwells beneath a ridge, it decompresses, crosses the solidus and begins to melt. Observations at the seafloor show that new crust is formed within a narrow (no more than 2 to 3 km) neovolcanic zone at the ridge crest, yet simple numerical models of mantle flow suggest that melting should occur over a much broader (> 50 km, perhaps hundreds of km wide) region of the mantle. This first order observation requires a mechanism to focus the melt towards the ridge, and places a strong constraint either on the actual pattern of upwelling within the melting region, or on the method of melt transport within the mantle.

Geochemical evidence requires rapid extraction of the melt from the melt source, suggesting an efficient transport network of melt through the overlying mantle. Geochemical data also place constraints on the depth range of melting, and suggest that most melting is initiated between depths of 50 and 100 km, with the melting column extending into the garnet stability field [e.g., *Salters and Hart*, 1989]. The presence of water may extend the melting to as deep as 120–150 km [*Hirth and Kohlstedt*, 1996]. A full review of the geochemical constraints on melting is given in *Kelemen et al.* [1997a].

If we think of the ridge in terms of a 2-dimensional (2-D) structure (i.e., one in which materials flow vertically and in the across-axis direction, but there is no transport of either energy or material in the along-axis direction), then we can consider two contrasting model classes that have been proposed to illustrate how melt is generated and how it is focussed towards the ridge crest. Passive flow models predict melting over a broad region of mantle (100–200 km wide at a depth of 60–70 km); while buoyant flow models, which incorporate the dynamic consequences of the melting, predict much narrower melt columns, tens of km wide [e.g., *Scott and Stevenson*, 1989; *Buck and Su*, 1989]. The difference between these types of models is straightforward. In passive flow models, the pattern of flow in the asthenospheric mantle beneath the ridge is controlled solely by the separation of the plates. Mantle upwells beneath the ridge and then turns to flow horizontally beneath the spreading oceanic plates. No additional driving forces for the flow are introduced into the model. An example of the resulting flow pattern is illustrated in Figure 2(a).

In reality there are significant additional forces that act on the overall pattern of mantle flow, of which probably the most important are the variations in density caused by partial melting. During partial melting, basaltic magma is produced from the original ultramafic mantle material. The melt has lower density (about 2900 kg m^{-3} compared to about 3300 kg m^{-3}) than the mantle source materials. Furthermore, and counter-intuitively, the depleted mantle rocks that are left behind as the residue from partial melting also have lower density than the original, undepleted asthenosphere. This is because the melting tends to partition heavier elements—notably iron—into the melt. As a consequence, the onset of melting leads to a significant reduction in overall density of the upwelling mantle column, and this in turn generates a significant buoyancy-driven component to the mantle flow pattern beneath a ridge.

In passive flow models, low melt fractions (around a few tenths of a percent) are predicted within a broad melting region extending 100 km or more off-axis (Figure 2(a)) and to depths of 50–60 km, within a roughly triangular (or 'tent-shaped') melt column that narrows upwards. When the buoyancy effects of melting are included in the model, the mantle flow becomes more complex. We can think of the resultant flow as resembling buoyancy-driven convective rolls beneath each flank of the ridge due to the density changes from melting, superimposed onto the passive flow resulting from plate separation [e.g., Figure 2(b)(c)(d); *Scott and Stevenson*, 1989; *Scott*, 1992]. The net effect is that more tightly focussed upwelling and melt generation is predicted beneath the ridge, with a depleted residual mantle produced just off-axis. Another important feature of this class of models is that the degree of focusing depends on spreading rate. At slow spreading rates, the buoyancy driven component of the flow is large compared to the plate-separation driven component, leading to strong focusing of upwelling and melting. However, at fast spreading rates, the plate-separation driven component dominates, so

that the resulting models are only slightly modified from those predicted by passive models.

Both the passive and the buoyancy-driven models presented here are over-simplistic. Other important factors to be considered include the effects on rheology of temperature [*Scott*, 1992], melt fraction [*Buck and Su*, 1989] and the presence of water [*Braun et al.*, 2000]. The relationships between upwelling rate, melt fraction, mantle and melt viscosities, temperatures and densities are sufficiently complex, recursive and poorly constrained to introduce significant uncertainties into such models. Nonetheless the interactions between these different factors almost certainly tend to concentrate both the upwelling and the melting (and hence high mantle temperatures, and a large thermal flux associated with the magma) into a narrower region beneath the axis than is predicted by the passive flow models.

In the preceding few paragraphs, we assumed that the flow pattern is 2-D—i.e., flow does not occur in the along-axis direction. In reality, ridges are highly segmented systems, and therefore we should expect mantle properties to vary in the along-axis dimension. A further consequence of buoyancy driven flow is that there is a tendency for any local increase in upwelling at any point *along* the ridge to lead to an increase in melting, and hence to an increase in buoyancy forces locally and hence to a reinforcement of the local increase in upwelling rate. In other words, upwelling has a tendency to become unstable in the along-axis direction, and to break up into a series of upwelling columns distributed along the ridge axis [e.g., *Whitehead et al.*, 1984]. This tendency is countered by the flow pattern imposed by plate separation, which is more 2-D [*Lin and Phipps Morgan*, 1992]. The balance between these two tendencies depends on three key factors. Firstly, how strong is the feedback between upwelling, melting and resulting buoyancy forces? The stronger this is, the greater the tendency for the flow to break up into narrow upwelling columns. Secondly, how fast are the plates separating? Faster spreading rates increase the relative importance of the passive component of flow relative to the buoyancy driven component. We expect fast spreading ridges to be more 2-D, and to have more sheet-like (i.e., invariant-along-axis) patterns of upwelling in the underlying mantle. Lastly, if we increase the viscosity of the mantle in our models, we reduce the amount of buoyancy-driven flow generated by a given density distribution; while simultaneously increasing the strength of forcing of the flow related to the passive response to plate separation. The pattern of upwelling predicted by models is therefore highly dependent on mantle viscosity. Higher viscosity promotes 2-D upwelling (sheets), while lower viscosity promotes 3-D upwelling (columns). These effects have been modelled and illustrated by (e.g.) *Phipps-Morgan and Forsyth* [1988], *Shen and Forsyth* [1992], and *Parmentier and*

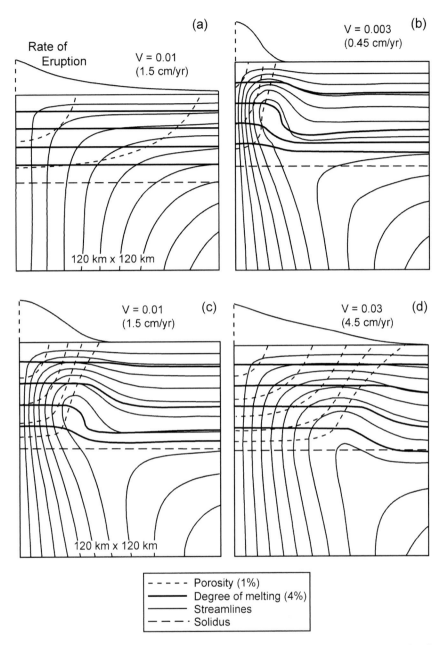

Figure 2. (a) An example of the pattern of mantle flow beneath a mid-ocean ridge, predicted by a passive flow model. Fine solid lines are mantle stream lines. Heavy lines are contours of the degree of partial melting which has taken place up to that point. Short-dashed lines indicate the instantaneous melt content. The long-dashed line indicates the location of the mantle solidus, i.e., the onset of partial melting. The curve (top) shows the predicted distribution of melt delivery to the base of the crust as a function of distance from the ridge axis. (b), (c), (d) Examples of the pattern of mantle flow predicted by models which include a component of buoyancy-driven flow. (b) for a half spreading rate of 4.5 mm a[-1]. (c) for a half spreading rate of 15 mm a[-1]. (d) for a half spreading rate of 45 mm a[-1]. From *Scott & Stevenson* [1989].

Phipps-Morgan [1990], some of whose results are illustrated in Figure 3.

While models of the type presented here can provide useful insights into mantle thermal structure, flow patterns and melt distribution, and indeed can help to explain our observations of ridge properties both along- and across-axis and as a function of spreading rate, remaining uncertainties mean that we cannot rely on the models to provide exact represen-

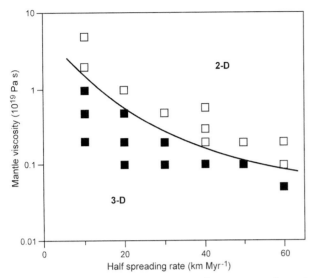

Figure 3. The transition from 2-D to 3-D flow patterns of mantle upwelling beneath a ridge, depending on spreading rate and upper mantle viscosity, predicted by numerical models. From *Parmentier & Phipps-Morgan* [1990].

tations of real processes. For example, the along-axis wavelength of 3-dimensional (3-D) upwelling in the models of Parmentier and Phipps-Morgan does not match the typical wavelength of ridge segmentation at slow spreading ridges [*Jha et al.*, 1994]. As yet there is no direct quantitative match between 3-D upwelling patterns inferred from numerical models, and segmentation observed at the sea floor.

3.2. Melt Focusing

There is a further important step between the generation of basaltic melt in the upwelling mantle beneath a ridge, and its delivery to a particular location at the base of the crust or into a magma chamber. That is the migration mechanism and path taken by the magma upwards through the mantle. Simple passive flow models do not focus melt into a sufficiently narrow region beneath the ridge crest to match observations, while even the more sophisticated flow models are unable to achieve this at anything other than the slowest spreading rates. Almost certainly this mismatch can be largely attributed to the non-vertical migration of melt once it has formed. The trajectory of each particle of magma between the point where it forms and separates from its host rock, and the point where it is delivered into the base of the crust, typically includes a significant horizontal component inwards towards the ridge axis (and possibly a component along axis as well). *Spiegelman & McKenzie* [1987] proposed that in the case of the passive flow models, a region of pressure drop immediately beneath the ridge axis is responsible for sucking melt inwards from far

off-axis. *Phipps-Morgan* [1987] proposed that the pattern of mantle flow leads to anisotropic (directionally dependent) permeability within the melting region, leading to further melt focusing. Another suggested focusing mechanism involves vertical melt migration to the base of the relatively impermeable lithosphere [*Sparks and Parmentier*, 1993; *Ghods and Arkhani-Hamed*, 2000] or crust [*Garmany*, 1989], at which point melt migrates up the slope of the impermeable layer base towards the ridge crest. *Buck and Su* [1989] predict that the additional decrease in mantle viscosity as a result of the presence of melt further influences the flow pattern of the magma, as well as of the solid mantle matrix.

Many observations, both geophysical and from ophiolites, suggest more complex mechanisms of melt flow than any of those considered above. Such models invoke branching networks of high-permeability channels in the mantle as the major transport mechanism for melt, rather than diffuse or porous flow. These types of model and the evidence for them are discussed in detail in *Kelemen et al.* [1997a; 2000].

Isotope geochemistry can provide some constraints on whether mantle melt transport occurs in focussed, high permeability conduits or by diffuse, porous flow. *Sims et al.* [2002] interpret an observed negative correlation between ^{226}Ra and ^{230}Th as evidence for a balance between these two flow mechanisms. *Jull et al.* [2002] have attempted to quantify this balance. They estimate that approximately 60% of the melt flux is transported by a porous flow mechanism, while 40% travels through high permeability channels that occupy only a small percentage of the mantle volume that undergoes melting. Mixing of the two melt populations is predicted to occur near the top of the melting column, where—at depths around 25 km—melt is transferred into channels. This depth, however, is not well constrained and channelized flow could extend deeper.

Recent numerical modelling by *Spiegelman et al.* [2001] has shown how it is possible to simulate a mantle flow regime in which melt transport at deeper levels is dominated by porous diffusion around grain boundaries, while at higher levels the upwelling melt becomes partitioned into a branching network of much higher permeability channels, separated by a low permeability matrix. This mechanism relies on the channels behaving as reactive features (i.e., chemical exchanges occur between the solid and melt phases within the channels), and on the presence of a vertical gradient of solubility of the solid within the melt.

Detailed mapping and chemical analysis of dunite channels in ophiolites [*Braun et al.*, 2002] indicate that these represent the preserved traces of a high permeability melt network of this kind. The network appears to have a riverine structure, converging on a region of high melt fraction in the uppermost mantle. It is fed by smaller branching tributaries that tap the deeper mantle source, through which melt percolates by porous flow [*Hart*, 1993; *Kelemen et al.*, 2000]. The solidified dunite channels continue to have substantially higher permeabilities than the host mantle, making it likely that these features are successively re-occupied as conduits for melt flow [*Zhu and Hirth*, 2003].

While we have outlined some of the existing and emerging models of melt migration and melt focussing, there are very few geophysical constraints to help discriminate in favour of any of them. This is largely because of a lack of co-ordinated large scale experiments that are able to provide resolution over the mantle depths critical to melt flow, but also because at the depths over which these small scale processes are occurring, geophysical resolution is inherently limited.

3.3. Complexity and Observations—MELT and Other Results

There have been few geophysical experiments that are able to map the distribution of melt in the mantle. A notable exception is the MELT study [*Forsyth*, 1992], which aimed to do just that at a super-fast spreading section of the southern East Pacific Rise (SEPR). The experiment featured a combination of electromagnetic (EM) and seismic methods [*MELT Seismic Team*, 1998; *Evans et al.*, 1999]. Both components involved deploying seafloor instruments for about a year. The EM experiment used the naturally occurring variations in the Earth's magnetic and electric fields to measure the electrical resistivity of the mantle to depths of around 200 km. The seismic experiment recorded signals from large earthquakes around the Pacific rim, which provided a variety of waveforms with which to constrain the seismic velocity and anisotropy in the mantle. Both electrical resistivity and seismic velocity are influenced by the presence of melt, and the experiment was designed to constrain the pattern of melt delivery from the underlying mantle to the ridge crest. The SEPR near 17°S was chosen as the site for this study because of its fast (~150 mm yr^{-1}) spreading rate and linear ridge morphology. These characteristics were expected to maximize the two-dimensionality of mantle flow. Despite this, the ridge has clear asymmetry both in spreading, with faster absolute plate motion on the Pacific plate resulting in a slow westward migration of the ridge axis with respect to the hot-spot reference frame, and also in terms of seafloor bathymetry, with less subsidence of the seafloor with distance from the ridge to the west than to the east [*Scheirer et al.*, 1998].

The primary result from seismic analyses is that melt generation and transport occurs over a broad region of the mantle, with no evidence for a narrow column of melt beneath the ridge that would be indicative of dominantly dynamic melt focusing. Asymmetry between the eastern and western sides of the ridge is seen. A zone of 1–2% melt is inferred in the top 100 km of mantle, extending some 150–200 km to the west of the ridge; but melting appears to shut off abruptly to the east of the ridge [*MELT seismic team*, 1998; *Forsyth et al.*, 1998; *Toomey et al.*, 1998]. Recent work has extended these results through more detailed analyses of the teleseismic P and S wave arrivals [*Hammond and Toomey*, 2003], and inversions of Rayleigh [*Forsyth and Dunn*, 2003] and short period Love waves that constrain structure within the uppermost 100 km of mantle [*Dunn and Forsyth*, 2003]. These latter inversions show that the asymmetry in structure extends upwards as far as the base of the crust, although it is less pronounced at these shallower depths (Plate 2). Also, the authors note that the transition from a broad melt bearing region deeper in the mantle to the narrow zone of crustal accretion occurs over a depth interval extending downwards for about 15 km from the base of the crust.

Modeling of the EM data also shows asymmetry in the underlying mantle, with higher conductivities to the west of the ridge, and with very little evidence for melt to the east [*Evans et al.*, 1999]. The transition laterally from conductive to resistive mantle appears to be quite sharp. Electrical resistivity in the mantle is influenced by the presence of water in the form of hydrogen dissolved in olivine [*Karato*, 1990], as well as by the presence of melt. Water plays a key role in the melting process: first, it can lower the solidus temperature facilitating more extensive melting. Second, it lowers viscosity [*Hirth and Kohlstedt*, 1996] and so impacts the mantle flow patterns beneath the ridge. Because water is preferentially partitioned into melt, there is a complex feedback between water content, flow patterns and melting [*Braun et al.*, 2000]. Off-axis, the MELT EM results show a transition downwards from a dry shallow mantle to a damp mantle below, at depths of about 60–80 km. Rayleigh wave models show a similar feature [*Forsyth et al.*, 2000]. Closer to the ridge, untangling regions of decreased resistivity due to the presence of water from those due to melt is harder. The MELT results suggest that the 60–80 km thickness of dry uppermost mantle off-axis represents that part of the sub-ridge mantle that underwent melting before being rafted away from the ridge crest. Anisotropic inversions of the EM data give further support to this interpretation, by showing lower resistivity in the direction of plate spreading within the 'damp' region of mantle [*Baba et al.*, in prep]. The resistivity of dry olivine is only weakly anisotropic, but the diffusion of hydrogen is much more efficient along its crystallographic a-axis. Lower resistivity in the direction of plate spreading can therefore be interpreted in terms of the effect of water in an anisotropic mantle, with the a-axis of olivine aligned with flow direction.

While there is some general agreement between the EM and seismic results, there are differences in the shapes of anomalous

GDH1 PLATE MODEL

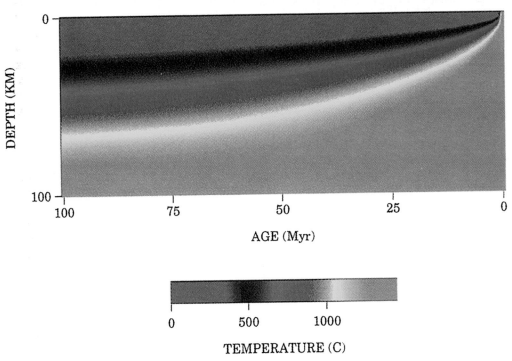

Plate 1. The basin-scale thermal structure of oceanic lithosphere, based on global compilations of bathymetry and heat flow data, and based on the GDH 1 model. From *Stein & Stein* [1996].

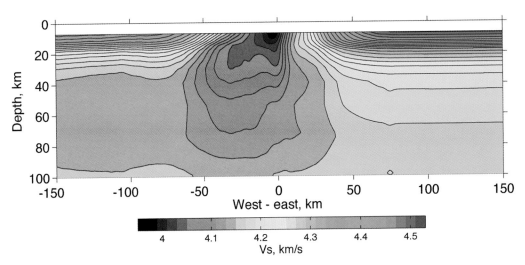

Plate 2. A cross-section of shear velocity from inversions of short period Love wave [*Dunn & Forsyth*, 2002] data from the MELT experiment. The inversion reveals an asymmetric velocity structure with slower velocities to the west of the ridge within the upper 60 km. This asymmetry extends upwards to the crust-mantle boundary, with the transition from a broad region of upwelling, melt bearing mantle, to the tightly focussed melt supply that feeds the ridge axis occurring over a ~15 km depth interval in the uppermost mantle. The region of low velocities extends to the base of the crust. To the east of the ridge, shear velocities increase rapidly.

regions of low seismic velocity and low electrical resistivity that have yet to be reconciled. In a broad sense, the geophysical results point to a passive flow model of melt generation within the mantle. Seismic modelling [*Hung et al.*, 2000] rules out the type of narrow, pipe-like low velocity feature that would be suggestive of a highly focussed melt supply. There is, however, a hint from Rayleigh wave models that melt is preferentially supplied along axis to the middle of the ridge segment, hinting at three dimensionality in melt supply, even at this super-fast spreading ridge [*Forsyth and Dunn*, 2003].

Seismic anisotropy inferred from splitting of SKS phases (i.e., shear waves that have been reflected at the core-mantle boundary), teleseismic P and S travel times and Rayleigh waves are all consistent with a mantle flow that contains an eastward component of pressure driven return flow from the South Pacific Superswell [*Toomey et al.*, 2002; *Conder et al.*, 2002]. SKS arrivals show greater splitting beneath the Pacific plate than to the east of the ridge [*Wolfe and Solomon*, 1998]. The SKS splits range in magnitude from about 2–3s west of the ridge to about 1s to the east. They are the result of anisotropic structure along the ray path between the core-mantle boundary and the seafloor. Beneath the 410 km discontinuity the mantle is believed to be isotropic, and so splitting is usually attributed to the fabric alignment of olivine in the upper mantle. How this anisotropy is distributed with depth in the upper mantle is not well constrained, however. Rayleigh wave velocities are able to constrain the depths of some of the anisotropy, and are azimuthally anisotropic in the uppermost 50 km of mantle, with the same fast direction as that seen in SKS analyses [*Forsyth and Dunn*, 2003].

The return flow hypothesis offers a means to explain the asymmetry in melt content on either side of the ridge, with the return flow bringing hotter mantle towards the ridge from the west. *Dunn and Forsyth* [2003] further argue that the apparent lack of melt immediately to the east of the ridge argues for a component of downwards flow here, as any upward migration of the mantle matrix should result in decompression melting—for which there is no evidence. One problem with the pressure driven return flow model is that while it explains most of the geophysical observations in the MELT area (asymmetric bathymetry, seismic velocities etc) it still does not adequately explain the larger shear wave splitting to the west of the ridge. The reason for this is that the predicted flow is almost uniformly from west to east, and so significant splitting should also be seen to the east of the ridge, where the flow is still expected to be eastwards, although at lower velocity. *Conder* et al. [2002] suggest that this discrepancy may be due to differences in the thickness of the anisotropic layer or to temperature differences not accounted for in their model. It is also possible that more complex flow patterns to the east, possibly including a component of downflow, remove some of

the fabric alignment on the eastern side of the ridge, producing the smaller amounts of anisotropy and splitting seen here compared to beneath the Pacific Plate. In any event, the upper mantle flow patterns in these emerging models are predominantly west to east, with only a small component of vertical flow. This is radically different from conventional ideas of upwelling beneath midocean ridges.

There is very little direct geophysical imaging of the mantle beneath slow spreading ridges. Gravity data [e.g., *Kuo and Forsyth*, 1988; *Lin et al.*, 1990; *Detrick* et al., 1995] and modelling [e.g., *Whitehead et al.*, 1984] suggest that melt delivery at slow spreading rates is strongly influenced by buoyancy driven effects and is likely to be highly segmented or diapiric in plan view. Seismic observations of thicker crust in the middle of slow spreading segments [e.g., *Tolstoy et al.*, 1993; *Purdy and Detrick*, 1986; *Hooft et al.*, 2000] also indicate that melt is either formed in the mantle primarily in mid-segment, or is channelled towards the central regions of the segment. Further evidence comes from a seismic tomography experiment at 35°N on the MAR, which shows sub-vertical pipe-like crustal velocity anomalies beneath the center of the ridge segment. These are interpreted as regions of lower crust that have recently been the locus of melt delivery [*Magde et al.*, 2000]. If this is typical of slow-spreading segments, then along-strike transport of melt at mid- to upper-crustal levels from the centers to the ends of segments is likely to be an important component of crustal construction.

Models which predict 3D upwelling and melt generation at mid-segment typically predict along-axis wavelengths that are longer than those of ridge segmentation on the MAR (150–400 km in the models compared to typical segment lengths of 50 km). More recent modelling of mantle Bouger anomalies and crustal thickness variations along portions of the MAR combined with models of dynamic flow have suggested a combination of upwelling that is driven by plate spreading and is largely 2-D but with three dimensional melt migration at the base of a sloping and impermeable lithosphere [*Sparks and Parmentier*, 1994; *Magde et al.*, 1997; *Magde and Sparks*, 1997]. Cooler thicker lithosphere at segment ends and thinner hotter lithosphere in mid-segment tends to focus melt into the mid-portions of the segment.

Seismic data from slow spreading ridges indicate that even directly beneath the ridge axis in the centers of magmatically active segments, the uppermost few km of the mantle has generally similar properties to off-axis mantle—i.e., there is no evidence (in the form of lower seismic velocity or absence of wide-angle reflections from the Moho) for a significant porous melt phase [e.g., *Navin et al.*, 1998]. Electromagnetic data that constrain upper mantle structure beneath a slow spreading ridge are sparse, but consistent with a transition from the pervasive distribution of a porous melt phase at depth to either unmelted uppermost mantle or melt restricted to highly channelized and dominantly vertical pathways, with the transition occurring at a depth of a few tens of km beneath the sea floor [*Heinson et al.*, 2000].

While most studies regard the ridges as isolated features, we have seen that across the SEPR return flow in the mantle from a feature several thousand kilometres away appears to have a substantial effect on the melting regime beneath the ridge. This should remind us that the ridges do not exist in isolation from the global mantle convection system. It also raises the questions of the extent to which the morphology of ridges—both in terms of their across axis shape and also in terms of their segmentation—are inherited from the underlying mantle flow; and the extent to which ridges themselves impart flow characteristics onto the mantle below. The MELT experiment suggests strongly that on a large scale the ridges are heavily influenced by the underlying mantle circulation.

3.4. Melt at the Crust-Mantle Boundary

Another way to approach the links between ridge segmentation and mantle flow is to look at the distribution of melt at the base of the crust and its relationship to ridge morphology at the seafloor. In this section we review further evidence for 3D melt distributions in the uppermost mantle beneath ridges, and suggestions that ridge segmentation is closely linked to magma supply from the mantle.

On the East Pacific Rise (EPR) near 9° 30′ N, seismic tomography [*Dunn and Toomey*, 1997; *Dunn et al.*, 2000] provides evidence for a 3D velocity anomaly at the base of the crust and in the uppermost few km of the mantle, which is interpreted in terms of the presence of a small melt fraction. The melt distribution in the mantle is segmented along axis in a fashion similar to that of the ridge crest. For example, the velocity anomaly pinches out northwards close to a deviation in axial linearity (deval) at 9° 35′ N, while the mantle midway between this and another deval at 9° 28′ N appears to be supplying abundant melt to the base of the crust.

Partial melt at 9° 30′ N is seen over a greater across-axis width in the uppermost mantle than in the crust [*Dunn et al.*, 2000], and a similar result has also been seen in the MELT area [*Dunn and Forsyth*, 2003]. A related observation is that compliance increases as far as 10 km off-axis at 9° 48′ N, suggesting a region of partial melt at the base of the crust this far off-axis [*Crawford and Webb*, 2002]. Further tantalizing geophysical evidence for melt at the base of the crust comes from unusual seismic phases recorded on the EPR at 12° 50′ N [*Garmany*, 1989]. The phase, which is a conversion from compressional to shear wave at the Moho (PmS), is rare and has not been reported elsewhere, although this is most likely due to limitations of other experimental geometries. The melt

distribution necessary for the generation of this phase is a melt sill, although the thickness of the layer is not well constrained, and could range from 10m to more than 100m. The sills are seen at a distance of around 20 km from the ridge axis. All of this evidence is consistent with a model in which some melt is supplied to the base of the crust over a fairly broad (at least 20 km across-axis) region of uppermost mantle; and in which focusing of melt occurs at the base of the crust and over a vertical distance of a few kilometres, reducing the width of the crustal neovolcanic zone to no more than 4–5 km across axis. One possible mechanism is that the base of the crust forms a sloping impermeable boundary to vertical melt migration, so that melt ponds at this boundary and flows upslope towards the ridge [*Sparks and Parmentier*, 1994].

3.5. Mantle Thermal Structure: Conclusions

Important progress has been made in recent years towards understanding the patterns of mantle flow, melt generation and melt migration that dictate the thermal regime that oceanic crust at ridges inherits from the underlying mantle. There remain, however, some considerable areas of uncertainty, related especially to the variation in behaviour with spreading rate; the relationship between segmentation at crustal levels and the pattern of flow in the mantle; the relative importance of focused upwelling and melting versus focused melt migration; and time variability in these systems, especially at slow spreading rates.

4. MAGMA CHAMBERS AND THE PLUTONIC CRUST

Geophysical and petrological evidence for the distribution and behaviour of crustal magma bodies at ridges was reviewed by *Sinton and Detrick* [1992]. In this section, we aim to bring some of the issues raised before and since that paper up to date. Melt within the crust can be identified by several geophysical methods. The most familiar technique is seismic reflection, which has been used to map midcrustal reflectors interpreted as the top of axial magma chambers (AMC) [*Detrick et al.*, 1987; *Kent et al.*, 1990; 1993]. The regions below these midcrustal reflectors were until recently not well constrained geophysically, although experiments using tomographic inversions of active source seismic arrivals [*Toomey et al.* 1990; *Dunn and Toomey* 1997; *Dunn et al.*, 2000] and compliance techniques [*Crawford and Webb*, 2002] have started to place constraints on melt distributions within the middle and lower crust, and also in the uppermost mantle immediately beneath the ridge. While a complete review of magma chamber structure and dynamics is beyond the scope of this paper, we will attempt to outline the first order constraints placed on melt within the crust by geophysical methods—and the implications that this has for ridge thermal structure.

4.1. Magma Body Properties at Fast and Intermediate Spreading Rates

There is a first order difference between the physical crustal structure of fast and slow spreading ridges. Fast and intermediate spreading ridges are characterised by readily detectable crustal magma bodies, while at slow spreading ridges such features are generally absent. Typical of the AMC reflectors seen at fast spreading ridges are those imaged between 9° and 10° N on the EPR, where the preferred model of the axial magma chamber consists of a narrow mid-crustal melt layer (approx 1500–2000m below the seafloor) with typical widths of 500m or so [Detrick et al., 1987; Kent et al., 1990, 1993; Babcock et al., 1998] (Plate 3). These features have been found along a high percentage of the EPR that has been surveyed, and have been followed along strike continuously for lengths of 10's of kms, but are also discontinuous on similar length scales, often related to segmentation at the seafloor [Detrick et al., 1993]. Similar features are also found on the SEPR [Detrick et al., 1993; Kent et al., 1994; Singh et al., 1999]. Detailed modelling of the reflections from these bodies shows that they are of limited vertical extent and they are interpreted to contain a fairly large melt fraction. Hussenoeder et al. [1996] studied reflection amplitudes from the southern EPR and concluded that the AMC reflector could be attributed to a 10–100m thick melt lens with a well defined roof, and a floor that was not completely molten. Singh et al. [1999] carried out full waveform inversions of reflections from the zone around an AMC on the SEPR. In some cases, shear wave velocities within the melt lens at the top of the AMC are zero, indicating a liquid composition. Elsewhere a short distance along axis the shear velocity is low but not zero, indicating that the contents of the 'melt lens' may vary between melt and mush on lengths scales of order 10 km. The melt lens is found to be only 50m thick, and has both a solid roof and floor.

Very similar AMC features have been documented at two back-arc ridges: the Valu Fa ridge in the Lau Basin [e.g., Collier and Sinha, 1990], and the East Scotia Ridge in the South Atlantic [e.g., Livermore et al., 1997]. MCS data from the intermediate spreading rate Juan de Fuca ridge shows an almost ubiquitous midcrustal reflector interpreted as the top of an AMC, even in the rifted Endeavour segment [Detrick et al., 2002]. At intermediate and fast spreading rates, crustal magma bodies are clearly very common if not ubiquitous, and share many similarities irrespective of tectonic setting.

The reflector beneath the Juan de Fuca ridge is (as on the EPR) less than 1 km wide. It is shallowest and brightest beneath the inflated central part of the Endeavour segment, confirming a link between melt supply and ridge morphology. This apparently extensive magma lens suggests that the driving mechanism for hydrothermal circulation at intermediate spreading rates is similar to that of fast spreading ridges, although the depth to the melt lens is greater on Endeavour (as it is at other intermediate spreading rate ridges) and the rifted morphology here is likely to result in a different permeability structure, potentially altering the way in which the fluids tap the heat source.

4.2. Thermal Structure of the Lower and Middle Crust

We do not fully understand how the lower part of the crust is formed at fast and intermediate spreading rates, and there are competing models to explain the process. These models embody different modes of melt storage and accumulation within the crust, and especially different depths at which melt crystallises to form gabbro. This region acts as a reservoir for a large amount of thermal energy, and its mode of formation and cooling has important consequences in terms of heat exchange between crust and hydrothermal fluids and on the maximum depth of circulation of the fluids. Two end member model classes use contrasting assumptions for where melt is crystallised and, hence, where latent heat is released. In the first class, melt crystallisation occurs only near the top of a crustal magma chamber (at or immediately below the AMC reflector melt lens), and the resulting gabbros are rafted down and away from the ridge crest with plate spreading [e.g., Phipps-Morgan and Chen, 1993; Henstock et al., 1993; Quick and Denlinger, 1993]. These are often referred to as 'Gabbro Glacier' models. They predict considerable straining of the lowermost gabbros, and this should result in specific foliation patterns that are not seen in ophiolites [Korenaga and Kelemen, 1998]. Models of the second type recognise that there are seismically mapped melt distributions at the base of the crust, and suggest that melt could be distributed throughout the lower crust (for which there is further evidence from compliance measurements [Crawford and Webb, 2002]) in the form of thin sills [Boudier et al., 1996; Kelemen et al., 1997b; Kelemen and Aharonov, 1998]. These models are also based on observations in ophiolites of modal layering of the entire gabbro sequence [e.g., Browning 1984]. Modelling suggests that in order to reproduce the chemical signals seen in ophiolites, the transport of melt from the base of the crust to the midcrustal magma chamber should be through some form of channelized or focused network rather than a diffuse porous flow network—which would result in extensive reaction and elimination of the observed chemical layering [Korenaga and Kelemen, 1998]. While proponents of both classes of models accept that there are likely to be significant amounts of melt within the lower crust at depths substantially below the

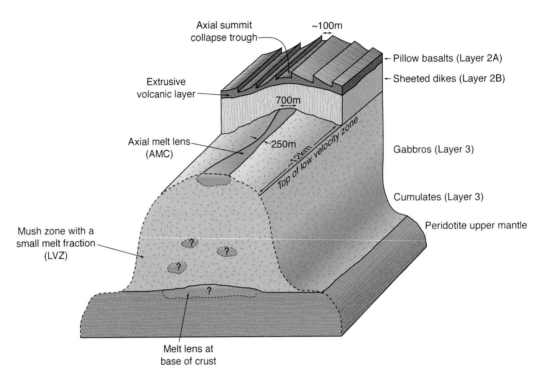

Plate 3. A composite model of crustal and upper-mantle structure at a fast spreading mid-ocean ridge based on the cartoon of *Kent et al.* [1990], and extended into the lower crust and upper-mantle based on more recent geophysical observations (discussed in the text). The upper crust consists of a surficial layer of fractured high porosity extrusives (layer 2A) which generally thickens with distance away from the ridge crest, reaching maximum thickness a few kilometers away. The extrusives are fed by dikes which form the seismic layer 2B. The Moho boundary and the boundary between seismic layers 2A and 2B are inferred from both reflection and wide-angle seismic results, while the boundary between Layers 2B and 3 is inferred from wide-angle results only. A key feature is the narrow and thin melt sill (AMC) typically seen as a bright seismic reflector at depths of 1–2 km below the seafloor. This AMC varies along axis in both width and depth, and also in the nature of the seismic reflection, suggesting that its composition varies between pure melt and a 'mush' with a high melt fraction. Below this sill is a region of low seismic velocities (LVZ) thought to contain a few percent partial melt. Although the distribution of melt within the LVZ remains largely unknown, there is increasing evidence that the melt is unevenly distributed, with local concentrations that may take the form of sills or lenses (shown by the red regions with question marks in the figure). The outer and lower edges of the 'mush' zone (seismic LVZ) are probably hot but completely solid. There is evidence for melt ponding at the base of the crust, although the size of the pond is not well constrained (see Plate 4). At the crust-mantle transition, tomography experiments on the EPR show a broadening of the hot, potentially melt bearing region suggesting that significant melt focussing may occur at the base of the crust (see Plate 5). Two important unknowns are: (a) Is the melt at levels deeper than the AMC crystallizing in-situ, or does most crystallization occur only after the melt has risen to the top of the system and come into close contact with hydrothermal circulation in layer 2? (b) Is the movement upwards of melt within the mid and lower crust—probably forming the local 'ponds' or 'sills' at the Moho or in the lower crust—mostly occurring by porous flow around grain boundaries within the larger mush zone, or by flow through high permeability channels which link and feed (and possibly drain) the local melt accumulations?

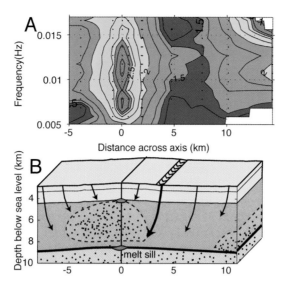

Plate 4. Compliance results from 9°50′N on the EPR [*Crawford & Webb*, 2002]. This relatively new technique measures the shear velocity structure of the crust and is able to constrain the structure of melt bearing regions within the lower crust and upper mantle. The top panel shows compliance as a function of frequency (low frequencies sense deeper structure) and distance across the ridge. High compliance values suggest low shear-wave velocities. The lower panel is an interpretive cartoon of the data. Evidence is seen for a broad region (6–8 km wide) containing a few percent melt underlying the melt lens. At 9°N, this melt body is asymmetric and extends further to the west of the ridge than to the east. There is also evidence for melt ponding at the base of the crust in a similar melt sill to to that seen seismically at the base of the sheeted dyke complex (i.e., the 'AMC' reflector).

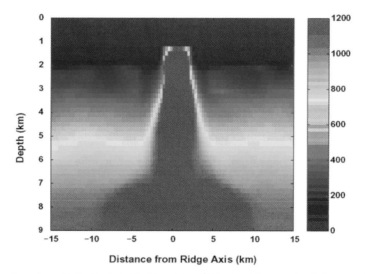

Plate 5. Crustal temperatures beneath the East Pacific Rise at 9°30′N inferred from a seismic tomography experiment [*Dunn et al.*, 2000]. The velocity structure responsible for these inferred temperatures featured similarly steep contours at the sides of the hot melt body running through the lower crust. These steep isotherms have been used to argue for extensive and deep penetrating hydrothermal circulation, that more efficiently removes heat from the magma chamber than would simple conductive cooling. Also significant is that the width of the temperature anomaly broadens in the upper mantle. This is interpreted as a broad region of melt that becomes focussed by the base of the crust into a narrower region through which melt accesses the crust.

AMC reflector, the argument centers primarily on whether this melt at deeper levels is crystallising—and if so, by what mechanisms the ensuing latent heat is removed from the mid or lower crust.

Constraints on the lower crust come from seismic refraction tomography and compliance methods which, at fast spreading ridges, show a large body with low melt fraction, thought to consist of a crystalline mush [*Sinton and Detrick*, 1992], underlying the melt lens that forms the AMC reflector. The melt fractions within this larger body (up to 6–8 km wide) are lower than in the melt sill at the top of the chamber, although there is conflicting evidence as to how this melt is distributed. At 9° 30′ N on the EPR melt fractions of at least 10% are inferred from tomography at a depth of 2 km below the seafloor. This is within the larger mush zone, which has a width of at least 4 km and possibly as great as 6 km [*Dunn et al.*, 2000]. Towards the base of the crust this melt fraction apparently decreases to around 1–2%, although this estimate could be as high as 10–11%, depending on how the melt is distributed (well connected or in isolated pockets). Compliance results from several sites along the EPR (9°–10° N) show evidence for abundant melt within the mid-crust [*Crawford and Webb*, 2002] (Plate 4). An asymmetric melt body of less than 8 km width is seen at 9° 48′ N centered beneath the rise axis. The body shifts to the west as it approaches the overlapping spreading center (OSC) at 9° 03′ N. The total amount of melt within this mush zone probably exceeds that in the mid-crustal melt lens seen with seismic reflection techniques. Slower shear wave velocities are measured in the upper half of the lower-crust, suggesting that more melt is located here than towards the base of the crust, although this may reflect differences in how the melt is interconnected, with less well connected melt at the base of the crust, in the same manner as suggested from tomography results.

Tomographic results from 9° 30′ N on the EPR show very steep iso-velocity contours at the sides of the low-velocity melt-bearing region, extending from beneath the AMC reflector to the base of the crust at distances of 2–3 km off-axis [*Dunn et al.*, 2000] (Plate 5). Conversions from velocity to temperature similarly result in steep sided isotherms on the flanks of the mid-crustal melt body. At the Juan de Fuca Ridge, micro-earthquake studies have shown deep focal mechanisms below the depth of the midcrustal reflector (to depths of 3.5 km) and so it has been proposed that hydrothermal circulation in this region is driven by a cracking front mining heat from hot, but solid crust, rather than by cooling of a crustal magma body [*Wilcock and Delaney*, 1996; *Wilcock et al.*, 2002].

One possible interpretation of these observations is that the middle and lower crust is cooled by deep penetrating hydrothermal circulation, mining heat from the sides of the magma chamber, resulting in near vertical isotherms. Modelling studies demonstrate that, given appropriate permeabilities within the lower crust, such a thermal structure is realisable [*Cherkaoui et al.*, 2003], and this provides support to the melt sill injection type models of lower crustal accretion. However the patterns of fluid flow required to remove heat rapidly from the lower crust predict significant off-axis seafloor venting of hydrothermal fluids, a phenomenon which has yet to be widely observed—although some evidence has been seen for hydrothermal sediment mounds and diffuse flow at scarp slopes some 25 km off axis at 9° 28′ N on the EPR [*Macdonald et al.*, 2002].

In summary, it is clear that the lower and mid crust beneath fast and intermediate spreading ridges can contain significant accumulations of melt—which may be distributed in discrete high porosity sills and channels, or may be more diffusely distributed as inter-granular porosity. At least some of this melt migrates upwards to maintain and replenish the high level axial melt lens that forms the prominent and characteristic seismic reflector, and from which much thermal energy is transferred into the overlying hydrothermal system. What is as yet unquantified is how much thermal energy is transferred into hydrothermal systems penetrating below the level of the melt lens; and hence how much heat is removed, and how much crystallization takes place, at these deeper levels.

Despite the uncertainty in the thermal and magmatic structure of the lower crust at ridges, there is clear agreement from all the seismic evidence (both reflection and wide angle) that the crust reaches its fully mature thickness, and a recognizable and sharp Moho boundary between crust and mantle forms, almost instantaneously at the ridge axis across a wide range of spreading rates [e.g., *Navin et al.* 1998; *Turner et al.* 1999; *Kent et al.* 2000]. Whatever the distribution of magma in the lower crust and the level at which it cools and crystallises, the segregation of crustal from mantle materials appears to occur over a distance that is no larger than the seismic resolution (on the order of a kilometre at the depth of the Moho) to either side of the ridge axis.

4.3. Crustal Magmatic Structure at Slow Spreading Rates

In contrast to widespread observations of magma bodies at fast and intermediate spreading centers, there has been a notable lack of similar observations at slow spreading ridges. The most comprehensively studied slow spreading ridge is the MAR. Numerous geophysical studies designed to look for crustal magma accumulations beneath the MAR have produced negative results. This is a crucial observation: most of the MAR axis is not underlain by any detectable magma body.

Further confirmation of this has come from a number of earthquake studies on the MAR. Earthquake focal depths at the MAR axis extend to depths of at least 10 km [e.g., *Wolfe et al.*, 1995], through the lower crust and into the uppermost mantle. This indicates a thermal regime in which temperatures are low enough (< 700°C) to allow brittle deformation throughout this depth range. This observation of a substantially thick, cool lithosphere at zero age is also consistent with rheological arguments for the origin of median valley topography at slow spreading rates [e.g., *Chen and Morgan*, 1990]. On the other hand, the presence of a well-developed layer 3 (at least beneath segment centers) [e.g., *Sinha and Louden* 1983] and the similarity in mean crustal thickness between the Atlantic and Pacific implies that, over time, a healthy supply of magma into the crust must be available.

Whether these observations reflect a first order difference in melt supply between fast and slow spreading ridges, such that at slower spreading rates melt supply is too infrequent to support a robust magma chamber, or whether they reflect enhanced hydrothermal cooling of the crust at slow spreading ridges, or whether they indicate fundamentally different mechanisms of melt transport and storage at slow spreading ridges (e.g., Figure 4) is not yet well understood.

The location of one possible melt body beneath the MAR is at 23° N—one of only a small number of known hydrothermally active segments. *Detrick et al.* [1990] carried out a seismic reflection survey here, similar to those which had revealed AMC reflectors beneath the East Pacific Rise. They reported that there was no evidence of a shallow AMC reflector, but the same data were subsequently reanalysed by *Calvert* [1995], and reinterpreted as presenting possible evidence of a melt lens reflection. Another important exception to the general absence of MAR magma bodies comes from the Reykjanes Ridge, where there is evidence both for a melt lens and for an underlying low seismic velocity region associated with partial melt [*Navin et al.*, 1998]. A comparison between MAR segments

allows us to note systematic differences between them. In some, the center of the median valley floor is occupied by a well developed axial volcanic ridge (AVR). These features are typically tens of km long, 2 to 3 km wide at their base, and of the order of 200 m high. Some median valleys with well developed AVRs also exhibit what has been described as 'hourglass' topography—the valley is narrowest and shallowest in the segment center, and both deepens and widens towards segment ends. Other median valley segments show different characteristics. Such segments have no discernable AVR. Instead, the valley floor is pervasively cut through by axis-parallel, extensional fault scarps. The median valley walls are typically higher and straighter, and there is evidence for much of the bathymetric relief being accommodated by a smaller number (sometimes only one) of larger offset, and sometimes low-angle, normal fault scarps. In such segments, there is little evidence of current or very recent magmatic activity. Instead, recent seafloor spreading appears to have been accommodated primarily by brittle, tectonic extension.

One possible explanation is that through time, each segment experiences cyclic changes in the rate at which magma is delivered from the mantle, giving rise to sea floor spreading through 'tectono-magmatic cycles' [e.g., *Parson et al.*, 1993]. Evidence for such cycles is, however, indirect. According to the tectono-magmatic cycle hypothesis, segments alternate between spreading in a fashion dominated by the injection of melt to construct new crust, and magmatically dormant periods when the lithosphere cools and spreading is accommodated by tectonic extension. The segments with well developed AVRs are considered representative of the magmatic phase of the cycle; while the tectonically-dominated segments are considered representative of the cold, amagmatic phase.

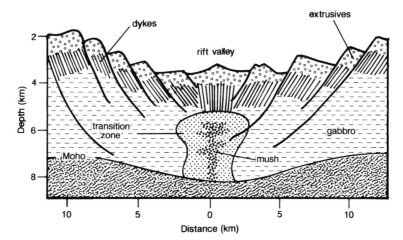

Figure 4. A contrasting cross-section through a possible model of the structure of a slow-spreading ridge axis. While explaining some observed features, there is considerably less agreement among geophysicists as to whether this model is substantially correct—see text, section 4.3. From *Sinton & Detrick* [1992].

The crustal magma body beneath the Reykjanes Ridge reported by *Navin et al.* [1998] and *MacGregor et al.* [1998] (on the basis of seismic and electromagnetic studies respectively) provides strong support for tectono-magmatic cycles [*Sinha et al.*, 1998]. The magma body lies beneath an AVR on the southern portion of the Reykjanes Ridge which was selected for study specifically because of morphological evidence for it being at the magmatic phase of any cycle. Although affected to some extent by the Iceland hot spot, it is far enough from the Iceland plume center for the axial topography to be characterised by a median valley, and for the thickness of the crust being formed to be normal for the MAR. The time-averaged melt flux into this part of the ridge is therefore no different from that at other slow spreading ridge segments. The magma body has many similarities to those found at higher spreading rates, but is, however, unusually large. *MacGregor et al.* [1998] estimated that the melt within the crustal reservoir represents at least 20,000 years' worth of crustal construction; but their thermal modelling showed that it is likely to become frozen in no more than about 1,500 years. It clearly, then, cannot be a steady state feature. A possible explanation is that the magma body underlies a segment that is at the peak of the magmatic stage of its life cycle, and that a rapid injection of melt from the mantle has just occurred. If so, this will be followed in the near future by a sharp decrease in the melt flux from the mantle. The melt body will solidify, and for the next few tens of thousands of years, sea floor spreading along this segment will occur predominantly by tectonic extension, until another influx of melt from the mantle starts the next magmatic cycle.

At 35°N on the MAR, wide angle seismic analysis and tomography has revealed evidence for anomalous velocities in the lowermost crust near the center of a major segment [*Canales et al.,* 2000; *Hooft et al.,* 2000; *Magde et al.,* 2000] These velocities are interpreted to indicate that lower crustal melt accumulations, possibly in the form of melt sills, play a major role in crustal construction at slow spreading rates. *Magde et al.* [2000] relate these lower crustal features both to patterns of melt delivery from the mantle, and to sub-axial, pipe-like, low velocity anomalies in the middle and upper crust that appear to represent significant bodies containing a small melt fraction. *Bonatti et al.* [2003] relate changes in crustal properties at the Vema transform, MAR to much longer-term cyclic variations in melt production rates in the mantle, with cycle times of the order of 3–4 Ma.

The tectono-magmatic cycle model for the MAR is broadly consistent with many of the geophysical and bathymetric observations. However, it is not yet clear how for example the observed petrological diversity of MAR basalts can be reconciled with it. The cyclic model also provides little help in explaining the extreme longevity of the TAG hydrothermal site, which lies a significant distance away from the neo-volcanic zone of the MAR spreading axis; and in particular the question of where the heat to drive hydrothermal output may have come from for a period of at least 20,000 years (albeit discontinuously) [*Lalou et al.*, 1993]. If taken to its extreme, the cyclic model would predict that all segments are essentially identical to each other, with the differences between them being attributable only to the stage of the cycle that each is at. In reality, there is evidence that the MAR may be more complex than this, with real and long-lived differences occurring between the properties of different segments.

Whether or not tectono-magmatic cycles are responsible for crustal construction processes at the MAR, a uniform feature of models invoked to explain the crustal, magmatic and thermal structures of slow spreading ridges is that they require melt to be delivered to crustal level from the underlying mantle in a fashion that is highly focused in space and probably in time: melt is delivered preferentially beneath the centers of individual spreading segments; and it may be delivered occasionally, and in batches, at intervals on the order of tens of thousands of years. More research is required to reconcile geophysical observations at crustal scale with analyses of the compositional variations in MAR rocks, and with modelling of mantle dynamics and especially of 3-D and 4-D patterns of focused upwelling or melt focusing beneath the Moho.

4.4. Thermal Boundary Layer Above the Magma Chamber: Interaction with Hydrothermal Circulation

Models of hydrothermal circulation generally include a boundary layer between the active regions of fluid circulation and the top of the axial magma chamber. Heat transport across this boundary layer is by conduction, and its thickness is a rate limiting parameter on the heat flux that can be extracted from a magma chamber [e.g., *Lowell and Germanovich*, 1997]. Geophysical constraints on the physical properties and dimensions of this important boundary layer are at first glance ambiguous, although this may reflect real differences in thermal structure among the different regions studied. The available high quality seismic velocity models have similar features. These are, starting from within the axial melt lens: a steep velocity gradient with velocities starting at those expected of the magma chamber and increasing upwards; a thin lid above the chamber with uniform high velocities; and an overlying region of depressed velocities relative to an expected linear velocity gradient with depth. The substantial differences are in the thicknesses of the units in the models in the different study areas, and the different interpretations that have been assigned to them.

Waveform analysis of reflections from around the AMC at 14° 10′ S on the SEPR shows a relatively thin region of steep velocity gradients (only about 25m thick) immediately above

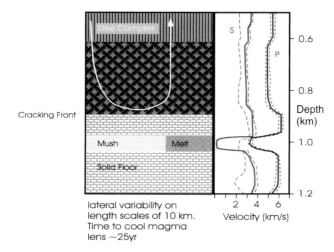

Cracking Front

lateral variability on
length scales of 10 km.
Time to cool magma
lens ~25yr

Figure 5. Results of waveform inversion from the SEPR [*Singh et al.*, 1999]. This model shows a thin low-velocity region associated with a melt sill. Overlying this body, which has both a solid roof and floor, is a steep velocity gradient capped by a ~70 m thick iso-velocity layer. This layer is interpreted to be largely uncracked and is thought to constitute the conductive thermal boundary layer above the AMC. A region of depressed compressional and shear velocities above this conductive boundary layer is interpreted as the lower reaches of hydrothermal circulation. In the text we discuss other models of seismic velocity in which different thicknesses of the conductive boundary layer are inferred. These differences most likely represent differences in the magmatic cycle of the various ridge segments studied. This region of the SEPR, with only a thin conductive boundary layer, will be susceptible to rapid cooling of the magma chamber, calculated by *Singh et al.* [1999] to occur on a 25–50 year time period.

or possibly part of the AMC, overlain by a 70m thick high velocity lid (Figure 5). This lid is interpreted by *Singh et al.* [1999] to be the conductive boundary layer at the base of hydrothermal circulation. High compressional and shear wave velocities indicate that it is probably unfractured. A 150–200m thick band of depressed velocities above this high velocity roof is interpreted to be the lower reaches of the hydrothermal circulation system. This assertion is based on the fact that decreases in both compressional and shear wave velocity are seen, consistent with an increase in porosity, but not with raised temperatures in an unfractured medium.

At 9° 30′ N on the EPR a 300–500m thick low-velocity region immediately above the AMC reflector is seen in tomographic inversions of compressional wave travel times (Plate 6) [*Toomey et al.*, 1994]. The largest anomaly is offset to the west of the ridge axis. Expanding spread refraction profiles (ESP) from 9° 30′ N also show a steep decrease in compressional velocity over about 250m immediately above the AMC reflector, but with relatively high shear wave velocities maintained [*Vera et al.*, 1990]. Their 1-D ESP model also shows a high-velocity layer about 100m thick, although it is 250m

above the AMC, compared to just 25–30m on the SEPR. *Toomey et al.* [1994] argue that the reduced velocities immediately above the AMC can be explained by elevated temperatures (250–450°C hotter), and that this thicker region constitutes the conductive boundary layer overlying a robust magmatic heat source.

A small aperture seismic tomography experiment focused on the ridge axis at 9° 50′ N shows evidence for along-axis variations in crustal thermal structure linked with a fourth order ridge axis discontinuity [*Tian et al.*, 2000]. North of 9° 52′ N, where hydrothermal venting is absent, upper crustal velocities are raised, interpreted to be the result of a cooler thermal structure. An along-axis temperature variation of about 300° C within the sheeted dike complex is invoked to explain the difference in velocities across the 9° 52′ N discontinuity. It is useful to note that the laboratory data on which the inferred crustal temperature differences are based were only made over the range from 20–500°C [*Christensen*, 1979]. Over this interval, basalt displays a fairly linear gradient in compressional velocity with increasing temperature. At higher temperatures, closer to melting point, samples of different rock types typically show more rapid decreases in velocity with increasing temperature. If basalt behaves in the same way then this would suggest that the differences inferred in temperature at the EPR are upper bounds—especially as temperatures towards the base of the conductive boundary layer will approach that of the magma chamber.

The primary difference between the SEPR and NEPR velocity models is the thickness of the steep velocity gradient immediately above the AMC. On the SEPR, this layer is thin and is capped by a 70m thick high velocity lid. In the Singh *et al.* model this velocity gradient is sufficiently close to the magma chamber that it is interpreted to be part of the AMC itself, with the thin iso-velocity lid constituting the conductive boundary layer. In the northern EPR model there is a much thicker layer containing a steep velocity gradient, and this layer is interpreted to be the conductive boundary layer. The cracking front, or the lower reaches of hydrothermal circulation, appears to be much closer to the AMC on the SEPR (within 100m) than it does at 9 °N on the EPR (300–500m). There are along strike variations in the thickness of this thermal boundary layer, although within the ridge segment at 9° 30′ N, where the low velocity layer is seen, there is no one-to-one correlation between the thickness of the layer and the location of active vents. However, both at 9° 30′ N and at 9° 50′ N, the thick layer is restricted to a 4th order ridge segment, bounded by discontinuities in the ridge axis, and venting is more active within this segment.

Further evidence for a thicker thermal boundary layer at 9°N comes from a series of micro-earthquakes observed by a cluster of ocean bottom seismometers. These have been interpreted as the result of thermal cracking induced by fluids mining heat from the conductive thermal boundary layer overlying the AMC [*Sohn et al.,* 1999a]. In this model, the earthquake locations roughly constrain the 450° C isotherm to a depth of approximately 1100m below the seafloor, and some 400–500m above the AMC reflector (Figure 6), consistent

with the interpretation of *Toomey et al.* [1994]. The earthquake swarm was followed some 4 days afterwards by a sharp rise in overlying vent fluid temperatures which reached a peak 11 days post-swarm before falling back to pre-swarm values [*Sohn et al.,* 1998a], apparently constraining the length of

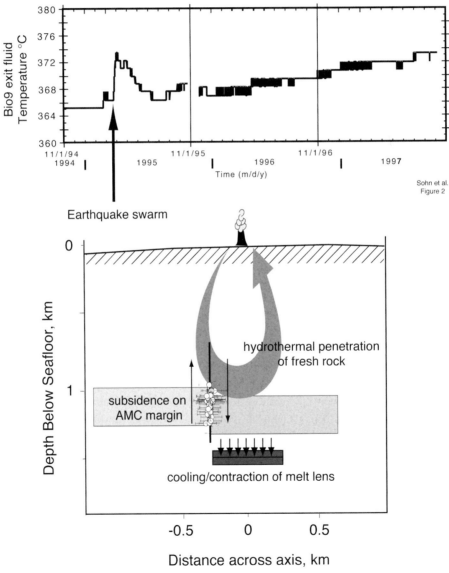

Figure 6. Results of a micro-earthquake study at 9°50′N on the EPR. A micro-earthquake swarm was seen at a depth of about 1100 m, some 400–500 m above the axial magma chamber, suggesting a thick conductive boundary layer at this ridge segment, consistent with other seismic observations as discussed in the text. Some 4 days after the earthquake swarm, vent fluid temperatures rose by about 8°C, apparently constraining the timescales of fluid transport from the cracking front to the seafloor [*Sohn et al.,* 1998, 1999].

time taken for fluids to travel from the cracking front to the seafloor.

Despite the different interpretations assigned to the different layers, it seems that there is actually little difference between the various velocity models except for the thicknesses of the units. The most likely explanation is that there is a much thicker thermal boundary layer at 9° 30′ N on the EPR compared to the SEPR. The AMC depths between the two locations are substantially different (1.6 km at 9° 30′ N compared to 1 km on the SEPR). These differences, both in terms of AMC depth and the thicknesses of the thermal boundary layers may simply reflect differences in the spreading rate between the two locations (rates at 9 °N on the EPR are about 65–75% of those of the SEPR south of the Garrett fracture zone [*Klitgord and Mammerickx*, 1982; *Rea*, 1981]), but also differences in the magmatic/hydrothermal state of the two settings.

Differences in the depth to the top of the AMC of the order of several hundred meters have also been reported from within single segments of EPR ridge crest [e.g., *Babcock et al.,* 1998; *Detrick* et al., 2002]. This suggests that, over time, magma intrusions might be injected to different depths, potentially breaking into the base of the dike sequence and possibly causing re-melting. *Coogan et al.* [2003] see evidence in ophiolites for the assimilation of crustal material into the magma chamber, and suggest that the injection of melt at new depths is the most likely cause.

Gillis and Roberts [1999] and *Gillis* [2002] have used field mapping of the lowermost sheeted dikes and uppermost gabbros in the Troodos ophiolite to develop a model of the boundary layer. Gabbros more than 50–100m into the sequence beneath the dikes do not show evidence of pervasive alteration associated with significant amounts of fluid flow, suggesting that circulation is mostly confined to the dikes and the uppermost gabbros. Field observations also constrain the highest temperatures of hydrothermal alteration to be around 700° C [*Gillis et al.*, 2001]. In this model, the eruption of a dike disrupts the conductive boundary layer, allowing fluids to penetrate deeper into the dike sequence and mine heat from hotter regions closer to the magma chamber. With time, the deeper cracks are re-sealed by precipitates and contraction due to cooling. If the magma chamber is efficiently cooled and not replenished, then the depth to the conductive boundary layer will increase downwards into the gabbroic section, allowing fluids to penetrate to these depths. Thus the thickness of the boundary layer will be dependent on the position of the ridge segment in a tectono-magmatic cycle, most critically the time since magma chamber replenishment and the time since the last diking event.

The lifetime of a thin melt lens will depend on how close the overlying hydrothermal circulation approaches it and, obviously, on the time scales for replenishment. From a thermal

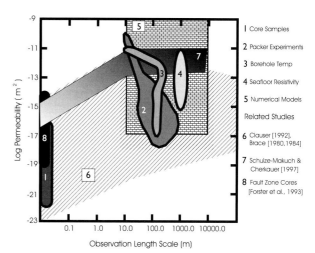

Figure 7. The scale dependence of crustal permeability, redrawn from *Fisher* [1998]. In general, the larger the scale of measurement, the larger the measured permeability, as regional scale faults and fissures are able to contribute to the fluid flow. Measurements made on core samples miss these contributions, making these analyses difficult to place in the context of ridge scale circulation.

viewpoint, a shoaling of the magma lens would thin the conductive boundary layer above the magma chamber and would speed up the rate at which heat is lost by conduction. The observed differences in thermal boundary layer thickness predict more rapid cooling on the SEPR than on the northern EPR where the conductive boundary layer is much thinner, allowing more heat to be extracted by conduction per unit time. *Singh et al.* [1999] calculate a period of 25–50 years to cool the melt lens seen on the SEPR, assuming that it is not replenished before then.

5. UPPER CRUST AND HYDROTHERMAL CIRCULATION

Within the upper crust, above the level of any axial magma body, the thermal regime is dominated by the circulation of hydrothermal fluids. This circulation is itself controlled by the temperature structure of the crust—in particular the presence and locations of any high temperature regions—and by the permeability, which controls the ease of fluid flow and the patterns of circulation [*Fornari and Embley*, 1995].

Permeability, the primary physical property that controls how fluids can circulate within the oceanic crust, is also the hardest to constrain geophysically. Permeability in the crust varies by more than 13 orders of magnitude, and its value has also been shown to be heavily dependent on the scale of observation [e.g., *Neuman*, 1994] (Figure 7). This is because, for instance, a measurement made on a drill core sample will include vesicular and inter-pore permeability on the micron-to-millimeter scale, but will not account for the influence of faults or cracks that may control circulation over the length scale of a hydrothermal convection cell. This makes the inference of appropriate permeabilities to use in models of circulation difficult, if not impossible, to obtain from laboratory scale measurements. There have, though, been substantial efforts in the hydrologic community to investigate relationships between permeability and observation length scale [e.g., *Neuman*, 1994; *Tidwell and Wilson*, 1997 and references therein].

The most direct measurements of permeability come from packer tests made in ODP drill holes. A full review of extant data and comparisons with estimates derived from other means is given in *Fisher* [1998]. With increasing depth and changing lithology within the upper crust, permeability decreases from values of around 10^{-12}–10^{-13}m^2 in the uppermost extrusive layer (2A) to around 10^{-17}m^2 in the sheeted dike section (Layer 2B). The values determined by the packer measurements assume an isotropic permeability, which may not be valid. They also span a range of length scales from 10s of meters up to several hundred metres, but may not sample enough of the crust to provide the representative values needed to understand hydrothermal flow. *Evans* [1994] demonstrated that measurements of electrical resistivity in tandem with seismic velocity measurements are able reduce the ambiguity in determining crustal porosity and its interconnectedness on the length scales pertinent to hydrothermal convection. Values of permeability inferred from that analysis were comparable to, but slightly higher than those from drilling measurements [*Fisher*, 1998], possibly indicative of the influence of larger scale fractures at the geophysical scale, although the analysis involved several key assumptions on the pore space geometry.

If we are to develop reliable models of fluid circulation and heat extraction, we need a better understanding of how our geophysical observables relate to permeability and critical flow parameters. However, this too is difficult because similar values of physical properties can be caused by different fluid distributions. For example, electrical conductivity is also a transport property and there have been many attempts to develop quantitative links between it and permeability [e.g., *Thompson et al.*, 1987; *Doyen*, 1988; *David*, 1993; *Schwartz et al.*, 1993; *Bernabe and Revil*, 1995; *Evans*, 1994]. However, the important pathways for electrical current flow through a rock may be different than those for fluid flow, because electrical current flow depends less critically on the size of the pathways. While fluid flow will be accommodated more readily though larger openings, current flow can take place through narrower networks and cracks [e.g., *David* 1993].

We can, though, make progress towards geophysically constraining two other highly important and interlinked crustal physical properties—namely, the porosity and the degree of interconnection between pore spaces. Higher porosities tend to be associated with higher permeabilities. The degree of interconnection is critical in terms of the ability of fluids to circulate. Highly interconnected pore systems will tend to be associated with high permeabilities and rapid convective circulation of fluids; whereas fluids trapped in isolated pore spaces will tend to have little impact on thermal structure. In the next section we discuss more recent approaches to jointly interpreting multiple physical properties made at similar length scales.

5.1. Geophysical Determinations of Physical Properties in the Upper Crustal Regime

The key parameters that geophysical studies of ridges can readily determine, and that can help to constrain pore space properties, are bulk density (from gravity), seismic velocity (most readily P-wave velocity) and electrical resistivity.

The bulk density of the seafloor, as measured by gravity methods, does not depend upon how fluid is distributed within the rock. Because of this, gravity measurements provide a robust estimate of seafloor porosity, albeit depth limited to the uppermost few tens of meters of seafloor in the case of recent gravity measurements made at the seafloor; or to averages over depth ranges of at least a kilometre, in the case of sea surface gravity data. There have been a growing number of seafloor gravity measurements which have sufficient precision to constrain the porosity of the uppermost seafloor. *Pruis and Johnson* [2002] review the data to date, together with the evidence for rapid decreases in porosity of the uppermost extrusives (Layer 2A) with age. Porosity estimates of very young crust vary widely, but values of 29–36% seem to be typical for very young lavas, decreasing to 10–15% by 1Ma. The increase in bulk density with age appears to be sufficient to explain observations of seismic velocity increase with age that *Wilkens et al.* [1991] similarly explained by closure of low aspect ratio cracks. Geological observations of heavily fractured rubbly pillow flows on the EPR agree with this model since, with time and additional eruptions that cover the rubble and fill voids and drain-back features, the bulk porosity is expected to decrease. While the number of data points on old crust is not large, the most rapid aging processes appear to occur over the first 0.5 Ma.

An additional parameter that can be extracted from seismic data is the degree of anisotropy, determined either from travel times [e.g., *Tong et al.*, 2004] or from shear wave splitting [e.g., *Jones et al.*, 1991]. Anisotropy studies indicate that the upper 1–2 km of the crust (Layer 2) at and close to ridge axes is weakly seismically anisotropic—with anisotropy values of typically 4% or less in the upper 1 km, and 2% or less between 1 and 2 km depth. There is no clear evidence for any anisotropy below 2 km depth. The orientation of the fast direction is consistent with the anisotropy being due to fractures within the sheeted dyke complex, preferentially aligned parallel to the spreading axis. However, as *Dunn and Toomey* [2001] point out, only a small proportion (<10%) of the total porosity in Layer 2 can be aligned in this fashion, since the bulk velocities indicate much higher overall porosities than those needed to generate the weak anisotropy. Thus the great majority of pore fluid within the upper crust is contained within pore spaces that have no strong alignment. What is less certain is whether the small proportion of aligned fractures result in higher permeabilities along-axis than across-axis, and so promote preferential along-axis fluid flow within the hydrothermal system. The available anisotropy data do not rule this out, but neither do they strongly support it.

The porosity in upper crustal rocks is inversely related to both P-wave velocity and electrical resistivity. The degree of interconnection between pore spaces is a second important factor in the relationship—increasing interconnection at constant porosity decreases both seismic velocity and resistivity. Fortunately, however, the two geophysical properties vary in rather different ways with the two hydrological properties. Electrical resistivity depends almost entirely on the interconnected parts of the pore system—since isolated pores are able to contribute very little to electrical conduction through the rock as a whole. Seismic velocity is slightly less sensitive to the degree of interconnection, and depends most strongly on overall porosity—since both isolated and interconnected pore spaces contribute (though to differing degrees) to the overall decrease in velocity. As a result of these behaviours, it is possible to make progress towards constraining the porosity of the upper crust and the degree of interconnection, provided that we have co-located P-wave velocity and electrical resistivity surveys, and provided that we apply appropriate geophysical effective medium modelling to the problem, in order to map the geophysical parameters into hydrological parameters.

Greer [2001] has developed an effective medium modelling approach specifically for this purpose, and has applied it to three contrasting sites on the mid-ocean ridge [*Greer et al., 2002; MacGregor et al., 2002*]. Greer's joint effective medium (JEM) modelling assumes that the pores consist of spheroidal inclusions with random orientations, so that pore distribution can be represented in terms of two parameters—porosity and pore aspect ratio. In contrast to previous studies, this approach uses identical probability arguments to explicitly account for the degree of interconnection of the pore spaces, in both the seismic and the electrical modelling (Figure 8). By making assumptions about the physical properties of both seawater and the basalt matrix (both well constrained by laboratory measurements), it is then possible to find a combination of porosity and pore aspect ratio which simul-

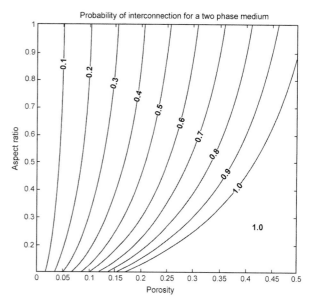

Figure 8. The probability of interconnection between pore spaces, for a two-phase medium containing randomly orientated spheroidal fluid inclusions. The probability ranges from 0 at low porosities and aspect ratios close to 1 (spherical), to 1 at high porosities and small aspect ratios. From *Greer et al.* [2002].

taneously fits both the observed P-wave velocity and electrical resistivity at any point in the crust (e.g., Plate 7).

Greer's first example of the application of this technique uses data from co-located wide angle seismic and controlled source electromagnetic (CSEM) studies of the axis of the EPR spreading center at 13° N [*Evans et al.,* 1991, 1994; *Greer,* 2001; *Harding et al.,* 1989; *Kappus et al.,* 1995; *Greer et al.,* 2002]. Convective numerical models of the circulation in hydrothermal systems [e.g., *Jupp and Schultz,* 2000] suggest that the zones of upwelling of hot seawater are spatially highly restricted, while zones of recharge with ambient seawater are much larger. Greer *et al.* therefore assume that the upper 1200 m of the crust (above the level of the small axial magma chamber) consists of a 2-phase medium, in which the properties of the solid matrix can be represented by those of mid-ocean ridge basalt (MORB) and those of the fluid by seawater at 3°C. Vertical profiles of porosity and pore aspect ratio for this location show several important features (Figure 9). Firstly, porosity within the upper 200 to 400 m of the crust is very high—greater than 20% at the seafloor—consistent with the findings from gravity and packer studies. Secondly, porosity decreases rapidly with depth, with the maximum gradient occurring just below the seafloor. Thirdly, pore aspect ratio also decreases rapidly with depth, indicating a transition from porosity dominated by relatively open void spaces immediately beneath the seafloor to one that is dominated by very thin fractures beneath 400 m depth. We can readily relate these changes to the inferred lithology of the rise axis. The

(a)
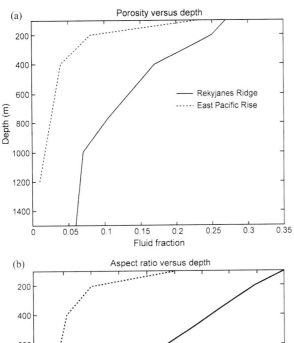

(b)
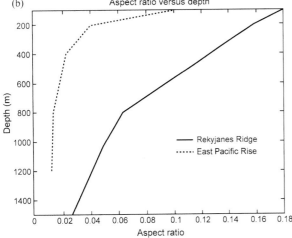

Figure 9. (a) Inferred fluid fraction versus depth and (b) inferred pore space aspect ratio versus depth, for the axes of the EPR at 13°N and the Reykjanes Ridge. From *Greer et al.* [2002].

upper 200 to 400 m of the sea floor consists of high-porosity (>20%) extrusive pillow basalts and lava flows, with many open sub-spherical pore spaces. Beneath layer 2A, in the intrusive sheeted dikes section of Layer 2B, porosity decreases dramatically to only around 4% at 400 m depth, and then more slowly to less than 2% at 1200 m depth. This coincides with a further decrease in crack aspect ratio, to order of 0.01 (i.e., disc-like, fluid filled void spaces that typically have a diameter-to-thickness ratio of 100).

Greer's method simultaneously determines the degree of inter-connection between pore spaces as a function of depth. The results from the EPR indicate that the porosity in the top 200 m is fully interconnected. It remains at least 80% interconnected down to 400 m depth, gradually decreasing below that, and reaching 40% interconnection at 1200 m depth, just above the expected level of any magma body.

A second combined seismic and CSEM dataset is available from the axis of the Reykjanes Ridge—part of the MAR—at 57° 45′ N [*Navin et al.*, 1998; *MacGregor et al.*, 1998; *Sinha et al.*, 1998]. Application of the same JEM method to the data reveals some broad similarities to the EPR results, but there are also some differences (Figure 9). Porosity immediately below the seafloor is greater than 25% at the Reykjanes Ridge. There is again a steep gradient in the porosity in the upper kilometre. A further similarity is that in both cases, crack aspect ratio decreases steadily with depth. At all depths, though, porosity beneath the Reykjanes Ridge is significantly higher than beneath the EPR, at least to a depth of 1.4 km. The transition between Layer 2A and Layer 2B is less clearly defined at the Reykjanes Ridge, so that the gradient of porosity in the upper 400 m is less steep. Also, the aspect ratio of the pore spaces is significantly closer to 1 at the Reykjanes Ridge than at the EPR. This effect may be related to the shallower water depth here, which allows out gassing of volatiles in the Reykjanes Ridge magmas to form vesicular lavas at the seafloor. The vesicles would be expected to be close to spherical, thus increasing the characteristic aspect ratio towards unity. The fact that aspect ratio remains higher than at the EPR all the way to 1.4 km depth suggests that some vesicles may be present to significant depths into the sheeted dike section, as well as in the extrusive Layer 2A.

At the Reykjanes Ridge, the pore spaces are 100% interconnected at the sea surface, and remain 70% or more interconnected at all depths down to 1400 m. In the depth range 800 to 1400 m, the probability of interconnection is significantly higher beneath the Reykjanes Ridge than it is beneath the EPR at the same depth. This means that the greater overall porosity beneath the Reykjanes Ridge is more than compensating for the higher aspect ratios (more rounded pore spaces) in terms of pore fluid connectivity. A highly interconnected pore fluid phase is present beneath the Reykjanes Ridge to depths of at least 2.5 km beneath the seafloor. This combined with the presence of a substantial crustal magma chamber at a depth of 2.5 km [*Sinha et al.*, 1998] makes it seem probable that this section of the ridge hosts a vigorous hydrothermal circulation system, even though no evidence of high temperature venting has been found here [*German et al.*, 1996; *German and Parson*, 1998].

The third example of co-located seismic and EM data comes from the intermediate spreading rate, back-arc Valu Fa Ridge (VFR) in the southern Lau Basin [*Turner et al.*, 1999; *MacGregor et al.*, 2001; *Day et al.*, 2001]. Here, both a wide-angle seismic tomography profile and a 2-D resistivity profile were obtained across the ridge axis, as part of an integrated study. Plate 8 shows a combined plot of electrical resistivity anomaly and seismic P-wave velocity anomaly, between the sea floor and a depth of 6 km, and extending 10 to 15 km to

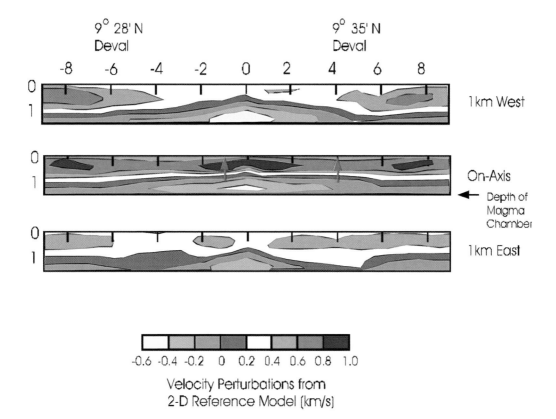

Plate 6. Ridge-parallel sections of anomalous seismic compressional wave velocity derived from a tomography experiment at 9°30′N on the EPR [*Toomey* et al., 1994]. The largest anomaly (-0.4 to -0.6 km/s) is 300–500m thick, occurs 1 km to the west of the ridge (top panel) and is consistent with a relative increase in temperature of about 250–450°C. The anomaly pinches out towards the two devals at 9°35′N and 9°28′N. The locations of known high temperature activity are denoted by the arrows in the center (axial) panel. The anomaly is interpreted as a thick conductive boundary layer above the AMC. How this thermal anomaly relates to present hydrothermal circulation is unknown. The asymmetry in thermal structure may indicate enhanced cooling to the east of the ridge.

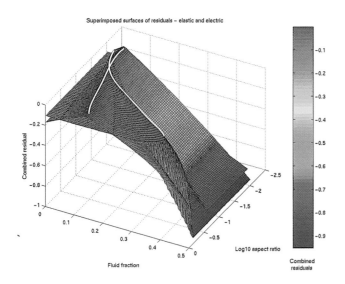

Plate 7. The electric and seismic fields of misfit between predicted and observed coincident geophysical data, superimposed, displaying the intersection of the individual lines of minimum misfit. Predictions and data are for a depth of 200 m beneath the EPR axis. From *Greer et al.* [2002].

each side of the ridge axis [*MacGregor et al.,* 2001]. In both cases, the 'anomaly' corresponds to the local deviation from a mean 1-D profile, derived from averages of off-axis results from the same study. The seismic data clearly pick out the layer 2A–2B boundary and the Layer 2–3 boundary. P-wave velocity anomalies are small throughout Layer 2, including in the axial region. The most prominent feature in the seismic tomography result is the large velocity anomaly located immediately beneath the Layer 2–3 boundary at the ridge axis. This corresponds to the large 'mush zone' region of partially crystallized material which underlies and flanks the crustal melt lens (also identified by seismic reflection data), which lies at a depth of 3 km beneath the ridge axis.

The electrical resistivity anomaly has a distinctly different pattern. The clearest feature is the marked asymmetry, with strongly negative anomalies seen at the axis and to the west, and positive anomalies to the east. The most pronounced resistivity anomaly is within Layer 2, and especially in Layer 2B, at depths of between 0.5 and 3 km beneath the sea floor. Here, both at the axis and to a distance of 10 km to the west, the crust is characterized by a pronounced low resistivity anomaly, with resistivities typically only a half or less of those that would be expected at a corresponding depth further off-axis or to the east (Figure 10).

To the east of the axis, the patterns of porosity, pore aspect ratio and pore-space interconnection are similar to those at the EPR and Reykjanes Ridge sites except that the axial magma body lies at greater depth. The probability of interconnection between pore spaces is between 85 and 100% at all depths down to 3 km.

Applying the JEM inversion procedure to the seismic and EM data either at or to the west of the VFR axis presents a problem. If the properties used to characterise the two-phase medium are those for the intermediate composition crustal rocks and ambient seawater at 3° C, it is not possible to find a solution in terms of porosity and pore aspect ratio. This indicates that the crust cannot be represented by a 2-phase mixture of basaltic andesite and ambient seawater. Since the data can be readily fit to the east of the axis, and since the P-wave velocity anomalies in Layer 2 are negligibly small, then the most likely scenario is that neither the rock properties nor the distribution of porosity and aspect ratio vary significantly across axis. If so, the reason for the failure to find a solution is that the pore fluids have a resistivity that is less than that of ambient seawater [*Greer*, 2001; *Greer et al.*, 2002]. It should be possible though to find a value of resistivity for the fluid phase at all depths and across-axis locations that can fit the JEM forward model and reproduce the observed bulk resistivity and seismic velocity.

The results of this rather different way of applying the JEM approach are shown in Figure 11. Figure 11(a) shows

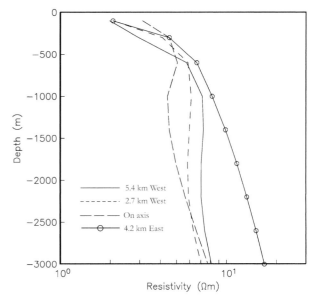

Figure 10. Profiles of electrical resistivity versus depth below the seafloor from points along the section of Plate 8. From *MacGregor et al.* [2002].

a set of three vertical profiles of pore fluid electrical conductivity, that are consistent both with the observed bulk resistivities and seismic velocities, and with an assumption that the aspect ratio vs. depth profile is similar to that determined off axis to the west at all locations on the cross-section. Since the conductivity of seawater is strongly dependent on temperature, the fluid conductivity profiles can be converted into equivalent temperature (for pore fluid of seawater composition) profiles (Figure 11(b)). In both figures the three profiles correspond to the ridge axis, and to positions 2.7 and 5.4 km to the west of the axis. Conductivity/temperature generally increase with depth, and decrease at any given depth with increasing distance from the axis.

MacGregor et al. [2002] discuss various possible origins for these anomalously low resistivities at and to the west of the VFR axis. Candidates include high temperature, seawater composition fluids (as in Figure 11); fluids of varying salinity, including highly saline brines in the lower parts of Layer 2, derived from phase separation in a hydrothermal circulation system and/or from a volatile phase exsolved from the magma; and extensive stock-work like mineralization by conductive mineral assemblages, pervading very large volumes of the upper crust that extend for at least several kilometres in all directions. They conclude that the most likely candidates are a combination of salinity and temperature variations.

The curves in Figure 11 are interesting in two respects. First, they indicate major changes in fluid properties with depth that are not required to explain the data from either the

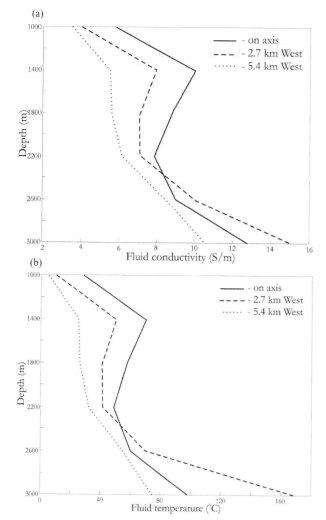

Figure 11. Profiles of (a) inferred pore fluid electrical conductivity, and (b) pore fluid temperature (assuming sea-water salinity), from the three sites on the section of Plate 8 and Figure 10, showing a highly non-linear increase in fluid properties with depth into Layer 2. From *MacGregor et al.* [2002].

EPR or Reykjanes Ridge sites. In both of those other cases, the available geophysical data are consistent with pore spaces throughout Layer 2 being filled with fluid that has ambient seawater properties (salinity and temperature). At the VFR, the geophysical data are not consistent with that model—but are consistent with a substantial increase in salinity and/or temperature with depth. The second important feature of the curves is that neither conductivity nor equivalent temperature increase monotonically with depth. At all three locations, three distinct depth zones are present. In the lower 800 m or so of Layer 2 (immediately above the magma or mush body), pore fluid conductivity increases rapidly with depth, reaching 4 to 5 times that of ambient seawater at the base of this layer. Above this, at about 1.4 to 2.2 km below the sea floor, conductivity either remains approximately constant or even decreases with depth. In the uppermost zone, fluid conductivity again increases rapidly with depth, while trending towards values typical of ambient seawater at the seafloor.

Vent fluids at high temperature seafloor sites close to the geophysical studies at the VFR show a range of salinities, from much less than seawater up to 30 wt% NaCl (ten times seawater) [*Lecuyer et al.*, 1999], so there is strong geo-chemical evidence for phase separation and the presence of both a fresh and a brine component in the hydrothermal system. Highly saline pore fluids, with densities exceeding that of ambient seawater even at elevated temperatures, could explain the persistence of the low resistivity anomaly in the geophysical data to such great distances to the west of the axis. However, neither a monotonic increase in salinity nor a monotonic increase in temperature with depth are able to explain the two clear inflections in the fluid-conductivity vs. depth curves in Figure 11—and especially the apparent downwards decrease in fluid conductivity over the 1.4 to 2.2 km depth range.

The inferred fluid conductivity profiles are more consistent with a two-layer hydrothermal convection system, in which phase separation of the hydrothermal fluids (and possibly exsolution from magma) generate a brine phase in the lower part of Layer 2, directly above the magma body. The steep gradient in fluid properties from about 2.2 km to 3 km depth would then represent a lower boundary layer, with the region above it corresponding to a convection cell in which fluids of a range of salinities and temperatures (including significant amounts of a hot, fresh phase) carry heat upwards. Much of the high salinity fluid would remain trapped by its high density within the lower boundary layer, where it would form in effect an isolated, lower convecting cell. Towards the top of the middle region, entrainment of increasing amounts of ambient seawater would be consistent with an upwards decrease in fluid resistivity. Thus the upper boundary layer above about 1.4 km depth would represent a region in which we might expect to see the observed gradient in properties towards those of ambient seawater at the seafloor.

The suggested model is far from unique, and remains highly speculative. Nonetheless the geophysical data, combined with appropriate effective medium modelling, now have sufficient resolution to enable us to identify and start to explore changes in the physical properties of hydrothermal fluids *in situ* within the crust, on length scales of hundreds of meters to kilometres. At least at the VFR, this reveals complexity in the fluid property distribution that tells us that a simple hydrothermal convection system (with spatially restricted upflow and much more pervasive seawater downflow everywhere else) is not compatible with the observations; and that phase separation is likely to play an important role not only in the geochemical aspects of hydrothermal circulation, but also in the physical patterns of fluid flow.

5.2. Other Geophysical Constraints on Hydrothermal Flow

Another approach to constraining modes of hydrothermal flow is to identify geophysical signals which delineate the pathways that fluids have taken as a result of alteration or some other impact on the rocks. A near-bottom magnetic survey around the main Endeavour vent field has revealed a series of circular magnetic lows centered beneath active and relict regions of venting [*Tivey and Johnson*, 2002] (Plate 9). These magnetic lows are interpreted as having been caused by high temperature vent fluids altering the magnetic minerals within the crust. The shape of the lows and the amplitude of the anomalies suggests that the fluid flow occurred in relatively narrow pipes, although the magnetic data alone cannot distinguish between an anomaly caused by a 40m diameter pipe and a 60m buried sphere. Furthermore, the magnetic signature comes predominantly from the upper-most extrusives, and there is no information on how deep the pipe might extend—although we know from thermal considerations that it must be at least several hundred meters into the seafloor. Other magnetic lows have been seen beneath the TAG and Lucky Strike hydrothermal mounds on the Mid-Atlantic Ridge [*Tivey et al.*, 1993; *Luis et al.*, 2000] and beneath Middle Valley on the Juan de Fuca Ridge [*Gee et al.* 2001]. At TAG, the magnetic low is also seen at the sea surface, and this has been interpreted as the signature of crustal thinning associated with activity on a detachment fault. The proximity of TAG to this fault (TAG sits on the hanging wall) supports the idea that crustal permeability here is controlled by faults that can reactivate pathways and maintain circulation for long periods of time [*Tivey et al.*, 2003]. At Lucky Strike, the hydrothermal activity is volcanically hosted at the center of the neovolcanic zone, and the magnetic anomaly appears to be due to demagnetization of the extrusive layer by hydrothermal alteration. The Lucky Strike anomaly may therefore be more similar to the Endeavour field anomalies, although spatially on a larger scale [*Luis et al.*, 2000]. In Middle Valley, which is a sedimented hydrothermal setting, heterogeneity in crustal magnetisation also points to varying spatial degrees of hydrothermal alteration [*Gee et al.*, 2001]. Similar high resolution surveys at fast spreading ridges, including 9° 50′N on the EPR [*Lee et al.*, 1996] have not found evidence for the same kind of localised upwelling. This suggests that the permeability structure of the intermediate spreading JDF and slow spreading MAR may be different to the EPR, and that the patterns of flow within the crust might also be different, as suggested by *Wilcock and Delaney* [1996], and which we discuss further in the next section.

5.3. How do Flow Patterns Depend on Ridge Morphology?

Most large-scale numerical models of ridge-axis circulation have considered flow through a porous medium [e.g., *Sleep et al.*, 1991; *Travis et al.*, 1991], driven by both 2D and 3D heat sources. These models feature flow which begins off-axis in a recharge zone, penetrates into the seafloor, and is drawn towards the ridge axis by magma-chamber driven convection. Hot fluids are vented in a narrow zone at the ridge-axis, as required by numerous seafloor observations. A magma chamber with finite length along-strike can also initiate along-strike flow, although if the medium is regarded as isotropic, then there will be no preference between along and across-strike flow.

Rosenberg et al. [1993] and *Haymon et al.* [1991] are advocates of along axis flow, at least for the deep-penetrating high temperature component of discharge. They propose models in which the deep penetrating fluids are confined to a narrow band beneath the axis, while more pervasive, but lower temperature, circulation occurs within the upper-most extrusives of Layer 2A. There is evidence from DSDP/ODP hole 504B and from ophiolites of alteration within dikes [e.g., *Alt et al.*, 1986; *Nehlig and Juteau*, 1988; *Nehlig*, 1993; *Gillis and Roberts*, 1999; *Gillis*, 2002], but both the porosity and permeabilities within the dike complex are low and possibly also anisotropic. *Wilcock and McNabb* [1996] used the spacing and alignment of vents on the Endeavour segment to argue for an anisotropic bulk permeability, with higher permeabilities in the along strike direction, and with primarily along strike flow of hydrothermal fluids.

The models proposed for segment scale circulation patterns differ widely in their predictions of recharge location, flow pattern and width of high temperature circulation zones. For example, there was discussion at the IRTI workshop of a model for Endeavour in which recharge occurs to the north of the segment, with along strike flow over a substantial distance at depths just above the magma chamber. Circumstantial evidence for this model comes from geochemical analyses of vent fluids which have elevated CH_4 and NH_4^+ concentrations—suggesting an organic (most likely sediment) contribution [*Lilley et al.*, 1993]. The closely spaced vents in the main Endeavour vent field would tap this source of hot fluid primarily by along strike flow, as suggested by *Wilcock and McNabb* [1996]. While there are no direct observations of how far fluids can flow along the ridge crest, measurements on the flanks of the JDF ridge suggest that off-axis transport can occur over distances greater than 20 km [*Davis et al.*, 1999].

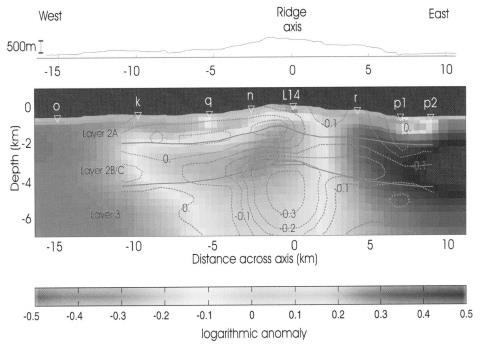

Plate 8. Anomalies in electrical resistivity (colors) and seismic p-wave velocity (gray contours) across the Valu Fa Ridge, Lau Basin, relative to a 1-D background profile [*MacGregor et al.*, 2000]. No significant seismic velocity anomaly is seen across the axis in layer 2, although the crustal magma body shows up clearly as a low-velocity region (and also as a seismic reflector). In contrast, the electrical resistivity result shows a strong and asymmetric conductive anomaly in the lower two-thirds of Layer 2 at and west of the axis, as well as decreased resistivities associated with the underlying magma body.

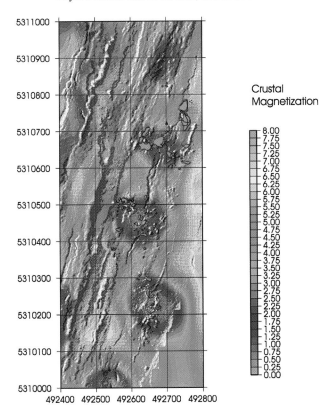

Plate 9. A map of the main vent field on the Endeavour segment. Bathymetric contours are shown with high resolution magnetic field data superimposed in color [*Tivey & Johnson*, 2002]. The regions of anomalously low magnetic field are interpreted as alteration pipes resulting from upflow of hot hydrothermal fluid, and have been dubbed "burnholes". These lows correspond to active and relict regions of venting (active vents are shown by the red patches), and suggested highly localised regions of focussed upwelling.

We do not know how changes in magma chamber geometry (width and depth) and changes in the fracturing of the overlying crust tie together with other tectonic factors to control the regional hydrothermal patterns and the partitioning of heat flux through different parts of the crust. *Wilcock and Delaney* [1996] discuss the implications of spreading rate and the subsequent changes in style of faulting on circulation patterns. The extensive normal faults found on the Endeavour segment of the JDF should provide high permeability access to deeper parts of the crust, and are presumably key to the regional patterns of hydrothermal circulation. This is in contrast to faster spreading ridges. EM data from 9° 50′ N on the EPR show a fairly uniform electrical resistivity structure along axis, even when areas of active venting are compared to less active regions [*Evans et al.*, 2002]. This suggests a relatively 2-D thermal structure here, with vent fluids tapping very localised cracks that extend deep into the crust. Sulphide chimneys at faster spreading ridges, including this section of the EPR, are small (a few meters), suggesting a short life-span commensurate with a restricted hot water supply that can become clogged with hydrothermal mineralization or closed through tectonism or volcanic flows.

In contrast, slower spreading ridges have fewer but larger sulphide outcrops. This has been interpreted to be due to the role of deep and long lived faults for providing circulation paths. Endeavour appears to have the characteristics of a fault controlled hydrothermal setting, and *Delaney et al.* [1992] note that the vents in the main Endeavour field appear to lie at the intersection of ridge parallel faults and oblique fissure sets.

5.4. Upper Crustal Thermal Structure: Conclusions

While we now have many observations, including the first few sets of constraints on how porosity and its interconnectivity are distributed within the crust between the seafloor and magma chamber level, the sub-seafloor hydrology and physics remain extremely poorly understood. Much of what we think we know is based on inference rather than observation by sub-seafloor geophysical methods. Key points that we should keep in mind as we investigate ridge hydrothermal systems further include:

- In some locations, the field geophysical evidence is consistent with circulation models in which high temperature upflow is restricted to a few small locations—so that almost all of the upper crust (Layer 2), even at or very close to the axis, might be expected to be permeated by relatively low temperature seawater recharge fluids. High temperatures in the upper 1 to 3 km of the crust would be restricted to the immediate environs of the crustal magma body, and to small and isolated high-T upflow zones.

- There are, however, locations where simple single-cell circulation models are not consistent with the geophysical data. Where phase separation plays a significant part in heat and fluid transport, both the circulation pattern and the analysis of the geophysical data will be more complex.

- There is some evidence that along-strike flow in circulation systems is more important than inflow at depth from off-axis regions. However, geophysical data suggest that it is hard to argue that Layer 2B (the sheeted dyke section) has significantly higher porosity and pore interconnectivity at the axis than at distances within a few km off-axis. Anisotropic permeability may be important here, but this has yet to be confirmed geophysically. While Layer 2B is weakly seismically anisotropic, the amount of anisotropy is consistent with only a small proportion of the low-aspect-ratio pore spaces having a preferential orientation, while the great majority of pore spaces may be randomly orientated. We should also bear in mind that much of the heat available to drive high temperature circulation is given up by the newly formed crust after, rather than before, it has crystallised—so if axis-parallel flow was particularly important, we might expect to see more, rather than fewer, high temperature venting systems at distances off-axis greater than the width of the axial magma body.

- The one major contrast in sub-seafloor hydrological properties of which we can be confident is the boundary between Layers 2A and 2B (or between extrusive basalt lavas and sheeted dikes, if we assume the geophysical-lithological correlation to be valid). This boundary occurs at depths of typically 200 to 600 m below the seafloor, and may be less sharp at some locations than at others. The high porosity and complete interconnection in Layer 2A mean that this layer is likely to be cold, and to be completely permeated by freely circulating ambient seawater. High temperature fluids entering this layer from below are likely to be very rapidly diluted by entrained seawater, unless the upflow region is sealed off from the surrounding parts of layer 2A by sub-seafloor hydrothermal precipitates. Below the 2A/2B boundary, porosities are significantly lower (though still quite highly interconnected), leading to quite different hydrological properties. Any satisfactory model for the circulation of fluids at ridges will need to take account of the importance of this hydrological, as well as geophysical/lithological, boundary.

Lastly it seems likely that relatively deeply penetrating and long-lived fault systems are more influential in controlling the patterns of circulation at slow spreading ridges than at fast spreading ridges. We need to better understand the significance of fault systems for overall hydrological properties at all spreading rates; but we must not forget that the brittle rheology of the upper and middle crust that gives rise to these fault systems is itself directly related to temperature structure—so there may be positive feedback mechanisms between faulting activity, fluid penetration and circulation and the thermal structure of the crust. We should not be surprised then by the correlation between ridge spreading rate and morphology and the pattern of high-temperature venting; but the exact nature of the linking mechanisms, let alone any ability to make quantitative predictions, still elude us.

6. HIGH LEVEL MAGMATIC SYSTEMS AND EVENTS

In considering the thermal structure of the oceanic crust at ridges, we are obliged to take account of the short-term excursions in thermal properties associated with eruptions and with the intrusion of dikes into the upper parts of the crust. There is now strong evidence that these events play an important role in modulating—and perhaps even dominating—the patterns of high-temperature hydrothermal circulation at ridges. While the deep circulation of hydrothermal fluids—extracting heat from mid- and lower-crustal magma bodies and recently crystallised plutonic units—may have properties that we can usefully model in terms of steady state structures and processes, we must think in terms of specific events if we are to understand the impact of higher level magma bodies on crustal thermal structure and hydrothermal circulation.

6.1. Diking Events and Upper Crustal Construction

Melt is transported from the magma chamber to the seafloor by diking events. We would expect that dike injection would occur more frequently close to the ridge axis, or above the melt source. Seismic reflection data (or more accurately long offset reflection data which measures rays turning within layer 2A) have been collected across many ridge segments, and show patterns of layer 2A evolution with distance away from the ridge crest that can be understood in terms of eruption patterns and cycles [e.g., *Harding et al.*, 1993]. *Hooft et al.* [1996] have produced a model of dike injection and subsequent off-axis flow of eruptive lava that is able to reproduce the commonly observed thickening of layer 2A with crustal age. In this model, the base of later 2A is the extrusive/dike contact. Dikes, which outcrop within the axial summit collapse trough (ASCT) or caldera (ASC) [*Fornari et al.*, 1998] are buried beneath a series of eruptions. In order to accurately match the patterns of 2A thickening away from the ridge crest, a bimodal distribution of lava flows is needed. About one half of the 2A thickness is constructed by flows confined to the ASC/ASCT. The remaining thickness is built up off-axis by flows which overflow the walls of the axial valley and flow off-axis, or by eruptions that take place off-axis to begin with. When 2A is fully developed, the uppermost portions will con-

sist of flows that were emplaced off-axis. The effect of flows draining down into the underlying sequence will tend to fill in voids and hence lead to decreasing porosity with depth.

With the advent of autonomous underwater vehicles has come the ability to generate detailed bathymetric maps at ridge crests, allowing detailed geologic mapping of lava flow morphology to be carried out. These maps have provided considerable insight into the mechanics of melt transport off-axis. They confirm many of the predictions of the Hooft *et al.* model by revealing flow tubes through which lava is transported off-axis. For example, a survey using the ABE (autonomous benthic explorer) vehicle equipped with high resolution micro-bathymetric mapping tools shows evidence of channelized melt transport and also of eruptions that flow from the ridge axis to distances of 2 km, covering existing lavas and contributing to the construction of layer 2A [*Schouten et al.*, 2002]. High resolution magnetic surveys using ABE and deep-towed magnetometers have also been used to constrain thicknesses of fresh eruptive sequences [*Tivey et al.*, 1998].

6.2. Eruptions and Transient Hydrothermal Events

The monitoring of earthquakes and seismic energy associated with mid-ocean ridge eruptions is difficult because of the small size of these events—generally well below the threshold for detection by the global seismic network. The declassification of data collected by the US Navy on its underwater sound monitoring hydrophone array (SOSUS) in the North East Pacific was a boon to researchers. The quasi-real-time data stream allowed several rapid response cruises to be mounted as eruptions were detected [*Fox et al.*, 1994]. Among these events were several major eruptions emanating from Axial seamount on the JDF [*Dziak and Fox*, 1999; Fox, 1999], a dike injection event on the Co-axial segment of the JDF [*Fox et al.*, 1995], and an eruption on Gorda Ridge [*Fox and Dziak*, 1999]. These eruptions were followed up by detailed seafloor monitoring including the installation of tilt meters [*Tolstoy et al.*, 1998], motion sensors [*Chadwick et al.*, 1999] and ocean bottom seismometers [*Sohn et al.*, 1998b; 1999b; *Tolstoy et al.*, 2002]. During the 1998 Axial eruption, for example, a seafloor pressure sensor recorded a drop of 3m immediately after an eruption, while another instrument was partially buried in a lava flow [*Fox*, 1999; *Fox et al.*, 2001]. These and other observations have allowed quantitative estimates to be ascribed to the durations, volumes of eruptions, and the along-axis extent of dike injection events. Observations continuing for long after eruptions have placed timescales on the cooling of dikes.

The northern portion of the Cleft and the Vance segments of the JDF were the sites of anomalous hydrothermal plumes, known as mega-plumes, which were detected in the mid-1980s

[*Baker et al.*, 1987; 1989]. These events had characteristics distinct from other hydrothermal plumes associated with the focussed discharge of black smoker fluid. The events were in some cases short lived, were at different depths in the water column to steady state black smoker plumes, had large positive temperature anomalies, contained a large amount of heat energy and had peculiar chemical signatures. Follow up surveys of the seafloor identified a large young sheet flow and a 17 km long linear series of pillow flows in the same area as the megaplumes [*Embley and Chadwick*, 1994; *Chadwick and Embley*, 1994]. The total volume of the lava flows is about 0.05 km³ and they are associated with a single fracture system. The root supply for the recent volcanism along this segment has been proposed to be a lateral dike intrusion from a buried magma supply situated below the sheet flow [*Embley and Chadwick*, 1994], and this diking event appears to be linked to the megaplume event. EM data collected at this site in 1994 showed raised crustal electrical conductivities consistent with a thermal anomaly in the uppermost few hundred metres of crust [*Evans et al.*, 1998], although the data coverage was insufficient to delineate the patterns and distribution of hot water.

A definitive link between diking and anomalous hydrothermal effluent was made at the Co-Axial segment of the JDF, where the SOSUS array detected a micro-earthquake swarm. Over a two day time period the locus of earthquakes migrated from the northern flank of Axial seamount to a distance of about 40 km to the north. Aftershocks continued for about 3 weeks [*Dziak et al.*, 1995; *Fox et al.*, 1995]. Subsequent seafloor observations found a site of hydrothermal effluent with mega-plumes larger than those from the Cleft [*Baker et al.*, 1995]. Eruptive material was only found at the northern end of the segment, although hydrothermal activity was found at the eruption site as well as at a site to the south where bacterial flocculent material was observed in the water column. Estimates of the amount of magma emplaced at Co-Axial are 5×10^6 m³ erupted and $120–180 \times 10^6$ m³ intruded [*Chadwick et al.*, 1995], smaller than that erupted at Cleft, by 1–2 orders of magnitude. Hydrothermal venting at both of these Co-Axial sites was extinct three years after the eruption [*Embley* et al., 2000].

A micro-earthquake study at Axial seamount which featured a very small aperture array of seismometers has been able to place tight constraints on the depths of earthquakes occurring immediately after the 1998 eruption [*Sohn et al.*, 1999b; 2004]. Activity was monitored over a 15 month period post eruption, during which time there was a decrease in the frequency of earthquakes. Hypocentral depths are around 700 m to 1 km, considerably shallower than the depth to the magma chamber which has been mapped seismically at 3 km [*West et al.*, 2001], and locations appear to correlate with subsurface magmatic features. For example, during the first 60 days post-

eruption, earthquake locations spatially correlate with seafloor observations of fresh lava flows. *Sohn et al.* [2004] assign a thermal contraction mechanism to the earthquakes, with the drop-off in activity related to an approximately 60 day time period for the intrusion to cool. The height of the events above the magma chamber is interpreted to indicate a sharp increase in temperature below about 1 km depth. *Sohn et al.* [2004] suggest that hydrothermal cooling within the caldera is, at present, limited to this upper 1 km of seafloor, although this depth may increase with time until the next eruption occurs and resets the thermal regime of the upper crust. Below 1 km, the rocks are sufficiently hot to deform aseismically.

The observations made at the Axial observatory and elsewhere on the JDF have provided fascinating insights into the eruption processes at ridge crests. They are only a taste of what is likely to happen in the future as observatory infrastructure is put in place at ridges, providing data bandwidth and power for instrumentation that will permit many arrays of sensors to be deployed for long periods of time, in turn allowing a continued presence at many locations along the ridge system.

7. AMAGMATIC HYDROTHERMAL SYSTEMS

There have been a number of discoveries of hydrothermal activity at slow and ultra-slow spreading ridges, where magmatic budgets are extremely low and where, in some cases, there is essentially no volcanic crust but rather unroofed mantle rocks are exposed at the seafloor. In addition to neovolcanic or axial valley venting, activity has also been found significant distances off-axis along the slow spreading MAR in sites that almost certainly cannot be influenced by the magma supply that forms the oceanic crust. The extent of hydrothermal activity in these settings appears to be greater than expected simply on the basis of the magmatic budgets of these segments, or from the simple predictive relationship between frequency of venting and spreading rate [*Baker et al.*, 1996]. This suggests that alternative sources of heat are being tapped by seawater, and several explanations for this alternative heat source have been proposed. In some cases it might represent a component of the lithospheric heat that would otherwise be lost through conductive cooling (as discussed in section 2) but which under normal circumstances is not available to hydrothermal removal, or in other cases, might be generated by chemical reactions between seawater and peridotite.

7.1. Off-Axis Venting on the MAR

An unusual vent field was found recently at a transform fault offset on the MAR, in 1.5 Ma crust. The field, known as the Lost City [*Kelley et al.*, 2001] is located on an inside-corner high at 30° N, between the spreading axis and the transform system, and is associated with a low angle detachment fault. Unlike the polymetallic sulphide compositions ubiquitous at axial vent sites, the Lost City chimneys are formed of carbonate. Furthermore, vent fluid temperatures are only 45–75° C. Geophysical evidence and geological sampling of this and other exposed footwall features at low-angle normal faults on the MAR, indicate that they consist of exposed mantle peridotite which has undergone extensive serpentinization as a result of exposure to seawater [Cann et al., 1997; Blackman et al., 1998]. The abundance of serpentinite, the carbonate chimney compositions and the low temperature of vent fluids are all consistent with exothermic reactions associated with serpentinization as a source of thermal energy driving the effluent heat flux.

In addition to this peridotite-hosted off-axis site, a number of the vent fields on the MAR are considerably removed from the neovolcanic zone and their positions appear to be controlled by intersecting fault patterns. TAG is one example of this [Kleinrock and Humphris, 1996], as is the Rainbow site which sits in a non-transform offset [German and Parson, 1998]. This tectonic control on hydrothermal activity was discussed above and has been proposed by Wilcock and Delaney [1996] as a primary reason why larger longer-lived sulphide bodies are found where active faults control the permeability structure. TAG has been active on and off for at least 20 Ka [Lalou et al., 1993], substantially longer than expected given a lack of an obvious thermal source.

7.2. Ultra-Slow Ridges

A number of hydrothermal plumes have been mapped on the South West Indian Ridge (SWIR), an ultra-slow spreading ridge [German et al., 1998]. The abundance of plumes is higher than expected for such a slow spreading segment, with 6 sites found along 400 km of ridge (in contrast, the MAR has an average known vent site density of 1 per 110 km [German and Parson, 1998]). Subsequent dredging of hydrothermal deposits from the SWIR has identified a peridotite hosted environment for the circulation. Bach et al. [2002] discuss the conditions that might drive heat export from the seafloor, given that magmatic heat does not seem to be a viable option. Two key controls are surmised to be exothermic reactions from serpentinization, and spatially restricted high permeability faults and cracks that limit where heat is mined but which allow heat to be mined over long periods of time resulting in large sulphide formations. Serpentinization reactions are thought to be able to heat water to anywhere between 25° C and 150°C, and so while they do not appear viable for high temperature venting, they do seem appropriate for these sites on the SWIR, and also for the Lost City vent field discussed in the previous section.

The Gakkel Ridge has the slowest spreading rate of any part of the global ridge system. Because of its high latitude, it is hard to access and has not been extensively studied. Bathymetric mapping has been completed using US Navy submarines and ice breaking vessels [e.g. Edwards et al., 2001], and dredge samples have been collected [Dick et al. 2003], but there is little sub-seafloor geophysical data to constrain the crust and mantle structure. The small amount of geophysical data that does exist suggests thin crust (1.9 to 3.3 km) with no gabbroic section [Jokat et al., 2003; Cochran et al., 2002]. Dick et al. [2003] suggest that ultra-slow spreading ridges are fundamentally different from slow spreading ridges, given the preponderance of exposed peridotite suggesting extremely limited and punctuated volcanic activity. Edmonds et al. [2003] report finding numerous water column plumes resulting from hydrothermal effluent. The Gakkel Ridge has an observed vent frequency of 9–12 in 1100 km, roughly the same as the MAR, but Gakkel has a spreading rate of 0.6–1.3 cm yr^{-1}, substantially slower than the MAR. The hydrothermal plumes appear to be co-located with centers of volcanic activity, suggesting a magmatic heat source. This relationship suggests a different thermal driving force mechanism to that proposed at the SWIR. Michael et al. [2003] argue that the strong focusing of magma delivery coupled with the intensity of faulting can account for the irregularly spaced but high levels of hydrothermal activity.

Because the Gakkel Ridge has undergone relatively little exploration, we have no idea of the longevity of these vent sites. We also have little idea of the role that the thin crust is able to take in facilitating extraction of heat from the mantle by seawater, although based on the observations of Edmonds et al. [2003] the venting observed does appear to be related to recent volcanic activity. At places where melt is injected into the crust, does the crustal structure make direct interaction between melt and seawater more likely?

The observation of high plume incidences at both the Gakkel Ridge and SWIR suggests that there may be a lower limit on the amount of heat removed from a ridge by hydrothermal means. Whether this limit is a serendipitous effect arising from many factors (crustal thickness and structure, patterns and frequency of melt injection, tectonic controls on fluid pathways) is open for debate, and is unlikely to be resolved until detailed structural images of the crust and mantle at these anomalous settings are obtained. In any event, these findings suggest that a significant amount of heat is lost to the oceans at slow and ultra-slow spreading ridges by reactions that are not directly linked to magmatic budget, but which may reflect other sources of heat that would either be lost through conduction or which is due to exothermic chemical reaction between sea water and the host rock.

8. CONCLUSIONS

On the scale of an oceanic lithospheric plate, the overall thermal structure of the crust is well constrained by geophysical data and plate cooling models (e.g., Plate 1). Over much of the oceans, the structure can be well represented by an approximately linear geothermal gradient associated with conductively dominated heat flow. Local deviations from this structure in the upper 1 to 2 km of the crust due to off-axis fluid circulation and advective heat transport are likely to be minor in temperature terms, but widely distributed, especially at crustal ages of less than 60 Ma.

As we approach the ridge axes, this picture changes, and uncertainties in thermal structure increase. A major source of the uncertainty is that as yet we have an incomplete understanding of the pattern of upwelling mantle flow beneath ridges, and of the migration routes and mechanisms of the resulting basaltic magmas that both act as vectors for much of the heat flow and provide the material for crustal construction. Since the thermal structure of the crust is in large part inherited from deeper-seated structures and processes in the mantle, a better knowledge of mantle kinematics would provide important boundary conditions for future attempts to reduce uncertainties in crustal properties. Aspects of mantle behaviour which contribute significant uncertainties include:

- To what extent is mantle circulation beneath ridges influenced by the larger scale convective flow in the upper mantle? Examples include the evidence of eastward flow of mantle material beneath the southern EPR from the MELT study; elevated mantle temperatures at sites of ridge-plume interactions; and ridges overlying regions of cold or down-welling mantle, for instance in the southern Indian Ocean.
- Is the flow pattern dominated by boundary conditions imposed by plate separation, or by buoyancy driven effects?
- Is the flow pattern dominantly two- or three-dimensional, and how does this vary with spreading rate?
- What factors are important in causing melt that is generated at depth over a region many tens of kilometres wide to be delivered to a neovolcanic zone that may be less than 1 km wide?
- How important is time variation in these flow and migration patterns?

Once the magma generated by decompression melting of the mantle has been delivered to the neovolcanic zone, it appears that the crust reaches its full thickness and a recognizable Moho boundary forms extremely rapidly—certainly within a kilometer or two of the spreading axis. Subsequent evolution of the crust and its thermal structure depend on a balance between magmatic thermal input, the size and location of any crustal magma body, and hydrothermal cooling. It is important to recognize that latent heat of crystallization of magma represents only a relatively minor component of the total heat flux, as is seen from Tables 2 and 4. Progressive cooling below the crystallization temperature and long-term cooling of the lithospheric mantle both provide far more heat energy that ultimately reaches the water column than the crystallization of the magma itself.

At fast and intermediate spreading rates, the axial crust contains substantial volumes of magma and of very recently crystallized plutonic rocks for much if not all of the time. A sill-like melt lens represents the top of this crustal magma distribution; and it is likely that the depth of the melt lens is controlled both by the available magmatic flux and by the permeability of the overlying crust, and hence the efficiency with which hydrothermal cooling can transport heat away. There appears to be a correlation between the depth of the melt lens and spreading rate, at least at fast and very fast spreading rates; but this correlation may break down at intermediate and slow rates. Beneath the melt lens, the distribution both of temperatures and of magma remains controversial; however, there is growing evidence both from ophiolite studies and from geophysical studies at the EPR that hydrothermal cooling may from time to time extend deep into the crust at short distances from the axis, allowing some of the melt in the lower regions of the crust to crystallize and cool at those depths.

While intermediate spreading rate ridges commonly have magma bodies within the axial crust that are similar to those at fast spreading rates, there is also geophysical evidence that tectono-magmatic cycles play an important role in modulating both the thermal structure and the magmatic construction of the crust, on time scales on the order of thousands of years. At slow spreading ridges, our knowledge of large crustal melt bodies is restricted to a sample of one. This is certainly grossly insufficient for us to extrapolate to the entire slow spreading ridge system; however, the evidence we do have is consistent with tectono-magmatic cycles playing a crucial role in crustal architecture and evolution at slow spreading rates. This would indicate that the crustal thermal structure of slow spreading ridges is highly variable through time, ranging from essentially cold periods in which sea floor spreading is accompanied by brittle rheology and extensional faulting, to hot periods in which crustal magma bodies similar to fast and intermediate spreading ridges may exist.

At slow and especially ultra-slow spreading ridges, there is now compelling evidence that substantial fractions of the seismically-defined 'crust' (i.e. above the Moho boundary) is composed not of crystallized basaltic magma from partial melting of the upper mantle, but of partially serpentinised mantle rocks—which outcrop in many locations. The composition of the crust is likely to be closely related to segmentation patterns. Along the MAR, for instance, it is likely that segment centers typically produce crust that is both thick (as much as 7 to 8 km) and is largely composed of basaltic material that has crystallized from mantle melt; whereas crust at segment ends is likely to be both thin (as little as 1 to 3 km) and dominated by serpentinized ultramafic rocks. Geological mapping of the seafloor at slow and ultra-slow sites remains patchy, while geophysical methods are not readily able to discriminate between these contrasting rock types in the sub-surface; so key aspects of this fundamental facet of crustal architecture at slow and ultra-slow spreading rates remain unquantified.

Above the level of the magma body, the long-term thermal structure is controlled by the porosity and permeability of the volcanic and sheeted dyke sections of layers 2A and 2B, and by the ability of seawater to penetrate, circulate, and convectively carry away heat both from latent heat of crystallization and from sub-solidus cooling. In layer 2B, porosity is typically quite low—between 1 and 4%—but the pore spaces are typically sheet or fracture-like features with low aspect ratio, and have a high degree of interconnection. A major permeability boundary separates layer 2B from layer 2A. In the volcanic layer 2A, porosities are much higher—typically above 20%—and there are abundant large void spaces which represent a fully interconnected network. Seawater can therefore flow very freely throughout this layer. It may be that the relatively low permeabilities encountered immediately beneath high-temperature vent sites are atypical of layer 2A, and are due to exceptional amounts of hydrothermally precipitated mineral deposits locally shutting off the otherwise high permeability. In this case, circulation pathways for high temperature fluids that reach the seafloor may be chimney-like in layer 2A. In general, fluid circulation between the seafloor and the base of layer 2A is likely to be qualitatively very different from circulation in the sheeted dike section.

While the long term thermal structure of the axial upper crust depends on the balance between magmatic thermal input and hydrothermal heat extraction, short term excursions from this structure due to eruption and diking events are important in terms of the shallowest high-temperature circulation cells, and hot water release events such as megaplumes.

At slow spreading ridges, temperatures low enough to correspond to brittle rheology (<~700°C) and to deformation by earthquakes and faulting extend, at least part of the time, throughout the crust and into the uppermost mantle, even at the ridge axis. Active faulting and the resulting high-permeability fracture pathways penetrating several kilometres deep into the axial lithosphere are likely to be highly influential in controlling the locations of long-lived circulation systems capable of delivering vent fluids at temperatures of several hundred degrees C. Such circulation can in turn influence the thermal structure, extending the range of brittle deformation and creating a positive feedback between fracturing and cooling.

Finally, observations in the Atlantic, Arctic and Indian Oceans show that magmatic heat is not the only driving mechanism for vigorous hydrothermal circulation. The hydration reaction of peridotite to serpentinite is exothermic, so penetration of seawater into fractured and unroofed mantle rocks may spontaneously lead to amagmatic intermediate or low temperature venting. The increase in volume that accompanies hydration of mantle rocks is likely to lead to further fracturing, and this may lead to self-sustaining circulation systems, as well as being influential in other aspects of slow spreading ridge tectonics.

Acknowledgments. We thank Jian Lin, Chris German and Mathilde Cannat for convening the InterRidge Theoretical Institute in Pavia and Sestri Levanti, Italy, September 2002; InterRidge, US RIDGE 2000 Program and the European Science Foundation for their organizational and financial support for the Institute; and all participants at the IRTI for the excellent and lively discussions there which inspired us to put this paper together. Rob Dunn, Mark Behn and Doug Toomey are thanked for particularly thorough and constructive reviews, and Damon Teagle is also thanked for comments on the manuscript. We thank K. Davis and J. Cook for assistance with the figures.

REFERENCES

Alt, J. C., J. Honnorez, C. Laverne, R. Emmerman, Hydrothermal alteration of a 1 km section through the upper oceanic crust, Deep Sea Drilling Project Hole 504B: Mineralogy, chemistry and evolution of seawater-basalt interactions, *J. Geophys. Res.*, 91, 10309–10335, 1986.

Bach, W. N. Banerjee, H. J. B. Dick, E. T. Baker, Discovery of ancient and active hydrothermal systems along the ultra-slow spreading Southwest Indian Ridge 10°–16°E, *Geochem. Geophys. Geosyst.*, 3, 10.1029/2001GC000279, 2002.

Babcock, J. M., A. J. Harding, G. M. Kent, J. A. Orcutt, An examination of along-axis variation of magma chamber width and crustal structure on the East Pacific Rise between 13 30N and 12 20N, *J. Geophys. Res.*, 103, 30451–30467, 1998.

Baker, E. T., G. J. Massoth, R. A. Feely, Cataclysmic venting on the Juan de Fuca Ridge, *Nature*, 329, 149–151, 1987.

Baker, E. T., W. Lavelle, R. A. Feely, G. J. Massoth, S. L. Walker, Episodic venting on the Juan de Fuca Ridge, *J. Geophys. Res.*, 94, 9237–9250, 1989.

Baker, E. T, G. J. Massoth, R. A. Feely, R. W. Embley, R. E. Thomson, Richard, B.J. Burd, Hydrothermal event plumes from the Coaxial seafloor eruption site, Juan de Fuca Ridge, *Geophys. Res. Letts.*, 22, 147–150, 1995.

Baker, E. T., Y. J. Chen, J. Phipps-Morgan, The relationship between near-axis hydrothermal cooling and the spreading rate of mid-ocean ridges, *Earth Planet. Sci. Letts*, 142, 137–145, 1996.

Bercovici, D., A simple model of plate generation from mantle flow, *Geophys. J. Int.*, 114, 635–650, 1993.

Bercovici, D., Generation of plate tectonics from lithosphere-mantle flow and void-volatile self-lubrication, *Earth Planet. Sci. Letts.*, 154, 139–151, 1998.

Bernabe, Y., A. Revil, Pore-scale heterogeneity, energy dissipation and the transport properties of rocks, *Geophys. Res. Letts.*, 22, 705–708, 1995.

Blackman, D. K., J. R. Cann, B. Janeson, D. K. Smith, Origin of extensional core complexes: Evidence from the Mid-Atlantic Ridge at Atlantis Fracture zone, *J. Geophys. Res.*, 103, 21315–21333, 1998.

Bonatti, E., M. Ligi, D. Brunelli, A. Cipriani, P. Fabretti, V. Ferranti, L. Gasperini, L. Ottolini, Mantle thermal pulses below the Mid-Atlantic Ridge and temporal variations in the formation of oceanic lithosphere. *Nature*, 423, 499–505, 2003.

Boudier, F., A. Nicolas, B. Ildefonse, Magma chambers in the Oman ophiolite: fed from the top or the bottom? *Earth Planet. Sci. Letts.*, 144, 239–250, 1996.

Braun, M. G., G. Hirth, E. M. Parmentier, The effects of damp melting on mantle flow and melt generation beneath midocean ridges, *Earth and Planet. Sci. Letts.*, 176, 339–356, 2000.

Braun, M.G., P.B. Kelemen, Dunite distribution in the Oman Ophiolite: Implications for melt flux through porous dunite conduits, *Geophys. Geochem. Geosyst.*, 3, 10.1029/2001GC000289, 2002.

Browning, P., Cryptic variation within the cumulate sequence of the Oman ophiolite: magma chamber depth and petrological implications, in: Gass, I. G., Lippard, S. J., Shelton, A. W. (Eds.), *Ophiolites and oceanic lithosphere*, Geol. Soc. Lond., Spec. Pub. 13, 71–82, 1984.

Buck, W. R., W. Su, Focused mantle upwelling below mid-ocean ridges due to feedback between viscosity and melting, *Geophys. Res. Lett.*, 16, 641–644, 1989.

Calmant, S., J. Francheteau, A. Cazenave, Elastic layer thickening with age of the oceanic lithosphere, *Geophys. J. Int.*, 100, 59–67, 1990.

Calvert, A. J. Seismic evidence for a magma chamber beneath the slow-spreading Mid-Atlantic Ridge. *Nature*, 377, 410–414, 1995.

Canales, J. P., R. S. Detrick, J. Lin, J. A. Collins, D. R. Toomey. Crustal and upper mantle seismic structure beneath the rift mountains and across a non-transform offset at the Mid-Atlantic Ridge (35°N). *J. Geophys. Res.*, 105, 2699–2719, 2000.

Cann, J. R., D. K. Blackman, D. K. Smith, E. McAllister, B. Janssen, S. Mello, E. Avgerinos, A. R. Pascoe, J Escartin, Corrugated slip surfaces formed at ridge-transform intersections on the Mid-Atlantic Ridge, *Nature*, 385, 329–332, 1997.

Cannat, M., Cann, J., Maclennan, J., Some hard rock constraints on the heat supply to mid-ocean ridges, *This Volume*, 2004.

Chadwick, W.W., R.W. Embley, Lava flows from a mid 1980s submarine eruption on the Cleft segment, Juan de Fuca Ridge. *J. Geophys. Res.*, 99, 4761–4776, 1994.

Chadwick, W. W., R. W. Embley, C. G. Fox, Sea Beam depth changes associated with recent lava flows, CoAxial segment, Juan de Fuca Ridge: Evidence for multiple eruptions between 1981–1993, *Geophys. Res. Letts.*, 22, 167–170, 1995.

Chadwick, W. W., R. W. Embley, H. B. Milburn, C. Meinig, M. Stapp, Evidence for deformation associated with the 1998 eruption of Axial Volcano, Juan de Fuca Ridge, from acoustic extensiometer measurements, *Geophys. Res. Letts.*, 26, 3441–3444, 1999.

Chen, Y., W. J. Morgan, Rift valley/no rift valley transition at mid-ocean ridges, *J. Geophys. Res.*, 95, 17571–17581, 1990.

Cherkaoui, A. S. M., W. S. D. Wilcock, R. A. Dunn, D. R. Toomey, A numerical model of hydrothermal cooling and crustal accretion at a fast spreading mid-ocean ridge, *Geochem Geophys. Geosyst.*, 4, 8615, 10.1029/2002GC000499, 2003.

Christensen, N. I., Compressional wave velocities in rocks at high temperatures and pressures, critical thermal gradients and crustal low-velocity zones, *J. Geophys. Res.*, 84, 6849–6857, 1979.

Cochran, J. R., G. J. Kurras, M. H. Edwards, B. J. Coaskley, The Gakkel Ridge: crustal accretion at extremely slow spreading rates, *Eos Trans AGU*, 83, 2002.

Collier, J., M. Sinha, Seismic images of a magma chamber beneath the Lau Basin back-arc spreading center. *Nature*, 346, 646–648, 1990.

Conder, J. A., D. W. Forsyth, E. M Parmentier, Astenospheric flow and asymmetry of the East Pacfic Rise, *J. Geophys, Res.*, 107, 10.1029/2001JB000807, 2002.

Coogan, L. A., N. C. Mitchell, M. J. O'Hara, Roof Assimilation at fast spreading ridges: an investigation combining geophysical, geochemical and field evidence, *J. Geophys. Res.*, 108, 10.1029/2001JB001171, 2003.

Crawford, W. C., S. C. Webb, Variations in the distribution of magma in the lower crust and at the moho beneath the East Pacific Rise at 9°–10°N, *Earth Planet. Sci. Letts.*, 203, 117–130, 2002.

David, C., Geometry of flow paths for fluid transport in rocks, *J. Geophys. Res.*, 98, 12267–12278, 1993.

Davis, E. E., D. S. Chapman, K. Wang, H. Villinger, A. T. Fisher, S. W. Robinson, J. Grigel, D. Probnow, J. Stein, K. Becker, Regional heat flow variations across the sedimented Juan de Fuca Ridge eastern flanks: constraints on lithospheric cooling and lateral heat transport, *J. Geophys. Res.*, 104, 17,675–17,688, 1999.

Day, A. J., C. Peirce, M. C. Sinha, Three-dimensional crustal structure and magma chamber geometry at the intermediate-spreading, back-arc Valu Fa Ridge, Lau Basin—results of a wide-angle seismic tomographic inversion. *Geophys. J. Int.*, 146, 31–52, 2001.

Delaney, J. R., V. Robigou, R. E. McDuff, M. K. Tivey, Geology of a vigorous hydrothermal system on the Endeavour segment, Juan de Fuca Ridge, *J. Geophys. Res.*, 97, 19663–19682, 1992.

DeMets, C. R. G. Gordon, D. F. Argus, S. Stein, Current plate motions, *Geophys. J. Int.*, 101, 425–478, 1990.

Detrick, R.S., P.Buhl, E.Vera, J.Orcutt, J.Mutter, J.Madsen, T.Brocher, Multi-channel seismic imaging of a crustal magma chamber along the East Pacific Rise. *Nature*, 326, 35–41. 1987

Detrick, R. S., J. C. Mutter, P. Buhl., I. Kim, No evidence from multichannel reflection data for a crustal magma chamber in the MARK area on the Mid-Atlantic Ridge, Nature, 347, 61–64, 1990.

Detrick, R. S., A. J. Harding, G. M. Kent, J. A. Orcutt, J. C. Mutter, P. Buhl, Seismic structure of the southern East Pacific Rise, *Science*, 259, 499–503, 1993.

Detrick, R. S., H. D. Needham, V. Renard, Gravity anomalies and crustal thickness variations along the Mid-Atlantic Ridge between 33°N and 40°N, *J. Geophys. Res.*, 100, 3767–3787, 1995.

Detrick, R. S., S. Carbotte, E. Van Ark, J. P. Canales, G. M. Kent, A.J. Harding, J. Diebold, M. Nedimovic, New Multichannel Seismic Constraints on the Crustal Structure of the Endeavour Segment, Juan de Fuca Ridge: Evidence for a Crustal Magma Chamber, EOS Trans AGU, 83(47), 2002.

Detrick, R. S., J. M Sinton, G. Ito, J. P. Canales, M. Behn, T. Balcic, B. Cushman, J. E. Dixon, D. W. Graham, J. J. Mahoney, Correlated geophysical, geochemical and volcanological manifestations of plume-ridge interaction along the Galapagos spreading center, Geochem., Geophys, Geosyst., 3, 10, 8501, 10.1029/2002GC000350, 2002.

Dick, H. J. B., J. Lin, H. Schouten, An ultraslow-spreading class of ocean ridge, Nature, 426, 405–412, 2002.

Doyen, P. M., Permeability, conductivity and pore geometry of sandstone, J. Geophys. Res., 93, 7729–7740, 1988.

Dumoulin, C., D. Bercovici, P. Wessel, A continuous plate-tectonic model using geophysical data to estimate plate-margin widths, with a seismicity-based example, Geophys. J. Int,, 133, 379–389, 1998.

Dunn, R. A., D. R. Toomey, Seismological evidence for three-dimensional melt migration beneath the East Pacific Rise, Nature, 388, 259–262, 1997.

Dunn, R. A., D. R. Toomey, S. C. Solomon, Three-dimensional seismic structure and physical properties of the crust and shallow mantle beneath the East Pacific Rise at 9° 30N, J. Geophys. Res., 105, 23537–23555, 2000.

Dunn, R. A., Toomey, D. R., Crack-induced seismic anisotropy in the oceanic crust across the East Pacific Rise (9° 30′ N). Earth Planet. Sci. Lett., 189, 9–17, 2001.

Dunn, R. A., D. W. Forsyth, Imaging the transition between the region of mantle melt generation and the crustal magma chamber beneath the southern East Pacific Rise with short period Love waves, J. Geophys. Res, 108, 2352, 10.1029/2002JB002217, 2003.

Dziak, R. P., C. G. Fox, A. E. Schreiner, The June–July 1993 seismoacoustic event at CoAxial segment, Juan de Fuca Ridge: Evidence for a lateral dike injection, Geophys. Res. Letts., 22, 135–138, 1995.

Dziak, R. P., C. G. Fox., The January 1998 earthquake swarm at Axial Volcano, Juan de Fuca Ridge: Hydroacoustic evidence of seafloor volcanic activity, Geophys. Res. Letts., 26, 3429–3432, 1999.

Edmonds, H. N., P. J. Michael, E. T. Baker, D. P. Connelly, J. E. Snow, C. H. Langmuir, H. J. B. Dick, R. Muhe, C. R. German, D. W. Graham, Discovery of abundant hydrothermal venting on the ultraslow-spreading Gakkel ridge in the Arctic Ocean, Nature, 421, 252–256, 2003.

Edwards, M. H, G. J. Kurras, M. Tolstoy, D. R. Bonnenstiehl, B. J. Coakley, J. R. Cochran, Evidence of recent volcanic activity on the ultraslow-spreading Gakkel Ridge, Nature, 409, 808–812, 2001.

Elderfield, H., A. Schultz, Mid-ocean ridge hydrothermal fluxes and the chemical composition of the ocean, Ann. Rev. Earth Planet. Sci., 24, 191–224, 1996.

Embley, R. W., Chadwick, W. W., Volcanic and hydrothermal processes associated with a recent phase of seafloor spreading at the northern Cleft segment: Juan de Fuca Ridge. J. Geophys. Res., 99, 4741–4760, 1994.

Embley, R. W., W. W. Chadwick, M. R. Perfit, M. C. Smith, J. R. Delaney, Recent eruptions on the CoAxial segment of the Juan de Fuca Ridge: Implications for mid-ocean ridge accretion processes, J. Geophys. Res., 105, 16501–16525, 2000.

Evans, R. L., S. C. Constable, M. C. Sinha, C. S. Cox, M.J. Unsworth, Upper-crustal resistivity structure of the East Pacific Rise near 13°N, Geophys. Res. Lett., 18, 1917–1920, 1991.

Evans, R. L., Sinha, M. C., S. C. Constable, M. J. Unsworth, On the electrical nature of the axial melt zone at 13°N on the East Pacific Rise, J. Geophys. Res., 99, 577–588, 1994.

Evans, R. L., Constraints on the large scale porosity of young oceanic crust from seismic and resistivity data. Geophys. J. Int., 119, 869–879, 1994.

Evans, R. L., S. C. Webb, R. N. Edwards, M. Jegen, K. Sananikone, Hydrothermal circulation at the Cleft-Vance overlapping spreading center: Results of a Magnetometric Resistivity survey, J. Geophys. Res., 103, 12,321–12,338, 1998.

Evans, R. L., P. Tarits, A. D. Chave, A. White, G. Heinson, J. H. Filloux, H. Toh, N. Seama, H. Utada, J. R. Booker, M. Unsworth, Asymmetric electrical structure in the mantle beneath the East Pacific Rise at 17°S, Science, 286, 756–759, 1999.

Evans, R. L., S. C. Webb and the RIFT-UMC Team, Crustal Resistivity Structure at 9°50 N on the East Pacific Rise: Results of an Electromagnetic Survey, Geophys. Res. Letts., 29, 10.1029/2001GL014106, 2002.

Fisher, A. T., Permeability within basaltic oceanic crust, Rev. Geophys., 36, 143–182, 1998.

Fornari, D. J., R. W. Embley, Tectonic and volcanic controls on hydrothermal processes at the mid-ocean ridge: An overview based on near-bottom and submersible studies. In Humphris, S. E., Zierenberg, R. A., Mullineaux, L. S., and Thomson, R. E. (eds.) Seafloor Hydrothermal Systems: Physical, Chemical, Biological, and Geological Interactions. Geophysical Monograph 91, American Geophysical Union, Washington, DC, 1–46, 1995.

Fornari, D. J., R. M. Haymon, M. R. Peerfit, T. K. P. Gregg, M. H. Edwards, Axial summit trough of the East Pacific Rise 9°–10°N: Geological characteristics and evolution of the axial zone on fast spreading mid-ocean ridges, J. Geophys. Res., 103, 9827–9855, 1998.

Forsyth, D. W., Geophysical constraints on mantle flow and melt generation beneath mid-ocean ridges, In: Mantle Flow and Melt Generation at Mid-Ocean Ridges, edited by J. P. Morgan, D. K. Blackman and J. M. Sinton, Geophys. Monogr. Ser. vol. 71, pp. 1–65, AGU, Washington, 1992.

Forsyth, D.W., S.C. Webb, L.M. Dorman, Y. Shen, Phase velocities of Rayleigh waves in the MELT experiment on the East Pacific Rise, Science, 280, 1235–1238, 1998.

Forsyth, D. W., S. C. Webb, L. Dorman, Y. Shen, Three-dimensional mantle structure beneath the East Pacific Rise in the MELT area from Rayleigh wave dispersion, in Second RIDGE workshop on mantle flow and melt generation beneath mid-ocean ridges, 2000.

Forsyth D. W., R. Dunn, Imaging the region of melt generation in the mantle beneath the East Pacific Rise in the MELT area with surface wave dispersion, Eos Trans AGU, paper T41C-0223, 2003.

Fox, C. G., R. P. Dziak, H. Matsumoto, A. E. Schreiner, Potential for monitoring low-level seismicity on the Juan de Fuca Ridge using military hydrophone arrays, *Mar. Technol. Soc. J.*, 27, 22–30, 1994.

Fox, C. G, W. E. Radford, R. P. Dziak, T-K Lau, H Matsumoto, A. E. Schreiner, Acoustic detection of a seafloor spreading episode on the Juan de Fuca Ridge using military hydrophone arrays, *Geophys. Res. Letts.*, 22, 131–134, 1995

Fox, C. G., In situ ground deformation measurements from the summit of Axial Volcano during the 1998 volcanic episode, *Geophys. Res. Letts.*, 26, 3437–3440, 1999.

Fox, C. G, R. P. Dziak, Internal deformation of the Gorda Plate observed by hydroacoustic monitoring, *J. Geophys. Res.*, 104, 17,603–17,616, 1999

Fox, C. G., W. W. Chadwick, R. W. Embley, Direct observations of a submarine eruption from a seafloor instrument caught in a lava flow, *Nature*, 412, 727–729, 2001.

Garmany J., Accumulations of melt at the base of young oceanic crust. *Nature*, 340, 628–632, 1989.

Gee, J. S., S. C. Webb, J. Ridgeway, H. Staudigel, M. A. Zumberge, A deep tow magnetic survey of Middle Valley, Juan de Fuca Ridge, *Geochem. Geophys. Geosyst.*, 1001GC000170, 2001.

German, C. R., L. M. Parson, B. J. Murton, and H. D. Needham, Hydrothermal activity and ridge segmentation on the Mid-Atlantic Ridge: a tale of two hot-spots? In: *Tectonic, Magmatic, Hydrothermal and Biological Segmentation of Mid-Ocean Ridges*, edited by MacLeod, C. J., P. A. Tyler, and C. L. Walker, *Geol. Soc. Spec. Pub.*, No. 118, 169–184, 1996.

German, C. R, L. M. Parson, Distributions of hydrothermal activity along the Mid-Atlantic Ridge: interplay of magmatic and tectonic controls, *Earth Planet. Sci. Lett.*, 160, 327–341, 1998.

German, C. R., E. T. Baker, C. Mevel, K. Tamaki & the FUJI Science Team, Hydrothermal activity along the southwest Indian ridge, *Nature*, 395, 490–493, 1998.

Ghods, A., Arkani-Hamed, J., Melt migration beneath mid-ocean ridges. *Geophys. J. Int.*, 140, 687–697, 2000.

Gillis, K. M., M. D. Roberts, Cracking at the magma-hydrothermal transition: evidence from the Troodos ophiolite Cyprus, *Earth Planet. Sci. Letts.*, 169, 227–244, 1999.

Gillis, K. M., K. Muehlenachs, M. Stewart, T. Gleeson, J. Karson, Fluid flow patterns in fast spreding East Pacific Rise crust exposed at Hess Deep, *J. Geophys. Res.*, 106, 26311–26329, 2001.

Gillis, K. M., The rootzone of an ancient hydrothermal system exposed in the Troodos Ophiolite, Cyprus, *J. Geol.*, 110, 57–74, 2002.

Greer, A. A., Joint interpretation of seismic and electromagnetic results, investigating zero age oceanic crust. Unpubl. PhD Thesis, University of Cambridge, 181 pp, 2001

Greer, A., M. Sinha, L. MacGregor, Joint effective medium modelling for coincident seismic and electromagnetic data and its application to studies of porosity structure at mid-ocean ridge crests. In: Singh S.C., barton, P. J. & Sinha, M. C. (Eds.), *Lithos Science Report—April 2002* (ISSN 1476-2706), University of Cambridge, pp 101–120, 2002.

Hammond, W. C. D. R. Toomey, Seismic velocity anisotropy and heterogeneity beneath the MELT region of the East Pacific Rise from analysis of P and S body waves, *J. Geophys. Res.*, 108, 10.1029/2002JB001789, 2003.

Harding, A. J., M. E. Kappus, J. A. Orcutt, E. E. Vera, P. Buhl, J. C. Mutter, R. S. Detrick, T. Brocher, The structure of young oceanic crust at 13°N on the East Pacific Rise from expanding spread profiles, *J. Geophys. Res.*, 94, 12163–12196, 1989.

Harding, A. J., G. M. Kent J. A. Orcutt, A multichannel seismic investigation of upper crustal structure at 9° N on the East Pacific Rise: implications for crustal accretion, *J. Geophys. Res.*, 98, 13,925–13,944, 1993.

Hart, S. R., Equilibration during mantle melting: a fractal tree model, *Proc. Natl. Acad. Sci.*, 90, 11914–11918, 1993.

Haymon, R. M., D. J. Fornari, M. H. Edwards, S. Carbotte, D. Wright, K. C. Macdonald, Hydrothermal vent distribution along the East Pacific Rise crest (9° 09′–54′ N) and its relationship to magmatic and tectonic processes on fast-spreading mid-ocean ridges. *Earth Planet. Sci. Lett.* 104, 513–534, 1991.

Heinson, G. S., S. C. Constable, A. White, Episodic melt transport at a mid-ocean ridge inferred from magnetotelluric sounding, *Geophys. Res. Lett.*, 2317–232-, 2000.

Henstock, T. J., A. W. Woods, R. S.White, The accretion of oceanic crust by episodic sill intrusion. *J. Geophys. Res.*, 98, 4143—4162, 1993.

Hirth, G., D. L. Kohlstedt, Water in the oceanic upper mantle: implications for rheology, melt extraction and the evolution of the lithosphere, *Earth Planet. Sci. Letts.*, 144, 93–108, 1996.

Hooft, E. E. E., H. Schouten, R. S. Detrick, Constraining crustal emplacement processes from the variation in seismic layer 2A thickness at the East Pacific Rise, *Earth Planet. Sci. Letts.*, 142, 289–309, 1996.

Hooft, E. E. E., R. S. Detrick, D. R. Toomey, J. A. Collins, J. Lin, Crustal thickness and structure along three contrasting spreading segments of the Mid-Atlantic Ridge, 33.5°–35° N, *J. Geophys. Res.*, 105, 8205–8226, 2000.

Hung, S.-H., D. W. Forsyth, D. R. Toomey, Can a narrow, melt-rich, low-velocity zone of mantle upwelling be hidden beneath the East Pacific Rise? Limits from waveform modelling and the MELT experiment, *J. Geophys. Res.*, 105, 2000.

Hussenoeder, S. A., J. A. Collins, G. M. Kent, R. S Detrick, and the TERA Group, Seismic analysis of the axial magma chamber reflector along the southern East Pacific Rise from conventional reflection profiling, *J. Geophys. Res.*, 101, 22087–22105, 1996.

Jha, K., Parmentier, E. M., Phipps Morgan, J., The role of mantle-depletion and melt-retetion buoyancy in spreading centre segmentation, *Earth Planet. Sci. Lett.*, 125, 221–234, 1994.

Jokat, W., O. Ritzmann, M. C. Schmidt-Aursch, S. Drachev, S. Gauger, J. Snow, geophysical evidence for reduced melt production on the Arctic ultraslow Gakkel mid-ocean ridge, *Nature*, 423, 962–965, 2003.

Jones, N., M. C. Sinha, R. B. Whitmarsh, Observations of split shear waves from young ocean crust. In *Shear Waves in Marine Sediments*, pp 345–352, J. M. Hovem, M. D. Richardson & R. D. Stoll (Eds.), Kluwer Academic Publishers, Dordrecht, 1991.

Jull, M., P. Kelemen, K. Sims, Consequences of diffuse and channelled porous melt migration on uranium series disequilibria, *Geochimica et Cosmochimica Acta*, 66, 4133–4148, 2002.

Jupp, T., A. Schultz, A thermodynamic explanation for black smoker temperatures, *Nature*, 403, 880–883, 2000.

Karato, S., The role of hydrogen in the electrical conductivity of the upper-mantle, *Nature*, 347, 272–273, 1990.

Kappus, M. E. A. J. Harding, J. A. Orcutt, A baseline for upper-crustal velocity variations along the East Pacific Rise at 13°N, *J. Geophys. Res.*, 100, 6143–6161, 1995.

Kelley, D. S., J. A. Karson, D. K. Blackman, G. L. Fruh-Green, D. A. Butterfield, M. D. Lilley, E. J. Olsen, M. O. Shrenk, K. K. Roe, G. T. Lebon, P. Rivizzigno and the AT-30 Shipboard Party, An off-axis hydrothermal bent field near the Mid-Atlantic Ridge at 30° N, *Nature*, 412, 145–149, 2001.

Kelemen, P. B., G. Hirth, N. Shimizu, M. Spiegelman, H. J. B. Dick, A review of melt migration processes in the adiabatically upwelling mantle beneath oceanic spreading ridges, *Phil. Trans. Roy. Soc. Lond.*, A, 355, 1–35, 1997a.

Kelemen, P. B. K. Koga, N. Shimizu, Geochemistry of gabbro sills in the crust/mantle transition zone of the Oman ophiolite: Implications for the origin of the oceanic lower cust, *Earth Planet. Sci. Letts.*, 146, 475–488, 1997b.

Kelemen, P. B., E. Aharanov, Periodic formation of magma fractures and generation of layered gabbros in the lower crust beneath oceanic spreading centers, in *Faulting and magmatism at mid-ocean ridges,* edited by Buck, W. R, P. T. Delaney, J. A. Karson and Y. Lagabrielle, pp267–289, AGU, Washington, 1998.

Kelemen, P. B., M. Braun, G. Hirth, Spatial distribution of melt conduits in the mantle beneath oceanic cpreading ridges: observations from the Ingalls and Oman ophiolites, *Geochem., Geophys., Geosytems.*, 1, 1999GC000012, 2000.

Kent, G. M., A. J. Harding, J. A. Orcutt, Evidence for a smaller magma chamber beneath the East Pacific Rise at 9° 30N, *Nature*, 344, 650–653, 1990.

Kent, G. M., A. J. Harding, J. A. Orcutt, Distribution of magma beneath the East Pacific Rise between the Clipperton transform and the 9° 19N deval from forward modelling of common depth point data, *J. Geophys. Res.*, 98, 13945–13969, 1993.

Kent, G. M. A. J. Harding, J. A. Orcutt, R. S. Detrick, J. C. Mutter, P. Buhl., Uniform accretion of oceanic crust south of the Garrett transform at 14° 15′S on the East Pacific Rise, *J. Geophys. Res.*, 99, 9097–9116, 1994.

Kent, G. M., S. C. Singh, A. J. Harding, M. C. Sinha, J. A. Orcutt, P. J. Barton, R. S. White, S. K. Bazin, R. W. Hobbs, C. H. Tong, J. W. Pye, Evidence from three-dimensional seismic reflectivity images for enhanced melt supply beneath mid-ocean ridge discontinuities. *Nature*, 406, 614–618, 2000.

Kleinrock, M. C., S. E. Humphris, Structural control on sea-floor hydrothermal activity at the TAG active mound. *Nature*, 382, 149–153, 1996.

Klitgord, K. D., J. Mammerickx, Northern East Pacific Rise: Magnetic anomaly and bathymetric framework, *J. Geophys. Res.*, 87, 6725–6750, 1982.

Korenaga, J., P. B. Kelemen, Melt migration through the oceanic lower crust: a constraint from melt percolation with finite solid diffusion, *Earth Planet. Sci. Letts.*, 156, 1–11, 1998.

Kuo, B.-Y., D. W. Forsyth, Gravity anomalies of the ridge-transform system in the South Atlantic between 31 and 34.5° S: Upwelling centers and variations in crustal thickness, *Mar Geophys. Res.*, 10, 205–232, 1988.

Lalou, C., J. L. Reyss, E. Brichet, M. Arnold, G. Thompson, Y. Fouquet, P. Rona. New age data for Mid-Atlantic Ridge hydrothermal sites: TAG and Snake Pit chronology revisited. *J. Geophys. Res.* 98, 9705–9713, 1993.

Lecuyer, C., M. Dubois, C. Marignac, G. Gruau, Y. Fouquet, C. Ramboz, Phase separation and fluid mixing in subseafloor back-arc hydrothermal systems: a microthermometric and oxygen isotope study of fluid inclusions in the barite-sulfide chimneys of the Lau Basin, *J. Geophys. Res.*, 104, 17911–17928, 1999.

Lee, S-M., S. C. Solomon, M. A. Tivey, Fine scale crustal magnetization variations and segmentation of the East Pacific Rise, 9°10′N–9°50′N, *J. Geophys. Res.*, 101, 22033–22050, 1996.

Lilley, M. D., D. A. Butterfield, E. J. Olson, J. E. Lupton, S. A. Macko, R. E. McDuff, Anomalous CH_4 and NH_4 concentrations at an unsedimented mid-ocean ridge hydrothermal system, *Nature*, 364, 45–47, 1993.

Lin, J., G. M. Purdy, H. Schouten, J-C. Sempere, C. Zervas, Evidence from gravity data for focused magmatic accretion along the Mid-Atlantic Ridge, *Nature*, 344, 627–632, 1990.

Lin, J., Phipps Morgan, J., The spreading rate dependence of three-dimensional mid-ocean ridge gravity structure, *Geophys. Res. Lett.*, 19, 13–16, 1992.

Livermore, R. A., A. Cunningham, L. Vanneste, R. Larter. Subduction influence on magma supply at the East Scotia Ridge, *Earth Planet. Sci. Letts.*, 150, 261–275, 1997.

Lowell, R. P., L. N. Germanovich, Evolution of a brine saturated layer at the base of a ridge crest hydrothermal system, *J. Geophys. Res.*, 102, 10,245–10,255, 1997.

Luis, J. F., J. M. Miranda, N. Lourenco, M. C. Sinha, Fine scale magnetisation in the Lucky Strike segment, *European Geophys. Soc., Geophys. Res. Abstracts*, 2, 2000.

Macdonald, K. C, R. M. Haymon, J. Blasius, S. Benjamin, Off-axis Hydrothermal Activity on the East Pacific Rise near 9°28′N, Faulted and Topographic Control of Hydrothermal Discharge? Eos Trans AGU., 83(47), 2002.

MacGregor, L. M., S. C. Constable, M. C Sinha, The RAMESSES experiment III: Controlled source electromagnetic sounding of the Reykjanes Ridge at 57°45′N. *Geophs. J. Int.*, 135, 773–789. 1998.

MacGregor, L., M. Sinha, S. Constable, Electrical resistivity structure of the Valu Fa Ridge, Lau Basin, from marine controlled source electromagnetic sounding. *Geophys. J. Int.*, 146, 217–236, 2001.

MacGregor, L., A. Greer, M. Sinha, C. Peirce, Properties of crustal fluids at the Valu Fa Ridge, Lau Basin their relationship to hydrothermal circulation, from joint analysis of electromagnetic and seismic data. In: Singh S.C., Barton, P.J. & Sinha, M.C. (Eds.),

Lithos Science Report—April 2002 (ISSN 1476-2706), University of Cambridge, pp 101–120, 2002.

Magde, L. S., D. W. Sparks, Three-dimensional mantle upwelling, melt generation and melt migration beneath segmented slow spreading ridges, *J. Geophys. Res.*, 102, 20,571–20,583, 1997.

Magde, L. S., D. W. Sparks, R. S. Detrick, The relationship between buoyant mantle flow, melt migration and gravity bull's eyes at the Mid-Atlantic Ridge between 33° and 35°N, *Earth Planet. Sci. Letts.*, 148, 59–67, 1997.

Magde, L. S., A. H. Barclay, D. R. Toomey, R. S. Detrick, and J. A. Collins, Crustal plumbing within a segment of the Mid-Atlantic Ridge, *Earth Planet. Sci. Letts.*, 175, 55–67, 2000.

MELT Seismic Team, Imaging the deep seismic structure beneath a mid-ocean ridge: the MELT experiment, *Science,* 280, 1215–1218, 1998.

Michael, P. J., C. H. Langmuir, H. J. B. Dick, J. E. Snow, S. L. Goldstein, D. W. Graham, K. Lehnert, G. Kurras, W. Jokat, R. Muhe, H. N. Edmonds. Magmatic and amagmatic seafloor generation at the ultra-slow spreading Gakkel Ridge, Arctic Ocean. *Nature*, 423, 956–961, 2003.

Muller, M. R., C. J. Robinson, T. A. Minshull, R. S. White, M. J. Bickle, Thin crust beneath Ocean Drilling Program borehole 735B at the Southwest Indian Ridge? *Earth Planet. Sci. Letts.*, 148, 93–107, 1997.

Muller, M. R., T. A. Minshull, R. S. White, segmentation and melt supply at the Southwest Indian Ridge, *Geology*, 27, 867–870, 1999.

Navin, D. A., C. Peirce, M. C. Sinha, The RAMESSES Experiment II: evidence for accumulated melt beneath a slow spreading ridge from wide-angle refraction and multichannel reflection seismic profiles. *Geophys. J. Int.*, 135, 746–772, 1998.

Nehlig, P., T. Juteau, Flow porosities, permeabilities and preliminary data on fluid inclusions and fossil thermal gradients in the crustal sequence of the Samail ophiolite (Oman), *Tectonophys.*, 151, 199–221, 1988.

Nehlig, P. Interactions between magma chambers and hydrothermal systems: oceanic and ophiolitic constraints, *J. Geophys. Res.*, 98, 19,621–19,633, 1993.

Neuman, S. P., Generalized scaling of permeabilities: validation and effect of support scale, *Geophys. Res. Letts.,* 21, 349–352, 1994.

Parmentier, E. M., J. Phipps-Morgan, Spreading rate dependence of three-dimensional structure in oceanic spreading centers, *Nature*, 348, 325–328, 1990.

Parson, L. M. *et al. (17 authors)* En echelon axial volcanic ridges at the Reykjanes Ridge: a life cycle of volcanism and tectonics. *Earth Planet. Sci. Lett.,* 117, 73–87, 1993.

Phipps-Morgan, J., Melt migration beneath mid-ocean spreading centers, *Geophys. Res. Letts.*, 145, 1238–1241, 1987.

Phipps-Morgan, J., D. W. Forsyth, Three-dimensional low and temperature perturbations due to a transform offset: Effects on oceanic crust and upper mantle structure. *J. Geophys. Res.*, 93, 2955–2966, 1988.

Phipps-Morgan, J., Y. J. Chen, The genesis of oceanic crust: magma injection, hydrothermal circulation and crustal flow, *J. Geophys. Res.*, 98, 6283–6297, 1993.

Pruis, M., H. P. Johnson, Age dependent porosity of young upper oceanic crust: insights from seafloor gravity studies of recent volcanic eruptions, *Geophys. Res. Letts.*, 29, 10.1029/2001GL013977, 2002.

Purdy, G. M., R. S. Detrick, Crustal structure of the Mid-Atlantic Ridge at 23° N from seismic refraction studies, *J. Geophys. Res.*, 91, 3739–3762, 1986.

Quick, J. E., R. P. Denlinger, Ductile deformation and the origin of layered gabbro in ophiolites, *J. Geophys. Res*, 98, 14,015–14,027, 1993.

Rea, D. K., Tectonics of the Nazca-Pacific divergent plate boundary, in Nazca Plate Crustal Formation and Andean Convergence, edited by L. D. Kulm, J. Dymond, E. J. Dasch, and D. M. Hussong, *Mem. Geol. Soc. Am.*, 154, 27–62, 1981.

Rosenberg, N. D., F. J. Spera, R. M. Haymon, The relationship between flow and permeability field in seafloor hydrothermal systems, *Earth Planet. Sci. Lett.*, 116, 135–153, 1993.

Salters, V. J. M, S. R. Hart, The hafnium paradox and the role of garnet in the source of mid-ocean ridge basalts, *Nature*, 342, 420–422, 1989.

Scheirer, D. S., D. W. Forsyth, M.-H. Cormier, K. C. Macdonald, Shipboard geophysical indications of asymmetry and melt production beneath the East Pacific Rise near the MELT experiment, *Science*, 280, 1221–1224, 1998.

Schouten H., M. Tivey, D. Fornari, D. Yoerger, A. Bradley, P. Johnson, M. Edwards, T. Kurokawa, Lava Transport and Accumulation Processes on EPR 9°27′N to 10°N: Interpretations Based on Recent Near-Bottom Sonar Imaging and Seafloor Observations Using ABE, Alvin and a new Digital Deep Sea Camera, *Eos Trans AGU*, 83(47), 2002.

Schwartz, L.M., N. Martys, D. P. Bentz, E. J. Garboczi, S. Torquato, Cross-property relations and permeability estimation in model porous media, *Physical Review E*, 48, 4584–4591, 1993.

Scott, D. R., D. J. Stevenson, A self-consistent model of melting, magma migration and buoyancy-driven circulation beneath mid-ocean ridges, *J. Geophys. Res.*, 94, 2973–2988, 1989.

Scott, D. R., Small scale convection and mantle melting beneath mid-ocean ridges, In: *Mantle Flow and Melt Generation at Mid-Ocean Ridges*, edited by J. P. Morgan, D. K. Blackman and J. M. Sinton, Geophys. Monogr. Ser. vol. 71, pp. 327–352, AGU, Washington, 1992.

Shen, Y., D. W. Forsyth, The effects of temperature and pressure dependent viscosity on three-dimensional passive flow of the mantle beneath a ridge-transform system, *J. Geophys. Res.*, 97, 19917–19728, 1992.

Singh, S. C., J. S., Collier, A. J. Harding, G. M. Kent, J. A. Orcutt, Seismic evidence for a hydrothermal layer above the solid roof of the axial magma chamber at the southern East Pacific Rise, *Geology*, 27, 219–222, 1999.

Sinha, M. C., K. E. Louden, The Oceanographer fracture zone—I. Crustal structure from seismic refraction studies. *Geophys. J. Roy. astr. Soc.*, 75, 713–736, 1983.

Sinha, M. C., S. C. Constable, C. Peirce, A. White, G. Heinson, L. M. MacGregor, D. A. Navin, Magmatic processes at slow spreading ridges: implications of the RAMESSES experiment at 57° 45′

N on the Mid-Atlantic Ridge. *Geophys. J. Int.*, 135, 731–745, 1998.

Sinton, J. M., R. S. Detrick, Mid-ocean ridge magma chambers. *J. Geophys. Res.,* 97, 197–216, 1992.

Sims, K. W. W, S. J. Goldstein, J. Blichert-toft, M. R. Perfit, P. Kelemen, D. J. Fornari, P. Michael, M. T. Murrell, S. R. Hart, D. J. DePaolo, G. Layne, L. Ball, M. Jull and J. Bender, Chemical and isotopic constraints on the generation and transport of magma beneath the East Pacific Rise, *Geochimica et Cosmochimica Acta*, 66, 3481–3504, 2002.

Sleep, N. H., Hydrothermal circulation, Anhydrite precipitation and thermal structure at ridge axes. *J. Geophys. Res.*, 96, 2375–2387, 1991.

Small, C., L. Danyushevsky, A plate kinematic explanation for mid-oceanic ridge depth discontinuities, *Geology*, 31, 399–402, 2003.

Sohn, R. A., D. J. Fornari, K. L. Von Damm, J.A. Hildebrand, S. C. Webb, Seismic and hydrothermal evidence for a cracking event on the East Pacific Rise crest at 9°50′N, *Nature*, 396, 159–161, 1998a.

Sohn, R. A., J. A. Hildebrand, S. C. Webb, Postrifting seismicity and a model for the 1993 diking event on the CoAxial segment, Juan de Fuca Ridge, *J. Geophys. Res.*, 103, 9867–9877, 1998b.

Sohn, R.A., J.A. Hildebrand, S.C. Webb, A microearthquake survey of the high-temperature vent fields on the volcanically active EPR (9 50N), *J. Geophys. Res.,* 104, 25,367–25,377, 1999a.

Sohn, R. A., W. C. Crawford, S. C. Webb, Local seismicity following the 1998 eruption of Axial Volcano, *Geophys. Res. Letts.*, 26, 3433–3436, 1999b.

Sohn, R. A., A. H. Barclay, S. C. Webb, Microearthquakes patterns following the 1998 eruption of Axial Volcano, Juan de Fuca Ridge: Mechanical relaxation and thermal strain, J. Geophys. Res. 109, 10.1029/2003JB002499, 2004.

Solomon, S. C., D. R. Toomey, The structure of mid-ocean ridges. *Annu. Rev. Earth Planet Sci.*, 20, 329–364, 1992.

Sparks, D. W., Parmentier, E. M., Phipps-Morgan, J., Three-dimensional mantle convection beneath a segmented spreading center: implications for along axis variations in crustal thickness and gravity, *J. Geophys. Res.*, 98, 21,977–21,995, 1993.

Sparks, D. W., Parmentier, E. M., The generation and migration of partial melt beneath oceanic spreading centers, in Magmatic systems, Ryan M.P. ed, *International Geophysics Series,* 57,55–76, 1994.

Spiegelman, M., D. McKenzie, Simple 2-D models for melt extraction at mid-ocean ridges and island arcs, *Earth Planet. Sci. Lett.*, 83, 137–152, 1987.

Spiegelman, M., P. B. Kelemen, E. Aharonov, Causes and consequences of flow organization during melt transport: the reaction infiltration instability in compactible media. *J. Geophys. Res.,* 106, 2061–2077, 2001

Stein, C. A., S.Stein. Constraints on hydrothermal heat flux through the oceanic lithosphere from global heat flow, *J. Geophys. Res.*, 99, 3081–3095, 1994.

Stein, S., C. A. Stein, Thermo-mechanical evolution of oceanic lithosphere; implications for the subduction process and deep earthquakes, in Subduction top to bottom edited by Bebout, G.E; Scholl, D.W, Kirby, S.H., Platt, J.P. *Geophysical Monograph*, vol.96, pp.1–17, 1996.

Thompson, A. H. A. J. Katz, C. E. Crohn, The microgeometry and transport properties of sedimentary rock, *Advances in Physics*, 36, 625–694, 1987.

Tian, T., W. S. D. Wilcock, D. R. Toomey, R. S. Detrick, Seismic heterogeneity in the upper crust near the 1991 eruption site on the EPR, 9°50′N. *Geophys. Res. Letts.*, 27, 2369–2372, 2000.

Tidwell, V. C., J. L. Wilson, Laboratory method for investigating permeability upscaling, *Water Resources Res.*, 33, 1607–1616, 1997.

Tivey, M. A., P. A. Rona, H. Schouten, Reduced crustal magnetization beneath the active sulphide mound, Tag hydrothermal field, Mid-Atlantic Ridge at 26°N, *Earth and Planet. Sci. Letts.*, 115, 101–115, 1993.

Tivey, M. A., H. P. Johnson, A. Bradley, D. Yoerger, Thickness of a submarine lava flow determined from near bottom magnetic field mapping by autonomous underwater vehicle, *Geophys. Res. Letts.*, 25, 805–808, 1998.

Tivey, M. A., H. P. Johnson, Crustal magnetization reveals subsurface structure of Juan de Fuca Ridge hydrothermal vent fields, *Geology*, 30, 979–982, 2002.

Tivey, M. A., H. Schouten, M. C. Kleinrock, A near bottom magnetic survey of the Mid-Atlantic Ridge axis at 26°N: Implications for the tectonic evolution of the TAG segment, *J. Geophys. Res.*, 108, 10.1029/2002JB001967, 2003.

Tolstoy, M., A. J. Harding, J. A.Orcutt, Crustal thickness on the Mid-Atlantic Ridge: bull's eye gravity anomalies and focused accretion. *Science*, 262, 726–729, 1993.

Tolstoy, M., S. C. Constable, J. A. Orcutt, H. Staudigel, F. K. Wyatt, G. Anderson, Short and long baseline tiltmeter measurements on Axial seamount, Juan de Fuca Ridge, *Phys. Earth. Planet. Ints.*, 108, 129–141, 1998.

Tolstoy, M., F. L. Vernon, J. A. Orcutt, F. K. Wyatt, Breathing of the seafloor: tidal correlations of seismicity at Axial Volcano, *Geology*, 30, 503–506, 2002.

Tong, C. H., White, R. S., Warner, M. R. & ARAD Working Group, 2004. Effects of tectonism and magmatism on crack structure in oceanic crust: a seismic anisotropy study. *Geology*, 32, 25–28.

Toomey, D. R., G. M. Purdy, S. C. Solomon, W.S.D. Wilcock, The three dimensional velocity structure of the East Pacific Rise near latitude 9°30′N. *Nature*, 347, 639–645, 1990.

Toomey, D. R., S. C. Solomon, G. M. Purdy, Tomographic imaging of the shallow crustal structure of the East Pacific Rise at 9°30′N, *J. Geophys. Res.*, 99, 24,135–24,157, 1994.

Toomey, D. R., W. S. D. Wilcock, S. C. Solomon, W. C. Hammond, J.A. Orcutt, Mantle seismic structure beneath the MELT region of the East Pacific Rise from P and S wave tomography, *Science*, 280, 1224–1227, 1998.

Toomey, D. R., W. S. D. Wilcock, J. A. Conder, D. W. Forysth, J. D. Blundy, E. M. Parmentier, W. C. Hammond, Asymmetric mantle dynamics in the MELT region of the East Pacific Rise, *Earth Planet. Sci. Letts.*, 200, 287–295, 2002.

Travis, B. J., D. R. Janecky, N. Rosenberg, Three-dimensional simulation of hydrothermal circulation at mid-ocean ridges, *Geophys. Res. Letts.*, 18, 1441–1444, 1991.

Turner, I. M., C. Peirce, M. C. Sinha, Seismic imaging of the axial region of the Valu Fa Ridge, Lau Basin—the accretionary processes of an intermediate back-arc spreading ridge. *Geophys. J. Int.*, 138, 495–519, 1999.

Vera, E. E., J. C. Mutter, P. Buhl, J. A. Orcutt, A. J. Harding, M. E. Kappus, R. S. Detrick, T. M.Brocher, The structure of 0- to 0.2 m.y. old oceanic crust at 9°N on the East Pacific Rise from expanded spread profiles. *J. Geophys. Res.*, 95, 15529–15556, 1990.

West M., W. Menke, M.Tolstoy, S. Webb, R. Sohn, Magma storage beneath Axial volcano on the Juan de Fuca mid-ocean ridge, *Nature*, 413, 833–836, 2001.

Whitehead, J., H. J. B. Dick, H. Schouten, A mechanism for magmatic accretion under spreading centers, *Nature*, 312, 146–148, 1984.

Wiens, D. A., S. Stein, Age dependence of oceanic intraplate seismicity and implications for lithospheric evolution. *J. Geophys. Res.,* 88, 6455–6468, 1983.

Wilcock, W. S. D., J. R. Delaney, Midocean ridge sulphide deposits: Evidence for heat extraction from magma chambers or cracking fronts? *Earth Planet. Sci. Letts.*, 145, 49–64, 1996.

Wilcock, W. S. D., A. McNabb, Estimates of crusal permeability on the Endeavour segment of the Juan de Fuca mid-ocean ridge, *Earth and Planet. Sci. Letts.*, 138, 83–91, 1996.

Wilcock, W. S. D., S. D. Archer, G. M Purdy, Microearthquakes on the Endeavour segment of the Juan de Fuca Ridge, *J. Geophys. Res.*, 107, 10.1029/2001JB000505, 2002.

Wilkens, R. H, G. J. Fryer, J. Karsten, Evolution of porosity and seismic structure of upper oceanic crust: Importance of aspect ratios, *J. Geophys. Res.*, 96, 17981–17995, 1991.

Wolfe, C. J., G. M. Purdy, D. R. Toomey, S. C. Solomon, Microearthquake characteristics and crustal velocity structure at 29° N on the Mid-Atlantic Ridge: The architecture of a slow-spreading segment. *J. Geophys. Res.*, 100, 24,449–24,472, 1995.

Wolfe, C. J., S. C. Solomon, Shear-wave splitting and implications for mantle flow beneath the MELT region of the East Pacific Rise, *Science*, 280, 1230–1232, 1998.

Zhu, W., G. Hirth, A network model for permeability in partially molten rocks, *Earth Planet. Sci. Letts.*, 212, 407–416, 2003.

Martin C. Sinha, Southampton Oceanography Centre, European Way, Southampton, SO14 3ZH, UK

Rob L. Evans, Department of Geology and Geophysics, Woods Hole Oceanographic Institution, Woods Hole, MA 02543, USA

The Rheology and Morphology of Oceanic Lithosphere and Mid-Ocean Ridges

R. C. Searle

Department of Earth Sciences, University of Durham, UK

J. Escartín

Laboratoire de Géosciences Marines (CNRS UMR7097), Institut de Physique du Globe de Paris, France

The rheology of oceanic lithosphere is primarily a function of temperature, the abundance and distribution of lithologies and fluids, and their mechanical properties. Rheology controls the overall strength and mode of deformation. Seafloor morphology is the surface expression of this deformation, modified by additional processes such as volcanism. Rheological models are key to interpreting both naturally deformed rocks as direct indicators of deformation conditions and the resulting morphology. Simple thermo-mechanical models have proven useful to study ridge processes, but are limited by lack of knowledge of lithospheric architecture, composition, and rheology. The mechanical properties of some components (olivine, dolerite, olivine plus melt, serpentinite) are reasonably known, but must be extended to other important materials such as alteration products and include the role of fluids and compositional variations. While the overall composition of oceanic lithosphere is relatively well known, particularly for fast-spreading ridges, the distribution and abundance of melt and alteration products is not. Though sparse, these weak phases can strongly control the overall strength, mode and localization of deformation. Thermo-mechanical models successfully reproduce observed axial relief and general faulting patterns. They provide plausible mechanisms of lithospheric behavior, but cannot constrain actual deformation processes. In particular, they must assume rheology, thermal structure, and composition and distribution of materials, and are non-unique. The most accurate constraints on rheology and deformation processes will come from study of naturally deformed rocks. This will guide the choice of the models used to interpret morphology and infer the detailed thermal structure under ridges.

1. INTRODUCTION

The subject of the first InterRidge Theoretical Institute in Pavia, Italy, 2002, was the "Thermal regime of oceanic ridges and dynamics of hydrothermal circulation". There is a fundamental, though complex, connection between ridge mor-

Mid-Ocean Ridges: Hydrothermal Interactions Between the Lithosphere and Oceans
Geophysical Monograph Series 148

phology and the rheology and thermal regime of the lithosphere, which we review here.

At the largest scale, the shape of the mid-ocean ridge itself is defined by the thermal contraction of the lithosphere created at the ridge [*Parsons and Sclater*, 1977]. The lithosphere can then be deformed continuously by elastic flexure in response to applied loads such as seamounts, ocean islands, fracture zones, and subduction zones [*Watts*, 1978]. The flexural response itself depends on the rheology of the lithosphere, which in turn is a function of lithospheric temperature [*McNutt*, 1984]. At a smaller scale, the lithosphere may be deformed discontinuously when a fault forms [*Atwater and Mudie*, 1968; *Ballard and Van Andel*, 1977]. The mechanics of faulting also depends strongly on the rheology of the rock, including its strength and the coefficient of friction on the fault, all of which may be strongly dependent on temperature, pressure and fluid content, among other parameters. Moreover, the loads resulting from faulting may cause further lithospheric flexure [*Bott*, 1996; *Buck*, 1988; *Weissel and Karner*, 1989]. The distribution, spacing, and size of faults may be controlled by and provide an indication of the lithospheric thickness and rheology [*Shaw*, 1992].

To fully describe and accurately model the processes occurring at mid-ocean ridges, therefore, we need a good understanding and parameterization of the rheology of the lithosphere. This can be approached by a combination of laboratory experiments and observations on actual deformed rocks. Alternatively, we might take observations on ridge morphology and other parameters, and attempt to invert them to determine the underlying rheology and temperature. However, given the complexities in the structures and processes involved, this link is still weak. In practice, a combined approach is necessary.

In this paper, we will review current experimental and field work on the rheology of the lithosphere or its components. We will also review the morphology of mid-ocean ridges and what can be inferred about from it concerning their rheology and underlying temperature structure.

2. RHEOLOGY OF THE OCEANIC LITHOSPHERE

2.1. The Strength of the Lithosphere

Experimental studies have revealed the physical mechanisms responsible for deformation in the lithosphere and provide constitutive equations to describe its mechanical behavior (see reviews by [*Kirby*, 1983] and [*Kohlstedt et al.*, 1995]). Early studies revealed that, to a first approximation, rocks deform in one of two ways: by brittle failure at low temperatures and pressures (shallow depths), and by ductile or plastic deformation at higher temperatures and/or pressures, e.g.,

[*Byerlee*, 1968; *Byerlee*, 1978; *Goetze*, 1978]. From experimental constraints on the strength of rocks at different temperature and pressure conditions, and assuming a simple lithology, a yield envelope for the whole lithosphere can be calculated, as we discuss below [*Brace and Kohlstedt*, 1980; *Goetze and Evans*, 1979; *Kirby*, 1983; *Kohlstedt et al.*, 1995] (Figure 1). These simple models provided a means to directly correlate rock properties with lithospheric thickness inferred from geophysical data (e.g., gravity, bathymetry, seismicity), but ignored the mode of deformation within the lithosphere and the rheological effects of parameters such as water or lithological heterogeneity.

In this approach, the mantle is modeled assuming the rheology measured for single crystals or aggregates of olivine crystals [*Chopra*, 1986; *Chopra and Paterson*, 1984; *Durham and Goetze*, 1977; *Durham et al.*, 1977; *Evans and Goetze*, 1979; *Kohlstedt and Goetze*, 1974]. For the oceanic crust, the rheology of diabase as determined in the laboratory [*Agar and Marton*, 1995; *Caristan*, 1982; *Mackwell et al.*, 1998; *Shelton and Tullis*, 1981] is assumed to apply to all rocks that compose the magmatic crust (i.e., gabbro, diabase and basalt), as these rocks are compositionally similar [*Goetze and Evans*, 1979; *Kohlstedt et al.*, 1995]. 'Classical' rheological models have assumed that the plastic flow law for the crust (diabase)

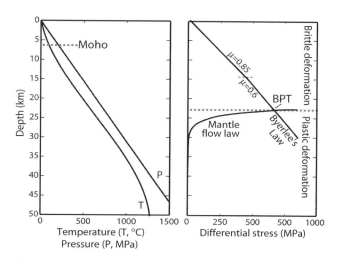

Figure 1. Left: variation in confining pressure (P) and temperature (T) with depth; right: corresponding yield strength envelope. The shallow levels deform in the brittle regime, where the maximum strength is given by Byerlee's friction law. The brittle domain overlies a plastic domain where deformation is accommodated by creep, and shows a fast decrease in the yield strength with increasing temperature and depth. The transition between the two regimes, where strength is maximum, corresponds to the brittle-plastic transition (BPT). The relative position of the Moho (6-km oceanic crustal thickness) is also indicated in the left panel. Calculations are for a strain rate of $\sim 10^{-15}$ s^{-1}.

is much weaker than that of the mantle (olivine), resulting in a decoupling weak lower crust under the ridge axis (Figure 2) [*Chen and Morgan*, 1990a; *Shaw and Lin*, 1996]. More recent experimental work shows that these models need to be revised, as the strength of dry crust may be similar to that of the mantle (Figure 3) [*Hirth and Kohlstedt*, in press; *Mackwell et al.*, 1998].

Being more complex, the continental crust is commonly modeled assuming a layered structure and using the rheology of quartz [*Blacic and Christie*, 1984; *Jaoul et al.*, 1984; *Kronenberg and Tullis*, 1984; *Mainprice and Paterson*, 1984], calcite [*Fredrich et al.*, 1989], and granite [*Tullis and Yund*, 1977]. More recently, experimental work has been extended to other rocks found in the lithosphere, such as serpentinite and serpentinized peridotite [*Escartin et al.*, 2001; *Escartin et al.*, 1997b; *Reinen et al.*, 1994], micas [*Mares and Kronenberg*, 1993; *Shea and Kronenberg*, 1992; *Shea and Kronenberg*, 1993], and feldspar [*Rybacki and Dresen*, 2000; *Tullis and Yund*, 1991], among other lithologies. In addition, numerous studies demonstrate that the presence of water [*Hirth and Kohlstedt*, 1996; *Karato et al.*, 1986; *Mainprice and Paterson*, 1984; *Tullis and Yund*, 1989], and melt [*Hirth and Kohlstedt*, 1995a; *Hirth and Kohlstedt*, 1995b], can have very signifi-

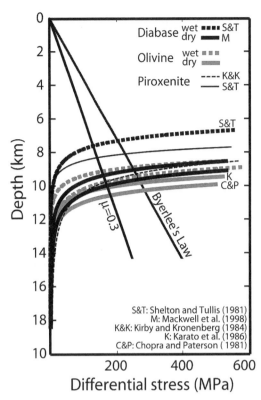

Figure 3. Plot of plastic flow laws for different lithologies and water content (curves) and predicted frictional strength in the brittle regime (straight lines) for a coefficient of friction for Byerlee's friction law ($\mu\sim0.85$) and for serpentinite ($\mu\sim0.3$) [Escartin et al., 1997b]. Flow laws are shown for diabase (thick black lines) [Mackwell et al., 1998; Shelton and Tullis, 1981], olivine (thick grey lines) [Chopra, 1986; Karato et al., 1986] and pyroxenite (thin black lines) [Kirby and Kronenberg, 1984; Shelton and Tullis, 1981].

cant effects on the rheology of the mantle and the crust, as discussed below.

2.2. Brittle Deformation

In the brittle regime, initial deformation is instantaneous and elastic. When the stresses exceed a yield point, brittle failure occurs. The maximum strength of the rock is dependant on the pressure, following a Coulomb law of the form:

$$\tau = c + \mu\sigma_n \qquad (1)$$

where τ is the shear stress, c is a material constant, μ is the coefficient of friction, and σ_n is the normal stress on the failing fault plane. As the lithosphere is assumed to be fractured, c is ~0 MPa, and the strength increases linearly with confining pressure and thus depth. [*Byerlee, 1978*] illustrated that for many rocks, strength is independent of composition. Thus a

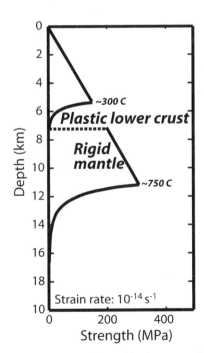

Figure 2. Yield strength envelope used in thermo-mechanical models, displaying a plastic lower crust decoupled from the mantle. These models have a brittle to plastic transition at a very low temperature (<400°C) [Chen and Morgan, 1990a]. These models used a wet diabase rheology for the plastic strength of the crust, which is now known to be unrealistic.

'universal friction law' (known as *Byerlee's friction law*) could be used:

$$c = 0 \text{ MPa}, \mu = 0.85, \quad \text{for } \sigma_n < 200 \text{ MPa}, \quad (2)$$

and

$$c = 60 \text{ MPa}, \mu = 0.6, \quad \text{for } \sigma_n > 200 \text{ MPa}. \quad (3)$$

While this parameterization of the strength in the brittle regime has been widely used to model the lithosphere, [Byerlee, 1978] and more recent experimental work has demonstrated that there can be important variations in the frictional characteristics of different materials. For example, the coefficient of friction for the serpentine polymorph lizardite, which is the most abundant form in the oceanic lithosphere, is $\mu \sim 0.35$ [Escartin et al., 1997b], and slightly serpentinized peridotites show a similar behavior to that of pure serpentinite [Escartín et al., 2001]. Serpentine may be an abundant component of the oceanic lithosphere, and therefore its presence can have a substantial weakening effect when incorporated into yield envelopes (Figures 1 and 3). As the modeled brittle strength of the lithosphere increases linearly with overburden pressure and thus depth, small variations in the coefficient of friction can result in large strength variations.

2.3. Plastic Deformation

Plastic deformation is accommodated by solid state creep [Goetze, 1978] according to:

$$\dot{\varepsilon} = A \sigma^n d^m e^{-Q/RT} \quad (4)$$

where $\dot{\varepsilon}$ is the strain rate, σ is the differential stress, d is the grain size, R is the universal gas constant, and A, Q, n and m are constants that are specific to each material and can be determined experimentally.

Because of the exponential term, the strain rate increases rapidly with increasing temperature for a given applied stress. This is equivalent to an exponential reduction in mantle strength with increasing depth, given a constant stress and strain rate (Figure 1). The crossing point between the frictional and the plastic strength is considered to mark the brittle-to-plastic transition (BPT), and is also where the strongest lithosphere is found at depth (Figure 1). The BPT is located close to the base of the mechanical lithosphere, which is defined as the depth at which the strength is a small fraction of the ambient stress-difference. The depth at which this occurs depends, among other factors, on the strain rate, and therefore varying lithospheric thicknesses may be expected at different time scales or for processes occurring at different rates. For

reasonable geological strain rate values (10^{-15}-10^{-18} s^{-1}) this transition occurs at ~750°C.

Plastic deformation can be accommodated by two creep mechanisms (Figure 4). The first, known as diffusion creep, is characterized by mobilization of atoms around grain boundaries and is limited by the rate of grain-boundary diffusion. When this mechanism operates, the strain rate varies linearly with stress ($n\sim1$). The second mechanism, known as dislocation creep, accommodates crystal deformation by the propagation of crystal defects (dislocations). In the dislocation creep regime, which tends to occur at larger grain sizes, the strain rate depends non-linearly on the applied stress ($n\sim3$) [*Karato et al.*, 1986]. Dislocation creep can also result in the formation of new grains and sub grains, reducing the overall grain size of the rock, and promoting a transition in creep

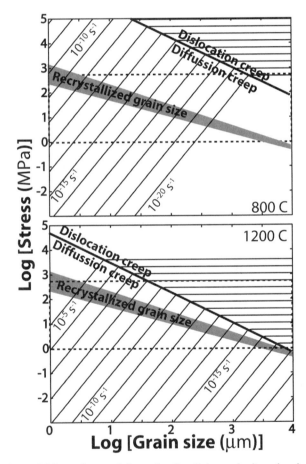

Figure 4. Maps of creep deformation for olivine mylonites, showing the variation in strain rate as a function of grain size and differential stress, for two temperatures as indicated. The bounds of geological stresses are indicated by the dashed lines. The shaded area gives the predicted grain size – stress relationship [Van der Wal, 1993]. Note that dislocation creep is activated at lower stresses and smaller grain sizes with increasing temperatures. After Jaroslow et al., [1996].

mechanism from dislocation creep to diffusion creep [*Hirth*, 2002; *Jaroslow et al.*, 1996].

2.4. The Effects of Composition, Grain Size, Water and Melt

Rock strength is determined by the composition of the rock (mineralogy), by the presence of fluids (e.g., water and melt), and by the stress and temperature conditions. While most experimental work is carried out in mono-mineralic rocks, natural rocks are composed of several mineralogical phases. [*Mackwell et al.*, 1998] showed that two types of diabase with different ratios of pyroxene and plagioclase have similar power law relationships (i.e., similar values for *n* and *Q*) but different strengths (as the measured *A* changed by a factor of 20). Figure 3 shows the variations in predicted strength for olivine, diabase and pyroxenite; depending on rock type, the thickness of the lithosphere could vary by more than 4 km near the mid-ocean ridge axis, with large variations of the maximum strength at the BPT. Therefore, knowledge of the composition and distribution of lithologies is necessary to construct accurate rheological models of the lithosphere, in addition to experimental and theoretical work on the rheology of polyphase materials [*Tullis and Yund*, 1991].

Water content is a second parameter that can strongly modify the overall strength of lithospheric materials (Figures 3 and 5) [Hirth, 2002; *Jaroslow et al.*, 1996; *Mei and Kohlstedt*, 2000a; *Mei and Kohlstedt*, 2000b]. In the oceanic mantle water can substantially weaken olivine by enhancing both dislocation and diffusion creep, with a transition occurring at 0.1–1 MPa for a grain size of 10 mm [*Karato et al.*, 1986]. Dewatering due to mantle melting below mid-ocean ridges may result in the formation of a strong, dry upper mantle layer ~60–70 km thick, with a viscosity more than an order of mag-

nitude larger than that of the underlying, wet mantle [*Hirth and Kohlstedt*, 1996]. A similar water-induced weakening is observed in diabase [*Mackwell et al.*, 1998]. Constraints on the water content of the lithosphere and mantle, and understanding of the processes responsible for hydration and dehydration, are needed in order to apply the experimentally-determined dependence of rheology on water content [*Hirth and Kohlstedt*, in press]. While the magmatic component of the oceanic crust (i.e., the melt-derived components: gabbro, diabase and basalt) may be nominally dry and therefore strong [*Hirth et al.*, 1998], other processes such as fracturing, water circulation and alteration will result in hydration and eventual weakening.

Grain size is a third parameter than can control the rheology of the lithosphere, and in particular the mode of localization of deformation. In undeformed rocks, grain size largely depends on the cooling history of the rock. However, deformation both during and after cooling also influences grain size [*Montesi and Hirth*, 2003; *Rutter and Brodie*, 1988; *Van der Wal*, 1993]. Studies on naturally deformed abyssal peridotite mylonites show two types of rock recording two conditions and modes of deformation [*Jaroslow et al.*, 1996]: medium to coarse-grained tectonites with equilibrium temperatures >755°C, and fine-grained mylonites with equilibrium temperatures of ~600°C. The first type is interpreted to record deformation associated with mantle upwelling, while the second one may be associated with localized ductile shear zones within the lithosphere that developed during extension and cooling. Grain-size reduction due to dislocation at the base of the lithosphere promoted a transition to the diffusion creep regime (Figure 4), further reducing the strength and favoring strain localization and long-lived faults. These results demonstrate that the mode of deformation and the strength of the lithosphere depend on the evolution of deformation with time, resulting in a complex and variable rheological structure that is not captured by the oversimplified yield strength envelopes commonly used (e.g., Figures 1 and 2).

Finally, the presence of melt can have an important effect on the strength of rocks [*Hirth and Kohlstedt*, 1995a; *Hirth and Kohlstedt*, 1995b; *Kohlstedt et al.*, 2000; *Renner et al.*, 2000]. An important strength reduction is documented in the diffusion creep regime at >5% melt content, while this reduction occurs at >4% melt for rocks deforming in the dislocation creep regime, associated with an increase of one order of magnitude in strain rate (Figure 6).

2.5. Semibrittle Deformation

As commonly accepted, all the models presented above are fundamentally oversimplified as they assume that deformation may only occur in either the brittle or the plastic deformation regime (Figure 1). Experimental work in numerous materi-

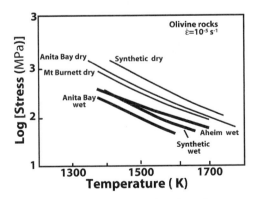

Figure 5. Variations in strength as a function of temperature for olivine with varying amounts of water. "Dry" olivine is stronger than "wet" olivine, but large variability is observed as the actual amount of water is not known. Modified from Evans and Kohlstedt, [1995].

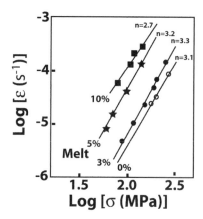

Figure 6. Effect of melt content on olivine rheology. An increase of about an order of magnitude in the strain rate at a constant stress is observed when the melt content increases from 3% to 5%. Modified from Hirth and Kohlstedt, [1995a].

als such as dry clinopyroxene [*Kirby and Kronenberg,* 1984] or feldspar aggregates [*Tullis and Yund,* 1992] demonstrate that a more complex behavior occurs in nature. Clinopyroxenites show both plastic and brittle deformation at moderate temperatures (600°C) and intermediate pressures (430–1190 MPa). Feldspars display a cataclastic ductile deformation field with some of the deformation accommodated by plasticity; the onset of deformation occurs at T>1000°C and P>1000 MPa with localized brittle faulting at T<300°C and P<500 MPa. Semibrittle behavior can be expected in heterogeneous materials that deform plastically at different pressure and temperature conditions (e.g., olivine, pyroxene and plagioclase in gabbros). This deformation regime is microstructurally complex and has been investigated in a limited number of lithologies. However, constitutive laws for the semibrittle regime are not available, and accurate constraints on the location of the plastic-semibrittle and semibrittle-brittle deformation regimes do not exist, though there are some "rules of thumb."

It is commonly accepted that localized brittle deformation occurs when the strength of the rock is larger than that predicted by Byerlee's friction law (Figure 7). Since Byerlee's law does not appear to be universal, this criterion may be improved using the friction law determined for each rock type [*Escartin et al.*, 1997b]. When the strength of the rock is lower than its frictional strength, deformation may be distributed (ductile, semibrittle regime). The onset of fully plastic behavior is assumed to occur when the strength of the rock equals that of the confining pressure (i.e., Goetze's criterion, Figure 7). The definition of the semibrittle regime may be further complicated in the case of heterogeneous materials, such as where brittle and plastic deformation of the different components may coexist under the same conditions [*Scholz*, 1988]. Incorporating a semibrittle deformation regime into rheolog-

ical models results in a profound modification of the yield strength envelope, with significant weakening of both the overall and the peak strength of the lithosphere, compared to 'classical' rheological models (Figure 8).

2.6. Limitations of Existing Rheological Models

Commonly used rheological models typically adopt numerous simplifications and assumptions, including the extrapolation of experimental results to natural conditions. For example, note the nine orders of magnitude difference between the experimental strain rates of Figures 5 and 6, and the "geological" rate used in the model of Figure 2. Nevertheless, these models have been successful at predicting and capturing some of the main observations and first-order processes taking place in the lithosphere, such as the brittle behavior of the upper lithosphere and crust and the resulting morphology at the ridge axis [*Chen and Morgan*, 1990a; *Chen and Morgan*, 1990b], the nature of faulting and the formation of abyssal hills at the seafloor [*Behn et al.*, 2002b; *Macdonald*, 1998], or the formation of detachment faults [*Lavier et al.*, 1999].

However, many of the key elements that characterize mid-ocean ridges in general, and slow-spreading ridges in particular, are not captured by current rheological models. In particular, these include: the three-dimensionality of tectonic structures near ridge discontinuities (as existing models are two-dimensional); the heterogeneity in composition of the oceanic lithosphere (mixture of gabbros, peridotite, serpentinite, and other rock types) as opposed to homogenous mod-

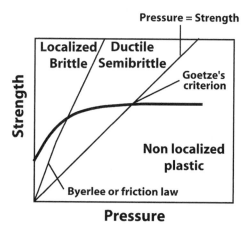

Figure 7. Schematic plot of strength of intact rock versus pressure with criteria to define the deformation regimes. Brittle deformation occurs when the strength of the rock exceeds that of the frictional strength (Byerlee's Law), while plastic deformation is initiated when the strength of the rock equals the pressure (Goetze's Criterion). The region between these two criteria may correspond to the semibrittle regime. After Evans and Kohlstedt, [1995].

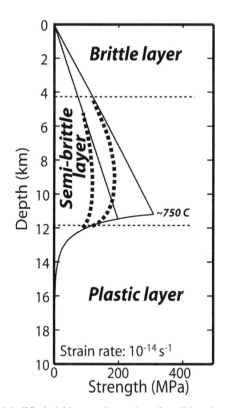

Figure 8. Modified yield strength envelope for a lithosphere exhibiting a field of semibrittle deformation. Both the overall and the peak stress can be significantly reduced compared with the 'classical' rheological model. Two models with differing friction laws (Byerlee's Law and the serpentinite friction law) are shown.

els; and feedback between the different active processes that will in turn affect the rheology of the lithosphere (e.g., hydrothermalism or volcanism, faulting and thermal regime). Some of these aspects, such as the composition of the lithosphere and distribution of lithologies, melt or water, are based on qualitative observations, and require additional high-resolution geophysical constraints to provide quantitative information of use for accurate rheological modeling. Additional experimental work to characterize the rheology of the less abundant lithologies is also required. Even if minor, the presence of a weak phase such as serpentinite or other phyllosilicates can substantially reduce the strength of the lithosphere, and influence the mode of strain localization [*Bos et al.*, 2000; *Escartin et al.*, 1997a]; the effects of these phases are not included in any of the existing mechanical models. Another key rheological parameter that is poorly constrained is temperature. For example, the presence of a magma chamber along portions of fast-spreading ridges, in combination with seismic data, can provide valuable information to construct complex but realistic thermal models under the ridge axis [*Dunn et al.*, 2000]. This information is not widely avail-

able, and additional high-resolution three-dimensional studies of lithospheric structure are required, particularly along slow-spreading ridges.

The study of actual deformed rocks offers some insight into and clues on the rheology and processes operating at depth, complementing both geophysical observations and numerical modeling. The comparison of deformed materials in nature and in the laboratory can provide important constraints on the mode of deformation of the lithosphere, and on the T-P conditions at which the rocks and deformation textures were formed. In particular, microstructural observations can be used to determine the mode of deformation (brittle vs. plastic and dislocation vs. diffusion creep regimes). This information can be combined with independent constraints on the temperature and pressure (e.g., geothermometry from mineral phase relationships) via the use of deformation mechanism maps such as that shown in Figure 4. Such studies can provide a deep understanding of the thermal state and history of the lithosphere, and of the deformation processes that operate at depth [*Jaroslow et al.*, 1996; *Yoshinobu and Hirth*, 2002]. These studies are also required to validate the extrapolation of experimental results from laboratory to natural conditions.

Having reviewed these studies of the rheology of ridge materials, we now consider the morphology of ridges, how it is controlled by the rheology, and what further may be learned from it concerning the rheology of the lithosphere.

3. THE THERMAL STRUCTURE OF OCEANIC LITHOSPHERE

To a first approximation, the temperature of the oceanic lithosphere can be calculated assuming conductive cooling with a constant mantle temperature at the base of the thermal plate [*Parsons and Sclater*, 1977]. Implicit in such models is the assumption that the lithosphere is rigid and does not convect on the timescale of thermal conduction. Such models reproduce the general seafloor subsidence with age (depth proportional to the square root of age) for ages <80 Ma. As horizontal heat conduction in the lithosphere is small when compared with the rate of horizontal advection by plate motion, the temperature at depth is mostly a function of age, so slow-spreading ridges display a more rapid increase in lithospheric thickness away from the spreading center than do fast-spreading ridges.

These models break down at ages >80 Ma, as they do not take into account sub-lithospheric convection, which slows plate cooling at large ages [*Parsons and McKenzie*, 1978; *Stein and Stein*, 1992]. They also break down in proximity of the ridge axis, as they do not accurately model the thermal effects due to the presence of an axial magma chamber, the latent heat of crystallization, or the advective cooling caused by hydrothermal circulation [*Chen and Morgan*, 1990a; *Davis and Lister*, 1974; *Henstock et al.*, 1993; *Lin and Parmentier*, 1989; *Phipps-Morgan* and *Chen*, 1993; *Wilson et al.*, 1988]. Current reviews on the thermal state near mid-ocean ridge axes are given by [*Sinha and Evans*, 2003] and [*Chen*, 2003].

Critical to the thermal state (and rheological structure) of the lithosphere under the ridge axis is the detailed pattern of hydrothermal circulation and the location and mode of emplacement and extrusion of magma in the crust. Constraints on the axial thermal structure may be more easily obtained and more accurate at fast-spreading ridges than at slow-spreading ones. In addition to a thicker lithosphere, slow-spreading ridges appear to show a heterogeneous crust that implies a non-steady state mode of magmatic accretion [*Cannat*, 1993; Dick et al., 2000], while the presence of ridge offsets suggests a three-dimensional thermal structure with along-axis variations [*Behn et al.*, 2002a; *Shaw*, 1992].

Neither the thermal structure nor the details of the hydrothermal cooling process are currently well constrained because of difficulties in sampling appropriate parts of the oceanic crust, or in geophysically imaging these areas or the effects of the processes. Geological constraints on the extent, depth, mode and temperature conditions of hydrothermal alteration are beginning to be obtained from ophiolites, e.g., [*Gillis and Roberts*, 1999]. [*Phipps-Morgan and Chen*, 1993] have modeled the general effect of hydrothermalism and show that the yield strength of the lithosphere and its effective elastic thickness depend on the balance between the rate of heat input by magma injection into the crust and the rate of hydrothermal cooling.

Two broad models exist for magma emplacement into the lower crust at fast-spreading ridges (Figure 9). In the "gabbro glacier" model, emplacement of magma takes place in a shallow sill-like magma chamber at the base of the sheeted dyke layer (corresponding to the seismically imaged axial magma chamber at fast-spreading ridges). Subsequent freezing and down- and outward movement of the crystallized residue constructs the lower crust [*Henstock et al.*, 1993; *Phipps-Morgan and Chen*, 1993; *Quick and Delinger*, 1993]. The more recent "multiple sill" model proposes magma injection in multiple lenses throughout the crustal section including near the Moho [*Crawford and Webb*, 2002; *Kelemen and Aharonov*, 1998; *MacLeod and Yaouancq*, 2000]. At present neither model appears to completely satisfy all the available geological and geophysical evidence.

Efforts are underway to better constrain the thermal state below the ridge axis by studying oceanic samples that record the interaction of hydrothermal fluids with the host rock, thus providing constraints on the thermal conditions at which such interaction occurred [*Coogan et al.*, 2002; *Manning and*

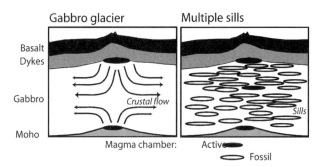

Figure 9. Models of crustal formation for a fast-spreading ridge. (Left) In the 'gabbro glacier' model most of the lower crust is produced from a high-level magma sill at the base of the dyke section by solid flow of the melt residue down and off-axis [Henstock et al., 1993; Phipps Morgan and Chen, 1993; Quick and Delinger, 1993]. A similar process may operate from a magma chamber at the base of the crust. (Right) In the 'many sills' model the crust is formed by emplacement of thin melt lenses into the crust at different levels between the Moho and a high-level magma chamber [Crawford and Webb, 2002; Kelemen and Aharonov, 1998]. These magma chambers may be emplaced independently [Kelemen et al., 2000], or as lateral extensions from a magmatic system feeding a high-level magma chamber from the upper mantle [MacLeod and Yaouancq, 2000].

MacLeod, 1996; *Manning et al.*, 2000], and from cooling rates inferred from grain sizes in ophiolites [*Garrido et al.*, 2001].

4. FLEXURE AND THE ELASTIC PROPERTIES OF THE LITHOSPHERE

Various authors have estimated the effective elastic thickness of the lithosphere by measuring its flexural deformation under loads such as seamounts, islands, ridges, fracture zones and trenches [*Caldwell et al.*, 1976; *Cazenave et al.*, 1980; *Watts*, 1978]. [*Goetze and Evans*, 1979] used experimental rock mechanical data to construct yield strength envelopes (see Figure 1) which could be used to study the elastic properties and bending of the lithosphere. This approach has been extensively used to obtain rheological information from flexural studies [*Kirby*, 1983; *McNutt and Menard*, 1982]. Recent compilations of results from numerous flexural studies are given by [*Watts and Zhong*, 2000] and [*Minshull and Charvis*, 2001] (Figure 10). These studies treat the lithosphere as a thin elastic plate overlying an inviscid substratum, and infer the flexural rigidity D from the shape of the deformed plate and the estimated load. The effective elastic thickness T_e is related to D:

$$D = E T_e^3 / 12 (1-v^2) \qquad (5)$$

where E is Young's modulus and v is Poisson's ratio.

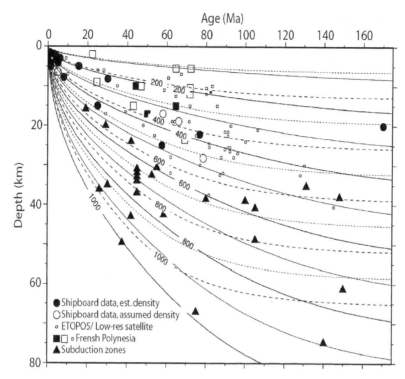

Figure 10. Comparison of the temperature structure of an oceanic plate with estimates of elastic thickness of the lithosphere from flexural studies (adapted from Minshull and Charvis, [2001]; see also references therein). The thermal models are those of Parsons and Sclater, [1977] (solid lines) and Stein and Stein, [1992] (dashed lines). Many of the flexural estimates give significantly lower values for lithospheric thickness than those that may be predicted by the thermal models.

In these studies, the envelope of all results shows an overall increase in the effective elastic thickness with age at time of loading, with most data bounded by the ~600°C isotherm (see Figure 10). This temperature is close to the predicted temperature corresponding to the brittle to plastic transition in classical rheological models (Figure 1). Many measurements, however, give elastic plate thickness estimates that are much smaller than the elastic thickness predicted by this isotherm (Figure 10).

Elastic thickness values reported for zero-age oceanic crust range from 0 to >10 km [*Bowin and Milligan*, 1985; *Cochran*, 1979; *Escartin and Lin*, 1998; *Kuo et al.*, 1986; *Madsen et al.*, 1984; *McKenzie and Bowin*, 1976; *Neumann et al.*, 1993; *Wang and Cochran*, 1993] . Estimates of the effective elastic thickness for the East Pacific Rise tend to be less than 4 km, while those for the Mid-Atlantic ridge range from less than 5 km to more than 10 km, consistent with a slightly thicker lithosphere under slow-spreading ridge axes. A value of 4 km is reported for the intermediate-spreading Juan de Fuca Ridge [*Watts and Zhong*, 2000]. However, estimates of elastic thickness at slow-spreading ridges may be biased by tectonic modification of both the thickness and seafloor morphology due to tectonic extension along the bounding rift faults [*Escartin and Lin*, 1998].

Lithospheric thickness may also be estimated seismically, e.g., by modeling surface-wave dispersion. Such methods yield lithospheric thickness estimates that are larger than those from flexural studies, fitting closer to the 1000°C isotherm [Leeds et al., 1974; Nishimura and Forsyth, 1989]. These discrepancies may be explained by the differences in the strain-rates involved: lithospheric flexure reflects deformation on the timescale of millions of years, while passage of seismic waves occurs on a time scale of seconds, emphasizing the strain-rate dependence of lithospheric rheology (see Figures 1 and 4).

5. THE THICKNESS OF THE SEISMOGENIC ZONE

Oceanic intraplate earthquakes occur over a broad range of depths, but a well-defined maximum depth is identifiable and increases with lithospheric age [*Wiens and Stein*, 1983] (Figure 11). This limit corresponds approximately to the 750°C isotherm, somewhat deeper in the Earth and at a higher temperature than that derived from flexural studies, but in agreement with the temperature of the brittle to plastic transition [*Bergman and Solomon*, 1990; *Chen and Morgan*, 1990a]. [*Wiens and Stein*, 1983] estimated seismogenic strain rates

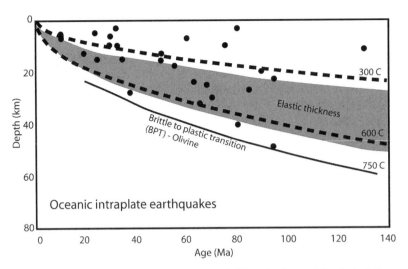

Figure 11. Oceanic intraplate earthquake depths as a function of lithospheric age. The dashed lines correspond to the 300°C and 600°C isotherms, and the brittle to plastic transition for olivine is indicated by the continuous line (~750°C). After Wiens and Stein, [1983].

from seismic moment release rates, concluding that they lie in the range of 10^{-18} to 10^{-15} s^{-1} for normal and highly active tectonic lithosphere, respectively. Assuming a dry olivine rheology and using values for the constants in the flow-law equations derived from laboratory rock mechanics experiments, they found that the maximum deviatoric stress that can be supported at 750°C is 20 MPa and 190 MPa at the lowest and highest strain rates, respectively. The critical depths at which these stresses are exceeded differ by only 5–10 km for these two strain rates.

A number of studies have estimated the depth of the seismogenic zone at mid-ocean ridges. Centroid depths of teleseismic events [*Huang and Solomon*, 1988] show a deepening with decreasing spreading rate, from less than 2 km depth at rates of >20 km Ma^{-1} to ~6 km for <5 km Ma^{-1} (Figure 12). A more detailed view of the distribution of seismicity is available from microseismicity studies using ocean-bottom seismometers that have been conducted on a limited number of mid-ocean ridge sites. Maximum hypocentral depths for the very slow-spreading South West Indian Ridge range from 6 to 10 km [*Katsumata et al.*, 2001]. Studies along the Mid-Atlantic ridge show a large variation in maximum hypocentral depth, both regionally and locally, ranging from <4 km to >10 km below seafloor, so that the deeper events occur within the upper mantle [*Barclay et al.*, 2001; *Kong et al.*, 1992; *Louden et al.*, 1986; *Tolstoy et al.*, 2002; *Toomey et al.*, 1988; *Toomey et al.*, 1985; *Wilcock et al.*, 1990; *Wolfe et al.*, 1995]. Some slow-spreading segments show seismicity limited to shallow levels, with maximum depths of ~4 km (e.g., OH1 segment [*Barclay et al.*, 2001]). This shallow limit on the seismicity was originally attributed to the commonly assumed weak nature of

the lower crust, although recent work suggests that the lower crust is strong [*Hirth et al.*, 1998] so this shallow seismicity most likely reflects a thin lithosphere at this location. Other slow-spreading segments, such as the 29°N area of the Mid-Atlantic Ridge, show hypocenter depths that vary from <6 km at the segment center to >9 km under the inside corner at the segment end [*Wolfe et al.*, 1995], thus reaching into the upper mantle (Figure 13). In most cases the seismicity tends to cluster near these maximum depths, with a zone of lesser activity in the upper crust and a relatively aseismic zone in between. Maximum hypocentral depths at medium and fast-spreading ridges tend to be smaller. They are on the order of 1.5–3.5 km at the Juan de Fuca Ridge [*Wilcock et al.*, 2002], 3–5 km at the fast-spreading East Pacific Rise near transform faults [*Lilwall et al.*, 1981], and <3 km away from transforms [*Tolstoy et al.*, 2002]. Variations in the maximum hypocentral depths therefore reflect the overall trends in elastic thickness inferred from flexural studies.

In the next sections we review some of the main observations regarding rift valley morphology, crustal thickness variations, and lithospheric composition as constrained from geophysical and geological investigations. These data are critical for constraining the composition of the lithosphere and the geometry of its different components, which in turn are required to construct more realistic and better constrained rheological models.

6. THE MEDIAN VALLEY AND THE AXIAL HIGH

An important characteristic of slow-spreading mid-ocean ridges is the median valley (Plate 1). Along many sections of

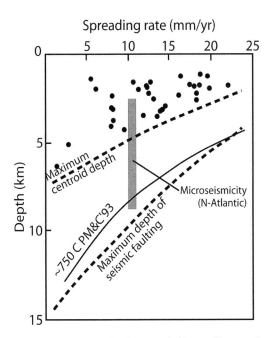

Figure 12. Plot of centroid depth versus half spreading rate for mid-ocean ridge axis earthquakes [Huang and Solomon, 1988]. As the centroid depth corresponds to the weighted centre of the focal volume, the inferred maximum depth of faulting is double this. These data are in agreement with the distribution of microseismicity along the northern Mid-Atlantic Ridge (grey box; see Barclay et al., [2001] and references therein). The solid line (PM&C'93) corresponds to the 750°C isotherm in the thermal models of Phipps Morgan and Chen [1993].

slow-spreading ridges there is an axial rift, several kilometers wide and 1–3 km deep, which is produced by stretching and necking of the lithosphere under horizontal tension as plates separate [Chen and Morgan, 1990a; Lin and Parmentier, 1989; Tapponnier and Francheteau, 1978]. At fast-spreading ridges the median valley disappears and is replaced by an axial high, which reflects dynamic viscous support and flexural bending of a thin plate over a hot axial region [Buck, 2001; Chen and Morgan, 1990b; Eberle and Forsyth, 1998; Madsen et al., 1984; Wang and Cochran, 1993]. Transitional morphologies are found at intermediate-spreading ridges, and other effects, such as the presence of hotspots, also play a role on ridge morphology, e.g., [Canales et al., 1997; Searle et al., 1998b].

Such a morphological transition is shown by simple rheological models that assume a brittle layer overlying power-law creep rheology, passive mantle upwelling driven by plate separation, and incorporating the effects of both hydrothermal cooling and the latent heat of crystallization [Chen and Morgan, 1990a; Chen and Morgan, 1990b]. Rift valley morphology is therefore a general indicator of the mechanical properties and overall thermal state of the lithosphere under

the ridge axis. These early rheological models assumed a weak lower crust deforming plastically and decoupled from the stronger upper mantle [Chen and Morgan, 1990a; Shaw and Lin, 1996]. This plastically deformed zone was thought to be associated with the aseismic zone often observed in microseismicity studies (Figure 13). The decoupling zone in these models is narrow at slow-spreading ridges: in the high-stress axial zone, the brittle upper crust exceeds its strength, fails, and is subsequently deformed plastically by the diverging ductile mantle (necking) to produce the rift valley. At faster spreading rates the axial region is hotter so the decoupling zone (ductile lower crust) is wider and exceeds the width of the zone of brittle failure. The thin and very weak axial lithosphere is thus decoupled from the mantle flow and, in these models, achieves almost perfect local isostatic equilibrium, producing an axial high, since there is little dynamic support for the topography.

The behavior of the [Chen and Morgan, 1990a] model depends critically on the crustal thickness. A thinner crust may reduce or totally remove the lower crustal decoupling zone, and the model then predicts a wider and deeper rift valley; this model has been invoked to explain the presence of an axial valley along the Australia–Antarctic Discordance [Chen and Morgan, 1990a; Hirth et al., 1998], where the mantle is inferred to be cold and to have a very low degree of melting [Weissel and Hayes, 1974]. By contrast, hotter mantle results in thicker crust, as observed near hotspots (e.g., Reykjanes Ridge [Bunch and Kennett, 1980]); the decoupling zone is then wider than the failure zone, and the lithosphere behaves like the fast-spreading case, producing an axial high [Chen and Morgan, 1990b]. This model was adopted and developed by [Neumann and Forsyth, 1993] and [Shaw and Lin, 1996] to

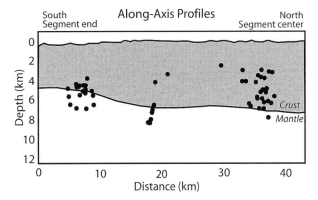

Figure 13. Distribution of microseismicity along the axis of a slow-spreading ridge segment (Mid-Atlantic Ridge, 29°N; see Plate 1 for location) over a 42 day period. Microseismicity clusters in three areas, at depths from 2 km to 9 km, mostly in the lower crust and upper mantle. Note that the mid-crust is relatively aseismic at depths ~ 5 km. After Wolfe et al., [1995].

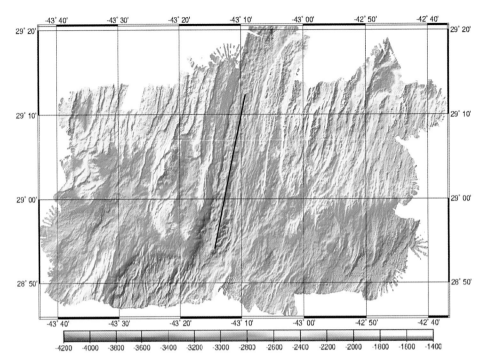

Plate 1. Shaded relief bathymetry over the southernmost two thirds of the 29°N segment of the Mid-Atlantic Ridge, illuminated from the northwest (Searle et al., [1998a] and unpublished data). Note the well-defined fault scarps facing towards the ridge axis, which outline both the median valley floor (centered at 29°02'N, 43°12'W), and the >1.5 km scarp at 28°55'N, 43°18'W that defines the inside corner high centered near 29°N and 43°20'W. The line corresponds to the axial section showing the microseismicity in Figure 13.

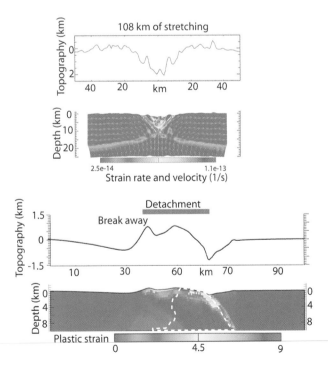

Plate 2. Numerical simulation of lithospheric stretching and strain localization. a: (top) Amagmatic stretching produces a seafloor topography that resembles that at slow-spreading ridges, with an axial valley and abyssal hills formed by many cross-cutting shear zones at the ridge axis [Buck and Poliakov, 1998]. b: (bottom) Detachment formation in numerical models that incorporate fault weakening. Under certain conditions deformation stabilizes for long periods of times and produces structures that have a similar geometry to the topography observed at oceanic detachments. The extent of the detachment is indicated by the grey bar, and the advection of mantle lithosphere can be tracked by the left-hand dashed white line, which indicates the present-day position of a marker originally at the base of the model [Lavier et al., 1999].

explain variations in seafloor morphology and faulting as a response to variations in the rheological structure along the ridge axis.

Recently this 'weak crust' model has been questioned on the basis of new experimental rheological data and geological arguments [*Hirth et al.*, 1998; *Hirth and Kohlstedt*, in press]. Recent experimental work on the rheology of dry diabase shows that this rock, at geologically relevant conditions, is much stronger than the wet 'diabase' reported in earlier experimental work, and has a similar strength to that of olivine [*Mackwell et al.*, 1998]. As water is a highly incompatible element in the mantle and present in small quantities, it is partitioned into the melt. As a consequence, both the mantle, having undergone a small amount of melting [*Hirth and Kohlstedt*, 1996], and the lower crust, formed by cumulate gabbros from which the melt (and hence the water) has been extracted, are nominally dry. Therefore, rheological models of the lower crust should adopt a 'dry' diabase rheology instead of the 'wet' one commonly used. Later hydration of the crust can occur due to circulation of fluids in the lithosphere, but this process requires the presence of fractures and interconnected porosity that is only possible in the brittle regime. This hydration will have consequences for the rheology of the brittle lithosphere (e.g., serpentinisation [*Escartin et al.*, 1997a]), but it is unlikely to affect the plastic, impermeable levels of the lithosphere. [*Hirth and Kohlstedt*, in press] suggest that in fact weak lower crust is less critical to these models than suggested by [*Chen and Morgan*, 1990a], and that increasing temperature may be the dominant effect rather than a weak lithology. Moreover, the isostatic balance of the axial high has also been questioned, and more recent work suggests that the high is regionally supported by dynamic viscous flow [*Eberle and Forsyth*, 1998] or by stresses in the brittle lithosphere [*Buck*, 2001].

The presence of a continuous magma chamber at fast-spreading ridges [*Babcock et al.*, 1998; *Vera et al.*, 1990] and its absence at slow-spreading ridges [*Detrick et al.*, 1990; *Lin et al.*, 2003] demonstrates that the rheological structure of fast and slow-spreading ridges are fundamentally different [*Poliakov and Buck*, 1998]. A magma lens at shallow crustal levels necessarily implies that the brittle layer above it is very thin and can be easily faulted or dissected by dykes. In contrast, the emplacement of discrete and ephemeral magma chambers in the thick lithosphere of slow-spreading ridges will result in large temporal variations in the rheological structure of the ridge axis. Over long periods of time, the thickness of the lithosphere at slow-spreading ridges can thus be assumed to be large and to vary gradually along the length of ridge segments, as indicated by the gradual variation in rift valley width and depth along axis. A thin lithosphere may be expected immediately after the emplacement of a magma chamber, but

such events must be limited both spatially and temporarily so as to maintain the axial rift valley.

7. MORPHOLOGY AND CRUSTAL ARCHITECTURE OF RIDGE SEGMENTS

The use of swath bathymetry, gravity and seismic studies along segmented slow-spreading ridges such as the Mid-Atlantic Ridge have revealed systematic along-axis variations from the segment ends to the segment center [*Detrick et al.*, 1995; *Hooft et al.*, 2000; *Hosford et al.*, 2001; *Kuo and Forsyth*, 1988; *Lin et al.*, 1990; *Magde et al.*, 2000; *Purdy et al.*, 1990; *Searle et al.*, 1998a; *Sempéré et al.*, 1995; *Thibaud et al.*, 1998] (Plate 1). Segments tend to be shallower and have a thicker crust at the segment center, which is considered to be an indication of focused magmatic accretion at the midpoint [*Lin et al.*, 1990; *Magde et al.*, 1997; *Tolstoy et al.*, 1993]. Crustal thickness variations along a segment can be as large as 7 km, with thicknesses of <3 km at the segment end increasing to >9 km at the center, e.g., [*Hooft et al.*, 2000]. Seismic data also show that the thickness variations occur primarily in layer 3 (V_p ~ 6.8 – 7.2 km/s), while the thickness of layer 2 remains relatively constant (Figure 14). Ultra slow-spreading ridges such as the South West Indian Ridge show a more extreme focusing of melt, with the construction of large central volcanoes at the centers of some segments while others are relatively magmatically starved [*Cannat et al.*, 1999; *Dick et al.*, 2003; *Fujimoto et al.*, 1999; *Michael et al.*, 2003]. Segments have a typical length of 40–90 km [*Schouten et al.*, 1985] and are generally sub-perpendicular to the direction of relative plate separation, although highly oblique segments are found near hotspots [*Abelson and Agnon*, 1997; *Searle et al.*, 1998b] and in extended regions of oblique spreading [*Taylor et al.*, 1994].

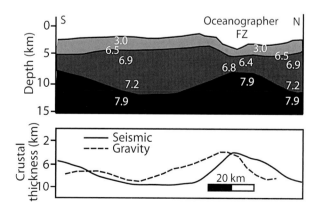

Figure 14. Along-axis variation of crustal thickness based on seismic observations (top) and gravity modeling (bottom). Note that the direction of the gravity scale is reversed. After Detrick et al., [1995].

Segments are laterally offset from each other by up to 30 km in "non-transform discontinuities" (NTDs), or by transform faults that normally accommodate larger offsets [*Fox et al.*, 1991; *Grindlay et al.*, 1991; *Grindlay et al.*, 1992]. NTDs leave wakes of depressed seafloor that show along-axis migration rates comparable to the spreading rate or higher [*Kleinrock et al.*, 1997], and that can lead to the lengthening and shortening of segments, and to their nucleation or extinction [*Gente et al.*, 1995; *Rabain et al.*, 2001; *Tucholke et al.*, 1997]. NTD migration along axis may be driven by changes in plate motion, differential variations in the relative magma supply or thermal state of adjacent segments [*Phipps Morgan and Sandwell*, 1994; *West et al.*, 1999]. In some cases "magmatically robust" segments, characterized by a shallow axis and narrower axial valley at the segment center, and showing evidence of voluminous volcanism, tend to grow at the expense of adjacent segments by the migration of the NTDs away from the segment center [*Gente et al.*, 1995; *Rabain et al.*, 2001; *Thibaud et al.*, 1998]. NTD migrations can also be driven by pressure gradients induced by topographic gradients across discontinuities [*Phipps Morgan and Parmentier*, 1985].

In the longer term, asymmetric spreading between adjacent spreading segments can vary the ridge offset, and promote a change from an NTD to a rigid, non-migrating transform fault or vice versa [*Grindlay et al.*, 1991]. The actual mechanism of NTD propagation is poorly understood, but requires the along-axis propagation of dykes and extensional faults into the crust formed at the adjacent segment across the segment boundary. Large offsets (>~30 km) will result in a thicker, cooler and therefore stronger lithosphere across a discontinuity, thus arresting the propagation of dykes and faults across it.

Similar plate-boundary segmentation is observed at both intermediate and fast-spreading ridges, but without the major morphological and crustal thickness changes observed at slow-spreading ridges. Crustal thickness variations along fast-spreading ridges and away from major transform offsets are of a much smaller amplitude (typically < 2 km) than those observed at slow-spreading ridges [*Bazin et al.*, 1998; *Canales et al.*, 1998; *Madsen et al.*, 1990]. The detailed morphology of the axial volcanic ridge does not directly correlate with the presence of a magma chamber, as some, but not all, discontinuities appear to be underlain by well-developed melt lenses [*Bazin et al.*, 2001; *Kent et al.*, 2000]. The presence of a melt lens above a zone of hot crust, possibly containing small amounts of melt, probably results in a weak crust that can deform plastically to accommodate variations in morphology and structure associated with ridge segmentation at the surface [*Bell and Buck*, 1992].

Slow-spreading ridges have characteristic tectonic patterns that demonstrate that both the seafloor morphology and the crust formed at the ridge axis undergo significant modification at the rift bounding walls [*Escartin and Lin*, 1998] (see Figures 14, 16). Along the ridge axis, the shallowest point in the rift valley and the thickest crust are located at the segment center, and the thinnest crust is found below the ridge discontinuities. Outside the rift valley, the shallowest points are commonly located in close proximity to the segment ends, at the inside corners of the ridge-transform or ridge-NTD intersections, while the thinnest crust is found under the elevated inside corners, where the topography must be dynamically supported [*Escartin and Lin*, 1995; *Rommevaux-Justin et al.*, 1997; *Tucholke et al.*, 1997]. The outside corners are commonly more subdued topographically and tectonically, indicating asymmetric tectonic processes and uplift [*Severinghaus and Macdonald*, 1988].

Ridge segmentation is generally agreed to be intimately associated with the pattern of melt delivery at mid-ocean ridges, although the ultimate cause of segmentation remains uncertain. While some models have suggested that segmentation may be controlled by focused mantle upwelling or mantle diapirs [*Lin and Phipps Morgan*, 1992; *Lin et al.*, 1990; *Whitehead et al.*, 1984], numerical modeling suggests that, for realistic viscosities, the characteristic size of diapirs exceeds the characteristic length of slow-spreading ridge segments [*Barnouin-Jha et al.*, 1997; *Sparks and Parmentier*, 1993]. Segmentation is more likely controlled by brittle processes in the lithosphere [*Macdonald et al.*, 1991b; *Macdonald et al.*, 1986; *Pollard and Aydin*, 1984], with some interaction and feedback with magmatic processes (e.g., focusing of melt to the center of segments at slow-spreading ridges). This feedback is supported by the apparent constant average melt supply from the mantle to three adjacent segments of different length along the Mid-Atlantic Ridge that otherwise display important differences in the absolute variations in crustal thickness along individual segments [*Hooft et al.*, 2000]. Initial melt focusing at the segment center can be achieved by melt migration along axis at the base of the lithosphere (following shallowly dipping isotherms), or by focusing within the lithosphere itself [*Magde and Sparks*, 1997; *Magde et al.*, 1997; *Sparks and Parmentier*, 1993]. The thinner crust at the ends of slow-spreading ridge segments results from along-axis dyke propagation from the segment center [*Hooft et al.*, 2000; *Lawson et al.*, 1996]. This crustal structure formed at the ridge axis, as mentioned above, is later modified [*Canales et al.*, 2000b; *Hosford et al.*, 2001] by extensional faulting along the rift valley walls [*Escartin and Lin*, 1998] as the crust is rifted off axis. The asymmetry in crustal thickness between inside and outside corners may result from initial asymmetric crustal accretion [*Allerton et al.*, 2000], possibly followed by asymmetric tectonic thinning [*Escartin and Lin*, 1998]. A summary of these processes is shown in Figure 15.

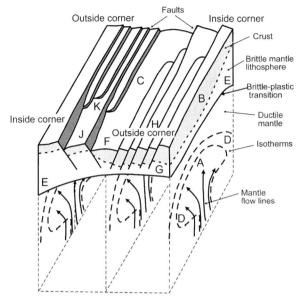

Figure 15. Block model of varying fault style along a slow-spreading ridge segment, modified from Shaw, [1992], and Shaw and Lin, [1993]. Crust is indicated by light grey shading, but the crust-mantle boundary is dashed to emphasise that it is unlikely to be a simple layered structure, but comprises a mixture of melt-derived rock (gabbro, diabase and basalt) and peridotite (see also Figure 16). Mantle wells up strongly under the segment centre (A), producing a high thermal gradient, enhanced melting and therefore thin lithosphere but thick crust (B). These conditions yield a weak lithosphere that deforms by closely-spaced, relatively low-amplitude faulting (C). Weaker mantle upwelling at segment ends (D) leads to lower thermal gradient, less melting, and therefore a thicker lithosphere and thinner crust (E). However, asymmetric crustal accretion at the segment end (F) leads to thicker crust under the outside corner [Allerton et al., 2000], which weakens the thermally thicker lithosphere there so that the BPT is shallower (G) than under the inside corner (E). Consequently the faulting style at the outside corner (H) is similar to that at the segment centre (C)[Escartin et al., 1999], while the stronger lithosphere at the inside corner can support much larger but more widely spaced faults (J). The different fault styles are accommodated by across-axis linking of faults ([Searle et al., 1998a], K). It is likely that the growth of large faults at the inside corner is also facilitated by weakening of the fault through serpentinisation [Escartin et al., 1997a].

8. LITHOLOGICAL STRUCTURE OF MID-OCEAN RIDGES

Geological observations and sampling of both fast- and slow-spreading ridges demonstrate that these have fundamental differences in the composition and architecture of the crust below the ridge axis and the mode of magmatic accretion. Sampling of normal oceanic crust formed along fast-spreading ridges yields basaltic rocks extruded at the seafloor

(upper seismic Layer 2). Lithologies that have been emplaced within lithosphere at deeper levels, such as diabase, gabbros (lower seismic layer 2 and layer 3 of the crust) and peridotite (mantle) are only found along transform faults or rift zones [*Früh-Green et al.*, 1996; *Karson*, 1998; *Karson et al.*, 2002; *MacLeod and Manning*, 1996; *Mével and Stamoudi*, 1996]. In these areas the pre-existing oceanic crust has been rifted and sections of the crust and deeper lithospheric levels exposed to the seafloor. Ocean drilling at site ODP 504B [*Shipboard Scientific Party*, 1993; *Shipboard Scientific Party*, 1995] , corresponding to intermediate-spreading crust, has revealed a 600 m thick layer of extrusive basalts, underlain by an ~200 m thick transitional zone in turn underlain by at least 1 km of sheeted dykes. Correlations of physical properties (e.g., seismic velocity) and recovered lithologies demonstrate that seismic velocity reflects variations in porosity rather than composition, as often assumed [*Detrick et al.*, 1994].

Sampling of slow-spreading ridges demonstrates that the simple layered structure is not correct for these ridges [*Cannat*, 1993; *Cannat*, 1996]. Here, outcrop of gabbro and peridotite (mostly serpentinized) at the seafloor is relatively common, and was recognized early on [*Aumento and Loubat*, 1971; *Auzende et al.*, 1989; *Bonatti and Hamlyn*, 1978; *Bonatti and Harrison*, 1976; *Dick*, 1989; *Engel and Fischer*, 1953; *Hekinian*, 1968; *Juteau et al.*, 1990; *Melson et al.*, 1966; *Prinz* et al., 1976]. These "deeper" rocks are most commonly found at the ends of slow-spreading segments and at inside corners, where the geophysically defined crust is thinner, and where tectonic processes appear to be more effective in exposing deeper lithologies. Tectonic models have been put forward in which the crust is extremely thin or altogether absent from inside corners (Figure 16), and deep lithologies are exposed along large-offset faults [*Blackman et al.*, 1998; *Cann et al.*, 1997; *Cannat*, 1993; *Cannat*, 1996; *Dick et al.*, 1991; *Karson*, 1998; *Karson and Lawrence*, 1997; *Kurewitz and Karson*, 1997; *Tucholke and Lin*, 1994].

Peridotite outcrops are not restricted to the ends of segments or ridge discontinuities, but may also be found along the centers of some segments [*Cannat*, 1993; *Cannat and Casey*, 1995; *Cannat et al.*, 1997; *Cannat et al.*, 1995; *Dick*, 1989; *Lagabrielle et al.*, 1998]. While it is thought that these segment-center peridotites may be more common at ultra-slow-spreading rates or ridges overlying unusually cold mantle, they are not restricted to such cases, and the precise conditions for their occurrence are not yet understood. These peridotite outcrops extend over several kilometers along the ridge axis, and are often capped by a thin layer of extrusive basalts. Although often referred to as "amagmatic", such crust may actually contain ~ 25% gabbro intruded into the peridotite [*Shipboard Scientific Party*, 2003]. Sections of ultra slow-spreading centers, such as the Gakkel ridge and South West Indian Ridge,

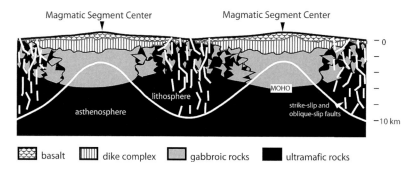

Figure 16. Along-axis variation in crustal thickness and lithology after Cannat et al., [1995]. Continuous white line represents the base of the lithosphere, approximately following an isotherm. White, sub-vertical lines represent mantle dykes. Note variations in crustal thickness and discontinuous nature of lower crust (gabbroic layer) at segment ends.

in addition to a geophysically thin crust, display ridge sections that correspond to tectonic stretching of the mantle lithosphere with little or no magmatism [*Coakley and Cochran*, 1998; *Cochran et al.*, 2003; *Dick et al.*, 2003; *Jokat et al.*, 2003; *Lin et al.*, 2003; *Michael et al.*, 2003].

These geological observations are inconsistent with a layered crustal model, as in most cases the fault scarps do not have sufficient throw to expose lower crustal and upper mantle levels. Instead, the geological constraints indicate that the crust can be compositionally heterogeneous [*Cannat*, 1993; *Cannat*, 1996; *Cannat et al.*, 1995], that transitions from a "magmatic" to a discontinuous or absent crust can occur along individual segments, and that the process of magmatic accretion is not continuous. Instead, magma is emplaced in discrete bodies in the thick, cold lithosphere, feeding the axial volcanism (Figure 16). The continuity and thickness of the crust will therefore depend on the relative supply of magma (which may itself vary with time and position) with respect to plate separation, and the seismic velocities that define the "crust" may be a complex function of composition, alteration and cracking of the lithosphere. Recent seismic studies at slow-spreading ridges have shown variations in crustal thickness and seismic velocities that are consistent with this compositionally heterogeneous lithospheric model [*Barclay et al.*, 1998; *Canales et al.*, 2000a; *Canales et al.*, 2000b]. The size and distribution of the different lithologies are still largely unconstrained, due both to the lack of sufficiently high resolution seismic data, and to the impossibility of distinguishing seismically between lithologies such as partially serpentinized peridotite and gabbro [*Carlson*, 2001; *Christensen*, 1972; *Horen et al.*, 1996].

Geological observations demonstrate that the crustal architecture and the processes responsible for its build up differ substantially from fast to slow-spreading ridges. Fast-spreading ridges display a homogeneous, layered crust that is formed with high melt supply and a relatively stable magmatic system (i.e., near-continuous axial magma chamber,

frequent eruptions, etc.) In contrast, slow-spreading ridges show a wide range of crustal composition, structure and thickness, with important variations both regionally and along individual segments. This complexity arises from a discontinuous mode of magma emplacement (discrete gabbro bodies), and from a wide variation in the magma supply to the ridge axis, ranging from no magmatism (e.g., extension by pure stretching of the mantle lithosphere such as at Gakkel Ridge and parts of the South West Indian Ridge [*Dick et al.*, 2003]), to well-developed and magmatically robust ridge segments that locally have crustal thicknesses exceeding 8 or 9 km (e.g., OH1 and Lucky Strike segments along the Mid-Atlantic Ridge [*Escartín et al.*, 2001; *Hooft et al.*, 2000], or other segments along the South West Indian Ridge [*Cannat et al.*, 1999]). Realistic rheological models of mid-ocean ridges must treat fast- and slow-spreading ridges separately, as these systems operate differently, and do not appear to represent end-members of an accretion process with gradual variations in magma supply. These fundamental differences relate both to the composition and thermal structure of the lithosphere, and to the time-dependence of the processes involved in magmatic accretion and lithospheric construction.

9. FAULTING AT MID-OCEAN RIDGES

Tensional stresses induced by plate separation result in disruption of the upper crust (brittle lithosphere) by normal faults that dissect the ocean floor [*Searle*, 1992]. On the slow-spreading Mid-Atlantic Ridge, most active faulting occurs within about 10–15 km of the ridge axis [*McAllister and Cann*, 1996; *McAllister et al.*, 1995; *Searle et al.*, 1998a]. The width of the active tectonic zone at fast-spreading ridges is less well constrained, though there is some evidence that it extends to ~30 km off-axis [*Macdonald*, 1998]. Variations in faulting patterns regionally (e.g., fast vs. slow-spreading ridges) and locally (along individual segments) can therefore provide insight into the rheological structure of the lithosphere and its spatial variations.

High-angle faulting is responsible for the formation of abyssal hill terrain, and estimates of tectonic strain at the seafloor indicate that <10–20% of the plate separation is taken up by such faulting [*Allerton et al.*, 1996; *Bohnenstiehl and Carbotte*, 2001; *Bohnenstiehl and Kleinrock*, 1999; *Carbotte and Macdonald*, 1994; *Carbotte et al.*, 2003; *Escartin et al.*, 1999; *Jaroslow et al.*, 1996; *Macdonald and Luyendyk*, 1977], while the rest must be taken up by magmatic emplacement or amagmatic accretion of mantle asthenosphere into the lithosphere. A similar value of ~10% tectonic strain has been obtained from seismic moment release studies in the case of slow-spreading ridges [*Solomon et al.*, 1988].

Faults are normally orthogonal to the spreading direction, except in oblique-spreading regions such as near hot spots and in proximity to NTDs. Faults identified in shipboard bathymetry and sonar data have a typical spacing on the order of 1–3 km, and lengths of tens of kilometers along the axial direction [*Cowie et al.*, 1994; *Macdonald et al.*, 1991a; *Searle*, 1984]. Faulting patterns at slow- and fast-spreading ridges differ substantially and reflect the fundamental differences in the structure and thermal state of the lithosphere under the axis in these two environments.

9.1. Normal Faulting at Slow-Spreading Ridges

Slow-spreading ridges are characterized by faults with throws that are an order of magnitude larger than those found at fast-spreading ridges (typically hundreds of meters to kilometers compared to <100 m). These faults are mostly facing the ridge axis, and produce a typical ridge-parallel abyssal hill terrain [*Bohnenstiehl and Kleinrock*, 2000; *Goff*, 1992; *Tucholke et al.*, 1997] with vertical relief of ~1 km. The traces of faults show spatial variations that appear to be systematically linked to the geometry of slow-spreading ridge segments [*Escartin et al.*, 1999; *Escartin and Lin*, 1995; *Searle et al.*, 1998a; *Shaw*, 1992; *Shaw and Lin*, 1993]. At the centers of "typical" slow-spreading ridge segments, faults tend to define a symmetrical axial valley with similar fault size and strain distribution at each flank (Figure 15). Mature faults have moderate throws (a few hundred meters at most) and are spaced 1–2 km apart. These faults grow from individual small faults that link to form larger structures [*Cowie and Scholz*, 1992; *Cowie et al.*, 1993; *Searle et al.*, 1999; *Searle et al.*, 1998a]. In contrast, segment ends are characterized by a marked asymmetry in topography and crustal thickness [*Escartin and Lin*, 1995; *Severinghaus and Macdonald*, 1988; *Tucholke and Lin*, 1994] that is associated with profound differences in fault patterns. It was early recognized that faults near segment ends had a larger throw and spacing that those

near the segment center [*Shaw*, 1992], but the overall tectonic strain does not seem to vary between segment end and segment center [*Escartin et al.*, 1999]. Asymmetry in tectonic strain can be important and extend along the whole length of a segment, and may be associated with a complementary asymmetry in magmatic accretion [*Allerton et al.*, 2000; *Escartin et al.*, 1999; *Searle et al.*, 1998a].

These variations in fault patterns have been interpreted to reflect broad variations in the overall strength of slow-spreading oceanic lithosphere, with a thicker lithosphere at the segment end than at the center [*Behn et al.*, 2002a; *Behn et al.*, 2002b; *Jaroslow*, 1996; *Shaw*, 1992; *Shaw and Lin*, 1996]. Other processes, such as fault weakening due to alteration of the mantle (e.g., via serpentinisation) can promote efficient strain localization and therefore influence faulting patterns observed at the seafloor. While these studies have provided some insight into the expected variations in lithospheric thickness along ridge segments, an accurate interpretation of fault patterns in the light of the thermal state of the lithosphere, its composition, and the mode of strain localization is still required.

Numerical models incorporating processes such as fault weakening have been successful in reproducing, in two dimensions, the broad characteristics of seafloor topography at slow-spreading ridges, with median valley and abyssal hills of the appropriate wavelength and height [*Buck and Poliakov*, 1998; *Poliakov and Buck*, 1998] (Plate 2a). These models incorporate an elastic-plastic-viscous layer, a temperature- and strain-rate-dependant brittle to plastic transition, and fault weakening by the reduction of cohesion of fault material. The models, which do not include any magmatic accretion, predict strain localization along lithospheric-scale faults. These faults advect mantle asthenosphere which is then accreted into the lithosphere. Models such as this demonstrate the importance of strain localisation in producing the observed ridge topography, and underline the importance of understanding the rheological properties that cause this to occur.

9.2. Detachment Faulting

Low-angle normal faults accommodating large amounts of displacement (detachment faults) have long been recognized in continental settings [*Davis and Lister*, 1988]. The presence of oceanic low-angle faults had been proposed early on [*Dick et al.*, 2000; *Dick et al.*, 1991; *Dick et al.*, 1981; *Karson and Dick*, 1983; *Mével et al.*, 1991] to explain the outcrop of basalt and gabbro, but the extent and geometry of the detachment fault surface was not defined. Oceanic detachments were first unambiguously identified on the Mid-Atlantic Ridge at 31°N [*Cann et al.*, 1997], as smooth, curved and sub-horizontal surfaces with corrugations ("mullions") parallel to the spreading direction. Numerous such structures have now been identified, surveyed and sampled along other sections of the Mid-Atlantic Ridge [*Escartin and Cannat*, 1999; *Escartin et al.*, 2003; *Fujiwara et al.*, 2003; *MacLeod et al.*, 2002; *Ranero and Reston*, 1999; *Reston et al.*, 2002; *Tucholke et al.*, 1998; *Tucholke et al.*, 2001], South West Indian Ridge [*Dick et al.*, 2000; *Searle et al.*, 2003], Central Indian Ridge [*Mitchell*, 1998], and back-arc basins [*Ohara et al.*, 2001] (Plate 3). Oceanic detachments tend to occur near ridge offsets (mostly transforms), but occurrences of detachments away from offsets have also been identified on the Mid-Atlantic Ridge near the Fifteen-Twenty fracture zone [*Escartin and Cannat*, 1999; *Fujiwara et al.*, 2003; *MacLeod et al.*, 2002].

The presence of oceanic detachments implies that, during periods of time of the order of 1 Ma or more, plate separation is accommodated mainly or entirely by pure tectonic extension in one flank of the ridge. Oceanic detachments have a limited along-axis extension, usually less than the length of the segment, so this highly asymmetric mode of plate separation must change abruptly from the detachment to adjacent areas. Neither the geometry of oceanic detachments (e.g., dip and depth of soling), nor the conditions of formation, deformation, and linkage to the adjacent seafloor, are properly understood. It has been proposed that these structures initiate as high-angle normal faults that then rotate by flexure [*Buck*, 1988; *Tucholke et al.*, 1998], while other models propose that the fault flattens at depth and becomes sub horizontal, as proposed for Basin and Range detachments [*Karson*, 1990; *Karson et al.*, 1987]. Oceanic detachments are proposed to root deeply near the brittle-plastic transition [*Tucholke et al.*, 1998], at shallower levels in melt-rich zones such as the axial magma chamber [*Dick et al.*, 2000], or within the brittle lithosphere at an alteration front [*Escartin and Cannat*, 1999; *Escartin et al.*, 2003; *MacLeod et al.*, 2002]. In all cases the root is placed at a rheological boundary that can localize deformation for long periods of time.

Oceanic detachments have been reproduced in numerical models that include the presence of strain softening to promote localization of deformation during long periods of time [*Lavier et al.*, 1999] (Plate 2b). In these models, the onset of detachment faulting occurs for a rheology in which fault strength is reduced to <10% of the total strength of the lithosphere. Such weakening mechanisms may be attained with alteration products such as serpentinite [*Escartin et al.*, 1997a] or talc, but these phases are only stable in the shallow, cold lithosphere, and not near the brittle-plastic transition, as required by the numerical models. More recent numerical modeling incorporating magmatic accretion has shown a transition from mostly lithospheric extension at low magma supply to formation of detachments and asymmetric magmatic accretion when the magma supply is ~50% of that corresponding to fully magmatic extension [*Buck et al.*, 2003; *Tucholke et al.*, 2003].

The presence of detachments throughout the oceanic crust, and their general, but not exclusive, association with transform offsets, NTDs, and the outcrop of serpentinized peridotites, indicates that their formation and development require specific conditions (e.g., of magma supply, temperature, lithospheric composition and geometry). It is also apparent that these conditions, while not unusual, are not found along all ridge segments or near all discontinuities. The existing numerical models, despite their limitations, support this qualitative interpretation; detachments form under certain rheologies that involve substantial weakening of the fault (allowing strain localization) and for a narrow range of magma supply to the ridge axis. Better understanding of the actual origin and conditions that lead to detachment fault development would provide important constraints on the thermal and rheological conditions at which they form.

9.3. Lithospheric Deformation at Depth

While most studies of tectonic strain at slow and fast-spreading ridges focus on faulting and deformation measurable at the seafloor as fault scarps [*Bohnenstiehl and Carbotte*, 2001; *Bohnenstiehl and Kleinrock*, 1999; *Cowie*, 1998; *Escartin et al.*, 1999], geological evidence demonstrates that additional mechanisms of extension and tectonic uplift are active under the axis of slow-spreading ridges [*Cannat and Casey*, 1995]. Serpentinized peridotites outcropping along rift valley bounding fault scarps in the immediate vicinity of the active axial volcanic zone have been uplifted from the base of the lithosphere (~750°C) to the seafloor. Rift valley wall faults have vertical throws of <2 km, and therefore tectonic uplift associated with these fault scarps cannot exhume rocks from such depths. *Cannat and Casey*, [1995] proposed the existence of a "tectonic lift" system of cross-cutting faults that allow the vertical ascent of rocks under the ridge axis from the base of the lithosphere to the seafloor. The nature and details of such uplift, required by the geological observations, is not constrained, though numerical models of pure stretching of the lithosphere [*Buck and Poliakov*, 1998] that simulate slow-spreading seafloor do show such patterns (Plate 2a). A similar mechanism is also required to operate along ridges with no magmatism, such as sections of the Gakkel Ridge [*Michael et al.*, 2003].

10. SUMMARY OF OBSERVATIONS: RHEOLOGICAL STRUCTURE OF SLOW AND FAST-SPREADING RIDGES

The cartoons in Plate 4 are intended to summarize, in two dimensions, some of the main observations described above for both fast- and slow-spreading ridges and their relevance to the fundamental differences in the rheology under the ridge axes. In addition, slow-spreading ridges in particular have further important along-axis variability.

Fast-spreading ridges (Plate 4a) have a homogeneous, compositionally layered crust and little variation in crustal thickness. A magma chamber at ~2 km depth is observed in numerous geophysical surveys, and additional magma chambers may be present in or below the crust. The thermal state of this type of ridge can be assumed to be quasi steady-state (when compared with that of slow-spreading ridges). The nature of the lower crust below the melt lens is not well constrained, but it is likely to be hot and may contain small amounts of partial melt, at least locally. Seismic data [*Hammond and Toomey*, 2003; MELT, 1998] suggest that the isotherms are steeply dipping at the flanks of the magma chamber, and that hydrothermal circulation is vigorous and reaching deep levels below the magma chamber adjacent to it. The brittle to plastic transition is likely to have a large amount of variation in depth at short distances from the axis, being controlled primarily by the presence of the axial magma chamber. We expect two areas of deformation; a near-axis one with development of smaller faults and dyking feeding axial volcanism, and off-axis growth of faults over the thicker lithosphere away from the melt lens.

The structure, composition and geometry of the lithosphere under the axis of slow-spreading ridges is not as well constrained, and probably has a large variability both in composition and temperature, varying both spatially (along and across-axis) and temporally (Plate 4b). The brittle lithosphere is, over large time scales, thicker than that at fast-spreading ridges, allowing the development of larger faults. Detachments accommodating large amounts of extension may form under certain rheological conditions that are yet to be determined. There is ample evidence of strain localization along ductile shear zones in the upper asthenosphere and lower mantle lithosphere at temperatures <700°C, that are frozen into the lithosphere and rafted off-axis. These shear zones may be connected with brittle zones at higher levels to allow the tectonic uplift of mantle rocks in close proximity to the ridge axis. Magma chambers are ephemeral, and may substantially alter the thermal structure and therefore the lithospheric rheology, with a transient position of the brittle to plastic transition. Other rheological boundaries may play a role in determining the mode of fault localization, such as the serpentinite alteration front associated with hydrothermal circulation and hydration of the upper lithosphere. Finally, the composition of the lithosphere may range from pure amagmatic (i.e., mantle lithosphere exposed at the seafloor) to a continuous magmatic crust formed by episodic intrusion and extrusion of magma.

Although constraints on the structure and composition of fast-spreading ridges are better than those for slow-spread-

ing ones, both systems require additional constraints to obtain realistic rheological models. These are in turn needed to provide further insights into processes that take place at these systems (e.g., faulting) and to provide a sound basis for the interpretation of geological observations such as faulting and naturally deformed rocks in terms of lithospheric rheology.

11. CONCLUSIONS

We have shown that, to a first order, the rheology of the lithosphere depends on its thermal state, and on the composition, abundance, distribution and mechanical properties of its components. Phases that are not abundant (such as alteration products) may play an important role in both the overall strength of the lithosphere and the mode of localization of the deformation, and need to be fully characterized and incorporated into rheological models. The final lithospheric rheology determines the mode of faulting, dyking and volcanic emplacement, and is ultimately responsible for the morphology of the seafloor and the distribution of faults that we observe. While simple thermo-mechanical models have provided great insight into important processes such as rift valley formation, interpretation of seafloor morphology and faulting in terms of lithospheric rheology has been hindered by a lack of actual constraints on the detailed thermal structure, composition and architecture of the lithosphere under the axis of both fast- and slow-spreading ridges.

Results from experimental rock mechanics have formed the basis for all rheological models commonly used, but require major extrapolation of the experimental results from laboratory to natural conditions. To date, we have gained a good understanding of the mechanical properties of olivine and olivine aggregates, and a body of experimental data exists for dolerite and serpentinite, including the important role of both melt and water: these are all important components of the oceanic lithosphere. This type of experimental work needs to be extended to other lithologies, such as the alteration products of the crust and mantle (i.e., amphibolites, talc, etc.), and the role of water, melt, and other parameters such as compositional variations in individual rock types need to be better characterized.

The main structure and composition of mid-ocean ridges, particularly of fast-spreading ones, is relatively well known thanks to detailed geophysical images of the crust and upper mantle. In the case of slow-spreading ridges, important remaining unknowns are the relative abundance, distribution and geometry of different lithologies (mainly gabbro, peridotite and serpentinite, but also the way in which dolerite and basalt are distributed at so-called "magmatic" and "amagmatic" crust); these are all required parameters to construct robust rheological models. In neither fast- nor slow-spread-

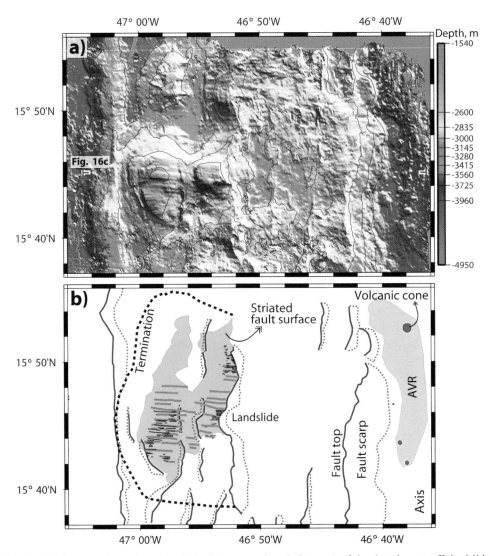

Plate 3. Shaded bathymetry (a, top) and geological interpretation (b, bottom) of the detachment off the Mid-Atlantic Ridge at 15°45'N. The detachment is sub-horizontal and gently curved in the direction of spreading, and has bathymetric corrugations at wavelengths of ~1 km. The exposed fault surface (green in b) is up to 12 km wide. These detachment surfaces also show lineations at shorter wavelengths in the deep-tow sonar images, and fault striations at rock outcrops. After Escartin et al., [2003] .

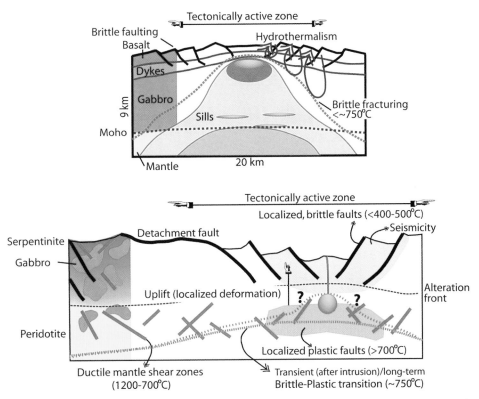

Plate 4. Schematic representation of the rheological structure of the lithosphere across the axis of fast- and slow-spreading ridges. While the fast-spreading system may show a steady magma chamber and the formation of a compositionally homogeneous crust, the slow-spreading system shows ephemeral magma chambers emplaced in the lithosphere and resulting in a heterogeneous composition. a: (top) Fast-spreading ridge. Coloring at left emphasizes simple layered lithological structure of crust formed at such ridges. Thin red lines represent isotherms, and thick dashed line is approximate 750°C isotherm representing brittle-plastic transition. Red areas represent crystal mush zones, grading into yellow representing magma chambers with more continuous melt. High-level sub-axial melt zone is quasi-continuous. Blue lines schematically represent hydrothermal circulation. b: (bottom) Slow-spreading ridge. Magmatism is highly discontinuous in space and time: orange-dashed line shows long-term brittle-plastic transition, while red-dashed line shows BPT immediately following transient magma injection. Melt is shown schematically grading from crystal mush zone (red) into connected magma (yellow), though the detailed geometry of the melt zone is unknown. Lithological structure may be highly heterogeneous. Shown on the left is an extreme case of "amagmatic" crust (typical of detachment zones, segment ends and parts of very cold or slow-spreading ridges) where the basaltic layer is absent; at other parts of slow-spreading ridges, and especially segment centers, there may be a basaltic carapace, possibly overlying a sheeted dyke layer (see Figure 16). Fine black dashed line schematically indicates the serpentinisation (peridotite alteration) front, which occurs at approximately 450°C and is probably limited by the depth of hydrothermal circulation. Upper crust is deformed by brittle faulting (black lines), while lower (and, locally, upper) crust and mantle are deformed in ductile shear zones. Hydrothermal activity (not shown) is probably highly episodic at slow-spreading ridges. See text for further discussion and explanation.

ing ridges are there yet good constraints on the distribution of melt and water. Finally, the thermal structure of fast-spreading ridges is relatively well determined, as the presence of the geophysically imaged magma lens imposes clear thermal constraints. In contrast, slow-spreading ridges probably display a thermal structure that varies substantially with time, as a consequence of emplacement of discrete magma bodies of unknown size and at unknown rates, but details are at present very poorly constrained.

To date, thermo-mechanical models have been successful in reproducing first-order observables such as the axial relief and general faulting patterns. These models can also reproduce structures such as oceanic detachments, although they do not fully explain all available geological observations, nor yet take into account parameters such as the compositional heterogeneity of the lithosphere. Magmatic accretion and its interplay with tectonism is an important but complex process that operates at mid-ocean ridges, and is just beginning to be incorporated into such models. The combined geological and theoretical study of the formation and evolution of detachments can provide valuable information on the rheological conditions at which these structures form, giving clues to the actual rheological state of different parts of the oceanic lithosphere.

Due to the numerous processes controlling the rheological structure of the lithosphere, there is no simple and direct correlation between fault structure and seafloor morphology on the one hand, and on the other the mechanical properties of the lithosphere at depth. A good understanding of this link will require further observations on the composition, thermal structure and processes operating near ridge axes as recorded by naturally deformed rocks, additional experimental work on the mechanical properties of different rock types that are present in the lithosphere, and the development of rheological models that capture the complexity of natural systems, including the interaction of magmatism and tectonism, the temporal variability of these processes, and the heterogeneous nature of the oceanic lithosphere, particularly at slow-spreading ridges.

Acknowledgments. This paper was developed from a presentation, delivered at the first InterRidge Theoretical Institute held in Pavia, Italy in September 2002, on the "Thermal Regime of Ocean Ridges and Dynamics of Hydrothermal Circulation." We want to acknowledge the stimulating presentations and discussions with the participants of this meeting that have made their way into this paper. In particular the comments, encouragement and persistence of Jian Lin made this paper possible. The paper was largely finalized while RCS was on sabbatical at Woods Hole Oceanographic Institution, and he is grateful for interactions with many colleagues there. The final paper benefited greatly from careful reviews by Roger Buck, Patience Cowie and Greg Hirth. Both authors gratefully acknowledge travel support from the European Science Foundation to Pavia, and RCS acknowledges support from the Leverhulme Trust for his stay at Woods Hole.

REFERENCES

Abelson, M., and A. Agnon, Mechanics of oblique spreading and ridge segmentation, *Earth and Planetary Science Letters, 148,* 405–421, 1997.

Agar, S. M., and F. C. Marton, Microstructural controls on strain localization in ocean diabases: Evidence from Hole 504B, in *Proceedings of the Ocean Drilling Program, Scientific Results*, edited by J. Erzinger, H. J. B. Dick, L. B. Stokking, and e. al., pp. 219–229, Ocean Drilling Program, College Station, TX, 1995.

Allerton, S., J. Escartin, and R. C. Searle, Extremely asymmetric magmatic accretion of oceanic crust at the ends of slow-spreading ridge-segments, *Geology, 28,* 179–182, 2000.

Allerton, S., R. C. Searle, and B. J. Murton, Bathymetric segmentation and faulting on the Mid-Atlantic Ridge, 24°00'N to 24°40'N, in *Tectonic, Magmatic, Hydrothermal and Biological Segmentation of Mid-Ocean Ridges, Geological Society of London, Special Publication 118*, edited by C. J. MacLeod, P. A. Tyler, and C. L. Walker, pp. 49–60, 1996.

Atwater, T., and J. D. Mudie, Block faulting on the Gorda Rise, *Science, 159,* 729–731, 1968.

Aumento, F., and H. Loubat, The Mid-Atlantic Ridge near 45°N. XVI. Serpentinized ultramafic intrusions, *Canadian Journal of Earth Science, 8,* 631–663, 1971.

Auzende, J.-M., D. Bideau, E. Bonatti, M. Cannat, J. Honnorez, Y. Lagabrielles, J. Malaveille, V. Mamaloukas-Frangoulis, and C. Mevel, Direct observation of a section through slow-spreading oceanic crust, *Nature, 337,* 726–729, 1989.

Babcock, J. M., A. J. Harding, G. M. Kent, and J. A. Orcutt, An examination of along-axis variation of magma chamber width and crustal structure on the East Pacific Rise between 13°30'N and 12°20'N, *Journal of Geophysical Research, 103 (B12),* 30371–30384, 1998.

Ballard, R. D., and T. H. Van Andel, Morphology and tectonics of the inner rift valley at lat. 36° 50'N on the Mid-Atlantic Ridge, *Geological Society of America Bulletin, 88,* 507–530, 1977.

Barclay, A. H., D. R. Toomey, and S. C. Solomon, Seismic structure and crustal magmatism at the Mid-Atlantic Ridge, 35°N, *Journal of Geophysical Research, 103 (B8),* 17827–17844, 1998.

Barclay, A. H., D. R. Toomey, and S. C. Solomon, Microearthquake characteristics and crustal Vp/Vs structure at the Mid-Atlantic Ridge, 35°N, *Journal of Geophysical Research–Solid Earth, 106 (B2),* 2017–2034, 2001.

Barnouin-Jha, K., E. M. Parmentier, and D. W. Sparks, Buoyant mantle upwelling and crustal production at oceanic spreading centers: On-axis segmentation and off-axis melting, *Journal of Geophysical Research, 102 (B6),* 11979–11989, 1997.

Bazin, S., A. J. Harding, G. M. Kent, J. A. Orcutt, C. H. Tong, J. W. Pye, S. C. Singh, P. J. Barton, M. C. Sinha, R. S. White, R. W. Hobbs, and H. J. A. Van Avendonk, Three-dimensional shallow crustal emplacement at the 9°03'N overlapping spreading center on the East Pacific Rise: Correlations between magnetization and tomographic images, *Journal of Geophysical Research, 106 (B8)*, 16101–16117, 2001.

Bazin, S., H. van Avendonk, A. J. Harding, J. A. Orcutt, J. P. Canales, and R. S. Detrick, Crustal structure of the flanks of the East Pacific Rise: Implications for overlapping spreading centers, *Geophysical Research Letters, 25 (12)*, 2213–2216, 1998.

Behn, M., J. Lin, and M. T. Zuber, Mechanisms of normal fault development at mid-ocean ridges, *Journal of Geophysical Research, 107 (B4)*, 10.1029/2001JB000503, 2002a.

Behn, M. D., J. Lin, and M. T. Zuber, A continuum mechanics model for normal faulting using a strain-rate softening rheology: implications for thermal and rheological controls on continental and oceanic rifting, *Earth and Planetary Science Letters, 202 (3–4)*, 725–740, 2002b.

Bell, R. E., and W. R. Buck, Crustal control of ridge segmentation inferred from observations of the Reykjanes Ridge, *Nature, 357*, 583–586, 1992.

Bergman, E. A., and S. C. Solomon, Earthquake swarms on the Mid-Atlantic Ridge: Products of magmatism or extensional tectonics?, *Journal of Geophysical Research, B, 95 (4)*, 4943–4965, 1990.

Blacic, J. D., and J. M. Christie, Plasticity and hydrolitic weakening of quartz single crystals, *Journal of Geophysical Research, 89 (B6)*, 4223–4239, 1984.

Blackman, D. K., J. R. Cann, B. Janssen, and D. K. Smith, Origin of extensional core complexes: Evidence from the Mid-Atlantic Ridge at Atlantis Fracture Zone, *Journal of Geophysical Research, B, 103*, 21,315–21,333, 1998.

Bohnenstiehl, D. R., and S. M. Carbotte, Faulting patterns near 19°30'S on the East Pacific Rise: Fault formation and growth at a superfast spreading center, *Geochemistry Geophysics Geosystems, 2*, art. no. 2001GC000156, 2001.

Bohnenstiehl, D. R., and M. C. Kleinrock, Faulting and fault scaling on the median valley floor of the trans-Atlantic geotraverse (TAG) segment, ~26°N on the Mid-Atlantic Ridge, *Journal of Geophysical Research–Solid Earth, 104 (B12)*, 29351–29364, 1999.

Bohnenstiehl, D. R., and M. C. Kleinrock, Evidence for spreading-rate dependence in the displacement-length ratios of abyssal hill faults at mid-ocean ridges, *Geology, 28 (5)*, 395–398, 2000.

Bonatti, E., and P. R. Hamlyn, Mantle uplifted blocks in the western Indian Ocean, *Science, 201*, 249–251, 1978.

Bonatti, E., and C. G. A. Harrison, Hot lines in the Earth's mantle, *Nature, 263*, 402–404, 1976.

Bos, B., C. J. Peach, and C. J. Spiers, Slip behaviour of simulated gouge-bearing faults under conditions favoring pressure solution, *Journal of Geophysical Research, 105 (B7)*, 16699–16717, 2000.

Bott, M. H. P., Flexure associated with planar faulting, *Geophysical Journal International, 126*, F21–F24, 1996.

Bowin, C. O., and J. Milligan, Negative gravity anomaly over spreading rift vcalleys: Mid-Atlantic Ridge at 26°N, *Tectonophysics, 113* (233–256), 1985.

Brace, W. F., and D. L. Kohlstedt, Limits on lithospheric stress imposed by laboratory experiments, *Journal of Geophysical Research, 85*, 6248–6252, 1980.

Buck, R. W., L. Lavier, and A. Poliakov, Modes of faulting at mid-ocean ridges, in *EGS-AGU-EUG Joint Assembly*, pp. EAE03-A-133333, EGS, Nice, France, 2003.

Buck, W. R., Flexural rotation of normal faults, *Tectonics, 7 (5)*, 959–973, 1988.

Buck, W. R., Accretional curvature of lithosphere at magmatic spreading centers and the flexural support of axial highs, *Journal of Geophysical Research, 106 (B3)*, 3953–3960, 2001.

Buck, W. R., and A. N. B. Poliakov, Abyssal hills formed by stretching oceanic lithosphere, *Nature, 392*, 272–275, 1998.

Bunch, A. W. H., and B. L. N. Kennett, The crustal structure of the Reykjanes Ridge at 59° 30'N, *Geophysical Journal of the Royal Astronomical Society, 61*, 141–166, 1980.

Byerlee, J. D., Brittle-ductile transition in rocks, *Journal of Geophysical Research, 73*, 4741–4750, 1968.

Byerlee, J. D., Friction of rocks, *Pure and Applied Geophysics, 116*, 615–626, 1978.

Caldwell, J. G., W. F. Haxby, D. E. Karig, and D. L. Turcotte, On the applicability of a universal elastic trench profile, *Earth and Planetary Science Letters, 31* (239–246), 1976.

Canales, J. P., J. A. Collins, J. Escartin, and R. S. Detrick, Seismic structure across the rift valley of the Mid-Atlantic Ridge at 23°20' (MARK area): Implications for crustal accretion processes at slow spreading ridges, *Journal of Geophysical Research–Solid Earth, 105 (B12)*, 28411–28425, 2000a.

Canales, J. P., J. J. Dañobeitia, R. S. Detrick, E. E. E. Hooft, R. Bartolomé, and D. F. Naar, Variations in axial morphology along the Galápagos spreading center and the influence of the Galápagos hotspot, *Journal of Geophysical Research, 102 (B12)*, 27341–27354, 1997.

Canales, J. P., R. S. Detrick, S. Bazin, A. J. Harding, and J. A. Orcut, Off-axis crustal thickness across and along the East Pacific Rise within the MELT area, *Science, 280*, 1218–1221, 1998.

Canales, J. P., R. S. Detrick, J. Lin, J. A. Collins, and D. R. Toomey, Crustal and upper mantle seismic structure beneath the rift mountains and across a nontransform offset at the Mid-Atlantic Ridge (35°N), *Journal of Geophysical Research–Solid Earth, 105 (B2)*, 2699–2719, 2000b.

Cann, J. R., D. K. Blackman, D. K. Smith, E. McAllister, B. Janssen, S. Mello, E. Avgarinos, A. R. Pascoe, and J. Escartin, Corrugated slip surfaces formed at ridge-transform intersections on the Mid Atlantic Ridge, *Nature, 385*, 329–332, 1997.

Cannat, M., Emplacement of mantle rocks in the seafloor at midocean ridges, *Journal of Geophysical Research, 98 (B3)*, 4163–4172, 1993.

Cannat, M., How thick is the magmatic crust at slow-spreading mid-ocean ridges?, *Journal of Geophysical Research, 101 (B2)*, 2847–2857, 1996.

Cannat, M., and J. F. Casey, An ultramafic lift at the Mid-Atlantic Ridge: Successive stages of magmatism in serpentinized peridotites from the 15°N region, in Mantle and Lower Crust Exposed in Oceanic Ridges and in Ophiolites, edited by R.L.M. Vissers, and A. Nicolas, pp. 5–34, Kluwer Academic Publishers, The Netherlands, 1995.

Cannat, M., Y. Lagabrielle, H. Bougault, J. Casey, N de Coutures, L. Dmitriev, and Y. Fouguet, Ultramafic and gabbroic exposures at the Mid-Atlantic Ridge: Geological mapping in the 15°N region, *Tectonophysics, 279*, 193–213, 1997.

Cannat, M., C. Mevel, M. Maia, C. Deplus, C. Durand, P. Gente, P.

Agrinier, A. Belarouchi, G. Dubuisson, E. Humler, and J. Reynolds, Thin crust, ultramafic exposures, and rugged faulting patterns at Mid-Atlantic Ridge (22°–24°N), *Geology, 23 (1)*, 49–52, 1995.

Cannat, M., C. Rommevaux-Jestin, D. Sauter, C. Deplus, and V. Mendel, Formation of the axial relief at the very slow spreading Southwest Indian Ridge (49°–69°), *Journal of Geophysical Research, 104 (B10)*, 22,825–22,843, 1999.

Carbotte, S. M., and K. C. Macdonald, Comparison of seafloor tectonic fabric at intermediate, fast, and super fast spreading ridges: Influence of spreading rate, plate motions, and ridge segmentation on fault patterns, *Journal of Geophysical Research, 99 (B7)*, 13,609–13,631, 1994.

Carbotte, S. M., W. B. F. Ryan, W. Jin, M.-H. Cormier, E. Bergmanis, J. Sinton, and S. White, Magmatic subsidence of the East Pacific Rise (EPR) at 18°14'S revealed through fault restoration of ridge crest bathymetry, *Geochemistry, Geophysics and Geosystems, 3 (1)*, doi:10.1029/2002GC000337, 2003.

Caristan, Y., The transition from high temperature creep to fracture in Maryland diabase, *Journal of Geophysical Research, 87*, 6781–6790, 1982.

Carlson, R. L., The abundance of ultramafic rocks in Atlantic Ocean crust, *Geophysical Journal International, 144 (1)*, 37–48, 2001.

Cazenave, A., B. Lago, K. Dominh, and K. Lambeck, On the response of the ocean lithosphere to seamount loads from Geos 3 satellite radar altimeter observations, *Geophysical Journal of the Royal Astronomical Society, 63 (233–252)*, 1980.

Chen, Y., and W. J. Morgan, A nonlinear rheology model for mid-ocean ridge axis topography, *Journal of Geophysical Research, 95*, 17583–17604, 1990a.

Chen, Y., and W. J. Morgan, Rift valley/no rift valley transition at mid-ocean ridges, *Journal of Geophysical Research, 95*, 17571–17581, 1990b.

Chen, Y. J., The thermal state of the oceanic crust, in *InterRidge Theoretical Institute: Thermal Regime of Ocean Ridges and Dynamics of Hydrothermal Circulation*, edited by C. German, J. Lin, R. Tribuzio, A. Fisher, M. Cannat, and A. Adamczewska, pp. 10, InterRidge, Tokyo, 2003.

Chopra, P. N., The plasticity of some fine-grained aggregates of olivine at high pressure and temperature, in *Mineral and rock deformation: Laboratory studies*, edited by H. E. Hobbs, and H. C. Heard, pp. 25–33, American Geophysical Union, Washington D. C., 1986.

Chopra, P. N., and M. S. Paterson, The role of water in the deformation of dunite, *Journal of Geophysical Research, 89*, 7861–7876, 1984.

Christensen, N. I., The abundance of serpentinites in the oceanic crust, *Journal of Geology, 80*, 709–719, 1972.

Coakley, B. J., and J. R. Cochran, Gravity evidence of very thin crust at the Gakkel Ridge (Arctic Ocean), *Earth and Planetary Science Letters, 162*, 81–95, 1998.

Cochran, J. R., An analysis of isostasy in the worlds oceans: 2. Mid-ocean ridge crests, *Journal of Geophysical Research, 84 (4713–4729)*, 1979.

Cochran, J. R., G. J. Kurras, M. H. Edwards, and B. J. Coakley, The Gakkel Ridge: Bathymetry, gravity anomalies, and crustal accretion at extremely slow spreading rates, *Journal of Geophysical Research, 108* (B2), 10.1029/2002JB001830, 2003.

Coogan, L. A., G. R. T. Jenkin, and R. N. Wilson, Constraining the cooling rate of the lower oceanic crust: a new approach applied to the Oman ophiolite, *Earth and Planetary Science Letters, 199* (1–2), 127–146, 2002.

Cowie, P. A., Normal Fault Growth in 3D in Continental and Oceanic Crust, in *Faulting and Magmatism at Mid-Ocean Ridges*, edited by W. R. Buck, P. T. Delaney, J. A. Karson, and Y. Lagabrielle, pp. 325–348, American Geophysical Union, 1998.

Cowie, P. A., A. Malinverno, W. B. F. Ryan, and M. H. Edwards, Quantitative fault studies on the East Pacific Rise: A comparison of sonar imaging techniques, *Journal of Geophysical Research, 99 (B8)*, 15205–15218, 1994.

Cowie, P. A., and C. H. Scholz, Displacement-length scaling relationship for faults: Data synthesis and discussion, *Journal of Structural Geology, 14*, 1149–1156, 1992.

Cowie, P. A., C. H. Scholz, M. Edwards, and A. Malinverno, Fault strain and seismic coupling on mid-ocean ridges, *Journal of Geophysical Research, 98*, 17,911–17,920, 1993.

Crawford, W. C., and S. C. Webb, Variations in the distribution of magma in the lower crust and at the Moho beneath the East Pacific Rise at 9–10°N, *Earth and Planetary Science Letters, 203* (1), 117–130, 2002.

Davis, E. E., and C. R. B. Lister, Fundamentals of ridge crest topography, *Earth and Planetary Science Letters, 21*, 405–413, 1974.

Davis, G., and G. Lister, Detachment faulting in continental extension: Perspectives from the southwestern US Cordillera, *Geological Society of America Special Paper, 218*, 133–159, 1988.

Detrick, R., J. Collins, and S. Swift, In situ evidence for the nature of the seismic layer 2/3 boundary in oceanic crust, *Nature, 370*, 288–290, 1994.

Detrick, R. S., J. C. Mutter, P. Buhl, and I. I. Kim, No evidence from multichannel reflection data for a crustal magma chamber in the MARK area on the Mid-Atlantic Ridge, *Nature, 347*, 61–64, 1990.

Detrick, R. S., H. D. Needham, and V. Renard, Gravity anomalies and crustal thickness variations along the Mid-Atlantic Ridge between 33°N and 40°N, *Journal of Geophysical Research, 100 (B3)*, 3767–3787, 1995.

Dick, H., J. Lin, and H. Schouten, An ultra-slow-spreading class of ocean ridge, *Nature, 426*, 405–412, 2003.

Dick, H. J. B., Abyssal peridotites, very slow spreading ridges and ocean ridge magmatism, in Magmatism in the Ocean Basins, edited by A. D. Saunders, and M. J. Norry, pp. 71–105, Geological Society, London, 1989.

Dick, H. J. B., J. H. Natland, J. C. Alt, W. Bach, D. Bideau, J. S. Gee, S. Haggas, J. G. H. Hertogen, G. Hirth, P. M. Holm, B. Ildefonse, G. J. Iturrino, B. E. John, D. S. Kelley, E. Kikawa, A. Kingdon, P. J. LeRoux, J. Maeda, P. S. Meyer, D. J. Miller, H. R. Naslund, Y. L. Niu, P. T. Robinson, J. Snow, R. A. Stephen, P. W. Trimby, H. U. Worm, and A. Yoshinobu, A long in situ section of the lower ocean crust: results of ODP Leg 176 drilling at the Southwest Indian Ridge, *Earth and Planetary Science Letters, 179 (1)*, 31–51, 2000.

Dick, H. J. B., H. Schouten, P. S. Meyer, D. G. Gallo, H. Bergh, R. Tyce, P. Patriat, K. T. M. Johnson, J. Snow, and A. Fisher, Tectonic evolution of the Atlantis II Fracture Zone, *Proceedings of the Ocean Drilling Program, Scientific Results, 118*, 359–398, 1991.

Dick, H. J. B., W. B. Thompson, and W. B. Bryan, Low angle faulting and steady-state emplacement of plutonic rocks at ridge-transform intersections, *EOS, Transactions, American Geophysical Union, 62*, 406, 1981.

Dunn, R. A., D. R. Toomey, and S. C. Solomon, Three-dimensional seismic structure and physical properties of the crust and shallow mantle beneath the East Pacific Rise at 9 degrees 30'N, *Journal of Geophysical Research–Solid Earth, 105 (B10)*, 23537–23555, 2000.

Durham, W. B., and C. Goetze, Plastic flow of oriented single crystals of olivine, 1, mechanical data, *Journal of Geophysical Research, 82 (36)*, 5737–5753, 1977.

Durham, W. B., C. Goetze, and B. Blake, Plastic flow of oriented single crystals of olivine. 2 Observations and interpretations of the dislocation structures, *Journal of Geophysical Research, 82 (36)*, 5755–5770, 1977.

Eberle, M. A., and D. W. Forsyth, An alternative, dynamic model of the axial topographic high at fast spreading ridges, *Journal of Geophysical Research, B, 103*, 12,309–12,320, 1998.

Engel, C. G., and R. L. Fischer, Granitic to ultramafic rock complexes of the Indian Ocean ridge system, western Indian Ocean, *Geological Society of America Bulletin, 86*, 1553–1578, 1953.

Escartin, J., and M. Cannat, Ultramafic exposures and the gravity signature of the lithosphere near the Fifteen–Twenty Fracture Zone (Mid-Atlantic Ridge, 14°–16.5°N), *Earth and Planetary Science Letters, 171*, 411–424, 1999.

Escartin, J., M. Cannat, G. Pouliquen, A. Rabain, and J. Lin, Crustal thickness of V-shaped ridges south of the Azores: Interaction of the Mid-Atlantic Ridge (36°–39°N) and the Azores hot spot, *Journal of Geophysical Research, 106 (B10)*, 21,719–21,735, 2001.

Escartin, J., P. A. Cowie, R. C. Searle, S. Allerton, N. C. Mitchell, C. J. MacLeod, and P. A. Slootweg, Quantifying tectonic strain and magmatic accretion at a slow-spreading ridge segment, Mid-Atlantic Ridge, 29°N, *Journal of Geophysical Research, B, 104*, 10,421–10,437, 1999.

Escartin, J., G. Hirth, and B. Evans, Effects of serpentinization on the lithospheric strength and the style of normal faulting at slow-spreading ridges, *Earth and Planetary Science Letters, 151*, 181–189, 1997a.

Escartin, J., G. Hirth, and B. Evans, Nondilatant brittle deformation of serpentinites: Implications for Mohr-Coulomb theory and the strength of faults, *Journal of Geophysical Research, 102 (B2)*, 2897–2913, 1997b.

Escartín, J., G. Hirth, and B. Evans, Strength of slightly serpentinized peridotites: Implications of the tectonics of oceanic lithosphere, *Geology, 29 (11)*, 1023–1026, 2001.

Escartin, J., and J. Lin, Ridge offsets, normal faulting, and gravity anomalies of slow spreading ridges, *Journal of Geophysical Research, 100 (B4)*, 6163–6177, 1995.

Escartin, J., and J. Lin, Tectonic modification of axial structure: Evidence from spectral analyses of gravity and bathymetry of the Mid-Atlantic Ridge flanks (25.5°–17.5°N), *Earth and Planetary Science Letters, 154 (1–4)*, 279–293, 1998.

Escartin, J., C. Mevel, C. J. MacLeod, and A. M. McCaig, Constraints on deformation conditions and the origin of oceanic detachments: The Mid-Atlantic Ridge core complex at 15°45'N, *Geochemistry Geophysics Geosystems, 4*, art. no. 1067, 2003.

Evans, B., and C. Goetze, The temperature variation of hardness of Olivine and its implication for polycrystalline yield stress, *Journal of Geophysical Research, 84 (B10)*, 5505–5524, 1979.

Evans, B., and D. L. Kohlstedt, Rheology of Rocks, in *Rock physics and phase relations: A handbook of physical constants*, edited by T. J. Ahrens, pp. 148–165, American Geophysical Union, Washington, D.C., 1995.

Fox, P. J., N. R. Grindlay, and K. C. Macdonald, The Mid-Atlantic Ridge (31°S–34°30'S): Temporal and Spatial Variations of Accretionary Processes, *Marine Geophysical Researches, 13*, 1–20, 1991.

Fredrich, J. T., B. Evans, and T. F. Wong, Micromechanics of the brittle to plastic transition in Carrara marble, *Journal of Geophysical Research, 94 (B4)*, 4129–4145, 1989.

Früh-Green, G. L., A. Plas, and L. N. Dell'Angelo, Mineralogic and stable isotope record of polyphase alteration of upper crustal gabbros of the East Pacific Rise (Hess Deep, Site 894), in *Proceedings of the Ocean Drilling Program, Scientific Results*, edited by C. Mével, K. M. Gillis, J. F. Allan, and P. S. Meyer, pp. 235–254, Ocean Drilling Program, College Station, TX, 1996.

Fujimoto, H., M. Cannat, K. Fujioka, T. Gamo, C. German, C. Mével, U. Münch, S. Ohta, M. Oyaizu, L. Parson, R. Searle, Y. Sohrin, and T. Yama-ashi, First Submersible Investigations of the mid-ocean ridges in the Indian Ocean, *InterRidge News, 8*, 22–24, 1999.

Fujiwara, T., J. Lin, T. Matsumoto, P. B. Kelemen, B. E. Tucholke, and J. F. Casey, Crustal evolution of the Mid-Atlantic Ridge near Fifteen–Twenty Fracture Zone in the last 5 Ma, *Geochemistry, Geophysics, Geosystems, 4*, article 1024, doi:10.1029/2002GC000364, 2003.

Garrido, C. J., P. B. Kelemen, and G. Hirth, Variation of cooling rate with depth in lower crust formed at an oceanic spreading ridge: Plagioclase crystal size distributions in gabbros from the Oman ophiolite, *Geochemistry, Geophysics, Geosystems, 2*, 2000GC000136, 2001.

Gente, P., R. A. Pockalny, C. Durand, C. Deplus, M. Maia, G. Ceuleneer, C. Mével, M. Cannat, and C. Laverne, Characteristics and evolution of the segmentation of the Mid-Atlantic Ridge between 20°N and 24°N during the last 10 million years, *Earth and Planetary Science Letters, 129 (1–4)*, 55–71, 1995.

Gillis, K. M., and M. D. Roberts, Cracking at the magma-hydrothermal transition: evidence from the Troodos Ophiolite, Cyprus, *Earth and Planetary Science Letters, 169*, 227–244, 1999.

Goetze, C., The mechanism of solid state creep, *Philosophical Transactions of the Royal Society of London, Series A, 288*, 99–119, 1978.

Goetze, C., and B. Evans, Stress and temperature in the bending lithosphere as constrained by experimental rock mechanics, *Geophysical Journal of the Royal Astronomical Society, 59*, 463–478, 1979.

Goff, J. A., Quantitative characterization of abyssal hill morphology along flow lines in the Atlantic Ocean, *Journal of Geophysical Research, 97*, 9183–9202, 1992.

Grindlay, N. R., P. J. Fox, and K. C. Macdonald, Second-order ridge axis discontinuities in the South Atlantic: Morphology, structure, and evolution, *Marine Geophysical Researches, 13*, 21–49, 1991.

Grindlay, N. R., P. J. Fox, and P. R. Vogt, Morphology and tectonics of the Mid-Atlantic Ridge (25°–27°30'S) from Sea Beam and magnetic data, *Journal of Geophysical Research, 97 (B5)*, 6983–7010, 1992.

Hammond, W. C., and D. R. Toomey, Seismic velocity anisotropy and heterogeneity beneath the Mantle Electromagnetic and Tomography Experiment (MELT) region of the East Pacific Rise from the analysis of P and S body waves, *Journal of Geophysical Research, 108 (B4)*, ESE1, doi:10.1029/2002JB001789, citation no. 2176, 2003.

Hekinian, R., Rocks from the mid-oceanic ridge in the Indian ocean, *Deep-Sea Research, 15*, 195–213, 1968.

Henstock, T., A. Woods, and R. White, The accretion of oceanic-crust by episodic sill intrusion, *Journal of Geophysical Research, 98 (B3)*, 4143–4161, 1993.

Hirth, G., Laboratory constraints on the rheology of the upper mantle, in *Plastic Deformation of Minerals and Rocks*, edited by H. R. Wenk, pp. 97–120, 2002.

Hirth, G., J. Escartin, and J. Lin, The rheology of the lower oceanic crust: Implications for lithospheric deformation at mid-ocean ridges, in *Faulting and Magmatism at Mid-Ocean Ridges, Geophysical Monograph 106*, edited by W. R. Buck, P. T. Delaney, J. A. Karson, and Y. Lagabrielle, pp. 291–303, American Geophysical Union, Washington, D.C., 1998.

Hirth, G., and D. L. Kohlstedt, Experimental constraints on the dynamics of the partially molten upper mantle 2. Deformation in the dislocation creep regime, *Journal of Geophysical Research, 100 (B8)*, 15,441–15,449, 1995a.

Hirth, G., and D. L. Kohlstedt, Experimental constraints on the dynamics of the partially molten upper-mantle—Deformation in the diffusion creep regime, *Journal of Geophysical Research–Solid Earth, 100 (B2)*, 1981–2001, 1995b.

Hirth, G., and D. L. Kohlstedt, Water in the oceanic upper mantle: implications for rheology, melt extraction and the evolution of the lithosphere, *Earth and Planetary Science Letters, 144*, 93–108, 1996.

Hirth, G., and D. L. Kohlstedt, Rheology of the upper mantle and the mantle wedge: a view from the experimentalists, in *The Subduction Factory*, edited by J. M. Eiler, in press.

Hooft, E. E. E., R. S. Detrick, D. R. Toomey, J. A. Collins, and J. Lin, Crustal thickness and structure along three contrasting spreading segments of the Mid-Atlantic Ridge, 33.5°–35°N, *Journal of Geophysical Research, B, 105*, 8205–8226, 2000.

Horen, H., M. Zamora, and G. Dubuisson, Seismic waves velocities and anisotropy in serpentinized peridotites from Xigaze ophiolite: abundance of serpentine in slow spreading ridges, *Geophysical Research Letters, 23 (1)*, 9–12, 1996.

Hosford, A., J. Lin, and R. S. Detrick, Crustal evolution over the last 2 m.y. at the Mid-Atlantic Ridge OH-1 segment, 35°N, *Journal of Geophysical Research, 106 (B7)*, 13,269–13,285, 2001.

Huang, P. Y., and S. C. Solomon, Centroid depths of mid-ocean ridge earthquakes: Dependence on spreading rate, *Journal of Geophysical Research, 93*, 13445–13477, 1988.

Jaoul, O., J. Tullis, and A. Kronenberg, The effect of varying water contents on the creep behavior of Heavitree quartzite, *Journal of Geophysical Research, 89 (B6)*, 4298–4312, 1984.

Jaroslow, G. E., The geological record of oceanic crustal accretion and tectonism at slow-spreading ridges, Ph.D. thesis, MIT/WHOI 97–09, 1996.

Jaroslow, G. E., G. Hirth, and H. J. B. Dick, Abyssal peridotite mylonites: Implications for grain-size sensitive flow and strain localization in the oceanic lithosphere, *Tectonophysics, 256 (1–4)*, 17–37, 1996.

Jokat, W., O. Ritzman, M. C. Schmidt-Aursch, S. Drachev, S. Gauger, and J. E. Snow, Geophysical evidence for reduced melt production on the Arctic ultraslow Gakkel mid-ocean ridge, *Nature, 423*, 962–965, 2003.

Juteau, T., E. Berger, and M. Cannat, Serpentinized, residual mantle peridotites from the M.A.R. median valley, ODP hole 670A (21°10'N, 45°02'W, leg 109): primary mineralogy and geothermometry, in *Proceedings of the Ocean Drilling Program, Scientific Results*, edited by R. Detrick, J. Honnorez, W.B. Bryan, and T. Juteau, pp. 27–45, College Station, Texas, 1990.

Karato, S.-I., M. S. Paterson, and J. D. Fitzgerald, Rheology of synthetic olivine aggregates: Influence of grain size and water, *Journal of Geophysical Research, 91 (B8)*, 8151–8176, 1986.

Karson, J. A., Seafloor spreading on the Mid-Atlantic Ridge: implications for the structure of ophiolites and oceanic lithosphere produced in slow-spreading environments, in *Proceedings of the Symposium on Ophiolites and Oceanic Lithosphere–TROODOS 87*, edited by J. Malpas, E. M. Moores, A. Panayyiotou, and C. Xenophontos, pp. 547–555, Geological Survey Department, Ministry of Agriculture and Natural Resources, Nicosia, Cyprus, 1990.

Karson, J. A., Internal structure of oceanic lithosphere: A perspective from tectonic windows, in Faulting and Magmatism at Mid-Ocean Ridges: *Geophysical Monograph 106*, edited by W. R. Buck, P. T. Delaney, J. A. Karson, and Y. Lagabrielle, pp. 177–218, American Geophysical Union, Washington, DC, 1998.

Karson, J. A., and H. J. B. Dick, Tectonics of ridge-transform intersections at the Kane Fracture Zone, *Marine Geophysical Researches, 6*, 51–98, 1983.

Karson, J. A., E. M. Klein, S. D. Hurst, C. E. Lee, P. Rivizzigno, D. Curewitz, and A. R. Morris, Structure of uppermost fast-spread oceanic crust exposed at the Hess Deep Rift: Implications for subaxial processes at the East Pacific Rise, *Geochemistry, Geophysics and Geosystems, 3*, 10.1029/2001GC000155, 2002.

Karson, J. A., and R. M. Lawrence, Tectonic setting of serpentinite exposures on the western median valley wall of the MARK area in the vicinity of site 920, *Proceedings of the Ocean Drilling Program, Scientific Results, 153*, 5–21, 1997.

Karson, J. A., G. Thompson, S. E. Humphris, J. M. Edmond, W. B. Bryan, J. R. Brown, A. T. Winters, R. A. Pockalny, J. R. Casey, C. A. C., G. Klinkhammer, M. R. Palmer, R. J. Kinzler, and M. M. Sulanowska, Along-axis variations in seafloor spreading in the MARK area, *Nature, 328*, 681–685, 1987.

Katsumata, K., T. Sato, J. Kasahara, N. Hirata, R. Hino, N. Takahashi, M. Sekine, S. Miura, S. Koresawa, and N. Wada, Microearthquake seismicity and focal mechanisms at the Rodriguez Triple Junction in the Indian Ocean using ocean bottom seismometers, *Journal of Geophysical Research–Solid Earth, 106 (B12)*, 30689–30699, 2001.

Kelemen, P. B., and E. Aharonov, Periodic formation of magma fractures and generation of layered gabbros in the lower crust beneath oceanic spreading ridges, in *Faulting and Magmatism at Mid-Ocean Ridges—Geophysical Monograph 106*, edited by W. R. Buck, P. T. Delaney, J. A. Karson, and Y. Lagabrielle, pp. 267–289, American Geophysical Union, Washington, D.C., 1998.

Kelemen, P. B., M. Braun, and G. Hirth, Spatial distribution of melt conduits in the mantle beneath oceanic spreading centers: Observations from the Ingalls and Oman ophiolites, *Geochemistry, Geophysics, Geosystems, 1*, 1999GC000012, 2000.

Kent, G. M., S. C. Singh, A. J. Harding, M. C. Sinha, J. A. Orcutt, P. J. Barton, R. S. White, S. Bazin, R. W. Hobbs, C. H. Tong, and J. W. Pye, Evidence from three-dimensional seismic reflectivity images for enhanced melt supply beneath mid-ocean-ridge discontinuities, *Nature, 406*, 614–618, 2000.

Kirby, S. H., Rheology of the lithosphere, *Reviews of Geophysics, 21 (6)*, 1458–1487, 1983.

Kirby, S. H., and A. K. Kronenberg, Deformation of clinopyroxene: Evidence for a transition in flow mechanisms and semibrittle behaviour, *Journal of Geophysical Research, 89 (B5)*, 3177–3192, 1984.

Kleinrock, M. C., B. E. Tucholke, J. Lin, and M. A. Tivey, Fast rift propagation at a slow-spreading ridge, *Geology, 25 (7)*, 639–642, 1997.

Kohlstedt, D. J., and C. Goetze, Low-stress high-temperature creep in olivine single crystals, *Journal of Geophysical Research, 74 (14)*, 2045–2051, 1974.

Kohlstedt, D. L., Q. Bai, Z.-C. Wang, and S. Mei, Rheology of partially molten rocks, in *Physics and Chemistry of Partially Molten Rocks*, edited by A. B. Thompson, pp. 3–28, Kluwer, 2000.

Kohlstedt, D. L., B. Evans, and S. J. Mackwell, Strength of the lithosphere: Constraints imposed by laboratory experiments, *Journal of Geophysical Research, 100 (B9)*, 17,587–17,602, 1995.

Kong, L. S. L., S. C. Solomon, and G. M. Purdy, Microearthquake characteristics of a Midocean Ridge Along-Axis High, *Journal of Geophysical Research, 97 (B2)*, 1659–1685, 1992.

Kronenberg, A., and J. Tullis, Flow strengths of quartz aggregates: grain size and pressure effects due to hydrolitic weakening, *Journal of Geophysical Research, 89 (B6)*, 4281–4297, 1984.

Kuo, B. Y., and D. W. Forsyth, Gravity anomalies of the ridge-transform system in the South Atlantic between 31 and 34.5°S: Upwelling centers and variations in crustal thickness, *Marine Geophysical Researches, 10 (3–4)*, 205–232, 1988.

Kuo, B. Y., D. W. Forsyth, and E. M. Parmentier, Flexure and thickening of the lithosphere at the East Pacific Rise, *Geophysical Research Letters, 13*, 681–684, 1986.

Kurewitz, D., and J. A. Karson, Structural settings of hydrothermal outflow: Fracture permeability maintained by fault propagation and interaction, *Journal of Volcanology and Geothermal Research, 79*, 149–168, 1997.

Lagabrielle, Y., D. Bideau, M. Cannat, J. A. Karson, and C. Mevel, Ultramafic-mafic plutonic rock suites exposed along the Mid-Atlantic Ridge (10°N–30°N). Symmetrical-asymmetrical distribution and implications for seafloor spreading processes, in *Faulting and Magmatism at Mid-Ocean Ridges: Geophysical Monograph no. 106*, edited by W. R. Buck, P. T. Delaney, J. A. Karson, and Y. Lagabrielle, pp. 153–176, American Geophysical Union, Washington, D.C., 1998.

Lavier, L. L., W.R . Buck, and A. N. B. Poliakov, Self-consistent rolling-hinge model for the evolution of large-offset low-angle normal faults, *Geology, 27 (12)*, 1127–1130, 1999.

Lawson, K., R. C. Searle, J. A. Pearce, P. Browning, and P. Kempton, Detailed volcanic geology of the MARNOK area, Mid-Atlantic Ridge north of Kane transform, in *Tectonic, Magmatic, Hydrothermal and Biological Segmentation of Mid-Ocean Ridges, Geol. Soc. London, Spec. Publ. 118*, edited by C. J. MacLeod, P. A. Tyler, and C. L. Walker, pp. 61–102, Geological Society, London, 1996.

Leeds, A. K., E. Kausel, and L. Knopoff, Variations of upper mantle structure under the Pacific Ocean, *Science, 186* (141–143), 1974.

Lilwall, R. C., T. J. G. Francis, and P. I. T., A microearthquake survey at the junction of the East Pacific Rise and the Wilkes (9°S) fracture zone, *Geophysical Journal of the Royal Astronomical Society, 66*, 407–416, 1981.

Lin, J., H. J. B. Dick, and H. Schouten, Evidence for highly focussed magmatic accretion along the ultra-slow Southwest Indian Ridge, *Eos Transactions*, American Geophysical Union, 84 (46, Fall Meeting Supplement), Abstract T11B-03, 2003.

Lin, J., and E. M. Parmentier, Mechanisms of lithospheric extension at mid-ocean ridges, *Geophysical Journal, 96*, 1–22, 1989.

Lin, J., and J. Phipps Morgan, The spreading rate dependence of three-dimensional mid-ocean ridge gravity structure, *Geophysical Research Letters, 19 (1)*, 13–16, 1992.

Lin, J., G.M. Purdy, H. Schouten, J.-C. Sempéré, and C. Zervas, Evidence from gravity data for focused magmatic accretion along the Mid-Atlantic Ridge, *Nature, 344*, 627–632, 1990.

Louden, K. E., R. S. White, C. G. Potts, and D. W. Forsyth, Structure and seismotectonics of the Vema Fracture Zone, Atlantic Ocean, *Journal of the Geological Society of London, 143*, 795–805, 1986.

Macdonald, K. C., in Faulting and Magmatism at Mid-Ocean Ridges, *Geophysical Monograph 106*, edited by W. R. Buck, P. T. Delaney, J. A. Karson, and Y. Lagabrielle, American Geophysical Union, Washington, D.C., 1998.

Macdonald, K. C., P. J. Fox, S. Carbotte, M. Eisen, S. Miller, L. Perram, D. Scheirer, S. Tighe, and C. Weiland, *The East Pacific Rise and its Flanks 8–17°N: History of Segmentation, Propagation and Spreading Direction Based on SeaMARC II and Seabeam Studies*, 1991a.

Macdonald, K. C., and B. P. Luyendyk, Deep-tow studies of the structure of the Mid-Atlantic ridge crest near 37°N (FAMOUS), *Geological Society of America Bulletin, 88*, 621–636, 1977.

Macdonald, K. C., D. S. Scheirer, and S. M. Carbotte, Mid-Ocean Ridges: Discontinuities, segments and giant cracks, *Science, 253*, 986–994, 1991b.

Macdonald, K. C., J.-C. Sempere, and P. J. Fox, Reply: The debate concerning overlapping spreading centers and mid-ocean ridge processes, *Journal of Geophysical Research, 91*, 10501–10510, 1986.

Mackwell, S. J., M.E . Zimmerman, and D. L. Kohlstedt, High-temperature deformation of dry diabase with application to tectonics on Venus, *Journal of Geophysical Research, 103* (B1), 975–984, 1998.

MacLeod, C. J., J. Escartin, D. Banerji, G. J. Banks, M. Gleeson, D. H. B. Irving, R. M. Lilly, A. M. McCaig, Y. Niu, S. Allerton, and

D. K. Smith, Direct geological evidence for oceanic detachment faulting: The Mid-Atlantic Ridge, 15°45'N, *Geology, 30*, 879–882, 2002.

MacLeod, C. J., and C. E. Manning, Influence of axial segmentation on hydrothermal circulation at fast-spreading ridges: insights from Hess Deep, in *Tectonic, hydrothermal and biological segmentation at Mid-Ocean Ridges*, edited by C. J. MacLeod, P. A. Tyler, and C. L. Walker, pp. in press, Geological Society, London, 1996.

MacLeod, C. J., and G. Yaouancq, A fossil melt lens in the Oman ophiolite: Implications for magma chamber processes at fast spreading ridges, *Earth and Planetary Science Letters, 176 (3–4)*, 357–373, 2000.

Madsen, J. A., R. S. Detrick, J. C. Mutter, P. Buhl, and J. A. Orcutt, A two- and three-dimensional analysis of gravity anomalies associated with the East Pacific Rise at 9°N and 13°N, *J. Geophys. Res., 95*, 4967–4987, 1990.

Madsen, J. A., D. W. Forsyth, and R. S. Detrick, A new isostatic model for the East Pacific Rise Crest, *Journal of Geophysical Research, 89*, 9997–10015, 1984.

Magde, L. S., A. H. Barclay, D. R. Toomey, R. S. Detrick, and J. A. Collins, Crustal magma plumbing within a segment of the Mid-Atlantic Ridge, 35°N, *Earth and Planetary Science Letters, 175*, 55–67, 2000.

Magde, L. S., and D. W. Sparks, Three-dimensional mantle upwelling, melt generation, and melt migration beneath segment slow spreading ridges, *Journal of Geophysical Research, B, 102 (9)*, 20,571–20,583, 1997.

Magde, L. S., D. W. Sparks, and R. S. Detrick, The relationship between buoyant mantle flow, melt migration, and gravity bull's eyes at the Mid-Atlantic Ridge between 33°N and 35°N, *Earth and Planetary Science Letters, 148 (1–2)*, 59–, 1997.

Mainprice, D. H., and M. S. Paterson, Experimental studies of the role of water in the plasticity of quartzites, *Journal of Geophysical Research, 89 (B6)*, 4257–4269, 1984.

Manning, C. E., and C. J. MacLeod, Fracture-controlled metamorphism of Hess Deep gabbros, site 894: Constraints on the roots of mid-ocean-ridge hydrothermal systems at fast-spreading centers, in *Proc. of the Ocan Drilling Program Sci. Res.*, edited by C. Mével, K. M. Gillis, J. F. Allan, and P. S. Meyer, pp. 189–212, Ocean Drilling Program, College Station, TX, 1996.

Manning, C. E., C. J. MacLeod, and P. Weston, Lower-crustal cracking front at fast-spreading ridges: Evidence from the East Pacific Rise and the Oman ophiolite, in *Ophiolites and the oceanic crust: New insights from field studies and the Ocean Drilling Program*, edited by Y. Dilek, E. M. Moores, D. Elthon, and A. Nicolas, pp. 262–272, Geological Society of America, Boulder, Colorado, 2000.

Mares, V. M., and A. K. Kronenberg, Experimental deformation of muscovite, *Journal of Structural Geology, 15 (9/10)*, 1061–1075, 1993.

McAllister, E., and J. Cann, Initiation and evolution of boundary wall faults along the Mid-Atlantic Ridge, 25–29°N, in *Tectonic, Magmatic, Hydrothermal and Biological Segmentation of Mid-Ocean Ridges*, edited by C. J. MacLeod, P. A. Tyler, and C. L. Walker, pp. 29–48, Geological Society Special Publication 118, London, 1996.

McAllister, E., J. Cann, and S. Spencer, The evolution of crustal deformation in an oceanic extensional environment, *Journal of Structural Geology, 17 (2)*, 183–199, 1995.

McKenzie, D. P., and C. O. Bowin, The relationship between bathymetry and gravity in the Atlantic Ocean, *Journal of Geophysical Research, 81 (1903–1915)*, 1976.

McNutt, M. K., Lithospheric Flexure and Thermal Anomalies, *Journal of Geophysical Research, 89 (B13)*, 11,180–11,194, 1984.

McNutt, M. K., and H. W. Menard, Constraints on Yield Strength in the Oceanic Lithosphere Derived from Observations of Flexure, *Geophysical Journal of the Royal Astronomical Society, 71 (2)*, 363–394, 1982.

Mei, S., and D. L. Kohlstedt, Influence of water on plastic deformation of olivine aggregates. 1. Diffusion creep regime, *Journal of Geophysical Research, 105 (B9)*, 21457–21469, 2000a.

Mei, S., and D. L. Kohlstedt, Influence of water on plastic deformation of olivine aggregates. 2. Dislocation creep regime, *Journal of Geophysical Research, 105 (B9)*, 21471–21481, 2000b.

Melson, W .G., V. T. Bowen, T. H. Van Andel, and R. Siever, Greenstones from the central valley of the Mid-Atlantic Ridge, *Nature, 209*, 604–605, 1966.

MELT, Imaging the deep seismic structure beneath a mid-ocean ridge: The MELT experiment, *Science, 280*, 1215–18, 1998.

Mével, C., M. Cannat, P. Gente, E. Marion, J.-M. Auzende, and J. Karson, Emplacement of deep crustal and mantle rocks on the west median valley wall of the MARK area (MAR, 23°N), *Tectonophysics, 190 (192–210)*, 1991.

Mével, C., and C. Stamoudi, Hydrothermal alteration of the upper-mantle section at Hess Deep, Proc. Ocean Drill. Prog., 147, 293–309, 1996.

Michael, P. J., C. H. Langmuir, H. J. B. Dick, J. E. Snow, S. L. Goldstein, D. W. Graham, K. Lehnert, G. J. Kurras, W. Jokat, R. Mühe, and H. N. Edmonds, Magmatic and amagmatic seafloor generation at the utraslow-spreading Gakkel ridge, Arctic Ocean, *Nature, 423 (26)*, 956–962, 2003.

Minshull, T., and P. Charvis, Ocean island densities and models of lithospheric flexure, *Geophysical Journal International, 145*, 731–739, 2001.

Mitchell, N. C., J. Escartin, and S. Allerton, Detachment faults at mid-ocean ridges, *EOS, Transactions of the American Geophysical Union, 79*, 127, 1998.

Montesi, L. G. J., and G. Hirth, Grain size evolution and the rheology of ductile shear zones: from laboratory experiments to post-seismic creep, *Earth and Planetary Science Letters, 211*, 97–110, 2003.

Neumann, G. A., and D. W. Forsyth The paradox of the axial profile: Isostatic compensation along the axis of the Mid-Atlantic Ridge?, *Journal of Geophysical Research, 98 (B10)*, 17,891–17,910, 1993.

Neumann, G. A., D. W. Forsyth, and D. Sandwell, Comparison of marine gravity from shipboard and high-density satellite altimetry along the Mid-Atlantic Ridge, 30.5°–35.5°S, *Geophysical Research Letters, 20* (15), 1639–1642, 1993.

Nishimura, C. E., and D. W. Forsyth, The anisotropic structure of the upper mantle in the Pacific, *Geophysical Journal, 96*, 203–229, 1989.

Ohara, Y., T. Yoshida, Y. Kato, and S. Kasuga, Giant megamullion in the Parece Vela Backarc Basin, *Marine Geophysical Researches, 22 (1)*, 47–61, 2001.

Parsons, B., and D. P. McKenzie, Mantle convection and the thermal structure of the plates, *Journal of Geophysical Research, 83*, 4,485–4,496, 1978.

Parsons, B., and J. G. Sclater, An analysis of the variation of ocean floor bathymetry and heat flow with age, *Journal of Geophysical Research, 82*, 803–827, 1977.

Phipps Morgan, J., and Y. J. Chen, Dependence of ridge-axis morphology on magma supply and spreading rate, *Nature, 364 (19 August)*, 706–708, 1993.

Phipps Morgan, J., and E. M. Parmentier, Causes and rate-limiting mechanism of ridge propagations: A fracture mechanics model, *Journal of Geophysical Research, 90*, 8603–8612, 1985.

Phipps Morgan, J., and D. T. Sandwell, Systematics of ridge propagation south of 30°S, *Earth and Planetary Science Letters, 121*, 245–258, 1994.

Phipps-Morgan, J., and Y. J. Chen, The genesis of oceanic crust: Magma injection, hydrothermal circulation, and crustal flow, *Journal of Geophysical Research, 98 (B4)*, 6283–6297, 1993.

Poliakov, A. N. B., and W. R. Buck, Mechanics of stretching elastic-plastic-viscous layers: Applications to slow-spreading mid-ocean ridges, in *Faulting and Magmatism at Mid-Ocean Ridges*, edited by W. R. Buck, P. T. Delaney, J. A. Karson, and Y. Lagabrielle, pp. 305–323, American Geophysical Union, Washington, D.C., 1998.

Pollard, D. D., and A. Aydin, Propagation and linkage of oceanic ridge segments *Journal of Geophysical Research, 89*, 10017–10028, 1984.

Prinz, M., K. Keil, A. Gree, A.M. Reid, E. Bonatti, and J. Honnorez, Ultramafic and mafic dredge samples from the equatorial mid-Atlantic ridge and fracture zones, *Journal of Geophysical Research, 81 (23)*, 4087–4103, 1976.

Purdy, G. M., J.-C. Sempere, H. Schouten, D. L. DuBois, and R. Goldsmith, Bathymetry of the Mid-Atlantic Ridge, 24°–31°N: A map series, *Marine Geophysical Researches, 12*, 247–252, 1990.

Quick, J. E., and R. P. Delinger, Ductile deformation and the origin of layered gabbro in ophiolites, *Journal of Geophysical Research, 98 (B8)*, 14015–14027, 1993.

Rabain, A., M. Cannat, J. Escartin, G. Pouliquen, C. Deplus, and C. Rommevaux-Jestin, Focused volcanism and growth of a slow spreading segment (Mid-Atlantic Ridge, 35°N), *Earth and Planetary Science Letters, 185 (1–2)*, 211–224, 2001.

Ranero, C. R., and T. J. Reston, Detachment faulting at ocean core complexes, *Geology, 27 (11)*, 983–986, 1999.

Reinen, L. A., J. D. Weeks, and T. E. Tullis, The frictional behavior of lizardite and antigorite serpentinites: experiments, constitutive models, and implications for natural faults, *Pure and Applied Geophysics, 143 (1/2/3)*, 318–358, 1994.

Renner, J., B. Evans, and G. Hirth, On the rheologically critical melt fraction, *Earth and Planetary Science letters, 181*, 585–594, 2000.

Reston, T. J., W. Weinrebe, I. Grevemeyer, E. R. Flueh, N. C. Mitchell, L. Kirstein, C. Kopp, and H. Kopp, A rifted inside corner massif on the Mid-Atlantic Ridge at 5°S, *Earth and Planetary Science Letters, 200 (3–4)*, 255–269, 2002.

Rommevaux-Justin, C., C. Deplus, and P. Patriat, Mantle Bouguer anomaly along an ultra slow-spreading ridge: Implications for accretionary processes and comparison with results from central Mid-Atlantic Ridge, *Marine Geophysical Researches, 19 (6)*, 481–503, 1997.

Rutter, E. H., and K. H. Brodie, The role of grain size reduction in the rheological stratification of the lithosphere, *Geologische Rundschau, 77 (1)*, 295–308, 1988.

Rybacki, E., and G. Dresen, Dislocation and diffusion creep of synthetic anorthite aggregates, *Journal of Geophysical Research, 105 (B11)*, 26017–26036, 2000.

Scholz, C. H., The brittle-plastic transition and the depth of seismic faulting, *Geologische Rundschau, 77 (1)*, 319–328, 1988.

Schouten, H., K. D. Klitgord, and J. A. Whitehead, Segmentation of mid-ocean ridges, *Nature, 317*, 225–229, 1985.

Searle, R. C., GLORIA survey of the East Pacific Rise near 3.5°S: tectonic and volcanic characteristics of a fast-spreading mid-ocean rise, *Tectonophysics, 101*, 319–344, 1984.

Searle, R. C., The volcano-tectonic setting of oceanic lithosphere generation, in *Ophiolites and their modern oceanic analogues*, edited by L. M. Parson, B. J. Murton, and P. Browning, pp. 65–80, Geological Society of London, 1992.

Searle, R. C., M. Cannat, K. Fujioka, C. Mevel, H. Fujimoto, A. Bralee, and L. Parson, The FUJI Dome: A large detachment fault near 64°E on the very slow-spreading southwest Indian Ridge, *Geochemistry, Geophysics, Geosystems, 4 (8)*, 9105, doi:10.1029/2003GC000519, 2003.

Searle, R. C., P. A. Cowie, S. Allerton, J. Escartin, P. Meredith, N. Mitchell, J. R. Cann, D. Blackman, E. McAllister, D. Smith, B. E. Tucholke, and J. Lin, Bending and Breaking the Ocean Floor, *BRIDGE Newsletter, 17*, 18–20, 1999.

Searle, R. C., P. A. Cowie, N. C. Mitchell, S. Allerton, C. J. MacLeod, J. Escartin, S. M. Russell, P. A. Slootweg, and T. Tanaka, Fault structure and detailed evolution of a slow spreading ridge segment: the Mid-Atlantic Ridge at 29°N, *Earth and Planetary Science Letters, 154 (1–4)*, 167–183, 1998a.

Searle, R. C., J. A. Keeton, S. M. Lee, R. Owens, R. Mecklenburgh, B. Parsons, and R. S. White, The Reykjanes Ridge: structure and tectonics of a hot-spot influenced, slow-spreading ridge, from multibeam bathymetric, gravity and magnetic investigations, *Earth and Planetary Science Letters, 160*, 463–478, 1998b.

Sempéré, J.-C., P. Blondel, A. Briais, T. Fujiwara, L. Geli, N. Isezaki, J. E. Pariso, L. Parson, P. Patriat, and C. Rommevaux, The Mid-Atlantic Ridge between 29°N and 31°30'N in the last 10 Ma, *Earth and Planetary Science Letters, 130 (1–4)*, 45–55, 1995.

Severinghaus, J. P., and K. C. Macdonald, High inside corners at ridge-transform intersections, *Marine Geophysical Researches, 9*, 353–367, 1988.

Shaw, P., Ridge segmentation, faulting and crustal thickness in the Atlantic Ocean, *Nature, 358*, 490–493, 1992.

Shaw, P. R., and J. Lin, Causes and consequences of variations in faulting style at the Mid-Atlantic Ridge, *Journal of Geophysical Research, 98 (B12)*, 21,839–21,851, 1993.

Shaw, W. J., and J. Lin, Models of ocean ridge lithospheric deformation: Dependence on crustal thickness, spreading rate, and segmentation, *Journal of Geophysical Research, 101 (B8)*,

17,977–17,993, 1996.

Shea, W. T., and A. K. Kronenberg, Rheology and deformation mechanisms of an isotropic mica schist, *Journal of Geophysical Research, 97 (B11)*, 15201–15237, 1992.

Shea, W. T., and A. K. Kronenberg, Strength and anisotropy of foliated rocks with varied mica contents, *Journal of Structural Geology, 15 (9/10)*, 1097–1121, 1993.

Shelton, G. L., and J. Tullis, Experimental flow laws for crustal rocks, *EOS, Transactions of the American Geophysical Union, 62*, 396, 1981.

Shipboard Scientific Party, Site 504, in *Proceedings of the Ocean Drilling Program, Initial Reports, vol 148*, edited by J. Alt, H. Kinoshita, and L. Stokking, pp. 27–121, Ocean Drilling Program, College Station, TX, 1993.

Shipboard Scientific Party, Ocean Drilling Program, Site 504B, Ocean Drilling Program, College Station, TX, 1995.

Shipboard Scientific Party, *Leg 209 Preliminary Report: Drilling Mantle Peridotite along the Mid-Atlantic Ridge from 14° to 16°N*, 6 May 2003–6 July 2003, pp. 100 + figures, Ocean Drilling Program, Texas A&M University, College Station TX 77845–9547, 2003.

Sinha, M., and R. L. Evans, Geophysical constraints on the thermal regime of the oceanic crust, in *InterRidge Theoretical Institute: Thermal Regime of Ocean Ridges and Dynamics of Hydrothermal Circulation*, edited by C. German, J. Lin, R. Tribuzio, A. Fisher, M. Cannat, and A. Adamczewska, pp. 18, InterRidge, Pavia, 2003.

Solomon, S. C., P. Y. Huang, and L. Meinke, The seismic moment budget of slowly spreading ridges, *Nature, 298*, 149–151, 1988.

Sparks, D. W., and E. M. Parmentier, The structure of three-dimensional convection beneath oceanic spreading centres, *Geophysical Journal International, 112*, 81–91, 1993.

Stein, C. A., and S. Stein, A model for the global variation in oceanic depth and heat flow with lithospheric age, *Nature, 359*, 123–129, 1992.

Tapponnier, P., and J. Francheteau, Necking of the lithosphere and the mechanics of slowly accreting plate boundaries, *Journal of Geophysical Research, 83*, 3955–3970, 1978.

Taylor, B., K. Crook, and J. Sinton, Extensional transform zones and oblique spreading centers, *Journal of Geophysical Research, B, 99 (10)*, 19,707–19,718, 1994.

Thibaud, R., P. Gente, and M. Marcia, A systematic analysis of the Mid-Atlantic Ridge morphology and gravity between 15°N and 40°N: Constraints of the thermal structure, *Journal of Geophysical Research, 103 (B10)*, 24201–24221, 1998.

Tolstoy, M., A. J. Harding, and J. A. Orcutt, Crustal thickness on the Mid-Atlantic Ridge—Bull's-eye gravity-anomalies and focused accretion, *Science, 262 (5134)*, 726–729, 1993.

Tolstoy, M., F. L. Vernon, J. A. Orcutt, and F. K. Wyatt, Breathing of the seafloor: Tidal correlations of seismicity at Axial volcano, *Geology, 30 (6)*, 503–506, 2002.

Toomey, D. R., S. C. Solomon, and G. M. Purdy, Microearthquakes beneath the median valley of the Mid-Atlantic Ridge near 23°N: Tomography and tectonics, *Journal of Geophysical Research, 93 (B8)*, 9093–9112, 1988.

Toomey, D. R., S. C. Solomon, G. M. Purdy, and M. H. Murray, Microearthquakes beneath the median valley of the Mid-Atlantic Ridge near 23°N: Hypocenters and focal mechanisms, *Journal of*

Geophysical Research, 90, 5443–5458, 1985.

Tucholke, B., J. Lin, and M. Kleinrock, Megamullions and mullion structure defining oceanic metamorphic core complexes on the Mid-Atlantic Ridge, *Journal of Geophysical Research, B, 103 (5)*, 9857–9866, 1998.

Tucholke, B., J. Lin, M. Kleinrock, M. Tivey, T. Reed, J. Goff, and G. Jaroslow, Segmentation and crustal structure of the western Mid-Atlantic Ridge flank, 25°25'–27°10'N and 0–29 m.y., *Journal of Geophysical Research, 102 (B5)*, 10,203–10,223, 1997.

Tucholke, B. E., W. R. Buck, L. Lavier, and J. Lin, Investigation of megamullion formation in relation to magma supply, *Geophysical Research Abstracts, 5*, EAE03-A-07925, 2003.

Tucholke, B. E., K. Fujioka, T. Ishihara, G. Hirth, and M. Kinoshita, Submersible study of an oceanic megamullion in the central North Atlantic, *Journal of Geophysical Research, B, 106*, 16,145–16,161, 2001.

Tucholke, B. E., and J. Lin, A geological model for the structure of ridge segments in slow spreading ocean crust, *Journal of Geophysical Research, 99*, 11,937–11,958, 1994.

Tullis, J., and R. Yund, Hydrolitic weakening of quartz aggregates: the effects of water and pressure on recovery, *Geophysical Research Letters, 16 (11)*, 1343–1346, 1989.

Tullis, J., and R. Yund, Diffusion creep in felspar aggregates: experimental evidence, *Journal of Structural Geology, 13 (9)*, 987–1000, 1991.

Tullis, J., and R. Yund, The brittle-ductile transition in felspar aggregates: an experimental study, in *Fault mechanics and transport properties of rock*, edited by B. Evans, and T. F. Wong, pp. 89–117, Academic Press, 1992.

Tullis, J., and R. A. Yund, Experimental deformation of dry Westerly granite, *Journal of Geophysical Research, 82 (36)*, 5705–5718, 1977.

Van der Wal, D., Deformation process in mantle peridotites with emphasis on the Ronda peridotite of SW Spain, *Geologica Ultraiectina, 102*, 1–180, 1993.

Vera, E. E., J. C. Mutter, P. Buhl, J. A. Orcutt, A. J. Harding, M. E. Kappus, R. S. Detrick, and T. M. Brocher, The Structure of 0- to 0.2-m.y.-old oceanic crust at 9°N on the East Pacific Rise from expanded spread profiles, *Journal of Geophysical Research, 95*, 15529–15556, 1990.

Wang, X., and J. R. Cochran, Gravity anomalies, isostasy, and mantle flow at the East Pacific Rise crest, *Journal of Geophysical Research, 98*, 19,505–19,532, 1993.

Watts, A., and S. Zhong, Observations of flexure and the rheology of oceanic lithosphere, *Geophysical Journal International, 142*, 855–875, 2000.

Watts, A. B., An analysis of isostasy in the worlds oceans: 1. Hawaiian-Emperor seamount chain, *Journal of Geophysical Research, 83 (5989–6004)*, 1978.

Weissel, J. K., and D. E. Hayes, The Australian–Antarctic discordance: new results and implications, *Journal of Geophysical Research, 79*, 2579–2587, 1974.

Weissel, J. K., and G. D. Karner, Flexural uplift of rift flanks due to mechanical unloading of the lithosphere during extension, *Journal of Geophysical Research, 94 (B10)*, 13,919–13,950, 1989.

West, B. P., J. Lin, and D. M. Christie, Forces driving ridge propagation, *Journal of Geophysical Research–Solid Earth, 104 (B10)*, 22845–22858, 1999.

Whitehead, J. A., H. J. B. Dick, and H. Schouten, A mechanism for magmatic accretion under spreading centres, *Nature, 312*, 146–148, 1984.

Wiens, D. A., and S. Stein, Age dependence of oceanic intraplate seismicity and implications for lithospheric evolution, *Journal of Geophysical Research, 88*, 6455–6468, 1983.

Wilcock, W. S. D., S. D. Archer, and G. M. Purdy, Microearthquakes on the Endeavour segment of the Juan de Fuca Ridge, *Journal of Geophysical Research, 107 (B12)*, EPM4, doi:10.1029/2001JB000505, 2002.

Wilcock, W. S. D., Purdy, and Solomon, Microearthquake evidence for extension across the Kane transform fault, *J. of Geophysical Research, B, 95 (10)*, 15,439–15,462, 1990.

Wilson, D. S., D. A. Clague, N. H. Sleep, and J. L. Morton, Implications of magma convection for the size and temperature of magma chambers at fast spreading ridges, *Journal of Geophysical Research, 93* (11,974–11,984), 1988.

Wolfe, C. J., G. M. Purdy, D. R. Toomey, and S. C. Solomon, Microearthquake characteristics and crustal velocity structure at 29°N on the Mid-Atlantic Ridge: The architecture of a slow-spreading segment, *Journal of Geophysical Research, 100 (B12)*, 24,449–24,472, 1995.

Yoshinobu, A. S., and G. Hirth, Microstructural and experimental constraints on the rheology of partially molten gabbro beneath oceanic spreading centers, *Journal of Structural Geology, 24 (6–7)*, 1101–1107, 2002.

R. C. Searle, Department of Earth Sciences, University of Durham, DH1 3LE, UK. (r.c.searle@durham.ac.uk)

J. Escartín, Laboratoire de Géosciences Marines (CNRS UMR7097), Institut de Physique du Globe de Paris, France. (escartin@ipgp.jussieu.fr)

Modeling the Thermal State of the Oceanic Crust

Yongshun John Chen[1]

Computational Geodynamics Laboratory, Department of Geophysics, Peking University, Beijing, China

New ocean crust is created every year along the mid-ocean ridge system, which is one of the most active plate boundaries on the Earth's surface. Studying the magmatic and tectonic processes of the construction of new oceanic crust at mid-ocean ridges is important for understanding the structure of oceanic crust and the evolution of the oceanic lithosphere. This review focuses on modeling the thermal state of the oceanic crust and its links to various seafloor observations such as ridge axis topography, gravity, and seismic crustal structure. Thermal modeling is important because the crustal thermal structure of a spreading ridge not only controls the rheology of the oceanic lithosphere and the ridge topography but also determines the style of oceanic crustal genesis, by a long-lived magma lens or by episodic magma intrusions.

1. INTRODUCTION

The 65,000-km-long mid-ocean ridge system is one of the most active plate boundaries on the Earth. Along this plate boundary, a chain of active volcanoes erupt more frequently then the volcanic eruptions on the continents. This largest and most volcanically active chain of mountains on the Earth's surface has continually produced the oceanic crust that covers about two thirds of the Earth's surface.

Plate divergence at mid-ocean ridges has induced mantle upwelling that resulted in partial melting of mantle rocks due to the "decompression melting" process. Melts separate from the upwelling solid residuum, and rise from tens of kilometers depths beneath the ridge. Some of the melt ascends all the way to the seafloor, producing extensive volcanic eruptions beneath the ocean surface and building the extrusive layer (basalt) of the oceanic crust (called layer 2). Most of the melt adds to the edges of the diverging plates and forms the intru-

sive layer (gabbro) of the oceanic crust (called layer 3). As the oceanic crust moves away from mid-ocean ridges, sediments gradually build up on top of the newly-formed crust, forming the layer 1 of oceanic crust.

The magmatic and tectonic processes associated with the construction of new oceanic crust at mid-ocean ridges have a profound impact on crustal structure and the evolution of oceanic lithosphere [e.g., *Lin*, 1992; *Chen*, 1996a], and therefore, they have been the subject of intense studies for the past few decades through international efforts such as the Inter-Ridge programs. In this review I focus on one important aspect of ocean ridge studies; modeling the thermal state of the oceanic crust, and its links to other processes at mid-ocean ridges such as dynamic rifting and the resulting topography, the oceanic crustal accretion, and the seismic structure of the oceanic crust. Understanding ridge thermal structure is important because it governs both the rheology of the oceanic lithosphere and the style of oceanic crustal accretion (that is, the interplay between tectonic and magmatic rifting processes). Almost all the seafloor observables such as the ridge-axis bathymetry, gravity, magnetics, seismic observations, rock composition, and hydrothermal vents are closely related to these crustal processes.

This paper was evolved from a presentation at the 1st Inter-Ridge Theoretical Institute at Pavia, Italy, 16–20 September 2002. The goal of this review article is to provide the readers

[1]was at College of Oceanic and Atmospheric Sciences, Oregon State University, Corvallis, Oregon

Mid-Ocean Ridges: Hydrothermal Interactions Between the Lithosphere and Oceans
Geophysical Monograph Series 148
Copyright 2004 by the American Geophysical Union
10.1029/148GM04

a brief summary of previous works on this topic and also to present an overview of current and ongoing thermal modeling efforts.

2. WHY STUDYING THE THERMAL STRUCTURE OF OCEANIC CRUST?

Numerous thermal models have been proposed for mid-ocean ridges since the recognition of the mid-ocean ridges as the active diverging plate boundary in late 60's [e.g., *Kusznir and Bott*, 1976; *Sleep*, 1975; 1978; *Kusznir*, 1980; *Reid and Jackson*, 1981; *Morton and Sleep*, 1985; *Phipps Morgan et al.*, 1987; *Phipps Morgan and Forsyth*, 1988; *Wilson et al.*, 1988; *Lin and Parmentier*, 1989; *Chen and Morgan*, 1990a,b; *Henstock et al.*, 1993; *Phipps Morgan and Chen*, 1993a; 1993b; *Quick and Denlinger*, 1993; *Eberle and Forsyth*, 1998a; 1998b; *Chen*, 2000; *Buck*, 2001; *Shah and Buck*, 2001]. It is now known that crustal thermal structure of a mid-ocean ridge, which is controlled mainly by the magma supply rate from the upwelling mantle, is important because (a) it controls the rheology of the oceanic lithosphere and thus the dynamic rifting process at the ridge and (b) it controls the presence or absence of a crustal magma chamber at a spreading ridge and the depth of such a magma chamber. In other words, the crustal thermal structure determines both ridge morphology and the style of oceanic crustal genesis by a long-lived magma lens or by episodic magma intrusion events [e.g., *Phipps Morgan and Chen*, 1993a; *Chen*, 2000].

2.1. Control on Lithospheric Rheology

A typical yield strength distribution with depth, often called the yield strength envelope, of a young oceanic lithosphere is shown in Figure 1. At shallow depths the yield strength is controlled by the frictional resistance along preexisting faults, which increases linearly with depth and is also known as the Byerlee's law [1978]. On the other hand, laboratory studies have shown that at high temperatures deeper into the earth's interior, rocks release the applied stress through creeping flow over a relatively long period of time [*Goetze*, 1978; *Brace and Kolstedt*, 1980; *Kirby*, 1983; *Kirby and Kronenberg*, 1987]. Therefore, the yield strength at depths is controlled by the power-law rheology for a creeping flow [e.g., *Brace and Kolstedt*, 1980]. While the frictional resistance along preexisting faults is independent of the rock type, crustal rocks usually have smaller rheological strength at high temperatures than mantle rocks. This would result in a weaker lower crust where temperature is relatively high. In summary, the total strength of an oceanic lithosphere, which is the depth integrated yield strength, depends on its thermal structure, crustal thickness, and the type of crustal rocks (Figure 1). Another factor that could

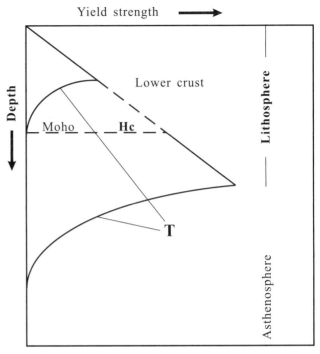

Oceanic lithosphere

Figure 1. Yield strength distribution with depth of a young oceanic lithosphere. Hc represents the thickness of the oceanic crust, which is about 6 km on average [*Chen*, 1992].

change the total strength significantly is the presence of water, that is, the wet versus dry rock rheology [*Hirth et al.,* 1998].

The rheological strength of the oceanic lithosphere at ridges depends strongly on its thermal structure. Oceanic crust is known to be simpler than the continental crust: it generally consists of the extrusive unit (the pillow basalts and the sheeted dikes) and the intrusive unit (the gabbros). It is believed that most of the oceanic crust is formed within the narrow neovolcanic zone at the ridge axis. Seismic observations have shown that the average oceanic crustal thickness is about the same for all the oceans independent of the spreading rate [*Chen*, 1992; *White et al.*, 1992], except at the recently surveyed ultra-slow spreading ridges where mantle peridotites are often directly exposed on the seafloor [*Dick et al.*, 2003]. Given these similarities, a variable that is likely to change the rheological strength of young oceanic lithosphere from one ocean to another is its thermal structure, particularly the thermal structure of the oceanic crust (Figure 1).

Previous dynamic modeling studies have shown that the rheological strength of the oceanic lithosphere dictates the style of rifting at mid-ocean ridges and therefore controls the ridge axis morphology [e.g., *Phipps Morgan et al.*, 197; *Lin and Parmentier*, 1989; *Chen and Morgan*, 1990a; 1990b; *Phipps Morgan and Chen*, 1993a; 1993b; *Shaw and Lin*,

1996]. Thus thermal modeling is an important constituent in any dynamic models of mid-ocean ridge processes because a ridge's thermal structure plays an important role in controlling the ridge axis morphology and other seafloor observations, which are the products of different styles of rifting at ridges.

For example, the contrasting difference in ridge axis morphology between a rift valley along the slow-spreading Mid-Atlantic Ridge (MAR) [e.g., *Macdonald*, 1986] and an axial high along the fast-spreading East Pacific Rise (EPR) [e.g., *Scheirer et al.*, 1996] has been attributed mainly to the differences in crustal thermal structure of the young oceanic lithosphere [*Chen and Morgan*, 1990a; 1990b]. The rift valley at the MAR was created by rifting a strong oceanic lithosphere [*Tapponnier and Francheteau*, 1978; *Phipps Morgan et al.*, 1987; *Lin and Parmentier*, 1989], which is resulted from a cold thermal structure at the slow-spreading ridge. The cold thermal structure is the combination of less frequent magma intrusions and vigorous hydrothermal cooling at a slow-spreading ridge. While heat is also brought up to the seafloor by vigorous hydrothermal cooling at the fast spreading EPR, the frequent magma supply (that keeps a long-lived magma lens) has created a rather hot lower crust, which in turns weakens the lithosphere. The weaker lower crust at a fast ridge effectively decouples the upper crust from the strong upper mantle lithosphere (Figure 2) and therefore, prevents the development of a rift valley at the ridge axis. An axial high at the fast-spreading EPR has been interpreted as either by "buoyant mantle conduit" model as being supported by a low-density conduit with small percent of melt present, which extends to a great depth in the mantle [*Madsen et al.*, 1984; 1990; *Wang and Cochran*, 1993; *Magde et al.*, 1995], or by "lower crustal drag" model that depends on viscous shear stresses generated by plate spreading at the axis [*Eberle and Forsyth*, 1998a; 1998b], or by flexural support model [*Buck*, 2001; *Shah and Buck*, 2001].

An earlier dynamic model [*Chen and Morgan*, 1990a; 1990b] examined the transition in the observed ridge axis topography, from rift valley to no rift valley, as a function of several variables including spreading rate, crustal thickness, and different thermal structures. Model calculations showed that the type of axial topography is controlled not only by the strong temperature dependence of the power law rheology of the oceanic lithosphere but also by the different rheological behaviors of the oceanic crust as compared to the underlying mantle. The success of this model in explaining the transition from rift-valley to no-rift-valley ridge axis topography confirmed the important role of the oceanic crust in shaping both the dynamic flow beneath a mid-ocean ridge and ridge axial topography [*Chen and Morgan*, 1990b]. It has been concluded that both the spreading rate difference and the crustal thickness variation are important factors in controlling ridge

axis dynamics [*Chen and Morgan*, 1990b; *Neumann and Forsyth*, 1993; *Shaw and Lin*, 1996].

2.2. Control on Magma Lens Depth

Another important aspect of the thermal structure of the oceanic crust is that it controls both the presence/absence of a crustal magma chamber at a spreading ridge and the depth of such a magma chamber. At ridges away from the influence

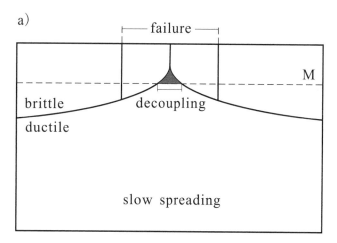

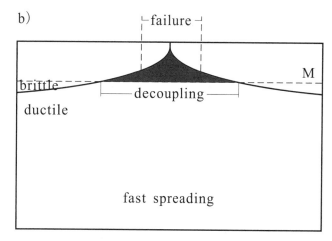

Figure 2. A conceptual model of a mid-oceanic ridge for slow and fast spreading rates, respectively [Chen and Morgan, 1990a]. (a) At a slow spreading ridge, the coupling between the viscous flow field and the strong lithospheric plate will cause a failure zone; an axial rift valley would be formed due to stretching/necking of the cold lithosphere. (b) At a fast spreading ridge, however, the hot lower crust beneath the ridge axis (dark shaded area) is rheologically weak and can effectively decouple the thin lithosphere of the upper crust (only 1–2 km thick) from underlying mantle flows. Thus the size of the hot lower oceanic crust with low ductile strength determines the transition from a well-developed rift valley at slow ridges to no-rift valley at fast ridges. (After *Chen and Morgan*, 1990a).

of a nearby hotspot, seismic studies have shown that the depth to the top of a magma chamber depends strongly on spreading rate [*Purdy et al.*, 1992], similar to the well-known strong spreading-rate dependence of the ridge axis morphology (a rift valley at slow ridges versus an axial high at fast ridges) [e.g., *Macdonald*, 1986; *Small and Sandwell*, 1989; *Malinverno*, 1991; *Small*, 1994; 1998].

Since both ridge axis morphology and the presence/absence of a magma chamber are correlated with spreading rate, it is reasonable to expect that ridge axis morphology is correlated with the presence/absence of a magma lens [*Phipps Morgan and Chen*, 1993b]. There seems indeed exist a good correlation between the presence of a magma lens with an axial high at the ridge axis and the presence of a rift valley and the absence of such a magma lens beneath the rift valley [*Phipps Morgan and Chen*, 1993b].

Model prediction (Figure 3) that assume the depth of the magma lens is controlled by crustal thermal structure match seismic observations of the magma chamber depths quite well [*Phipps Morgan and Chen*, 1993a; *Chen*, 2000]. These models, the details of which are discussed in a later section, show a relatively hot oceanic crust, usually created at a fast-spreading ridge, which allows the presence of a long-lived magma lens. On the other hand, a cold oceanic crust, often created at slow spreading ridges, cannot host a long-lived magma chamber and instead, the ocean crust is formed by episodic magma intrusion events. Therefore, the crustal thermal structure determines the style of the crustal accretion at a spreading ridge [e.g., *Chen*, 2000].

3. RIDGE THERMAL MODELS

3.1. Observations at Ocean Ridges and Ophiolites on Land

Ophiolite studies have provided us with detailed information of the oceanic crustal structure and reasonable interpretation of the seismically determined layering structure of the oceanic crust. For example, the well-studied Oman ophiolite, which is believed to be created at a fast spreading ridge [*Nicolas*, 1989], has been mapped in the field in great details and information obtained there has significantly influenced the ridge studies for the past two decades.

Crustal accretion models developed from early ophiolite studies required a crustal-size magma chamber at a mid-ocean ridge [e.g., *Cann*, 1974; *Smewing*, 1981]. However, this view has changed significantly because of seismic observations at mid-ocean ridges since 1980s. Multi-channel seismic reflection studies (the technology became available in early 80s) on the fast-spreading East Pacific Rise have found a rather small magma body about 1–2 km beneath the ridge crest, which is narrow (~1 km wide) and thin (few hundreds

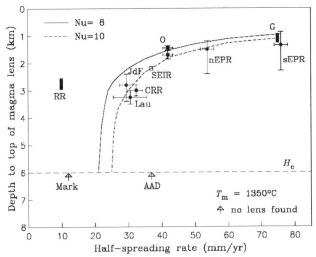

Figure 3. The calculated depth to top of magma lens (defined as depth of the 1200°C isotherm at the ridge axis) for a reference mantle temperature $T_m = 1350$°C as a function of spreading rate, together with seismic observations [*Chen*, 2000]. The two curves show the results of a suite of numerical experiments with Nu = 8 (solid line) and 10 (dotted line), respectively, while all other parameters are held constant. Data sources: G, East Pacific Rise south of the Garret transform [*Tolstoy et al.*, 1997]; O - the EPR north of Orozco transform [*Carbotte et al.*, 1998]; SEIR - the Southeast Indian Ridge [*Tolstoy et al.*, 1995]; CRR - the Costa Rica Rift [*Mutter*, 1995]; RR - the Reykjanes Ridge at 57°43′N [*Sinha et al.*, 1997]; AAD - the Australian-Antarctic Discordance [*Tolstoy et al.*, 1995]; MARK, Mid-Atlantic Ridge at the Kane Fracture Zone [*Detrick et al.*, 1990]; JdF, Juan de Fuca Ridge [*Morton et al.*, 1987; *Rohr et al.*, 1988]; Lau, Lau Basin [*Collier and Sinha*, 1990]; sEPR and nEPR, southern and northern East Pacific Rise, respectively [*Detrick et al.*, 1987; *Purdy et al.*, 1992].

of meters thick) in cross-axis section, and it also seems to be continuous along the axis [*Detrick et al.*, 1987; *Mutter et al.*, 1988; *Harding et al.*, 1989; 1993; *Vera et al.*, 1990; *Kent et al.*, 1990]. This thin melt lens is underlain by a broader (~ 6 km wide and 2–4 km thick) low velocity region of hot rock that may include ~3–5% of melt (Figure 4). The depth to the magma lens was reported to be greater at intermediate spreading Juan de Fuca Ridge (~3 km) [*Morton et al.*, 1987; *Rohr et al.*, 1988] than that along the fast spreading East Pacific Rise (1–2 km) [*Detrick et al.*, 1987; *Mutter et al.*, 1988]. While strong seismic reflectors of a magma lens can be traced over tens of kilometers along the EPR, such seismic signatures had not been detected at the slow-spreading Mid-Atlantic Ridge [*Detrick et al.*, 1990], except at one location of the hotspot influenced Reykjanes Ridge [*Sinha et al.*, 1997].

Motivated by these exciting seismic studies at fast- and slow-spreading ridges and field observations at the Oman ophiolite, three modeling studies were published in Journal of

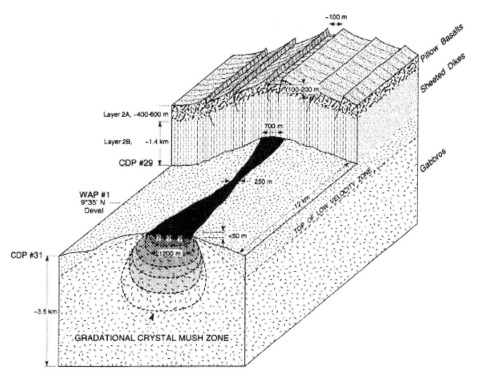

Figure 4. Seismic crustal structure of a fast spreading ridge derived from reflection and refraction seismology. (After *Kent et al.*, 1993).

Geophysical Research simultaneously in 1993 for the crustal genesis at mid-ocean ridges [*Henstock et al.*, 1993; *Phipps Morgan and Chen*, 1993a; *Quick and Denlinger*, 1993]. As an extension of an earlier model of Sleep [1975] by including a crustal-accretion lens in the upper crust, these models focused only on the crustal accretion process within the crust, that is, what could happen after the melt ascends to the mid-crustal level. Nevertheless, these models are successful in explaining a number of observations and therefore, have made a strong impact on the ridge studies for the past decade.

It has shown [*Phipps Morgan and Chen*, 1993a; 1994] that magma injection into a small magma lens at the base of the sheeted-dike complex has the potential to accommodate both the gabbroic layering patterns seen in the Oman ophiolite record and the small magma lens observed in seismic studies at fast-spreading ridges (Figure 5). Assuming the depth of the magma lens is controlled by a freezing temperature (say 1200°C), the model successfully predicted the observed pattern in the variation of the magma lens depth with spreading rate, that is, the deepening of the magma lens with decreasing spreading rate [*Purdy et al.*, 1992] and the sharp transition from a mid-crustal level lens to no lens for half-spreading rates less than ~20 mm/yr [*Detrick et al.*, 1990] (Figure 3). These models have been recently extended to include the effect of melt production in the upper mantle [*Chen*, 2000], which will be briefly described next.

3.2. Current Model Formulation and Results

The crustal genesis model presented here was built on our current understanding that it is the interplay between the magma supply to the crust and the heat loss to the seafloor through hydrothermal cooling that controls the ridge thermal structure, the crustal accretion process, and the ridge axis topography (Figure 6). Magma supply rate depends on both spreading rate and mantle temperature [e.g., *Phipps Morgan*, 1987; *Shen and Forsyth*, 1992; *Su and Buck*, 1993; *Chen*, 2000] although other factors such as the lateral heterogeneity in the mantle source (geochemical variations) could also affect the magma supply rate to the crust [e.g., *Shen and Forsyth*, 1995; *Niu and Hékinian*, 1997; *Niu et al.*, 1996; 2001; *Korenaga and Kelemen*, 2000]. The faster the ridge is spreading the more mantle is upwelling through the partial melting region beneath the ridge axis, and therefore, the more melt is produced and supplied to the crust per unit time. Similarly the hotter the mantle beneath the ridge the more melt is generated from a larger partial melting region with higher content of partial melt [e.g., *McKenzie and Bickle*, 1988; *Shen and Forsyth*, 1995; *Chen*, 1996b; *Niu*, 1997].

The basic model geometry is shown in Figure 7; it includes the effects of melt production at depths of ~20–60 km within upwelling mantle, melt migration to shallow level of ~1–3 km depth beneath the seafloor, and melt injection into upper

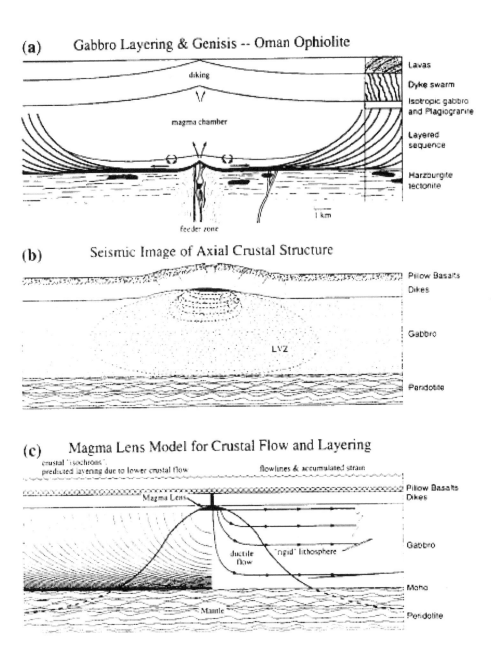

Figure 5. (a) Schematic model of a mid-ocean ridge spreading center derived from Oman ophiolite studies [Smewing, 1981]. As shown on the right side, isotropic gabbros just beneath the sheeted dike complex grade into gabbros with a (weakly developed) near vertical dip which becomes more developed and more shallowly dipping deeper into the gabbro section. This dip structure was used by Smewing [1981] to infer that gabbro layering reflects cumulate deposition on the floor of the large magma chamber sketched here. (b) Seismic model of magma chamber structure for the East Pacific Rise. Molten magma is concentrated in a lens that is approximately 1 km wide that resides at the base of the sheeted dike complex. Beneath the magma lens and extending to mid-crustal depths is a broader region of rock at elevated temperatures that contains a few percent partial melt. (c) Schematic theoretical model showing the flowlines, accumulated strain, and layering predicted by lower crustal flow away from a shallow injection lens at the sheeted dike/gabbro interface [Phipps Morgan and Chen, 1993a]. The left side of the figure shows predicted layering "isochrons" generated from crustal flow away from a magma lens at the base of the sheeted dike complex. The right hand side shows several typical crustal flow lines and the accumulated strain along each flowline. Strain is most intense in the lowermost part of the gabbro section. (After Phipps Morgan et al., 1994).

Factors (u, Tm, Nu) controlling ridge-axis thermal structure

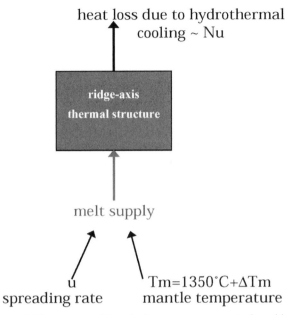

Figure 6. Chart summarizing the three-parameter control on ridge-axis thermal structure.

crust accompanied by the resulting crustal flow. The model is briefly described next and its most important features are outlined; details of the model could be found in *Chen* [2000].

Here the total melt production rate in the upwelling mantle is calculated from a numerical model for a passive spreading center [*Chen*, 1996b]. Each combination of the assumed mantle temperature, Tm, and half spreading rate, u, yields a calculated maximum crustal thickness H_c^* (which equals the total melt production rate divided by spreading rate). A model parameter β is introduced to denote the fraction of total melt produced in the upwelling mantle that eventually migrates to the crustal level to create the new oceanic crust, i.e., $H_c = \beta \cdot H_c^*$, where H_c is the crustal thickness. The remaining fraction of the melt that does not make to the crustal level, i.e., $(1-\beta) \cdot H_c^*$, is assumed to have frozen in the upper mantle (Figure 7).

Seismic observations over the world's ocean basins indicated that, on average, the oceanic crust has a constant thickness for half-rates greater than 10 mm/yr [e.g., *Reid and Jackson*, 1981; *Chen*, 1992; *White et al.*, 1992], below which the average crustal thickness appears to decrease with spreading rate [*e.g., Cochran et al.*, 2003; *Dick et al.*, 2003]. Using this observation as a constraint, the value of β is adjusted in the model calculations to produce a 6-km crust for a reference mantle temperature of 1350°C at a given spreading rate.

This simplified approach yields a linear increase in crustal thickness with mantle temperature [*Chen*, 2000].

When the melt reaches the crust level beneath the ridge axis, it is injected into a narrow vertical dike zone and a thin horizontal magma lens (Figure 7) [*Phipps Morgan and Chen*, 1993a]. The crystallization temperature (here chosen to be 1200°C) determines the depth of the lens at the ridge axis. However, a steady-state magma lens is not permitted in the crust if the entire crust beneath the ridge axis is cooler than 1200°C. In that case, melt solidification within the narrow vertical dike zone is assumed to be the only mode of crustal accretion.

Hydrothermal cooling is treated by an enhanced thermal conductivity, K_c^*, within the crustal region penetrated by seawater circulation (Figure 7), which is assumed to occur within the cold and brittle crust (< 600°C). Here the enhanced thermal conductivity is parameterized as $K_c^* = Nu \cdot K_c$, where K_c is the actual thermal conductivity of the crust and the Nusselt number (Nu) is defined as the ratio of convective to conductive heat transfer within a hydrothermal convection system (see *Phipps Morgan and Chen*, 1993a for details of the model).

In summary the key elements and results of the current thermal models [*Phipps Morgan and Chen*, 1993a; *Chen*, 2000] are the following:
(a) Collection (or ponding) of the ascending melt in the crust is controlled by a freezing temperature, assumed at 1200°C.
(b) Gabbros (layer 3) are formed by crystallization from the melt lens.
(c) Magma lens depth variation with spreading rate is due to the different thermal structures at different spreading ridges, and as a consequence,
(d) The model predicts a sharp transition from the presence of a magma lens to the absence of such a lens in the crust at intermediate spreading rates (Figure 3).

Model predictions of the magma lens depth (defined as the depth of the 1200°C isotherm beneath the ridge axis) are shown in Figure 3 for a reference mantle temperature of 1350°C and Nu = 8 (dashed line) and Nu = 10 (solid line), respectively. The predicted trend matches the seismic observations quite well: the magma lens deepens from ~1–1.8 km at fast ridges (40–75 mm/yr) to ~3–3.5 km at intermediate ridges (~30 mm/yr), and the sharp transition from a deeper lens at 30 mm/yr to no crustal magma lens < 20 mm/yr.

4. DISCUSSIONS

4.1. Sensitivity at Intermediate Spreading Rates

Both the observations and model calculations show a rapid transition from a crustal magma lens style of (hot) crustal accretion at fast-spreading ridges to a cold crustal accretion

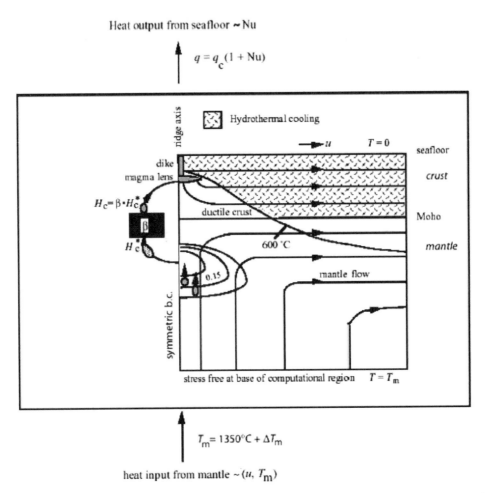

Figure 7. Sketch of key factors controlling thermal equilibrium at mid-ocean ridges [*Chen*, 2000]. Magma supply is controlled by potential mantle temperature, Tm, and half spreading rate, u. Steady-state temperature field near the ridge axis is determined by thermal equilibrium between heat conduction and advection, release of latent heat as melt solidifies in a vertical narrow dike zone and a horizontal magma lens, and hydrothermal circulation, represented here as enhanced thermal conductivity, in the shallow crust.

environment (absence of such a magma lens) at slow-spreading ridges (Figure 3). It shows that the interplay between the magma supply and the hydrothermal cooling is at a critical state at intermediate spreading rates and therefore, any perturbations in either magma supply or hydrothermal cooling could cause significant variations in thermal structure and the magma lens depth. This is further illustrated in Figure 8, where the changes of the magma-lens depth relative to the lens depth at a reference ridge with Nu = 8 and Tm = 1350°C as a function of mantle temperature variations are shown at three spreading rates. (Recall that higher mantle temperatures result in higher magma supply rates to the crust.) At each spreading rate, the left vertical bar shows decreasing in magma lens depth (negative lens-depth change) for ΔTm = 30°C, and the middle bar shows the deepening (positive lens-depth change) for ΔTm = −30°C.

These model calculations show that ridges with intermediate spreading rates have the highest sensitivity to mantle-temperature variations. For example, the change of magma-lens depth at u = 37 mm/yr is about a factor of 3 larger than that at a faster ridge with u = 55 mm/yr while it is about a factor of 3 smaller than that at a slower ridge with u = 30 mm/yr. In particular, the model did not predict a crustal magma lens in the crust for a ridge spreading at u = 30 mm/yr if the mantle beneath the ridge is 30°C cooler than the reference value of 1350°C.

The conclusion from these model calculations that intermediate spreading ridges have the highest sensitivity to mantle temperature variations (or to magma supply fluctuations) is consistent with field observations along several intermediate spreading ridges including the Southeast Indian Ridge

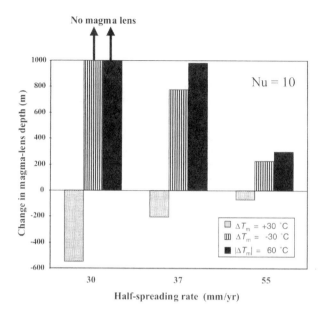

Figure 8. Sensitivity of magma-lens depth to mantle temperature variations at ridges spreading at intermediate and fast rates. (After *Chen*, 2000).

[*Tolstoy et al.*, 1995; *Cochran et al.*, 1997; 2002; *Ma and Cochran*, 1996; 1997; *Sempere et al.*, 1997], the Juan de Fuca and Gorda Ridges [*Hooft and Detrick*, 1995], and the Galápagos Spreading Center [*Canales et al.*, 1997; 2002; *Detrick* et al., 2002]. Along these intermediate spreading ridges, noticeable transitions in ridge-axis topography from an axial high to a rift valley were observed and were attributed to variations in mantle temperature due to a nearby hotspot or a relatively cold mantle anomaly.

Recently the intermediate spreading Galápagos Spreading Center in the East Pacific was not only mapped in great details by multi-beam bathymetry but was also surveyed by multi-channel seismic experiments. Multi-beam mapping along the Galápagos Spreading Center indicated an eastward transition in ridge-axis bathymetry from a rift valley west of the 95.5°W propagating rift tip to an axial high topography towards the ridge segment near 91°W where it is closest to the Galápagos hotspot [*Canales et al.*, 1997; 2002]. This transition correlates with both progressive eastward increase in crustal thickness, decrease in residual mantle Bouguer gravity anomaly, and decrease in axial magma chamber depth [*Canales et al.*, 2002; *Detrick et al.*, 2002]. Application of the thermal models to the Galápagos Spreading Center illustrates a threshold effect in axial topography and the model results (Figure 9) demonstrate that the along-axis variations are primarily caused by increased magma supply from the influence of the nearby Galápagos hotspot [*Chen and Lin*, 2004].

4.2. Effect of Mantle Plume on Hydrothermal Cooling at Ridges

When a mantle plume is located sufficiently close to a mid-ocean ridge, its mass- and heat-flux can influence significantly the style of lithospheric accretion at the ridge axis [*Schilling*, 1973; 1985; 1991; *Ribe et al.*, 1995; *Ribe*, 1996; *Ribe and Delattre*, 1998; *Lin*, 1998; *Ito et al.*, 1997; 2003; *Ito*, 2001]. An excellent review on the observational and theoretical studies of the mantle plume-mid-ocean ridge interaction can be found in a recent publication [*Ito et al.*, 2003]. Next we focus on the influence of the on-axis Iceland hotspot on the heat-flux of the ridge axis, inferred from the water-column plume observations, which have become a standard field operation to systematically look for hydrothermal vents along a ridge [*Baker et al.*, 1994; 2001; *Baker and Urabe*, 1996; *German et al.*, 1994; 1996a; 1996b].

The observed global increase in plume incidence with spreading rate is summarized in Figure 10 [*Baker et al.*, 1996] and the plume incidence is defined as the fraction of the ridge segment length overlain by hydrothermal plumes or vent fields [*Baker*, 1996]. Thus it represents an average assessment of the hydrothermal activity on a segment scale. *Chen and Phipps Morgan* [1996] showed that the hydrothermal heat output per kilometer along the ridge axis represented from this data set

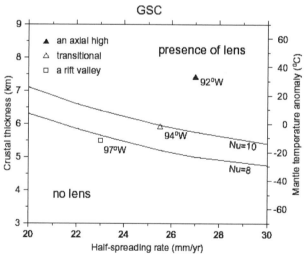

Figure 9. Modeling results showing the transition from the presence to absence of a magma lens within the crust as a function of spreading rate (horizontal axis), curstal thickness (left axis), and mantle temperature anomaly (right axis). Data are from three locations of the GSC [*Detrick et al.*, 2002]: 92°W with an axial high morphology and a relatively shallow magma lens (filled triangle); 94°W with a rifted high or transitional morphology and a relatively deep magma lens (open triangle); 97°W with an axial valley and no magma lens (open square). (After *Chen and Lin*, 2004).

is consistent with the linear increase in the heat loss from hydrothermal cooling at ridge axis predicted from thermal models.

In a recent study [*Chen*, 2003], model calculations have shown that a magma chamber at depths similar to those at EPR could be present at the inflated Reykjanes Ridge within a reasonable range of crustal thickness if hydrothermal cooling is a factor of 2–4 less than the global average as shown in Figure 10. These modeling results and the similarities in many ridge characteristics between the EPR and the hotspot-influenced Reykjanes Ridge have led *Chen* [2003] to suggest that hydrothermal cooling at the Reykjanes Ridge is not as effective as at elsewhere along the Mid-Atlantic Ridges due to the influence of the Iceland mantle plume although this link has not been investigated in details to-date.

The only available observation of the hydrothermal activity at the Reykjanes Ridge is the water column, vent plume survey along over 600 km of the Reykjanes Ridge reported *by German et al.* [1994]. Only one vent site was discovered, which is translated into an extremely low plume incidence, shown as "RR" in Figure 10. Although other factors could also contribute to this finding [*German and Parson*, 1998], one cannot rule out the interpretation that this unique observation does indeed reflect an anomalous low level of overall hydrothermal activity currently operating at this hotspot-influ-

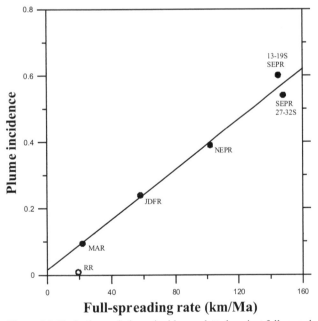

Figure 10. Hydrothermal plume incidence plotted against full spreading rate. Plume incidence (solid squares) is the fraction of ridge crest length overlain by hydrothermal plumes. The solid line is least-squares fit to these global data. The anomalously low plume incidence of the Reykjanes Ridge (RR) is shown as an open circle. (The figure is modified from *Baker et al.*, 1996).

enced ridge. Vent plume observations at another on-axis mantle plume, the St Paul and Amsterdam hotspot also indicated a very low plume incidence along the ridge that is at or near the hotspot relative to the ridge further away from the hotspot [*Baker*, personal communication].

The intensity of hydrothermal cooling mainly depends on the permeability structure of the oceanic lithosphere at the ridge axis, particularly the upper few kilometers of the crust [e.g., *German and Parson*, 1998; *Baker et al.*, 2001]. It is also known that the rheology of the oceanic lithosphere is strongly temperature dependent. The oceanic lithosphere of the Reykjanes Ridge could be much weaker than elsewhere of the MAR because of the elevated crustal temperatures from the extra heat input of the Iceland mantle plume. Thus one additional aspect of the Iceland hotspot influence on the Reykjanes Ridge is the reduced permeability of the oceanic lithosphere, possibly resulting from both reduced number of fractures and fissures near the seafloor and faulting activities, and from the limited depth extent of faults and fractures as well. Whether this interpretation is valid or not requires more detailed field surveys at the hotspot-influenced ridges, particularly at the Reykjanes Ridge. For example, as suggested by *Chen* [2003], future studies similar to the RAMESSES experiments [*Sinha et al.*, 1997] at the inflated Reykjanes Ridge north of 59°N could resolve the issues that if a magma chamber is present beneath this portion of the ridge and what is the depth range of the magma chamber.

A detailed geological/hydrothermal investigation of the East Pacific Rise between the Orozco Transform Fault at 15°20'N and the Rivera Transform Fault at 18°30'N was conducted to access the relative influence of magma budget and permeability structure at the segment scale [*Baker et al.*, 2001]. It was found that the hydrothermal plumes over the inflated 16°N segment with robust magma supply were weaker than that long the narrower, rifted 17°N segment with subdued magma supply. It was interpreted by *Baker et al.* [2001] that the local permeability structure here controls the hydrothermal activity. The 16°N segment with little faulting on the seafloor might have its hydrothermal activity suppressed by widespread volcanic flows and the elevated crustal temperatures. Activity on the 17°N segment, on the other hand, may be tectonically enhanced, with seawater circulating through faults to a deeper depth.

4.3. Debate on Gabbros Accretion

Alternative models for the formation of the gabbroic layer have been proposed recently based primarily on the field observations from Oman ophiolite [*Boudier et al.*, 1996; *Kelemen et al.*, 1997; *Kelemen and Aharonov*, 1998; *Korenaga and Kelemen*, 1998]. Essentially these models require more than one melt lenses in the crust and suggest that significant portion of the lower gabbros is formed from these melt bodies below the seismically observed magma lens. One major problem of these in-situ crystallization models is that they require extensive cooling of the lower crust beneath the top melt lens to remove the considerable latent heat resulted from in-situ crystallization in order to prevent significant heating up the lower crust [*Chen*, 2001], which would have not been consistent with seismic observations [e.g., *Dunn and Toomey*, 1997; *Dunn et al.*, 2000]. This should not be a problem if the sills seen in Oman ophiolite are injected episodically into the crust off axis [*MacLeod and Yaouancq*, 2000].

A new approach has been developed for calculating the cooling rate of the lower oceanic crust based on the closure temperature of down-temperature diffusive exchange of Ca from olivine to clinopyroxene in gabbros from the Oman ophiolite [*Coogan et al.*, 2002]. The results have shown that the cooling rates for the lower oceanic crust decrease with depth, rapidly in the upper third of the gabbros but more slowly in the lower two thirds, which leads to several orders of magnitude variation in cooling rate between the top and bottom gabbros. This depth distribution of cooling rate has suggested more efficient heat removal from the shallow magma lens than that from deeper in the crust, which is consistent with the thermal models with one magma lens at the seismically observed AMC depth [*Phipps Morgan and Chen*, 1993a; *Chen*, 2000].

Moreover, it also has been shown that the depth of the top magma lens is not sensitive to the difference in distributing a fraction of the crystallization of the gabbros below the lens (Figure 11) [*Chen*, 2001]. Thus the major results presented here will probably not be affected by the on-going debate on the accretion styles of the gabbros.

4.4. Steady-State Versus Transient Crustal Accretion

It is now widely accepted that the crustal accretion on a segment scale is steady-state or quasi steady-state at fast-spreading ridges as indicated by the smooth ridge topography and the presence of a magma lens that can be traced tens of kilometers along the axis. The magmatic construction of the crust at slow-spreading ridges is episodic since the cold crust can not permit the presence of a long-lived magma chamber and the dominant tectonic extension there created an axial rift valley and the rugged ridge flanks bathymetry.

This hypothesis is consistent with the gravity observations: the slow-spreading Mid-Atlantic Ridge is characterized by distinct "Bulls-eye" shaped gravity lows associated with large zero-age-seafloor-depth variations, suggesting large along-axis variations in crustal thickness/magma supply rate [*Lin et al.*, 1990; *Detrick et al.*, 1995], whereas the fast spreading

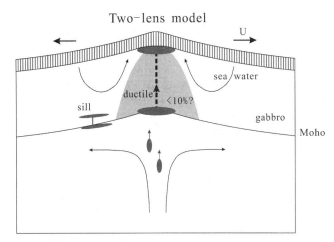

Figure 11. Sketch of the two-lens model for a fast spreading East Pacific Rise. While the melt supply to the shallow melt lens keeps the gabbros beneath the ridge axis at high temperatures, hydrothermal circulation of cold seawater rapidly cools the gabbros a few kilometers off the axis. Melt could pond at a deep reservoir at or above the Moho but most of the melt eventually reached the shallow melt lens. The gabbros in layer 3 are mostly formed from the top melt lens whereas a small percentage (< 10%) of the gabbros could be crystallized directly from the deeper reservoir at the Moho. Alternatively, a small portion of the gabbros could be added by off-axis sill intrusions. (After *Chen*, 2001).

East Pacific Rise is associated with much smaller along-axis variations and smooth topography [*Lin and Phipps Morgan*, 1992].

Recently a new type of mid-ocean ridge, the ultra-slow spreading ridge was proposed based on the discovery along the western portion (9°–25°E) of the Southwest Indian Ridge (SWIR), where gravity observations suggested the crust is almost absent, magnetic anomaly is extremely weak, and abundant mantle peridotites are directly exposed on the seafloor [*Dick et al.*, 2003]. These data, plus the data collected during earlier cruises along the east portion (49°–69°E) of the SWIR [*Cannat et al.*, 1999] and the data collected along the Gakkel Ridge under the Arctic Ocean [*Cochran et al.*, 2003], suggested that ultra-slow ridges, including both the SWIR and the Gakkel Ridge, could sustain steady-state amagmatic lithospheric accretion and the magma accretion could be extremely focused both spatially and temporally. It was hypothesized that such extreme focusing in melt production and magma accretion in time is caused by a thick and cold lithosphere at ultra-slow ridges, which could suppress melting columns significantly and prevented the development of a steady-state partial melting region beneath the ridge [*Lin et al.*, 2003].

The steady-state nature of the current thermal models prevents investigation of time-dependent crustal accretion process at a slow-spreading ridge and the transient behaviors of magma intrusion/eruption and rifting dynamics at mid-ocean ridges. Extension of current thermal models into time-dependent models would, therefore, enable the exploration of what time scales can appropriately be used to parameterize dike intrusions above the gabbroic sections as a time-averaged, steady-state process despite the inherently time-dependent nature of individual dike-intrusion episodes. In addition, the magma chamber might be fed by episodic melt intrusion from underlying mantle, or the melt might be supplied laterally by along-axis flow between connected magma chambers within a ridge segment [*Hooft et al.*, 2000; *Magde et al.*, 2000]. The current two-dimensional models limit our investigation of how the crustal accretion occurs along a segment where mantle upwelling is focused or where significant along-axis crustal flow takes place.

Finally, the most poorly understood model parameter in current thermal models is the Nu, which describes the efficiency of the hydrothermal cooling at a ridge [e.g., *Phipps Morgan et al.*, 1987; *Chen and Morgan*, 1990b; *Neumann and Forsyth*, 1993; *Shaw and Lin*, 1996; *Eberle and Forsyth*, 1998; *Chen*, 2000]. However, with more data becoming available, theoretical studies of hydrothermal circulation systems at mid-ocean ridges will provide important constraints on this crustal accretion variable.

5. SUMMARY

Modeling the thermal state of the oceanic crust is important in understanding both the dynamic rifting processes along the longest active plate boundary on the Earth and the construction of the oceanic crust that covers two thirds of the earth's surface. Previous studies have demonstrated that the crustal thermal structure of a ridge not only controls the rheology of the oceanic lithosphere and thus the dynamic rifting process at the ridge but also controls both the presence or absence of a crustal magma chamber at a spreading ridge as well as the depth of such a magma chamber. In other words, it determines the style of oceanic crustal genesis, by a long-lived magma lens or by episodic magma intrusions. Therefore, thermal modeling should be an important constituent of any dynamic models of mid-ocean ridges and the modeling results provide important links between various seafloor observations such as ridge axis topography, gravity, seismic crustal structure and etc. and controlling variables of the ridge processes such as spreading rate, mantle temperature (or magma supply rate), and hydrothermal cooling.

One of the most important modeling results is that intermediate spreading ridges have the highest sensitivity to either the mantle-temperature variations or hydrothermal cooling fluctuations. This prediction is consistent with observations from recent research cruises along several intermediate-spreading ridges including the Southeast Indian Ridge and the Galápagos Spreading Center.

Future improvements of current thermal models should include incorporating (a) time-dependent capability to investigate the episodic magmatic/tectonic spreading at slow spreading ridges, (b) better approach of simulating the hydrothermal cooling as a consequence of detailed theoretical studies of hydrothermal circulation systems at mid-ocean ridges, and (c) extension to three-dimensional models for investigation of along-axis flow from a mantle plume or from the focused magma supply beneath a slow spreading ridge.

Acknowledgments. I thank Chris German, Jian Lin, and P. Lindsey for their invitation for my presentation at the 1st InterRidge Theoretical Institute at Pavia, Italy, 16–20 September 2002. Their efforts of both organizing this first Theoretical Institute and editing this special volume are greatly appreciated. Reviews by Jian Lin, Yaoling Niu, and two anonymous referees helped improving the manuscript. Travel to the 1st InterRidge Theoretical Institute at Pavia, Italy was partially supported by US RIDGE 2000 Office. Partial support for writing this paper came from both the National Science Foundation of China and the Peking University.

REFERENCES

Baker, E. T., Geological indexes of hydrothermal venting, *J. Geophys. Res.*, *101*, 13,741–13,753, 1996.

Baker, E. T., and T. Urabe, Extensive distribution of hydrothermal plumes along the superfast spreading East Pacific Rise, 13°30′–18°40′S, *J. Geophys. Res., 101*, 8685–8697, 1996.

Baker, E. T., R. A. Feely, M. J. Mottl, F. T. Sansone, C. G. Wheat, J. A. Resing, and J. E. Lupton, Hydrothermal plumes along the East Pacific RIse, 8°40 to 11°50′N: Plume distribution and relationship to the apparent magmatic budget, *Earth Plan. Sci. Letts., 128*, 1–17, 1994.

Baker, E. T., Y. J. Chen, and J. Phipps Morgan, Relationship between hydrothermal cooling and the spreading rate of mid-ocean ridges, *Earth Plant. Sci. Lett, 142*, 137–145, 1996.

Baker, E.T., M.-H. Cormier, C. H. Langmuir, and K. Zavala, Hydrothermal plumes along segments of contrasting magmatic influence, 15°20′–18°30′N, East Pacific Rise: Influence of axial faulting, *Geochem. Geophys. Geosyst., 2*, 2000GC000165, 24 September 2001.

Boudier, F., A. Nicolas, and B. Idefonse, Magma chambers in the Oman ophiolite: Fed from the top or from the bottom? *Earth Planet. Sci. Lett., 144*, 239–250, 1996.

Brace, W. F., and D. L. Kolstedt, Limits on lithosphere stress imposed by laboratory experiments, *J. Geophys. Res., 85*, 6248–6252, 1980.

Buck, W. R., Accretional curvature of lithosphere at magmatic spreading centers and the flexural support of axial highs, *J. Geophys. Res., 106*, 3953–3960, 2001.

Byerlee, J. D., Friction of rocks, *Pure Appl. Geophys., 116*, 615–626, 1978.

Canales, J. P., J. J. Danobeitia, R. S. Detrick, E. E. E. Hooft, R. Bartolomé, and D. Naar, Variations in axial morphology along the Galápagos Spreading Center and the influence of the Galápagos hotspot, *J. Geophys. Res., 102*, 27,341–27,354, 1997.

Canales, J. P., G. Ito, R. S. Detrick, and J. Sinton, Crustal thickness along the western Galápagos spreading center and the compensation of the Galápagos hotspot swell, *Earth Planet. Sci. Lett., 203* , 311–327, 2002.

Cann, J. R., A model for oceanic crustal structure developed, *Geophys. J. R. Astron. Soc., 39*, 169–187, 1974.

Cannat, M., C. Rommevaux-Jestin, D.,Sauter, C. Deplus, and V. Mendel, Formation of the axial relief at the very slow spreading Southwest Indian Ridge (49° to 69°E), *J. Geophys. Res., 104*, B10, 22,825–22,843, 1999.

Carbotte, S., C. Mutter, J. Mutter and G. Ponce-Correa, Influence of magma supply and spreading rate on crustal magma bodies and emplacement of the extrusive layer: Insights from the East Pacific Rise at lat 16°N, *Geology, 26*, 455–458, 1998.

Chen, Y. J., Oceanic crustal thickness versus spreading rate, *Geophys. Res. Lett., 19*, 753–756, 1992.

Chen, Y. J., Dynamic of the Mid-Ocean Ridge plate boundary: Recent observations and theory, *Pure and Applied Geophys., 146*, 621–648, 1996a.

Chen, Y. J., Constraints on the melt production rate beneath the mid-ocean ridges based on passive flow models, *Pure and Applied Geophys., 146*, 590–620, 1996b.

Chen, Y. J., Dependence of crustal accretion and ridge axis topography on spreading rate, mantle temperature, and hydrothermal cooling, *in Ophiolites and Oceanic Crust: New Insights from Field Studies and the Ocean Drilling Program*, edited by Y. Dilek, E. M. Moores, D. Elthon, and A. Nicolas, pp. 161–179, Geological Society of America Special Paper 349, Boulder, Colorado, 2000.

Chen, Y. J., Thermal effects of gabbro accretion from a deeper second melt lens at the fast spreading East Pacific Rise, *J. Geophys. Res., 106*, 8581–8588, 2001.

Chen, Y. J., Influence of the Iceland mantle plume on crustal accretion at the inflated Reykjanes Ridge—Magma lens and low hydrothermal activity? *J. Geophys. Res., 108(B11)*, 2524, 10.1029/2001JB000816, 2003.

Chen, Y., and W. J. Morgan, Rift valley/no rift valley transition at mid-ocean ridges, *J. Geophys. Res., 95*, 17,571–17,581, 1990a.

Chen, Y., and W. J. Morgan, A nonlinear-rheology model for mid-ocean ridge axis topography, *J. Geophys. Res., 95*, 17,583–17,604, 1990b.

Chen, Y. J. and J. Lin, High sensitivity of ocean ridge thermal structure to changes in magma supply: the Galapagos Spreading Center, *Earth and Planetary Science Letters, 221*, 263–273, 2004.

Chen, Y. J., and J. Phipps Morgan, The effect of magma emplacement geometry, spreading rate, and crustal thickness on hydrothermal heat flux at mid-ocean ridge axes, *J. Geophys. Res., 101*, 11,475–11,482, 1996.

Cochran, J. R., J.-C. Sempéré, and SEIR Scientific Team, The Southeast Indian Ridge between 88°E and 118°E: gravity anomalies and crustal accretion at intermediate spreading rates, *J. Geophys. Res., 102*, 15,463–15,487, 1997.

Cochran, J. R., G.J. Kurras, M. H. Edwards, and B. J. Coakley, The Gakkel Ridge: Bathymetry, gravity anomalies, and crustal accretion at extremely slow spreading rates, *J. Geophys. Res., 108*, B2, 10.1029/2002JB001830, 2003.

Coogan, L. A., G. R. T. Jenkin, and R. N. Wilson, Constraining the cooling rate of the lower oceanic crust: a new approach applied to the Oman ophiolite, *Earth Planet. Sci. Lett., 199*, 127–146, 2002.

Detrick, R. S., J. P. Madsen, P. E. Buhl, J. Vera, J. Mutter, J. Orcutt, and T. Brocker, Multichannel seismic imaging of an axial magma chamber along the East Pacific Rise between 4°N and 13°N, *Nature, 326*, 35–41, 1987.

Detrick, R. S., J. Mutter, P. E. Buhl, and I. I. Kim, No evidence from multichannel reflection data for a crustal magma chamber in the MARK area on the Mid-Atlantic Ridge, *Nature, 347*, 61–64, 1990.

Detrick, R. S., H. D. Needham, and V. Renard, Gravity anomalies and crustal thickness variations along the Mid-Atlantic Ridge between 33°N and 40°N, *J. Geophys. Res., 100*, 3767–3787, 1995.

Detrick, R. S., J. M. Sinton, G. Ito, J. P. Canales, M. Behn, T. Blacic, B. Cushman, J. E. Dixon, D. W. Graham, and J. J. Mahoney, Correlated geophysical, geochemical and volcanological manifestations of plume-ridge interaction along the Galápagos Spreading Center, *Geochem. Geophys. Geosyst., 3(10)*, 8501, doi:10.1029/2002GC000350, 2002.

Dick, H. J. B., J. Lin, and H. Schouten, An ultraslow-spreading class of ocean ridge, *Nature, 426*, 405–412, 2003.

Dunn, R. A., and D. R. Toomey, Seismological evidence for three-dimensional melt migration beneath the East Pacific Rise, *Nature, 388*, 259–262, 1997.

Dunn, R. A., D. T. Toomey, and S. C. Solomon, Three-dimensional seismic structure and physical properties of the crust and shallow mantle beneath the East Pacific Rise at 9°30′N, *J. Geophys. Res., 105*, 23,537–23,555, 2000.

Eberle, M. A., and D. W. Forsyth, An alternative, dynamic model of the axial topographic high at fast-spreading ridges, *J. Geophys. Res., 103*, 12,309–12,320, 1998a.

Eberle, M. A., and D. W. Forsyth, Evidence from the asymmetry of fast-spreading ridges that the axial topographic high is due to extensional stresses, *Nature, 394*, 360–363, 1998b.

German, C. R., and L. M. Parson, Distributions of hydrothermal activity along the Mid-Atlantic Ridge: interplay of magmatic and tectonic controls, *Earth and Planetary Science Letters, 160*, 327–341, 1998.

German, C. R., J. Briem, C. Chin, M. Danielsen, S. Holland, R. James, A. Jonsdottir, E. Ludford, C. Moser, J. Olafsson, M. R. Palmer, and M. D. Rudnicki, Hydrothermal activity on the Reykjanes Ridge: the Steinaholl vent-field at 63°06'N, *Earth Planet. Sci. Lett., 121*, 647–654, 1994.

German, C. R., L. M. Parson, B. J. Murton, and H. D. Needham, Hydrothermal activity and ridge segmentation on the Mid-Atlantic Ridge: A tale of two hot-spots? *Geol. Soc. Spec. Pub., 118*, 169–184, 1996a.

German, C. R., L. M. Parson, and HEAT Scientific Team, Hydrothermal Exploration at the Azores Triple-Junction: Tectonic control of venting at slow-spreading ridges?, *Earth Planet. Sci. Lett., 138*, 93–104, 1996b.

Goetze, C. The mechanism of creep in olivine, *Philos. Trans. R. Soc. London, Ser. A, 288*, 99–119, 1978.

Harding, A. J., J. A. Orcutt, M. E. Kappus, E. E. Vera, J. C. Mutter, P. Buhl, R. S. Detrick, and T. M. Brocher, Structure of young oceanic crust at 13°N on the East Pacific Rise from expanding spread profiles, *J. Geophys. Res., 94*, 12,163–12,196, 1989.

Harding, A. J., G. M. Kent, and J. A. Orcutt, A multichannel seismic investigation of upper crustal structure at 9°N on the East Pacific Rise: Implications for crustal accretion, *J. Geophys. Res., 98*, 13,925–13,944, 1993.

Henstock, T. J., A. W. Woods, and R. S. White, The accretion of oceanic crust by episodic sill intrusion, *J. Geophys. Res., 98*, 4143–4154, 1993.

Hirth, G., J. Escartin, and J. Lin, The rheology of the lower oceanic crust: Implications for lithospheric deformation at mid-ocean ridges, in *Faulting and Magmatism at Mid-Ocean Ridges, Geophys. Monogr. Ser.*, vol. 106, edited by W. R. Buck, et al., pp. 291–323, AGU, Washington, D. C., 1998.

Hooft, E. E. E, and R. S. Detrick, Relationship between axial morphology, crustal thickness, and mantle temperature along the Juan de Fuca and Gorda Ridges, *J. Geophys. Res., 100*, 22,499–22,508, 1995.

Hooft, E. E., R. S. Detrick, D. R. Toomey, J. A. Collins, and J. Lin, Crustal and upper mantle structure along three contrasting spreading segments of the Mid-Atlantic Ridge, 33.5°N–35°N, *J. Geophys. Res., 105*, 8205–8226, 2000.

Ito, G., Reykjanes 'V'-shaped ridges originating from a pulsing and dehydrating mantle plume, Nature, 411, 681–684, 2001.

Ito, G., J. Lin, and C. W. Gable, Interaction of mantle plumes and migrating midocean ridges: Implications for the Galápagos plume-ridge system, *Journal of Geophysical Research, 102*, 15403–15417, 1997.

Ito, G., J. Lin, and D. Graham, Observational and theoretical studies of the dynamics of mantle plume-mid-ocean ridge interaction, *Review of Geophysics, 41*, 1017, 10.1029/2002RG000117, 2003.

Kelemen, P. B., and E. Aharanov, Periodic formation of magma fractures and generation of layered gabbros in the lower crust beneath oceanic spreading ridges, *in Faulting and Magmatism at Mid-Ocean Ridges*, edited by W. R. Buck, P. T. Delaney, J. A. Farson, and Y. Lagabrielle, Geophysical Monograph, 106, pp. 267–289, American Geophysical Union, Washington D.C., 1998.

Kelemen, P., K. Goga, and N. Shimizu, Geochemistry of gabbro sills in the crust/mantle transition zone of the Oman Ophiolite: Implications for the origin of the oceanic lower crust, *Earth Planet. Sci. Lett., 146*, 475–488, 1997.

Kent, G. M., A. J. Harding, and J. A. Orcutt, Evidence for a smaller magma chamber beneath the East Pacific Rise at 9°30′N, *Nature, 344*, 650–653, 1990.

Kirby, S. H., Rheology of the lithosphere, *Rev. Geophys., 21*, 1458–1487, 1983.

Kirby, S. H., and A. K. Kronenberg, Rheology of the lithosphere: Selected topics, *Rev. Geophys., 25*, 1219–1244, 1987.

Korenaga, J. and P. B. Kelemen, Melt migration through the oceanic lower crust: a constraint from melt percolation modeling with finite solid diffusion, *Earth Planet. Sci. Lett., 156*, 1–11, 1998.

Korenaga, J. and P. B. Kelemen, Major element heterogeneity in the mantle source of the North Atlantic igneour province, *Earth Planet. Sci. Lett., 184*, 251–268, 2000.

Kusznir, F. D., Thermal evolution of the oceanic crust: its dependence on spreading rate and effect on crustal structure, *Geophys. J. R. astr. Soc., 61*, 1677–1681, 1980.

Kusznir, F. D., and M. H. P. Bott, A thermal study of oceanic crust, *Geophys. J. R. astr. Soc., 47*, 83–95, 1976.

Lin, J., The segmented Mid-Atlantic Ridge, *Oceanus, 34*, 9–16, 1992.

Lin, J., Hitting the hotspots, *Oceanus, 41*, 34–37, 1998.

Lin, J., and E. M. Parmentier, Mechanisms of lithospheric extension at mid-ocean ridges, *Geophys. J. R. Astrono. Soc., 96*, 1–22, 1989.

Lin, J., and J. Phipps Morgan, The spreading rate dependence of three-dimensional mid-ocean ridge gravity structure, *Geophysical Research Letters, 19*, 13–16, 1992.

Lin, J., H. J. B. Dick, and H. Schouten, Evidence for highly focused magmatic accretion along the ultra-slow Southwest Indian Ridge, *EOS Trans. AGU, 84 (46)*, Fall Meet. Suppl., Abstract T11B-03, 2003.

Lin, J., G. M. Purdy, H. Schouten, J.-C. Sempere, and C. Zervas, Evidence from gravity data for focused magmatic accretion along the Mid-Atlantic Ridge, *Nature, 344*, 627–632, 1990.

Ma, Y. and J. R. Cochran, Transitions in axial morphology along the Southeast Indian Ridge, *J. Geophys. Res., 101*, 15,849–15,866, 1996.

Ma, Y. and J. R. Cochran, Bathymetry roughness of the Southeast Indian Ridge: Implications for crustal accretion at intermediate spreading rate mid-ocean ridge, *J. Geophys. Res., 102*, 17,697–17,711, 1997.

Macdonald, K. C., The crest of the Mid-Atlantic Ridge: Models for crustal generation processes and tectonics, in *The Geology of North America: The Western North Atlantic Region, vol. M*, edited by P. Vogt and B. Tucholke, pp. 51–68, Geol. Soc. of Am., Boulder, Colo., 1986.

MacLeod, C. J., and G. Yaouancq, A fossil melt lens in the Oman ophiolite: Implications for magma chamber processes at fast spreading ridges, *Earth Planet. Sci. Lett., 176*, 357–373, 2000.

Madsen, J. A., D. W. Forsyth, and R. S. Detrick, A new isostatic model fro the East Pacific crest, *J. Geophys. Res., 89*, 9997–10,015, 1984.

Madsen, J. A., R. S. Detrick, J. C. Mutter, P. Buhl, and J. C. Orcutt, A two- and three-dimensional analysis of gravity anomalies associated witht the East Pacific Rise at 9°N and 13°N, *J. Geophys. Res., 95*, 4967–4987, 1990.

Magde, L. S., R. S. Detrick, and the TERA Group, Crustal and upper mantle contribution to the axial gravity anomaly at the southern East Pacific Rise, *J. Geophys. Res., 100*, 3747–3766, 1995.

Magde, L. S., A. H. Barclay, and D. R. Toomey, Crustal magma plumbing within a segment of the Mid-Atlantic Ridge 31–34°S, *Earth Planet. Sci. Lett., 198*, 59–67, 2000.

Malinverno, A., Inverse square-root dependence of mid-ocean ridge flank roughness on spreading rate, *Nature, 352*, 58–60, 1991.

McKenzie, D. P., and M. J. Bickle, The volume and composition of melt generated by extension of the lithosphere, *J. Petrology, 29,* 625–679, 1988.

Morton, J. L., and N. H. Sleep, A mid-ocean ridge thermal model: Constraints on the volume of axial hydrothermal flux, *J. Geophys. Res., 90,* 11,345–11,353, 1985.

Morton, J. L., N. H. Sleep, W. R. Normark, and D. H. Tomkins, Structure of the southern Juan de Fuca Ridge from seismic reflection records, *J. Geophys. Res., 92,* 11,315–11,326, 1987.

Mutter, C. Z., Seismic and hydrosweep study of the western Costa Rica Rift, *EOS, Trans. Amer. Geophys. Union, 76,* F595, 1995.

Mutter, J. C., G. A. Barth, P. Buhl, R. S. Detrick, J. Orcutt, and A. Harding, Magma distribution across ridge-axis discontinuities on the East Pacific Rise from multichannel seismic images, *Nature, 336,* 156–158, 1988.

Nicolas, A., *Structures of Ophilites and Dynamics of Oceanic Lithosphere,* Kluwer Academic Publishers, 1989.

Niu, Y., Mantle melting and melt extraction processes beneath ocean ridges: Evidence from abyssal peridotites, *J. Petrology, 38,* 1047–1074, 1997.

Niu, Y., and R. Hékinian, Spreading rate dependence of the extent of mantle melting beneath ocean ridges, *Nature, 385,* 326–329, 1997.

Niu, Y., G. Waggoner, J. M. Sinton & J. J. Mahoney, Mantle source heterogeneity and melting processes beneath seafloor spreading centers: The East Pacific Rise 18°–19°S, *J. Geophys. Res., 101,* 27,711–27,733, 1996.

Niu, Y., D. Bideau, R. Hékinian & R. Batiza, Mantle compositional control on the extent of melting, crust production, gravity anomaly and ridge morphology: a case study at the Mid-Atlantic Ridge 33–35°N, *Earth Planet. Sci. Lett., 186,* 383–399, 2001.

Neumann, G. A., and D. W. Forsyth, The paradox of the axial profile: Isostatic compensation along the axis of the Mid-Atlantic Ridge?, *J. Geophys. Res., 98,* 17,891–17,910, 1993.

Phipps Morgan, J., Melt migration beneath mid-ocean spreading centers, *Geophys. Res. Lett., 14,* 1238–1241, 1987.

Phipps Morgan, J., and Y. J. Chen, The genesis of oceanic crust: Magma injection, hydrothermal circulation, and crustal flow, *J. Geophys. Res., 98,* 6283–6297, 1993a.

Phipps Morgan, J., and Y. J. Chen, Dependence of ridge-axis morphology on magma supply and spreading rate, *Nature, 364,* 706–708, 1993b.

Phipps Morgan, J., and Y. J. Chen, Reply, *J. Geophys. Res., 99,* 12,031–12,032, 1994.

Phipps Morgan, J., E. M. Parmentier, and J. Lin, Mechanisms for the origin of mid-ocean ridge axial topography: Implications for the thermal and mechanical structure of accretion plate boundaries, *J. Geophys. Res., 92,* 12,823–12,836, 1987.

Purdy, G. M., L. S. L. Kong, G. L. Christeson, and S. Solomon, Relationship between spreading rate and the seismic structure of mid-ocean ridges, *Nature, 355,* 815–817, 1992.

Quick, J. E., and R. P. Denlinger, Ductile deformation and the origin of layered gabbro in ophiolites, *J. Geophys. Res., 98,* 14,015–14,027, 1993.

Reid, I. D., and H. R. Jackson, Oceanic spreading rate ands crustal thickness, *Mar. Geophys. Res., 5,* 165–172, 1981.

Ribe, N., The dynamics of plume-ridge interaction: 2. Offridge plumes, J. Geophys. Res., 101, 16,195–16,204, 1996.

Ribe, N., and W. L. Delattre, The dynamics of plume-ridge interaction, 3: The effects of ridge migration, Geophys. J. Int., 133, 511–518, 1998.

Ribe, N., U. R. Christensen, and J. Theissing, The dynamics of plume-ridge interaction, 1: Ridge-centered plumes, Earth Planet. Sci. Lett., 134, 155–168, 1995.

Rohr, K. M. M., B. Milkereit, and C. J. Yorath, Asymmetric deep crustal structure across the Juan de Fuca Ridge, *Geology, 16,* 533–537, 1988.

Shah, A. K., and W. R. Buck, Causes for axial high topography at mid-ocean ridges and the role of crustal thermal structure, *J. Geophys. Res., 106,* 30,865–30,879, 2001.

Shaw, W. J., and J. Lin, Models of ocean ridge lithospheric deformation: Dependence on crustal thickness, spreading rate, and segmentation, *J. Geophys. Res., 101,* 17,977–17,993, 1996.

Scheirer, D. S., K. C. Macdonald, D. W. Forsyth, S. P. Miller, D. J. Wright, M.-H. Cormier, and C. M. Weiland, A map series of the southern East Pacific Rise and its flanks, 15°S to 19°S, *Mar. Geophys. Res., 18,* 1–12, 1996.

Schilling, J.-G., Iceland mantle plume: Geochemical study of Reykjanes Ridge, Nature, 242, 565–571, 1973.

Schilling, J.-G., Upper mantle heterogeneities and dynamics, Nature, 314, 62–67, 1985.

Schilling, J.-G., Fluxes and excess temperatures of mantle plumes inferred from their interaction with migrating midocean ridges, *Nature, 352,* 397–403, 1991.

Sempéré, J.-C., J. R. Cochran, and SEIR Scientific Team, The Southeast Indian Ridge between 88°E and 118°E: variations in crustal accretion at constant spreading rates, *J. Geophys. Res., 102,* 15,489–15,505, 1997.

Shen, Y., and D. W. Forsyth, The effects of temperature and pressure dependent viscosity on three-dimensional passive flow of the mantle beneath a ridge-transform system, *J. Geophys. Res., 97,* 19,717–19,728, 1992.

Shen, Y., and D. W. Forsyth, Geochemical constraints on initial and final depths of melting beneath mid-ocean ridges, *J. Geophys. Res., 100,* 2211–2237, 1995.

Sinha, M. C., D. A. Navin, L. M. MacGregor, S. Constable, C. Peirce, A. White, G. Heinson, and M. A. Inglis, Evidence for accumulated melt beneath the slow-spreading Mid-Atlantic Ridge, *Phil. Trans. Roy. Soc. A., 355,* 233–253, 1997.

Sleep, N. H., Formation of oceanic crust: Some thermal constraints, *J. Geophys. Res., 80,* 4037–4042, 1975.

Sleep, N. H., Thermal structure and kinematics of the mid-ocean ridge axis, some implications to basaltic volcanism, *Geophys. Res. Lett., 5,* 426–428, 1978.

Small, C., A global analysis of mid-ocean ridge axial topography, *Geophys. J. Int., 116,* 64–84, 1994.

Small, C., Global systematics of mid-ocean ridge morphology, *in Faulting and Magmatism at Mid-Ocean Ridges,* edited by W. R. Buck, P. T. Delaney, J. A. Farson, and Y. Lagabrielle, Geophysical

Monograph, 106, pp. 1–25, American Geophysical Union, Washington D.C., 1998.

Small, C., and D. T. Sandwell, An abrupt change in ridge axis gravity with spreading rate, *J. Geophys. Res., 94*, 17,383–17,392, 1989.

Smewing, J. D., Mixing characteristics and compositional differences in mantle-derived melts beneath spreading axes: Evidence from cyclically layered rocks in the ophiolite of North Oman, *J. Geophys. Res., 86*, 2645–2659, 1981.

Su, W., and W. R. Buck, Buoyancy effects on mantle flow under mid-ocean ridge, *J. Geophys. Res., 98*, 12,191–12,205, 1993.

Tapponnier, P. and J. Francheteau, Necking of the lithosphere and the mechanics of slowly accreting plate boundaries, *J. Geophys. Res., 83*, 3955–3970, 1978.

Tolstoy, M., A. J. Harding, J. A. Orcutt, and J. Phipps Morgan, A seismic refraction investigation of the Australian Antarctic Discordance and neighboring South East Indian Ridges: Preliminary results, *Eos Trans. AGU, 76*, S275, 1995.

Vera, E. E., J. C. Mutter, P. Buhl, J. A. Orcutt, A. J. Harding, M. E. Kappus, R. S. Detrick, and T. M. Brocher, The structure of 0- to 0.2-m.y.-old oceanic crust at 9°N on the East Pacific Rise from expanded spread profiles, *J. Geophys. Res., 95*, 15,529–15,556, 1990.

Wang, X., and J. R. Cochran, Gravity anomalies, isostasy, and mantle flow at the East Pacific Rise crest, *J. Geophys. Res., 98*, 19,505–19,531, 1993.

White, R. S., D. McKenzie, and R. K. O'Nions, Oceanic crustal thickness from seismic measurements and rare earth element inversions, *J. Geophys. Res., 97*, 19,683–19,715, 1992.

Wilson, D. S., D. A. Clague, N. H. Sleep, and J. L. Morton, Implications of magma convection for the size and temperature of magma chambers at fast spreading ridges, *J. Geophys. Res., 93*, 11,974–11,984, 1988.

Yongshun John Chen, Computational Geodynamics Laboratory, Department of Geophysics, School of Earth and Space Sciences, Peking University, Beijing 100871, China, johnyc@pku.edu.cn

Some Hard Rock Constraints on the Supply of Heat to Mid-Ocean Ridges

Mathilde Cannat[1], Joe Cann[2], and John Maclennan[3]

In this paper, we examine many aspects of how heat from melts and from the mantle is transferred to the oceanic lithosphere at mid-ocean ridges, focusing on constraints that can be derived from rock specimen and geological observations at mid-ocean ridges and in ophiolites. These range from experimental constraints on the specific heat and latent heat of crystallisation of oceanic crustal materials, to geological constraints on how melt is distributed throughout the axial lithosphere, and partitioned between gabbros, dykes, and lava. We also discuss oceanic magma chamber processes and the episodicity of magma input at fast and slow-spreading ridges, and test four different ways of estimating the melt supply to mid-ocean ridges (seismic crustal thickness; regional axial depth; sodium and REE contents in MORB). We find that these four methods yield consistent results at East Pacific Rise sites, but that determination of the melt (and heat) supply to the crust of slow-spreading ridges is not as straightforward and requires a good integration of local scale geological, geophysical, and geochemical data.

Many of our conclusions concern the tectonically and lithologically complex slow-spreading ridges. We discuss these ridges under two crustal end-members: a magmatically-dominated, segment centre type, end-member, and a magma-poor, segment end type, end-member. In this magma-poor end-member mantle-derived peridotite must be a significant source of heat, and tectonic heat advection is expected as deeply-derived rocks are tectonically uplifted through the axial lithosphere and emplaced into the crust. We estimate that the total heat supply per unit area of young crust in this magma-poor end-member setting should not be much less than the heat supply per unit area of young crust formed in magmatically-dominated segment centres. This heat would not, however, be available in the same way to axial hydrothermal systems. We also show that there is substantial evidence for partial crystallisation of magma in the form of gabbros (and for the production of latent heat of crystallisation) at sub-crustal depth below slow-spreading ridges.

[1]Laboratoire de Géosciences Marines, CNRS-Institut de Physique du Globe de Paris, France
[2]School of Earth Sciences, University of Leeds, United Kingdom
[3]School of GeoSciences, University of Edinburgh, United Kingdom

Mid-Ocean Ridges: Hydrothermal Interactions Between the Lithosphere and Oceans
Geophysical Monograph Series 148
10.1029/148GM05

1. INTRODUCTION

The thermal regime of mid-ocean ridges (i.e., the temperature distribution within the axial lithosphere) is the result both of the supply of heat from the mantle and of its removal by conduction and advection into the ocean. The supply of heat to mid-ocean ridges is fundamentally controlled by the volume and distribution of melts supplied to the axial lithosphere, both in space and in time, and by the partition of heat between melts and solid hot rock. Conversely, the distribution of melt at the axis, and where it crystallises, are strongly influenced by the thermal regime of the ridge. Axial thermal gradients control the minimum depth at which permanent or semi-permanent melt bodies can be maintained [*Chen and Morgan*, 1990; *Phipps Morgan and Chen*, 1993], and the extent to which they may crystallise on axis [*Chen*, 2001; *Cherkaoui et al.*, 2003; *Maclennan et al.*, 2004].

In this paper we discuss constraints on the heat supply to mid-ocean ridges that can be derived from the study of rock specimens and from geological observations at mid-ocean ridges and in ophiolites. This discussion focuses on melts as principal carriers of heat to the ridge axis, but also covers the question of the contribution of heat from serpentinisation and other hydration reactions. We do not cover the major issue of how heat is removed from the axis by hydrothermal circulation. We do, however, discuss to some extent how differences in axial melt supply and distribution could affect the efficiency of ridge hydrothermal systems.

We review first the range of possible lithological structures in the crust and upper mantle as a function of spreading rate, and the distribution of heat supply that follows from these structures. We then evaluate the experimentally determined values of the heat capacity and latent heat of fusion or crystallisation for mantle and basaltic rocks, and the heat emitted during hydration reactions, and quantify the space and time averaged heat supply to mid-ocean ridges as a function of spreading rate. We next address the constraints that mid-ocean ridge basalt (MORB) compositions can provide on how much melt is supplied to mid-ocean ridges, and the question of the depth at which melts crystallise within the axial region. There follow sections dealing with the construction of the lower crust at fast and slow ridges, with the use of magma geochemistry to estimate the ratio of cumulates to lavas plus dykes at slow-spreading ridges, and with the timescales of episodicity of magma supply at fast and slow-spreading ridges from a variety of indicators.

2. LITHOLOGICAL STRUCTURE OF THE OCEAN CRUST AS A FUNCTION OF SPREADING RATE

2.1. Fast-Spreading Ridges

A consensus, following from the Penrose Conference on Ophiolites in the early 70s [Penrose Conference on Ophiolites, 1972], has developed that the lithological structure of the crust at fast-spreading ridges is similar to that of major ophiolites such as Troodos and Oman, capped by a unit of extrusives that overlies a unit of sheeted dykes. Beneath the sheeted dykes is a plutonic section that overlies ultramafics and the

upper mantle. In such a structure the heat supply is dominated by upward advection of magma, with heat coming both from latent heat of crystallisation of the melt and from the specific heat released as the solid rock cools. Though the ophiolitic comparison has yet to be tested directly by drilling, there are many observations that indirectly favour it, such as the presence of sheeted dykes at ODP Hole 504B [*Anderson et al.*, 1982; *Adamson*, 1985] and in the walls of Hess Deep [*Francheteau et al.*, 1992; *Karson et al.*, 1992; *Karson*, 2002]. Also in favour of a widespread unit of sheeted dykes at fast-spreading ridges is the narrow zone of fissuring observed along the crest of the East Pacific Rise [*Crane*, 1987; *Wright et al.*, 1995], which, if interpreted as fissures above dykes that failed to reach the surface, would require that those dykes form the central part of a sheeted dyke unit [*Kidd*, 1977].

The gabbroic-plutonic nature of the lower crust at the East Pacific Rise (EPR) is strongly suggested by the seismic identification of a thin sill-like axial magma chamber at a depth corresponding to the base of the sheeted dykes in the ophiolitic model [*Detrick et al.*, 1987; *Vera et al.*, 1990; *Detrick et al.*, 1993a; *Kent et al.*, 1993], and by the demonstration of a thick zone of low seismic velocity interpreted as a crystal mush zone beneath the axial magma chamber [*Harding et al.*, 1989; *Toomey et al.*, 1990; *Dunn et al.*, 2000]. Construction of the lower crust in such a magma-rich environment would indicate a lower crust of gabbroic composition, and gabbroic rocks have been recovered from equivalent levels in the Hess Deep [*Francheteau et al.*, 1990; *Hekinian et al.*, 1993; *Coogan et al.*, 2002a]. The simplified structure of the spreading centre producing such a basaltic crust is shown in Figure 1. The 6 km thickness of this crust would derive both latent heat and specific heat from basaltic magma and its solidified equivalents. The contribution of specific heat from the mantle would be small close to the spreading axis.

The distribution of the heat supply to the axial lithosphere and its availability for driving hydrothermal circulation is very different between the three lithological layers of the crust.

The extrusives lose all of their specific and latent heat to the ocean water almost as soon as they are erupted, and do not contribute to deep hydrothermal circulation. The sheeted dykes in Hole 504B were encountered at a depth of about 650 metres [*Adamson*, 1985; *Becker and et al.*, 1989]. Seismic reflection surveys close to the EPR ridge axis show a reflector (the Layer 2A-2B reflector) at about that depth [*Hooft et al.*, 1996 ; *Carbotte et al.*, 1997] which has been interpreted as coming from a rather abrupt lava-sheeted dyke boundary. If that identification is correct, then the lava unit is typically a few hundred metres thick.

Most dykes in an ophiolitic dyke swarm have well-defined chilled margins on each side, and the crystal size of dyke centres in ophiolites is related to dyke thickness, not to depth

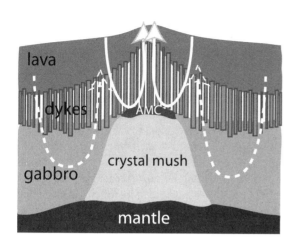

Figure 1. Idealized section across a ridge axis, based on the interpretation of seismic data from the fast-spreading East Pacific Rise (after *Sinton and Detrick* [1992]). Vertical dimension of sketch: ~7 km; horizontal dimension: ~12 km. Respective thicknesses of the lava, dykes and gabbro units are not to scale. Melt extracted from the mantle crystallises as gabbros, either in the AMC (Axial Magma Chamber) only [*Phipps Morgan and Chen*, 1993; *Chen*, 2001], or also in the crystal mush domain underneath the AMC [*Boudier et al.*, 1996; *Kelemen and Aharanov*, 1998; *MacLeod and Yaouancq*, 2000; *Cherkaoui et al.*, 2003; *Maclennan et al.*, 2004]. Arrows : near-axis hydrothermal cells.

in the sheeted dyke unit [*Kidd*, 1977]. The 504B ODP drill hole yields similar results [*Iturrino et al.*, 1996]. This indicates that dykes must be intruded into a well-cooled environment, losing both latent and specific heat to axial hydrothermal systems over a short period of time. *Jaeger* [1957] calculated that solidification of a dyke intruded into a well-cooled medium would be complete in $0.01D^2$ years, where D is the dyke thickness in metres. For a typical 1.5 m wide ophiolitic dyke, this corresponds to ~ 8 days. A 10 m wide dyke would solidify after about a year. The dykes would cool rapidly to ambient temperature; for a 1.5 m dyke this cooling would be complete after about one month, and for a 10 m dyke after about 3.5 years. If dykes are intruded in swarms, as happened at Krafla in Iceland [*Tryggvason*, 1980], then the temperature locally might be somewhat raised for a few years, but in general the heat from dyke intrusion must be dissipated and transferred to axial hydrothermal systems very rapidly after intrusion.

The thickness of the sheeted dykes is more difficult to estimate than that of the extrusives. ODP Hole 504B drilled about 1100 m of sheeted dykes without coming to plutonics [*Alt et al.*, 1996], so the unit must be thicker than that at this site. In ophiolites, estimates of about 1.5 km have been made [*Pallister*, 1981; *Lippard et al.*, 1986], but these estimates are hard to justify precisely in the absence of horizontal markers in the dyke complex. The depth of the axial magma chamber

at fast-spreading ridges gives an indirect estimate of the maximum thickness of sheeted dykes, since the sheeted dykes must represent the cover of the magma chamber at the spreading axis, perhaps thinned if the axial magma chamber intrudes the base of the dykes as it does in places in the Troodos ophiolite [*Gillis*, 2002]. Seismic estimates of the depth of the magma chamber [*Harding et al.*, 1989; *Kent et al.*, 1993; *Tolstoy et al.*, 1997; *Babcock et al.*, 1998] give values of 1 to 2 km for the fast-spreading EPR.

Given the above estimates for the thickness of the overlying units, the plutonic unit must be about 4 kilometres thick. Because of this greater thickness, it carries a large proportion of the latent and specific heat supply to the ridge axis. How this heat is removed from the plutonic unit is a matter of controversy, since it is bound up with the different models for the formation of the lower crust at fast-spreading ridges. This question will be discussed in Section 6.

Close to a fast-spreading axis, there are therefore three temperature domains of importance: the upper crust, including extrusives and sheeted dykes, about 2 kilometres thick, cooled to temperatures of less than ~ 400°C (upper bound for the temperature of black smoker type hydrothermal fluids); a very thin thermal boundary layer, possibly only a few tens of metres thick separating the upper crust and the axial magma chamber [*Nehlig*, 1993; *Lister*, 1995]; and the lower crust, close to magmatic temperatures at the axis.

Along-axis crustal thickness variations of up to 2.5 km have been documented in the 9°–11°N region of the EPR [*Barth and Mutter*, 1996 ; *Canales et al.*, 2003]. However, the small along-axis mantle Bouguer gravity anomaly (MBA) gradients [*Madsen et al.*, 1990; *Wang et al.*, 1996], and the very gentle along-axis topographic slopes (typically 1 m/km or less at the EPR) suggest that magma is distributed evenly along fast-spreading segments. This discrepancy between seismic crustal thickness estimates, and the gravity and topographic record suggests that regions of thicker crust may be underlain by higher density upper mantle, possibly containing a smaller melt fraction [*Wang et al.*, 1996 ; *Canales et al.*, 2003]. Along-axis variations in the melt content of the uppermost mantle have recently been proposed based on seismic tomography data [*Dunn et al.*, 2001]. If these account for the discrepancy between seismic and gravity results, the melt supply, integrated over the crust and uppermost mantle, would not vary significantly along-axis the fast-spreading East Pacific Rise. An even along axis distribution of melt would suggest either that there is no significant focusing of melt in the mantle towards segment centres (as represented in Figure 2a), or that magma is efficiently redistributed along axis at crustal or upper mantle level [*Bell and Buck*, 1992 ; *Lin and Phipps Morgan*, 1992].

2.2. Slow-Spreading Ridges

The topography of slow-spreading ridges is much more rugged than at fast-spreading ridges, and a major axial valley is present, up to 1.5 km deep and with a floor several kilometres wide. Thermal modelling [*Chen and Phipps Morgan*, 1996], studies of hypocentre depths [*Francis and Porter*, 1973; *Wolfe et al.*, 1995; *Toomey et al.*, 1988], and active seismic surveys [*Fowler*, 1976; *Detrick et al.*, 1990; *Magde et al.*, 2000] suggest that crustal magma chambers at the axis are likely to be isolated in both space and time. A large body of partially

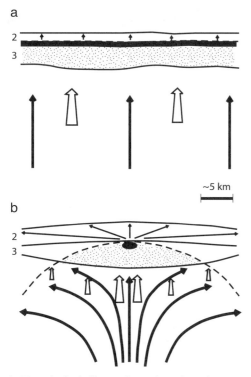

Figure 2. Hypothetical along-axis sections through (a) a segment of fast-spreading crust, and (b) a magmatically-dominated segment of slow-spreading crust (see text). Same approximate scale, no vertical exaggeration. Solid arrows are mantle flow lines. In these sketches, mantle flow is drawn as 2-dimensional in the fast-spreading case, and 3-dimensional in the slow-spreading case (see text for discussion). Open arrows are melt delivery trajectories in the mantle. In these sketches, it is inferred that melt delivery is uniform along axis in the fast-spreading case, while more melt is delivered to segment centres at slow-spreading ridges (see text for discussion). The numbers 2 and 3 refer to seismic layers 2 and 3, and the thin arrows in layer 2 show possible melt trajectories in the upper crust. The horizontal thick line in (a) represents the axial magma chamber, and the black ellipse in (b) represents a non permanent magma chamber at the centre of the slow-spreading segment. The stippled region of layer 3 is that part of layer 3 at or close to the melting point of basalt. The ends of the slow-spreading segment in (b) may contain a substantial fraction of variably serpentinised peridotite.

magmatic material has been unambiguously imaged only once at the axis of a slow-spreading ridge [*Sinha et al.*, 1997; *Navin et al.*, 1998]. Lower crustal seismic velocities suggesting melt fractions of a few percent have been modelled under the Snake Pit hydrothermal vent field at 23°20'N on the MAR [*Canales et al.*, 2000a]. The structure of slow-spreading ridges is also apparently more complex than that of fast-spreading ridges, to judge from the variety of rock types that outcrop close to the spreading axis [*Dick*, 1989; *Cannat*, 1993]. And determining the deep structure of slow-spreading ridges is made difficult by the apparent absence of major ophiolitic bodies that have the lithological complexity seen on the ocean floor.

Large along-axis variations in axial depth and crustal thickness [*Kuo and Forsyth*, 1988; *Lin et al.*, 1990; *Tolstoy et al.*, 1993; *Detrick et al.*, 1995; *Hooft et al.*, 2000] indicate that there are significant differences between the crust formed at segment centres and at segment ends of slow-spreading ridges. Most slow-spreading segment centres have well-developed axial volcanic ridges and a characteristic volcanic small-scale topography. On the other hand, segment ends (and in some places whole segments) commonly show weaker volcanic activity, and outcrops of gabbros and serpentinised peridotite [*Cannat et al.*, 1995]. In thermal terms, it is therefore useful to discuss slow-spreading ridges under two crustal end-members, keeping in mind that many sites may be intermediate between these two end-members. At ultra-slow-spreading ridges such as the Southwest Indian and Gakkel ridges, the dichotomy between these two end-members may be more marked: a greater proportion of the seafloor appears to be composed of gabbros and peridotite outcrops than at the Mid-Atlantic Ridge [*Dick*, 1989; *Hellebrand et al.*, 2002; *Dick et al.*, 2003; *Seyler et al.*, 2003], and the volcanic constructions are up to 1000 m high, suggesting much greater variability in lava thickness than at the Mid-Atlantic Ridge [*Cannat et al.*, 1999; *Cochran et al.*, 2003].

2.2.1. Magmatically-Dominated, Segment Centre End Member. Crustal thickness in a magmatically-dominated slow-spreading ridge segment decreases from segment centres to segment ends [*Lin et al.*, 1990; *Tolstoy et al.*, 1993; *Hooft et al.*, 2000], indicating that melt supply is focused towards the segment centre (Figure 2b). As proposed by *Lin et al.* [1990], there may be diapiric instabilities in the mantle beneath the ridge. Melt focusing could also, however, be a consequence of the thermal and mechanical effects of ridge axis discontinuities: because of these effects, mantle flow beneath slow-spreading ridges is expected to be three-dimensional (which does not mean necessarily diapiric), and the axial thermal regime should vary along axis, with a thicker lithosphere at ridge segment ends [*Phipps Morgan and Forsyth*, 1988]. It has been proposed that melt focusing toward slow-spreading segment centres is due to along-axis melt migration in a permeable horizon at the sloping base of the axial lithosphere [*Magde et al.*, 1997].

The floor of the median valley at segment centres is often occupied by an axial volcanic ridge a few hundred metres high and a few kilometres wide, rising to an along-axis peak close to the centre of the segment. Away from this peak, along-axis topographic gradients can reach tens of metres per kilometre, similar to those of the Puna Ridge (the offshore extension of the East Rift Zone of Kilauea volcano; [*Smith and Cann*, 1999]). Since we know that the axis of the Puna Ridge is fed by dykes propagating from the magma chamber beneath Kilauea, 50–100 km away, it is likely that the along-axis slopes on the Puna Ridge are characteristic of rifts fed by dyke intrusion [*Fialko and Rubin*, 1998]. Similar along-axis slopes in the Mid-Atlantic Ridge (MAR) can thus be taken to represent sections of the ridge axis fed remotely from a central magma chamber, suggesting that the upper crust of magmatically-dominated spreading segments is at least partially constructed by intrusion of dykes along the axis of the ridge from magma chambers beneath the segment centres [*Smith and Cann*, 1999].

A crustal magma chamber as shown in Figure 2b would persist only as long as high magma fluxes to the segment centre are maintained. In that case, the crust at the segment centre would include units of extrusives, dykes and gabbroic plutonics, similar to those of fast-spreading ridges. Heat would be supplied at the segment center in much the same way as at fast-spreading ridges, but along-axis conduction and 3D hydrothermal convection would be expected to increase the cooling rate. The structure of the crust at the centre of magmatically-dominated slow-spreading ridge segments would not, however, be as systematically layered as that of the major ophiolites. As discussed above, many dykes at slow-spreading ridges probably propagate along-axis from isolated magma chambers (Figure 2b), and feed eruptions along axial volcanic ridges. Within one axial volcanic ridge, dykes probably form sheeted dyke units up to a few kilometers wide, with the width depending on the duration of volcanic activity along this ridge, similar to the rift zones in the extinct volcanoes of the Hawaiian islands [*Walker*, 1987]. These rift zones may overlap and interact with each other, forming sheeted dyke units of plurikilometric extension (as reported from the wall of the Vema Transform Fault in the Equatorial Atlantic; [*Auzende et al.*, 1989]), but a continuous layer of sheeted dykes is unlikely.

The height of axial volcanic ridges in slow-spreading ridge segments is a local measure of minimum extrusive thickness, and these are commonly 200 m high, and occasionally as high as 500 m [*Smith and Cann*, 1999]. A lava thickness of 500 m was recorded in the only ODP hole (395A) to penetrate

into dykes on a slow-spreading ridge [*Aumento et al.*, 1977]. As mentioned before, axial volcanic ridges can be up to 1000 m high at ultra-slow ridges [*Cannat et al.*, 1999; *Cochran et al.*, 2003].

If it is hard to say much that is definite about dykes in these magmatically-dominated slow-spreading segment centres, it is even harder to write meaningfully about the role of gabbro in their structure, as gabbros have not yet been sampled from such settings. Basalt geochemistry gives some clue (see Section 8), but there is great ambiguity about the interpretation of seismic experiments, for example, since, as is discussed below, partly-serpentinised peridotite may make up a substantial part of the seismic crust towards segment ends.

2.2.2. Magma-Poor, Segment-end End Member. Magma focusing toward slow-spreading segment centres leads to magma depletion at segment ends. Dyke swarms propagating from the segment centre could feed lava flows at segment ends, perhaps associated with underlying diabase bodies. These may intrude into, or be intruded by, gabbro plutons and dykes representing melt supplied from the underlying mantle. Drilling at ODP site 735B (Southwest Indian Ridge) has shown that such end of segment gabbro bodies may be > 1.5 km thick and formed by repeated melt injections ([*Natland and Dick*, 2001]; see section 7).

Hypocentre depths of up to 8 km at segment ends of the MAR indicate that the crust is underlain, right on axis, by cooled lithospheric mantle [*Toomey et al.*, 1988; *Kong et al.*, 1992; *Wolfe et al.*, 1995]. In this configuration (Figure 3), magma bodies emplaced at crustal depths are intrusive into colder material, and the possibility exists for gabbro bodies to crystallise in the sub-axial mantle lithosphere [*Cannat*, 1996].

The ends of many slow-spreading ridge segments exhibit a seafloor morphology that is different from that of the volcanically dominated segment centres, and serpentinised peridotites are often found to outcrop. Such exposures are widespread on ultra-slow-spreading ridges [*Dick*, 1989; *Hellebrand et al.*, 2002; *Seyler et al.*, 2003] and have also been reported from the centre of some slow-spreading ridge segments [*Cannat et al.*, 1997c]. Outcrops of serpentinised peridotites are intermixed with basalts and gabbros and characteristically occur in areas of thin seismic crust (< 4-5 km).

The seismic signature of these thin crust domains is distinct from that of thicker crust at nearby segment centres: the high velocity gradient layer 2 has near normal thicknesses (1.5 to 3 km), while layer 3 is thin or absent [*Detrick et al.*, 1993b; *Tolstoy et al.*, 1993; *Minshull et al.*, 1998; *Canales et al.*, 2000b]. The relatively constant thickness of layer 2 despite the large changes in layer 3 suggests that layer 2 thickness,

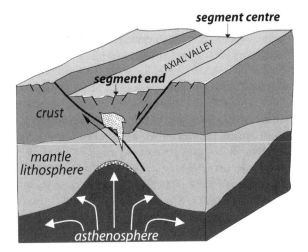

Figure 3. Idealized block diagram of a slow-spreading ridge (half) segment. Vertical dimension of sketch: ~15 km; horizontal dimensions: ~25 km. Hypocenter depths up to 8 km at the MAR [*Toomey et al.*, 1988; *Kong et al.*, 1992; *Wolfe et al.*, 1995] indicate that the seismic crust (dark grey) is underlain by cooled, lithospheric mantle (light grey). Segment end (front) has thinner crust and thicker axial lithosphere than segment centre (back). Melt extracted from the asthenospheric mantle (black) may form long lasting magma bodies (dotted grey) only in the lowermost lithosphere. Shallower magma bodies (also dotted grey) should be transient and intrude into colder tectonised material that may include variably serpentinised peridotites.

at least in these thin crust settings, is controlled more by porosity and alteration than by lithology [*Muller et al.*, 1999; *Canales et al.*, 2000b]. Thin crust domains also commonly have relatively high velocity gradients throughout the entire crustal section. This pattern is consistent with serpentinisation of a large ultramafic crustal component at rates that decrease with depth [*Detrick et al.*, 1993b; *Canales et al.*, 2000b]. Material interpreted as partially serpentinised peridotite, with seismic velocities higher than normal lower oceanic crust (Vp ~ 7.2 km/s), but lower than typical upper mantle velocities (Vp > 7.6 km/s), is also sometimes present at the base of the crust [*Minshull et al.*, 1991], or in the uppermost mantle [*Canales et al.*, 2000b]. Seismic data from slow-spread crust are therefore consistent with the presence of a significant component of variably serpentinised ultramafics in thin crust domains. There is, however, no indication for large degrees of serpentinisation at depths greater than about 5 km, even in the valley of large offset transform faults, where extensive tectonism should most favor deep water penetration [*Minshull et al.*, 1998]. Furthermore, segment ends in slow-spreading environments commonly have seismic crustal thicknesses < 3 km [*Tolstoy et al.*, 1993; *Wolfe et al.*, 1995; *Muller et al.*, 1999; *Canales et al.*, 2000b]. This

suggests that pervasive serpentinisation at ridge axes, even in these highly tectonised settings, generally does not exceed 3 km. *Wilcock and Delaney* [1996] argued that volume increase during serpentinisation should fill the porosity created by thermal contraction and tectonism, so that the permeability in ultramafic rocks should be low. This could restrict pervasive serpentinisation of large volumes of peridotite to the vicinity of frequently reactivated faults, and of cooling gabbro or diabase intrusions.

In the magma-poor slow-spreading end-member, specific heat of peridotite released during cooling as the mantle rises towards the surface, and the heat released by the exothermic hydration of peridotite to serpentinite should contribute to the axial thermal budget. Heat would be extracted by conduction and by advection in hydrothermal fluids circulating along faults and cracks [*Lister*, 1974]. Latent and specific heat would also be derived from the crystallisation and cooling of gabbro and diabase bodies intruded into the peridotite.

3. QUANTIFICATION OF SOURCES OF HEAT IN MANTLE AND OCEAN CRUST

In this section we first examine experimental results that pertain to the specific and latent heats of both basalt and mantle, and to the heat released during the transformation of peridotite to serpentinite.

3.1. Heat Capacity and the Specific Heat of Mantle and Basaltic Materials

Heat capacity (Cp) is a measure of the capacity of a material to store or release heat as temperature varies : for a given pressure, it is the slope of the heat content vs. temperature curve. In crystalline materials, this slope decreases at decreasing temperature and is zero at T= 0 °K. In melts, Cp does not vary significantly with temperature and is ~ 1600 J kg^{-1} K^{-1} for basaltic compositions [*Lange and Navrotsky*, 1992]. Cooling basaltic melt by 100°C should therefore release ~0.16 MJ/kg.

Experimentally determined values of Cp for the main mineral constituents of mantle peridotites (forsterite, enstatite and diopside) are similar (Figure 4). The Cp of mantle peridotites should therefore be only marginally dependent on modal composition, and vary between ~ 1300 J kg^{-1} °K^{-1} at temperatures > 1000°C, and ~ 800 J kg^{-1} K^{-1} at 0°C (Figure 4). Basaltic rocks (gabbros, diabase and basalts) also contain significant modal Ca-rich plagioclase. The Cp of pure anorthite is close to that of diopside (Figure 4). Cp of basaltic rocks should thus be only marginally lower than that of mantle rocks, at a given temperature.

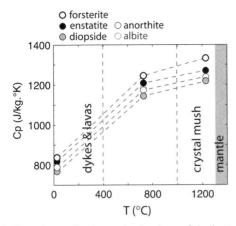

Figure 4. Experimentally determined values of the heat capacity (Cp) for forsterite, enstatite, diopside, anorthite and albite [*Robie et al.*, 1978; *Berman*, 1988; *Richet and Fiquet*, 1991]. Temperature domains labelled as dykes and lavas, crystal mush, and mantle correspond to the range of temperature inferred to prevail on axis in the corresponding domains of Figure 1 (see text).

3.2. Latent Heat of Crystallisation of Basaltic Melts, Latent Heat of Mantle Melting

The latent heat released by basaltic melts upon crystallisation is a major component of the heat budget of mid-ocean ridges. The latent heat of mantle melting is also an important parameter in thermal calculations: for a given mantle potential temperature, input of higher values of the latent heat of melting results in calculation of lower maximum and mean degrees of melting, hence a smaller melt supply to the ridge axis (e.g., [*Langmuir et al.*, 1992]). Table 1 presents experimentally determined values of the enthalpy change upon melting, or latent heat of melting (Hf), for the principal mineral constituents of mantle peridotites (forsterite, enstatite and diopside), for Fe olivine and Na clinopyroxene end-members

(fayalite and jadeite), and for Ca and Na end-members of the plagioclase series. These latent heat values have been measured in each case at the mineral's liquidus temperature (T_1). They can be corrected back to a common temperature, for example the inferred liquidus temperature of a melting mineral assemblage, using the following formula [*Hess*, 1992], that corresponds to dHf /dT~ 400 J kg^{-1} K^{-1} :

$$H_f(T) = \int_{T_1}^{T} \Delta Cp \, dT \qquad (1)$$

where ΔCp is the heat capacity change on fusion.

Variation of latent heat with pressure can be estimated using the following formula [*Hess*, 92], that corresponds to dHf /dP~ 28 10^3 J kg^{-1} GPa^{-1} :

$$H_f(P) = H_f(1) + \int_{1}^{P} \Delta V (1 - \Delta \alpha) \, dP \qquad (2)$$

where ΔV is the change of volume on melting, and $\Delta \alpha$ is the change in thermal expansion on melting.

Experimental measurements of latent heat performed directly on mineral assemblages are very close to values calculated from latent heat values of the individual mineral components, weighted by their normative proportions, and corrected to the assemblage's liquidus temperature [*Kojitani and Akaogi*, 1997]. This demonstrates that mixing enthalpies of melts are small, and allows for direct determination of a melt's latent heat of crystallisation, using its normative mineralogy. For example, latent heat calculated at 1 atm. and 1523°K for a composition inferred to represent primary melts generated by melting of the mantle at pressures of 1.1 GPa (composition a in Table 2), is ~ 0.5 MJ kg^{-1} [*Kojitani and Akaogi*, 1997]. This inferred composition is not very different from that of a little fractionated MORB (composition b in

Table 1. Experimentally determined values of the heat of fusion (Hf) for selected minerals. Tf : melting temperature at 1 atm. Hf at 1523°K: Hf value recalculated at 1523°K using the formula proposed by *Hess* [1992]. References : (1) *Navrotsky et al.* [1989] ; (2) *Richet et al.* [1993] ; (3) *Orr* [1953] ; (4) *Richet and Bottinga* [1986] ; (5) *Stebbins et al.* [1984] ; (6) *Richet and Bottinga* [1984] ; (7) *Stebbins et al.* [1983].

MINERAL		Tf (°K)	Hf (J/g)	Hf at 1523°K	ref.
forsterite	Mg2SiO4	2163	810	554	(1)
forsterite	Mg2SiO4	2174	1009	748	(2)
fayalite	Fe2SiO4	1490	452	439	(3)
enstatite	MgSiO3	1834	750	625	(4)
diopside	CaMgSi2O6	1665	639	582	(5)
jadeite	NaAlSi2O6	1100	293	462	(6)
anorthite	CaAl2Si2O8	1830	492	369	(6)
albite	NaAlSi3O8	1373	240	300	(7)

Table 2. Melt composition in oxide weight fractions. a- Inferred composition of primary melts generated by melting of the mantle at a pressure of 1.1 Gpa [*Kojitani and Akogi*, 1997]. b- compositionof MAR basalt sample ALV-2004-3-1 [*Grove et al.*, 1992].

	(a)	(b)
SiO2	47.4	49.6
Al2O3	19.0	16.3
FeO	7.3	8.65
MgO	11.0	9.13
CaO	12.5	11.7
Na2O	2.7	2.6

Table 2). The latent heat of crystallisation of basaltic melts in the crust should therefore be ~ 0.5 MJ kg^{-1}.

Determination of the latent heat of mantle melting is less straightforward because melting does not proceed to 100%, so that the composition of melt products differs substantially from that of the peridotite that is melting. The latent heat value of 0.75 MJ kg^{-1} calculated by *Hess* [1992] for melting of a forsterite, enstatite and diopside assemblage at 1 GPa and 1700 K is therefore most probably too large. The 0.59 MJ kg^{-1} value calculated at 1 GPa and 1523 K (solidus temperature) by *Kojitani and Akaogi* [1997] for partial melting of a peridotite yielding melts of composition (a) in Table 2, is probably a better choice, although it should be corrected to slightly higher temperatures (between solidus and liquidus temperatures), corresponding to the inferred mean degree of mantle melting beneath mid-ocean ridges (5% to 20%; [*Dick et al.*, 1984; *Klein and Langmuir*, 1987]). Our preferred value for the latent heat of mantle melting at a mid-ocean ridge, integrated over the thickness of the melting column, is therefore ~ 0.6 MJ kg^{-1}.

3.3. Heat of Reaction from Serpentinisation and Basalt Hydration

The hydration of peridotite to serpentinite is an exothermic reaction, as is the hydration of ferromagnesian minerals in basaltic rocks to chlorite and other metamorphic phases. Serpentine minerals are only stable below about 400°C, and serpentinisation rates are probably fastest at about 250°C [*Martin and Fyfe*, 1970]. Heat released by serpentinisation of peridotite is ~ 0.25 MJ kg^{-1} [*Fyfe and Lonsdale*, 1981], which is about half of the latent heat released by crystallisation of basalt, and thus potentially an important source of heat at mid-ocean ridges. Significant heating can, however, be generated by serpentinisation only if fluid flow rates are low, and if serpentinisation rates are very high [*Lowell and Rona*, 2002]. As discussed in *Seyfried et al.* [this volume],

this means that heat from serpentinisation probably does not contribute significantly to high-temperature hydrothermal circulation, as in black smoker systems. However heat from serpentinisation may drive hydrothermal circulation at temperatures of tens of degrees Celsius [*Seyfried et al.*, this volume]. Heat derived from hydration of ferromagnesian minerals in basalt is of the same order as that from serpentinisation (data from *Robie et al.* [1978]), but the proportion of hydrated minerals in basaltic rocks is small, and the contribution of heat from this source to hydrothermal circulation is likely not to be important.

3.4. Availability of Heat From the Axial Regions of Mid-Ocean Ridges

We now estimate the heat flux per kilometre of mid-ocean ridge crest as a function of spreading rate and crustal lithology, under a number of different assumptions.

From the discussion above, we take the latent heat of basalt as 0.5 MJ kg^{-1} and the specific heat for basaltic melt and for basalts as 1600 J kg^{-1} K^{-1} and 1100 J kg^{-1} K^{-1}, respectively. The latent heat is assumed to be released at a temperature of 1150°C (crystallisation interval for basaltic melts: 1250 to 1050°C ; [*Sinton and Detrick*, 1992]). Specific heat from basaltic melts is assumed to be released over the 1250° to 1150°C cooling interval, yielding 0.16 MJ/kg. If basalt is cooled from its crystallisation temperature (~ 1150°C) to 0°C, 1.42 MJ kg^{-1} of specific heat will be released, giving a total heat released of 1.92 MJ kg^{-1}. If instead it is cooled only to 400°C (an upper temperature limit for high-temperature hydrothermal fluids), 0.98 MJ kg^{-1} of specific heat will be released giving a total heat released over that temperature interval of 1.48 MJ kg^{-1}.

With a specific heat of 1200 J kg^{-1} K^{-1} (Figure 4), peridotite can supply 1.38 MJ kg^{-1} of specific heat between 1150°C and zero (or 0.9 MJ kg^{-1} between 1150°C and 400°C). In addition 0.25 MJ kg^{-1} may be released from exothermic serpentinisation at around 250°C [*Seyfried et al.*, this volume], giving a total of 1.63 MJ kg^{-1} if it is 100% serpentinised, and cooled all the way to 0°C. Only 0.9 MJ kg^{-1} is released if the rock is cooled from 1150°C to 400°C.

At fast-spreading ridges, the heat that is available close to the axis is essentially that from 6 km of basaltic crust. The specific and latent heats together yield a heat output per kilometre of:

$$h\, s\, \rho (Cp_1 \Delta T_1 + Hf + Cp_2 \Delta T_2) 10^{-3} \text{MW/km} \qquad (3)$$

where h is the thickness of the basaltic layer in metres, s the full spreading rate in m/s, ρ the rock density in kg/m^3, Hf the latent heat in J kg^{-1} K^{-1}, Cp$_1$ the specific heat of basaltic melt and Cp_2 the specific heat of basalt in J kg^{-1}, ΔT_1 the average cooling interval for melts (1250–1150°C), and ΔT_2 the average cooling interval for basaltic rocks.

This relationship can be used to estimate the heat available to drive high-temperature axial hydrothermal systems at a fast-spreading ridge. Extrusive lavas (estimated thickness: 600 m; Section 2.1) cool to 0°C, losing most of their heat directly to seawater. Dykes (estimated thickness: 1400 m) crystallise and cool from magmatic to hydrothermal temperatures at the ridge axis. For a temperature drop ΔT_2 of 1150 to 400°C, a rock density of 2.8 10^3 kg/m^3, and a spreading rate of 100 mm/yr, this yields ~19.4 MW per km of ridge length. Gabbros (estimated thickness 4 km) crystallise on or near axis, producing ~18.6 MW km^{-1} in the form of latent heat, and ~6 MW/km in the form of melt specific heat. The extent to which fast-spreading ridges gabbros cool due to near axis hydrothermal circulations is a matter of current debate ([*Chen*, 2001; *Cherkaoui et al.*, 2003; *Nicolas et al.*, 2003; *Maclennan et al.*, 2004] ; see section 6). Two extreme cases bracket heat supply estimates [*Mottl*, 2003]: - hydrothermal cooling is restricted to the top of the upper axial magmatic lense (heat from dykes + melt specific heat and latent heat from gabbros = ~44 MW km^{-1}); - or extensive hydrothermal cooling of the whole gabbro section to hydrothermal temperatures occurs near axis (providing an additional ~31 MW km^{-1} to axial hydrothermal vents). The latter case provides an absolute upper estimate of 75 MW km^{-1} for the heat available to axial hydrothermal systems at a fast (100 mm/yr) spreading ridge.

As noted by *Mottl* [2003], near axis cooling of the entire crust to hydrothermal temperatures is probable at slow-spreading ridges. As mentioned above, the thickness of extrusive and dyke units in magmatically-dominated slow-spreading segment centres is not well known. Using the fast-spreading ridge values for these variables, as well as for h, ρ, Cp_1, Cp_2, ΔT_1 and Hf, and with a temperature drop ΔT_2 of 1150–400°C, yields an estimate of ~18.6 MW km^{-1} (~4.8 MW km^{-1} from the dykes, and ~13.8 MW km^{-1} from the gabbros) for the heat available to axial hydrothermal systems at the centre of a magmatically-dominated slow (25 mm/yr) spreading segment. Seismic data suggest that the crust at magmatically-dominated MAR segment centres is actually thicker than the average 6 km [*Tolstoy et al.*, 1993; *Hooft et al.*, 2000]. With a crustal thickness of 8 km (and a lava thickness of 600 m), the heat available at the centre of such a segment would be ~25.5 MW/km.

In the slow-spreading, magma-poor end-member, rising peridotites will also be sources of heat. The maximum depth of 8 kilometres for earthquakes in these parts of the ridge suggests that the axial lithosphere is brittle to that depth, and is therefore cooled at least below 700°C. The seismic veloc-

ity structure (see Section 2.2) suggests that serpentinisation is not pervasive until the peridotite arrives within ~3 km of the surface, and is not complete except in the upper kilometre or so. The total heat supplied by peridotites in an hypothetical completely amagmatic end-member case can be approximated as :

$$(Z\, s\, \rho\, C\rho\, \Delta T + z\, s\, \rho\, f\, H)/10^3 \text{ MWkm}^{-1} \qquad (4)$$

where s is the full spreading rate in m/s, Z is the thickness of peridotite cooled into the brittle lithosphere in metres, ρ is the peridotite density in kg/m^3, z the thickness of pervasively serpentinised peridotite in metres, f the mean fraction serpentinised, H the heat emitted by the serpentinisation reaction in J kg^{-1} K^{-1}, Cp the specific heat of peridotites in J kg^{-1}, and ΔT the mean temperature drop.

If Z is 8 km, ΔT is 600°C, z is 3 km, ρ is 3000 kg/m^3, and f is 25%, this gives a heat supply of 16 MW km^{-1} (of which only ~0.5 MW km^{-1} would come from serpentinisation reactions) at a ridge spreading at 25 mm/yr. This is not much less than the heat available from the basaltic crust in magmatically-dominated segment centres at the same spreading rate. This heat may be accessible to axial hydrothermal systems (faults and cracks could channel hydrothermal fluids down to the lower brittle lithosphere; [*Lister*, 1974]). However, it is unlikely that a thin thermal boundary layer would develop in this amagmatic end-member setting as it presumably does when magmatic bodies are present [*Lister*, 1995].

4. USING BASALT COMPOSITION, SEISMIC CRUSTAL THICKNESS, AND REGIONAL AXIAL DEPTH TO CONSTRAIN THE MELT SUPPLY TO MID-OCEAN RIDGES

Thermal models of ridge axes are generally constructed on the assumption that the volume of melt supplied to the ridge for each increment of plate spreading is enough to build a magmatic layer equivalent to the seismically defined crustal layer [*Sleep*, 1975; *Morton and Sleep*, 1985; *Phipps Morgan et al.*, 1987]. Seismic crustal thickness and melt thickness are therefore commonly taken as equivalent. This hypothesis was developed in the 1970s on the basis of a comparison between ophiolitic geological sequences, and the seismic structure of the oceanic crust [*Participants to the Penrose Conference on Ophiolites*, 1972; *Cann*, 1974]. It is still consistent with our understanding of the geology of fast-spreading ridges [*Sinton and Detrick*, 1992]. At slow-spreading ridges, however, it is challenged by the evidence for emplacement of variably serpentinised peridotites in the crust [*Dick*, 1989; *Cannat*, 1993], and by petrological data suggesting deep melt partial crystallisation in the lithospheric part of the sub-axial mantle

[*Flower*, 1981; *Tormey et al.*, 1987; *Grove et al.*, 1992]. Seismic crustal thickness at slow-spreading ridges may thus incorporate a non-magmatic component of variably serpentinised peridotite, and miss magmatic rocks that may have remained trapped in the upper mantle [*Cannat*, 1996].

In addition to seismic crustal thickness, three other data sets have been used to estimate melt thickness at mid-ocean ridges: $Na_{8.0}$ and rare-earth geochemistry of MOR basalts, assuming passive mantle upwelling [*Klein and Langmuir*, 1987; *White et al.*, 1992a] ; and regional axial depth, assuming isostatic compensation [*Klein and Langmuir*, 1987]. Contrary to the seismic method, these three approaches yield melt thickness estimates that are independent of how melt is distributed between the crust and upper mantle. These three approaches are, however, based on variables that integrate the combined effects of mantle composition, temperature, degree of melting, and dynamics. As such, these approaches are, also contrary to the seismic approach, strongly model-dependent.

MORB composition, once corrected from partial crystallisation effects, reflects the combined effects of the composition of the mantle source, the degree of mantle melting, and the dynamics of mantle and melt fluxes beneath the ridge [*Klein and Langmuir*, 1987; *McKenzie and Bickle*, 1988; *Forsyth*, 1992; *Langmuir et al.*, 1992]. These parameters also determine the volume of melt that is supplied to the ridge for each increment of plate spreading. Given some simplifying assumptions, MORB composition can thus be used to estimate (independently of seismic crustal thickness) how much melt is supplied to a ridge region.

Klein and Langmuir [1987] modelled this relation for MORB sodium content, over a range of mantle temperatures, using a simplified tent-shaped region of mantle melting, with passive mantle flow, complete melt extraction, perfect focusing of melts toward the ridge axis, and a uniform sodium content in the mantle at the onset of melting. This modelled relation fits the global trend of MORB $Na_{8.0}$ (sodium content corrected for the effect of fractional crystallisation back to a common MgO content of 8 wt%) vs. seismic crustal thickness [*Klein and Langmuir*, 1987]. However, the seismic crustal thickness vs MORB $Na_{8.0}$ global trend identified by *Klein and Langmuir* [1987] relies mostly on its thin and thick crust end-members (the ultra-slow-spreading Cayman Rise and Arctic Ridge, and the Kolbeinsey and Reykjanes ridges associated with the Iceland hot spot) and is not well defined for most EPR and MAR locations.

The relationship between seismic crustal thickness and melt thickness has also been approached through the study of MORB Rare Earth Elements (REE) contents, using similar simplifying assumptions for melting parameters and for the shape and dynamics of the melting region [*McKenzie and*

O'Nions, 1991; *White et al.*, 1992a]. The average melt thickness derived from such REE inversions is, given analytical and model uncertainties, of the same order as the average thickness of the seismic crust, for ridges that spread faster than 15 mm/yr, away from hot spot influence [*White et al.*, 1992a]. Very few of the REE inversion results and seismic thickness data in the global study of *White et al.* [1992a] were, however, acquired for the same locations, so that it is not possible to use this data set to test for a correlation between calculated melt thickness and seismic crustal thickness.

Axial depths, if averaged over a sufficiently long portion of the ridge to assume isostatic compensation, can be interpreted as reflecting the density structure of the crust and mantle beneath the study area. Regional axial depth variations can thus be interpreted in terms of relative changes of melt thickness, and of mantle temperature and composition [*Klein and Langmuir*, 1987]. The amplitude of these inferred changes, however, is dependent on compensation depth (a larger compensation depth translates into smaller relative changes in mantle temperature and composition). In addition, mantle melting models are needed to relate mantle temperature and composition to melt thickness, in order to interpret regional axial depth variations in terms of melt thickness variations. *Klein and Langmuir* [1987] modelled the axial depth vs. melt thickness trend, and the axial depth vs. predicted MORB $Na_{8.0}$ trend, for a range of mantle temperatures, using the same melting model they used to invert MORB major element data (with a constant compensation depth of 200 km). They showed that the global axial depths and MORB $Na_{8.0}$ data sets very closely fit this predicted trend. This global fit, and the agreement between global averages of seismic crustal thicknesses and of melt thicknesses calculated from MORB compositions [*Klein and Langmuir*, 1987; *White et al.*, 1992a], are indirect arguments that: 1) the simple (passive) mantle dynamics and melting models used in these calculations are valid, at least at the regional scale; and 2) that the global average of seismic crustal thickness is a good estimate of the melt supplied to the global mid-ocean ridge system per increment of plate separation [*Klein and Langmuir*, 1987; *Langmuir et al.*, 1992].

4.1. Seismic Crustal Thickness, MORB Composition, and Axial Depth in Nine East Pacific Rise and Mid-Atlantic Ridge Regions

New seismic data have been acquired at the EPR and MAR in recent years, particularly in MAR locations where the existence of variably serpentinised peridotites in the crust is well documented. These new data frequently include along-axis profiles and therefore allow for seismic crustal thicknesses

Table 3. Average seismic crustal thickness (av. SC), average MORB $Na_{8.0}$ content (av. $Na_{8.0}$), melt thickness (MREE; * for segment OH1; ** for segment OH3) and percent fractionation (%F) calculated from MORB REE contents, and average axial depth (av. depth), for twelve regions of the East Pacific (EPR) and Mid-Atlantic (MAR) ridges. See text and Appendix A for explanation. References : (1) *Barth and Mutter* [1996]; (2) *Canales et al.* [2003] ; (3) *Canales et al.* [1998]; (4) *Fowler* [1976]. Av. SC calculated from gravity model [*Escartin et al.*, 2001] ; (5) *Hooft et al.* [2000] ; (6) *Purdy and Detrick* [1986] ; (7) Detrick and Collins pers. comm. [1998]. Av. SC calculated from gravity model [*Escartin and Cannat*, 1999]; (8) *Tolstoy et al.* [1993] ; (9) *Reynolds* [1995] ; (10) *Batiza and Niu* [1992] ; (11) *Sinton et al.* [1991] and Sinton, pers. comm. [1999] ; (12) *Stakes et al.* [1984] and Langmuir, pers. comm. [1998] ; (13) Langmuir, pers. comm. [1998] ; (14) *Bryan et al.* [1981] ; (15) *Bryan et al.* [1981]; *Humphris et al.* [1990]; and *Reynolds and Langmuir* [1997]; (16) *Bougault et al.* [1988] ; (17) *Michael et al.* [1994]; (18) *Castillo et al.* [2000]; (19) *Stakes et al.* [1984] ; (20) *Niu et al.* [2001] ; (21) *Lawson et al.* [1996].

RIDGE REGION	noted	av. S.C. (km)	ref.	av.$Na_{8.0}$ (%)	std	ref.	M(REE) (km)	rms	%F	ref.	av. depth (m)	std
EPR 12-12.7°N	EPR12N	5-6.5	(1)	2.60	0.18	(9)					2759	102
EPR 9.5-9.7°N	EPR9N	6.8	(2)	2.63	0.07	(10)	8.1	0.3	0.34	(10)	2577	16
EPR 14-18.5°S	EPR14S	5.7-6.3	(3)	2.66	0.14	(11)					2668	58
MAR 36.3-37.3°N	MAR37N	6.1	(4)	2.17	0.08	(12)	10.7	0.3	0.19	(18)	2751	202
MAR 33.5-35.5°N	MAR34N	5.5	(5)	2.37	0.12	(13)	*12.7	0.3	0.20	(19)	3025	420
							**11.5	0.4	0.30	(19)	3025	420
MAR 23.8-24.7°	MAR24N			2.97	0.04	(14)	6.6	0.3	0.07	(20)	3287	756
MAR 22.8-23.7°N	MAR23N	6.8	(6)	2.98	0.12	(15)	6.6	0.3	0.33	(14)	3876	487
MAR 15.4-16.4°N	MAR16N	5.4	(7)	3.05	0.04	(16)	5.1	0.6	0.11	(16)	3576	528
MAR 32-33.5°S	MAR33S	5.9	(8)	2.79	0.07	(17)					3038	375

to be averaged over the same ridge length as chemical data for MORB samples. In Table 3, we present a compilation of seismic crustal thickness, MORB $Na_{8.0}$ values, melt thickness from REE inversion, and regional axial depth, for three regions of the EPR, and six regions of the MAR.

Where published along-axis profiles of Moho depth are available, we averaged the seismic crustal thicknesses over the MORB sampling region. Where such profiles are not available, crustal thicknesses are either bracketed (smallest and largest published Moho depth for the MORB sampling region), or averaged over the MORB sampling region using gravity-derived crustal thickness models, having checked that these models fit available Moho depth data. Average MORB $Na_{8.0}$ values for each ridge region are calculated from glass data stored in the RIDGE Geochemical Database [*Lehnert et al.*, 2000]. We corrected these data for interlaboratory differences in microprobe calibration, excluding samples with high K/Ti ratios (enriched MORBs). Details on these calculations are given in Appendix A. High K/Ti samples represent a significant part of the dataset only in one region (MAR34N; see Table A1 in Appendix A), where they yield an average $Na_{8.0}$ similar to that of samples with lower K/Ti ratio (see Appendix A). Melt thicknesses from REE inversion are calculated from published glass and whole rock data, as described in

Appendix B. Finally, average axial depths are calculated from shipboard bathymetry over each MORB sampling region.

Our first observation is that average seismic crustal thicknesses are very similar in the nine studied regions. They range between ~ 5.5 km in the MAR34N and MAR16N regions, and to ~ 6.8 km in the EPR9N and MAR23N regions (Table 3). The limited spread observed in the EPR and MAR seismic crustal thicknesses plotted in Figure 14 of *Klein and Langmuir* [1987] is thus even more reduced using more recent data, and there is no correlation between MORB $Na_{8.0}$ and seismic crustal thickness in our dataset (Figure 5a).

The seismic crustal thickness vs. REE-derived melt thickness diagram in Figure 5b shows the same lack of correlation, with REE-derived melt thicknesses more than twice the seismic crustal thickness in the MAR37N (FAMOUS area) and MAR34N regions south of the Azores. The REE-derived melt thickness for the EPR9N region is similar to the global average value for non hot spot locations [*White et al.*, 1992a], and ~1 km thicker than the seismic crustal thickness. The REE-derived melt thicknesses for the EPR12N, MAR16N and MAR23N (MARK area) regions are less than the global average value for non hot spot locations [*White et al.*, 1992a], and similar to seismic crustal thicknesses (Figure 5b). The correlation between $Na_{8.0}$ and melt thickness calculated from REE over the 8 areas for which both are available (Figure 6),

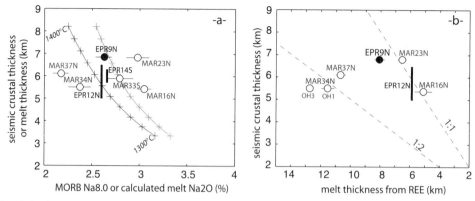

Figure 5. a- Seismic crustal thickness vs MORB $Na_{8.0}$ contents in Mid-Atlantic (open circles) and East Pacific (filled circles) locations listed in Table 3. Thick black lines show range of seismic crustal thicknesses determined in the EPR12N and EPR14S regions. Thin lines : ± 1 standard deviation for MORB $Na_{8.0}$ values in each region (Table 3). Thin black line with crosses: model melt Na_2O content vs. melt thickness for a range of mantle temperatures at 100 km depth (from 1300°C to 1400°C, cross every 10°C), calculated with simplified model of mantle melting (Appendix C) for an initial mantle sodium content of 0.27%. Thin grey line with crosses: corresponding calculated melt $Na_{8.0}$ contents (see text). b- Seismic crustal thickness vs. melt thickness calculated from MORB REE contents in Mid-Atlantic (open circles) and East Pacific (filled circles) locations listed in Table 3. Thin lines : ± 1 rms error for MREE values in each region (Table 3).

suggests that both geochemical methods are sensitive to the same features of the melt generation.

We used a simplified, tent-shaped melting regime [*Langmuir et al.*, 1992], to calculate the predicted melt Na_2O content vs melt thickness trend shown in Figures 5a and 6 (using mantle melting parameters listed in Appendix C). The position of this trend, and the corresponding mantle temperature values, are strongly model-dependent [*Forsyth*, 1992; *Langmuir et al.*, 1992; *Plank and Langmuir*, 1992]. Modelled melt Na_2O contents correspond to the calculated sodium content of primitive melts extracted from the melting regime, not to melts fractionated to a MgO content of 8%. Melt $Na_{8.0}$ contents predicted for low pressure fractional crystallisation are higher by ~ 0.1-0.2% (Figures 5a and 6); melt $Na_{8.0}$ contents predicted for high pressure fractional crystallisation could be up to 0.5% higher than modelled Na_2O values [*Langmuir et al.*, 1992]. Predicted MORB $Na_{8.0}$ vs regional axial depths, and melt thickness vs regional axial depths trends (Figure 7) can be calculated using the same simplified mantle melting model as above, provided that an arbitrary reference ridge depth is set for a given mantle temperature (we used 3000 m for a mantle temperature of 1360°C, corresponding in our melting model to a melt thickness of 6 km).

MORB $Na_{8.0}$ and REE contents measured in the Atlantic at the MAR34N and MAR37N areas correspond to predicted melt thicknesses about twice the seismic crustal thickness (Figures 5 and 6). This discrepancy could be explained in at least three ways: (a) A thickness of several kilometres of melt did not reach the crust, and was trapped in the mantle. But regional axial depths should then be shallower (they should

plot close to the predicted melt Na_2O vs axial depth trend in Figure 7a); (b) The mantle beneath these areas was twice as depleted in Na and REE as the depleted mantle source assumed for the modelling. But this is contradicted by isotopic and trace element evidence for a relative enrichment of the mantle source in the region [*Schilling*, 1975; *White and Schilling*, 1978; *Dosso et al.*, 1999]; (c) The MAR37N and MAR34N ridge regions have sampled melts from mantle

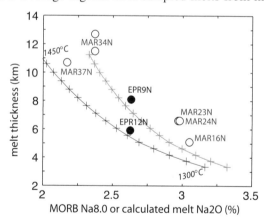

Figure 6. Melt thickness calculated from MORB rare earth elements contents vs MORB $Na_{8.0}$ contents in Mid-Atlantic (open circles) and East Pacific (filled circles) locations listed in Table 3. Thin black line with crosses: melt thickness vs. melt Na_2O content for a range of mantle temperatures at 100 km depth (from 1300°C to 1450°C, cross every 10°C), calculated with simplified model of mantle melting (Appendice C) for an initial mantle sodium content of 0.27%. Thin grey line with crosses: corresponding calculated melt $Na_{8.0}$ contents (see text).

entrained along axis from the Azores plume, which would have experienced prior melting and melt extraction further north beneath the Azores Platform. Our objective here is not to examine this last hypothesis further, but to make the point that melt thicknesses inferred from MORB compositions in these two near hot spot regions do not agree with seismic crustal thicknesses.

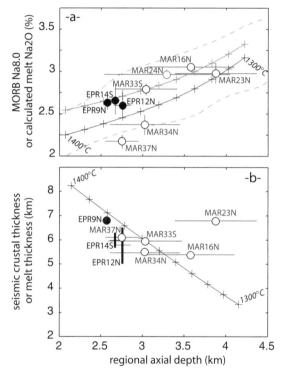

Figure 7. a- MORB $Na_{8.0}$ contents vs regional axial depth in Mid-Atlantic (open circles) and East Pacific (filled circles) locations listed in Table 3. Thin lines : ± 1 standard deviation for axial depths and MORB $Na_{8.0}$ values in each region (Table 3). Thin black line with crosses: melt Na_2O content vs. average axial depth for a range of mantle temperatures at 100 km depth (from 1300°C to 1400°C, cross every 10°C), calculated with simplified model of mantle melting (Appendice C) for an initial mantle sodium content of 0.27%. Thin grey line with crosses: corresponding calculated melt $Na_{8.0}$ contents (see text). Dashed grey line shows outline of global MORB $Na_{8.0}$ vs regional axial depth trend in Figure 15 of *Langmuir et al.* [1992]. b- Seismic crustal thickness vs regional axial depth in Mid-Atlantic (open circles) and East Pacific (filled circles) locations listed in Table 3. Thick lines show range of seismic crustal thicknesses determined in the EPR12N and EPR14S regions. Thin lines : ± 1 standard deviation for axial depths in each region (Table 3). Thin black line with crosses: melt thickness vs. regional axial depth for a range of mantle temperatures at 100 km depth (from 1300°C to 1400°C, cross every 10°C), calculated with simplified model of mantle melting (Appendice C) for an initial mantle sodium content of 0.27%.

With the exception of these two near hot spot regions, MORB $Na_{8.0}$ and regional axial depths in Table 3 plot near the predicted melt $Na_{8.0}$ vs axial depth trend in Figure 7a. The three EPR regions plot at the shallow and sodium poor end, the MAR24N and MAR33S regions at intermediate axial depths and MORB sodium contents, and the MAR23N and MAR16N regions at the deep and sodium rich end (Figure 7a). Seismic crustal thicknesses in the three EPR regions, and in the MAR24N and MAR33S regions, plot near the predicted melt thickness vs. regional axial depth trend in Figure 7b. MORB composition, axial depth, and seismic crustal thickness in these three EPR regions, and in the MAR24N and MAR33S regions, therefore do fit predicted trends, suggesting, as noted before, both that the mantle dynamics and melting models behind these trends are reasonable, and that seismic crustal thickness in these regions does provide a good estimate of the melt supply per increment of plate separation.

By contrast, seismic crustal thicknesses in the MAR23N region, and to a lesser extent in the MAR16N region, are too large, given MORB $Na_{8.0}$ contents, to fit the predicted melt thickness vs. MORB $Na_{8.0}$ trend (Figure 5a). In Figure 7b, the MAR23N region, and the MAR16N region to a lesser extent, also plot away from the predicted melt thickness vs. regional axial depth trend. These two regions of the MAR are also where the most extensive outcrops of variably serpentinised mantle-derived peridotites have been documented [*Karson et al.*, 1987; *Cannat et al.*, 1995; *Cannat et al.*, 1997c], and it is possible that partly-serpentinised peridotites make up a significant part of the seismic crust there. We discuss this further in section 4.2.

4.2. The Effect of Variable Volumes of Serpentinised Peridotites in Slow Spread Crust

We now proceed to examine the effect of the presence of serpentinite in the crust on estimates of melt thickness. Figure 8 shows the results of modelling the interrelationship between mantle temperature, percent of serpentinised peridotite in the crust, seismic crustal thickness and $Na_{8.0}$ content of MORB, showing how two ridge regions may have similar seismic crustal thicknesses, but distinct axial depths and MORB $Na_{8.0}$ contents, if the seismic crust in the deeper region contains a significant proportion of serpentinised peridotites. Calculated axial depths depend strongly on the choice of isostatic compensation depth (200 km in Figures 7 and 8), and on the average density of serpentinised peridotites in the crust. In Figure 8, we assumed that they have the same average density as basaltic crustal materials (2700 kg/m³). According to grain density measurements on abyssal peridotites, this corresponds to an average degree of serpentinisation of about 75% [*Miller*

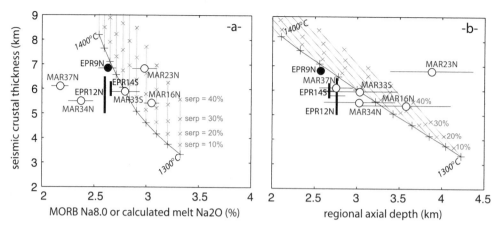

Figure 8. Effect of a serpentinised peridotite component in the crust on MORB $Na_{8.0}$ (a) and on regional axial depths (b). a- Plotted values for seismic crustal thicknesses and MORB $Na_{8.0}$: same as in Figure 5a. Black curve with crosses: calculated melt $Na_{8.0}$ contents for mantle temperatures at 100 km depth between 1300°C and 1400°C (cross every 10°C); grey lines with crosses: predicted effect of serpentinised peridotite crustal component (0% to 40% of seismic crustal thickness). b – Plotted values for seismic crustal thicknesses and regional axial depths: same as in Figure 7b. Black curve with crosses: calculated melt $Na_{8.0}$ contents for mantle temperatures at 100 km depth between 1300°C and 1400°C (cross every 10°C); grey lines with crosses: predicted effect of serpentinised peridotite crustal component (0% to 40% of seismic crustal thickness).

and Christensen, 1997]. If the average serpentinisation of mantle-derived material at crustal levels is less, the overall density of the crust, and thus the isostatic axial depth, should be greater.

The seismic crustal thickness vs MORB $Na_{8.0}$ content (Figure 8a), and the seismic crustal thickness vs average axial depth (Figure 8b) diagrams show that a 5.4 km thick seismic crust with a high MORB $Na_{8.0}$ content and a deep seafloor in the MAR16N region could be explained if serpentinised peridotites make up 15 to 20 % of the seismic crust there. Serpentinised peridotites make ~ 40% of the outcrops explored by submersible in the axial valley walls in this region [*Cannat et al.*, 1997c]. Although it is not clear how this translates into regional proportions of serpentinised peridotite in the crust, proportions of 15-20%, corresponding to melt thicknesses of 4.3-4.6 km (average seismic crustal thickness of 5.4 km minus 15 to 20% variably serpentinised material), appear possible. The presence of a variably serpentinised peridotite component in the crust could thus reconcile the geochemical, seismic and axial depth melt supply indicators in the MAR16N region.

The diagram of seismic crustal thickness vs. MORB $Na_{8.0}$ content (Figure 8a) shows that a 6.8 km-thick seismic crust with a high MORB $Na_{8.0}$ content in the MAR23N region could be explained if variably serpentinised peridotites there make up about 30% of the seismic crust. This proportion may, however, be somewhat too high, given that surveys have suggested that outcrops of serpentinised peridotites in the MAR23N region are common over ~25% of the seafloor

(in crust with age of 1 to 10 myr; [*Cannat et al., 1995]).* With an average crustal thickness of 4 km modelled from gravity data for areas of common ultramafic outcrops in the region [*Maia and Gente*, 1998], and assuming that intermixed gabbros and basalts represent only 20% of the crust in these areas, we estimate the proportion of variably serpentinised peridotite in the crust of the MAR23N region to be about 9% only. This is not enough to reconcile seismic crustal thickness and MORB $Na_{8.0}$ content in Figure 8a. In addition, the average axial depth in the MAR23N region is much too high to fit the calculated trends in Figure 8b. We are thus faced with a ridge region that does not fit global trends of MORB composition, axial depth and seismic crustal thickness.

Why this is so would deserve a more detailed study that is beyond the scope of this paper. We note, however, that seismic data acquired along an across-axis profile at 23°20'N (the latitude of the Snake Pit hydrothermal vent field) offers a possible path to explore, involving episodic injections of magma from the mantle over periods of 400 to 800 kyrs [*Canales et al.*, 2000a]. This across-axis seismic profile shows that the axial crust at 23°20'N is 6.2 km-thick, while the average crustal thickness over the whole profile (crustal ages up to ~ 2.5 myrs) is only 4.9 km. The average along-axis thickness of the seismic crust in the MAR23N region could thus be greater now than in the recent past, reflecting a recent increase in magma supply. MORB compositions, on the other hand, may reflect melt generation processes in the mantle that could vary over longer, or different time periods.

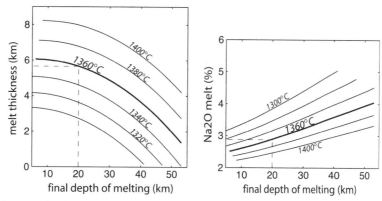

Figure 9. Melt thickness and melt Na$_2$O content calculated as a function of the final depth of melting (ie the depth to the top of the melting regime), using a simplified model of mantle melting (Appendice C), with an initial mantle sodium content of 0.27%, and a range of mantle temperatures at 100 km depth (from 1300°C to 1400°C). Dotted lines show that cessation of melting of "normal temperature" mantle (in this model ~ 1360°C at 100 km) at 20 km rather than at the base of the crust has little effect on resulting melt thickness, but that predicted melt sodium contents are significantly higher.

5. SUB-CRUSTAL MELT CRYSTALLISATION AT SLOW-SPREADING RIDGES

Earthquakes at depths of 8 km beneath segment ends in the MAR [*Toomey et al.*, 1988; *Kong et al.*, 1992; *Wolfe et al.*, 1995] show that temperatures at those depths are well below that of the basalt solidus, and therefore that rising melt may crystallise below the Moho [*Tormey*, 1987; *Grove et al.*, 1992; *Cannat*, 1996; *Sleep and Barth*, 1997]. Melting of the mantle should also cease at depth beneath the crust, with a reduction in the volume of melt produced. Among the implications of sub-crustal melt crystallisation are: 1. Geochemical estimates of the melt thickness contributed to a slow-spreading ridge may be systematically in error due to the retention of melt in the mantle, and to the differences in basalt fractionation behaviour at high and low pressure; 2. Less heat will be supplied to the crust because latent and specific heat from melts that crystallise below the crust will be supplied instead to the sub-axial mantle lithosphere.

However, not all sub-crustal crystallisation has a direct impact on the thermal structure of the spreading axis. Because mantle temperatures at the top of the melting column are expected to be less than liquidus temperatures for magnesium-rich melts [*Sinton and Detrick*, 1992], it is possible that these melts crystallise ultramafic cumulates below the Moho at fast, EPR-type ridges. In terms of heat supply, this shallow sub-crustal fractionation is simply equivalent to having a greater melt thickness than that inferred from seismic observations. In addition, at all ridges there may be wall-rock reaction between slowly-rising melt and the enclosing peridotite [*Kelemen et al.*, 1990; *Dick and Natland*, 1996], leading to crystallisation of olivine from the melt and dissolution of pyroxene from the wall rock. In this case, though the melt crystallises in part, the crystallisation is balanced by wall-rock mineral dissolution.

5.1. The Effect of Ending Mantle Melting at Depth Beneath the Crust

Cessation of mantle melting at depth beneath the crust at slow-spreading and cold ridges [*Langmuir et al.*, 1992; *Shen and Forsyth*, 1995] is a predictable consequence of having a thick axial lithosphere. Thermal models of slow-spreading ridges [*Lin and Parmentier*, 1989; *Phipps Morgan and Chen*, 1993; *Bown and White*, 1994] suggest that temperatures corresponding to the cessation of melting are reached at depths <15-20 km, except in the vicinity of large offset transforms, where the cessation of melting may occur at 40 km or more [*Detrick et al.*, 1995; *Magde et al.*, 1997]. As shown by *Langmuir et al.* [1992], cessation of mantle melting beneath the crust reduces the maximum extent of mantle melting, and should lead to an increase of the melt sodium contents (Figure 9b), but it does not reduce melt thickness in proportion. The cause of this lack of proportionality is that the mantle that rises above the final depth of melting (i.e., into the axial lithosphere) has melted by the maximum amount. As a result, the predicted decrease in melt thickness is moderate for final depths of melting < 20-30 km, except for the melting of unusually cold mantle (ie. for very shallow depths of initial melting; Figure 9a).

5.2. The Effect of Crystallizing Part of the Melt at Depth Below the Crust at Slow-Spreading Ridges

Melt crystallisation at sub-crustal pressures (> 0.2 GPa) should have an effect on MORB composition, because the temperature of appearance of clinopyroxene in the crystallised assemblage increases with pressure [*Bender et al.*, 1978; *Presnall et al.*, 1979], so that the abundances of most oxides in a melt fractionated at higher pressure to 8% MgO differ from those produced by fractionation at lower, crustal pressures [*Weaver and Langmuir*, 1990; *Langmuir et al.*, 1992]. Sodium and iron contents in the residual melt at 8% MgO, in particular, are expected to be higher if fractionation occurred at sub-crustal pressures. Neglecting a substantial sub-crustal melt fractionation effect while using MORB sodium contents to constrain the extent of mantle melting, would thus lead to underestimates of the melt thickness. Substantial sub-crustal melt fractionation should not, however, affect regional axial depths, and thus cannot account for the position of the MAR16N and MAR23N regions in Figures 5a and 7b.

5.3. Extent of Sub-Crustal Melt Fractionation and the Proportion of Deeply-Crystallised Magmatic Rocks in Slow Spread Crust

The extent to which sub-crustal fractionation occurs in slow-spreading ridge settings depends on the efficiency of melt transport in the axial region. If melts travel rapidly from the final depth of mantle melting to the base of the crust, the extent of sub-crustal fractionation will be minimal. At the other extreme, modelled by *Sleep and Barth* [1997], sub-crustal melt fractionation is allowed to buffer the temperature of the host mantle to the liquidus temperature of residual melts. This end-member model predicts that for very slow-spreading rates (or very large rates of axial heat loss), no melt actually reaches the crust, because all the melt that is extracted from the mantle is used up to maintain melt liquidus temperatures in the uppermost axial mantle [*Sleep and Barth*, 1997]. In a more realistic setting, taking the possibility for fast upward melt transport into account, sub-crustal melt crystallisation should be less extensive.

As illustrated in Figure 10, crystallisation of a significant volume of melt beneath the crust in a thick axial lithosphere, end of segment-type environment, neither requires that the crust will contain only a small proportion of magmatic rocks, nor that the uppermost mantle in the resulting off-axis lithosphere will contain large volumes of magmatic rocks.

Two factors should control the depth distribution of magmatic rocks in the resulting off-axis lithosphere: the geometry of solid state material flow lines on axis; and the width of the domain where most axial magmatic activity occurs. Both factors are very poorly constrained at present. Solid material flow lines in Figure 10 fit the requirement that mantle-derived rocks are brought up into the crust, but are clearly oversimplified. Displacements in the upper axial lithosphere at slow-spreading ridges occur along faults. In order to better

understand material flow lines in this tectonised context, we need to know the geometry of these faults at depth, and the kinematics of brittle-ductile and ductile flow in the deeper levels of the axial lithosphere.

Figure 10a is a schematic representation of a case where axial magmatic activity occurs in a domain that is very narrow with respect to the domain of solid material axial upflow. In this end-member case, all magmatic rocks that have crystallised below the crust are ultimately emplaced at crustal levels. Sub-crustal fractionation of these melts thus does not modify the relative proportions of magmatic and mantle-derived rocks in the crust of off-axis domains. It does, however, have an effect on the heat distribution on axis, because part of the melt that is supplied to the ridge releases latent and specific heat beneath the crust. The other configuration, shown in Figure 10b, has a domain of magmatic activity that is broader with respect to the domain of solid material axial upflow. It has similar implications in terms of axial heat distribution, but part of the deeply crystallised magmatic rocks remain in the uppermost mantle of off-axis domains.

5.4. Evidence for Sub-Crustal Melt Crystallisation at Slow-Spreading Ridges

In this section, we review observations made on rock samples from slow-spreading ridges that can help constrain the extent and depth of sub-crustal melt fractionation.

5.4.1. Liquid Lines of Descent of MORB. The slope of MORB liquid lines of descent (LLDs) in MgO variation diagrams depends on the crystallisation assemblage: olivine, olivine + plagioclase, or olivine + plagioclase + clinopyroxene. As mentioned before, the clinopyroxene saturation temperature has a stronger dependence on pressure, than that of olivine or plagioclase [*Bender et al.*, 1978; *Presnall et al.*, 1979]. Higher pressure favours the earlier crystallisation of clinopyroxene, leading in turn to the earlier appearance of the trend of decreasing Ca/Al with increasing fractionation of the liquid [*Weaver and Langmuir*, 1990; *Langmuir et al.*, 1992]. Crystallisation paths for different melt starting compositions and at different pressures can be calculated from experimentally determined phase equilibria [*Tormey et al.*, 1987; *Longhi and Pan*, 1988; *Grove et al.*, 1992; *Kinzler and Grove*, 1992; *Yang et al.*, 1996]. Resulting major element LLDs can then be compared to measured LLDs of MORB suites to infer their crystallisation pressure. Experimental data can also be used to determine the evolution of mineral-melt partition coefficients with pressure and to calculate model LLDs [*Weaver and Langmuir*, 1990; *Langmuir et al.*, 1992]. These approaches involve large analytical and experimental imprecision (e.g., [*Grove et al.*, 1992]). In addition, these methods yield only one frac-

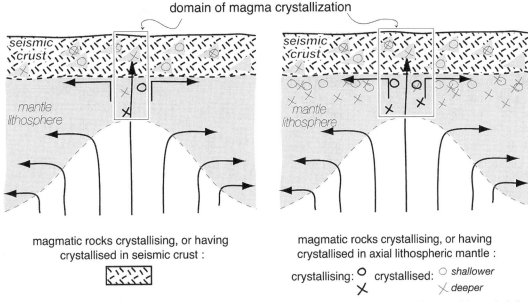

a - all deeply crystallised magmatic rocks are tectonically incorporated into the crust

b - some deeply crystallised magmatic rocks remain in the uppermost mantle

domain of magma crystallization

Figure 10. Idealized sections across a slow-spreading ridge axis. Sub-crustal crystallisation of some of the melt delivered from the mantle is shown to occur (black crosses and open dots) in the two sketches. In (a), the axial domain of magma crystallisation is narrow relative to the width of the axial deformation zone. All products of sub-crustal crystallisation (grey crosses and open dots), with their ultramafic host rocks, are thus tectonically emplaced into the crust. As it leaves the axis, the newly formed mantle lithosphere thus does not contain magmatic rocks. In (b), the axial domain of magma crystallisation is wider relative to the width of the axial deformation zone. Some of the products of sub-crustal crystallisation (grey crosses and open dots), with their ultramafic host rocks, are thus tectonically emplaced into the crust, while some are emplaced into the upper lithospheric mantle. As it leaves the axis, the newly formed mantle lithosphere thus does contain magmatic rocks. Vertical dimension of sketches: ~20 km; horizontal dimension: ~40 km. Arrows show hypothetical flow lines for solid material in the axial domain.

tionation pressure, while MORBs may commonly experience polybaric fractionation (e.g., [*Langmuir et al.*, 1992]). For example, melt that has fractionated clinopyroxene at depth can then ascend to shallower levels where it no longer lies on the olivine + plagioclase + clinopyroxene saturation boundary, and crystallises olivine + plagioclase only until it reaches the olivine + plagioclase + clinopyroxene cotectic again at that lower pressure. If it does reach this lower pressure cotectic, this melt will yield a low estimated presssure; if it erupts before reaching this cotectic, it will yield an intermediate estimated pressure. Pressure estimated with the LLDs methods are therefore minimum pressures for the onset of fractionation, and maximum pressures for the completion of fractionation [*Michael and Cornell*, 1998].

Estimated fractionation pressures from LLDs for basalts from the MAR are commonly between 0.3 and 0.6 GPa, corresponding to sub-crustal depths. This is the case in the Kane Fracture Zone region (MAR23N ; [*Tormey et al.*, 1987; *Grove et al.*, 1992]), in the region of the TAG hydrothermal site at 26°N [*Meyer and Bryan*, 1996], in the AMAR and FAMOUS area (MAR37N ; [*Grove et al.*, 1992]), and in the Mohns-Knipovitch, 10°N–20°N, and 31°S–46°S regions [*Michael and Cornell*, 1998]. This is also the case for the ultra-slow-spreading Mid-Cayman Rise [*Grove et al.*, 1992] and Southwest Indian Ridge [*Michael and Cornell*, 1998], and for the intermediate-spreading but anomalously deep and thin-crusted ridge in the Australian Antarctic Discordance Zone [*Michael and Cornell*, 1998]. In contrast, most fractionation pressures estimated for MORB suites from the fast-spreading EPR at 9°N [*Michael and Cornell*, 1998], 10°N–12°N [*Grove et al.*, 1992; *Yang et al.*, 1996], and 27°S–35°S [*Michael and Cornell*, 1998] are less than 0.2 GPa, corresponding to crustal depths. As mentioned before, sub-crustal estimated fractionation pressures do not mean that no further fractionation occurred in the crust, but indicate that shallow crystallisation events did not proceed to complete re-equilibration. Such high crystallisation pressure estimates are therefore an indication against an extended stay of melts in crustal magma chambers. Crustal estimated fractionation depths at the EPR, on the other hand, do not mean that melts did not start crystallizing at greater depths, but indicate extensive shallow fractionation, consistent with the geophysical evidence for crustal magma lenses.

It is worth noting that high crystallisation pressure estimates for MAR basalts concern samples collected both at the centre, and at the ends of slow-spreading ridge segments. Significant sub-crustal melt crystallisation would therefore not be a characteristic only of the magma-poor, segment-end type settings. This would indicate that the hot thermal regime envisioned for the magmatically-dominated slow-spreading ridge segment centre in Figure 2b, with sub-solidus temperatures for basaltic melts being reached only in the upper crust, is not common. A similar conclusion can be reached from seismic evidence that active crustal magma chambers are uncommon at the MAR (see Section 2.2).

The main disadvantage of the LLD approach for estimating crystallisation pressures is that the method only holds if all of the samples have parental melts of similar compositions. Another possible drawback is that processes other than fractional crystallisation (e.g., *in situ* crystallisation, melt mixing) may influence the major element trends observed in MORB sample suites. In some cases it may be possible to check for action of such processes using trace element and isotopic compositions in addition to major elements.

The method of *Yang et al.* [1996] is not prone to such difficulties, and we used it to estimate the pressure of crystallisation of samples from the AMAR-FAMOUS region of the MAR (MAR37N; [*Stakes et al.*, 1984]). *Yang et al.* [1996] parameterised MORB experimental data and found that the molar fraction of Mg, Ca and Al in melts that are saturated with olivine, plagioclase and augite can be given as a function of pressure and of the Na, K, Ti, Si and Fe molar contents in the melt. The key assumption that has to be made when using this method is that the melt is saturated in olivine, clinopyroxene and plagioclase. In common with the LLD approach, this method also suffers from uncertainty due to analytical imprecision. We checked that the major element trends found in AMAR-FAMOUS samples with <8.5 wt% MgO are controlled by crystallisation of all three phases, using the principal component approach described in detail by *Maclennan et al.* [2001b]. Samples that lie along this crystallisation trend also contain phenocrysts of these phases. Figure 11 shows that these samples have the composition of melts saturated with olivine, clinopyroxene and plagioclase at pressures of 0.1 to 0.4 GPa, equivalent to depths of 3 to 12 km. Therefore, substantial sub-crustal crystallisation takes place. Some samples that appear to have crystallised beneath the crust have MgO contents as low as 7.5 wt%, indicating that they have lost 50% or more of their original liquid mass during sub-crustal crystallisation.

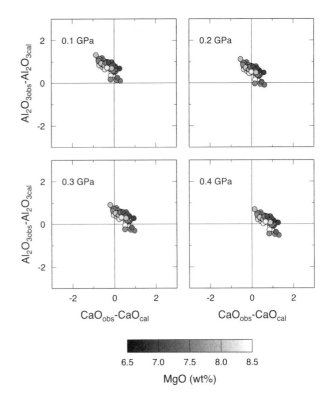

Figure 11. *Yang et al.* [1996] fractional crystallisation model applied to MORB samples from the AMAR-FAMOUS region [*Stakes et al.*, 1984]. Only samples with MgO < 8.5 %, found to be in equilibrium with an olivine+clinopyroxene+plagioclase assemblage, are represented. The x-axis shows the difference between the observed and predicted CaO contents at a given pressure, and the y-axis shows the difference in Al2O3 contents. The colour of the circle shows the MgO content of the sample. Samples that appear to have crystallised at pressures > 0.2 Gpa (depths > 6 km) have MgO contents as low as 7.5 wt%, indicating that they have lost 50% or more of their original liquid mass during sub-crustal crystallisation.

An extensive study of crystallisation depths at mid-ocean ridges has been published by *Michael and Cornell* [1998], who compared the results of the *Yang et al.* [1996] barometer with a number of other barometers, including that of *Danyushevsky et al.* [1996]. The presence of water is known to influence the crystallisation behaviour of basalt [*Michael and Chase*, 1987; *Danyushevsky*, 2001] and for some water-rich basalts, such as those found in back-arc basins, water has been shown to play an important role. While the *Yang et al.* [1996] barometer is based on the results of anhydrous experiments, *Danyushevsky et al.* [1996] also accounted for the effect of water. The strong 1:1 correlation between the *Yang et al.* [1996] and *Danyushevsky et al.* [1996] results shown by *Michael and Cornell* [1998] indicates that the modest water contents of most MORB samples has little influence on the barometric results.

5.4.2. Clinopyroxene and clinopyroxene-melt thermo-barometry The jadeite ($NaAlSi_2O_6$) content of clinopyroxene that crystallises from a melt has a strong dependency on pressure because the partial molar volume of jadeite is significantly smaller than the partial molar volume of Na and Al oxides in the melt [*Robie et al.*, 1978; *Lange and Carmichael*, 1987]. The clinopyroxene-melt thermobarometer of *Putirka et al.* [1996] is based on an experimental investigation of the jadeite-diopside/hedenbergite exchange between clinopyroxene and basaltic melts at a range of pressures and temperatures. Clinopyroxene phenocrysts do occur in MORBs, but are frequently not in equilibrium with the surrounding groundmass. This is a serious limitation to the use of the *Putirka et al.* [1996] thermobarometer in ridge settings. Equilibrium pressures can in principle also be estimated from the crystal structure of clinopyroxene phenocrysts alone, based on the respective volumes of the unit crystallographic cell, and of crystallographic site M1 [*Nimis and Ulmer*, 1998]. This barometer does, however, also require the clinopyroxene compositions used to be representative of equilibrium compositions. In addition, it is poorly calibrated for the range of pressures that is most relevant to mid-ocean ridge studies (0-0.8 GPa; [*Nimis and Ulmer*, 1998]).

Maclennan et al. [2001b] applied the *Putirka et al.* [1996] thermobarometer to carefully selected samples from the Krafla area in Iceland and determined crystallisation pressures between 0.3 and 0.9 GPa. In Figure 12, we have applied the same approach to clinopyroxene phenocrysts and basalt glass compositions published for the FAMOUS and Kane regions of the MAR [*Bryan et al.*, 1981; *Stakes et al.*, 1984]. Melt-clinopyroxene compositional pairs that were clearly out of equilibrium were removed from the dataset by filtering out any pairs where the calculated clinopyroxene-melt K_{dFe-Mg} lay outside the range 0.27-0.33. If the melt and clinopyroxene are the products of crystallisation of parental melts of similar compositions, then this filtering limits the errors in pressure estimates due to disequilibrium to 0.1 GPa [*Maclennan et al.*, 2001b]. Given the ± 0.14 GPa calibration error calculated by *Putirka et al.* [1996], and the disequilibrium error, the results are consistent with MORB fractionation pressures estimated independently for these regions using the LLD or *Yang et al.* [1996] methods. These results indicate that sub-crustal crystallisation, at pressures of 0.2 GPa or more, does occur at slow-spreading ridges.

5.4.3. Melt crystallisation products in mantle peridotites.
Crystallisation products of basaltic magmas are common in abyssal peridotites : many samples contain veins of variably altered gabbroic material [*Bloomer et al.*, 1989; *Cannat et al.*, 1992; *Cannat and Casey*, 1995; *Constantin et al.*, 1995; *Dick and Natland*, 1996; *Cannat et al.*, 1997b; *Tartarotti et al.*,

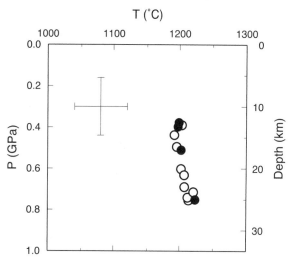

Figure 12. Pressure (depth) and temperature of crystallisation estimated with the *Putirka et al.* [1996] thermobarometer for clinopyroxene phenocrysts and basalt glass compositions published for the FAMOUS (closed circles; [*Stakes et al.*, 1984]) and Kane regions of the MAR (open circles; [*Bryan et al.*, 1981]). Estimated errors (error bars in upper left corner of figure) on pressures are ± 1.4 kbar, and on temperatures, ± 30° C [*Putirka et al.*, 1996].

2002]. Plagioclase peridotites, which represent a significant proportion of the peridotites dredged from the ocean floor, most likely result from impregnation with basaltic melts [*Dick*, 1989]. Abyssal peridotites also commonly contain magmatic clinopyroxene, as microscopic impregnations or as macroscopic lenses [*Seyler et al.*, 2001].

Figure 13 illustrates the textural and compositional variability of magmatic rocks in peridotites drilled in the MAR axial valley wall at 23°N (ODP Leg 153). Gabbroic veins up to 35 cm-thick, but mostly less than 10 cm-thick are found all throughout the 142 m of recovered core, and elongated lenses of magmatic clinopyroxene (Figure 13C) are present in ~25% of the core [*Cannat et al.*, 1997b]. Clinopyroxene in these CPX lenses are magnesium-rich (Figure 13D) and have characteristic impregnation textures, filling intergranular spaces in the host peridotite. One thick clinopyroxene-bearing troctolite and gabbro dyke, with magnesium-rich clinopyroxene and calcium-rich plagioclase, has distinct dunite screens a few tens of centimetres thick [*Shipboard Scientific Party*, 1995; *Niida*, 1997]. These could represent residual dunite after extensive melt-rock interactions [*Kelemen*, 1990], or cumulate olivine precipitated from a magnesium-rich melt, or a combination of both. Other veins range from clinopyroxene-plagioclase (Figure 13B), to orthopyroxene-bearing, and to zircon and apatite-bearing gabbros. These veins span the full range of plagioclase An content vs. clinopyroxene Mg# measured in kilometer-sized gabbroic bodies drilled at slow [*Cannat et*

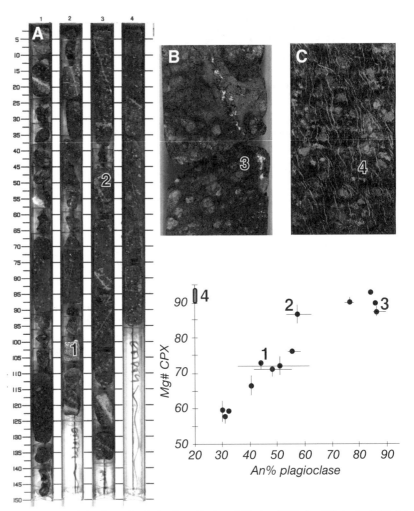

Figure 13. Gabbroic veins in serpentinised mantle-derived peridotites drilled at ODP site 920 in the Mid-Atlantic Ridge 23°N region (after *Cannat et al.* [1997b]). A- Core 153–920D–18R. This 4.5 m long core, representing 10 m of hole, contains an accumulated thickness of gabbroic veins of ~ 0.4 m. B- Interval 153–920D–14R-2, 109–120cm. Thin and irregular vein containing An-rich plagioclase and high Mg# (Mg/(Mg+Fe) clinopyroxene; C- Interval 153–920B–12R-2, 108–118 cm. Clinopyroxene lenses appear as small grey patches, elongated sub-horizontally, sub-parallel to the crystal plastic spinel foliation in this sample. Larger grey patches are partly serpentinised orthopyroxenes from the peridotite host. Thin and steep light grey later serpentine microcracks are also visible in this sample. D- Average Mg# and anorthite content of clinopyroxene and plagioclase in gabbroic veins from ODP site 920. Numbers refer to gabbroic veins and clinopyroxene lenses identified in A, B and C (CPX lenses in C do not contain plagioclase). Massive gabbros drilled at ODP sites 921–924 to the north of site 920 show the same wide range of composition, corresponding to increasing (from the Mg and An-rich right side of diagram) fractional crystallisation of a basaltic melt.

al., 1997a; *Ross and Elthon*, 1997], and ultra slow-spreading ridges [*Bloomer et al.*, 1991; *Ozawa et al.*, 1991; *Dick et al.*, 2000]. The host peridotite commonly bears evidence (iron, sodium and titanium enrichment, nickel depletion) for reaction with a percolating melt over distances of a few millimetres to a few centimetres of the veins [*Cannat et al.*, 1997b].

The existence of gabbroic veins in residual peridotites that have been sampled at or near the seafloor is not in itself an indication that melt crystallisation occurred at sub-crustal depths. Peridotites that now outcrop in axial valley walls have come up from the melting mantle, through the axial lithospheric mantle, and into the crust. Gabbroic veins now sampled in these rocks could have been emplaced at any time during this uplift (Figure 10). Extensive melt-rock interactions in the peridotite near most gabbroic veins indicate, however, that melt emplacement occurred in peridotite that was still very hot [*Cannat et al.*, 1997b]. A fortiori, the interstitial texture of clinopyroxene lenses, and the dunite screens near irregular plagioclase-clinopyroxene veins, indicate melt emplacement in peridotite that was close to solidus temperatures. Plagioclase peridotites, not found in the core from ODP leg 153, but common at other ridge locations [*Dick*, 1989], must similarly have formed in near solidus temperature conditions. Such high temperature conditions are unlikely at crustal depths at slow-spreading ridges, particularly in thin-crust, end of segment-type settings. A significant proportion of the magmatic lenses and veins found in outcrops of residual peridotites at slow-spreading ridges therefore probably crystallised at sub-crustal depths.

6. CONSTRUCTION OF THE LOWER CRUST AT FAST-SPREADING RIDGES

The mechanisms by which the lower crust is formed at fast-spreading ridges are currently a matter of debate. There are two current end-member models, the gabbro glacier model [*Dewey and Kidd*, 1977; *Phipps Morgan and Chen*, 1993] in which all of the lower crust is created by crystallisation in a shallow axial magma chamber (Figure 14a), and the sill intrusion model [*Kelemen et al.*, 1997; *Kelemen and Aharanov*, 1998], in which the crust is created in a series of stacked sills of which the shallow axial magma chamber is merely the highest and last (Figure 14b). A hybrid model has also been developped, with the upper gabbros formed as part of a gabbro glacier, and the lower gabbros crystallising in situ in a series of sills [*Boudier et al.*, 1996; *Maclennan et al.*, 2004].

The gabbro glacier and sill intrusion models have very different thermal implications. In the gabbro glacier model, the whole of the latent heat of the crystallisation of the lower crust (0.5 MJ/kg) is supplied near the roof of the axial magma chamber and is thus directly available to drive high temperature hydrothermal convection within the sheeted dykes. In the sill-intrusion model, only a small fraction of the latent heat of crystallisation is supplied near the roof of the axial magma chamber. In this model the hydrothermal circulation must penetrate much deeper in the crust to extract the same amount of heat. The difference between the two models is thus significant in determining the depth of penetration of

high-temperature fluids close to the spreading axis, and the location of zones of high-temperature water-rock interaction.

The principal problem with the gabbro glacier model is that it does not fit with field and laboratory observations made on the gabbros of the Oman ophiolite. This ophiolite is used for comparison because of its exceptionally large and well-exposed section of gabbro, its magma geochemistry close to MORB, and its structure which is close to that expected for oceanic crust formed at fast-spreading rates. In the gabbro section of the Oman ophiolite there is a distinction between an upper section, made of weakly banded gabbro, and a lower section in which most of the gabbro show strong igneous layering, with variation between layers in the proportions of igneous phases [*Pallister and Hopson*, 1981 ; *Boudier et al.*, 1996; *Kelemen et al.*, 1997; *Korenaga and Kelemen*, 1997; *MacLeod and Yaouancq*, 2000]. The gabbro glacier model gives no explanation for this observation, since all of the lower crust is considered to have formed by the same process from the shallow axial magma chamber. In addition, the gabbro glacier model predicts that gabbros must be very intensely deformed at temperatures close to basalt melting point, since layers of cumulate that start the same width at the axial magma chamber (about 1 km), must be extended and warped to fill the full 4 km thickness of the lower crust. Exposures in the Oman ophiolite suggest that the lower crust there is not deformed by as much as would be required to fulfil the glacier model requirements [*Nicolas et al.*, 1988; *Boudier et al.*, 1996].

In its simplest form, the sill intrusion model predicts that primary magma is injected into the crust from the mantle, and rises through the crust in a stack of sills, crystallising as it rises, to build the lower crust. This requires that crystal fractionation is distributed throughout the lower crust, and that the magma, and the crystals forming from it, should become more and more evolved from the bottom to the top of the lower crust. The lavas and dykes should have the composition of the evolved magma that fills the uppermost sill, the axial magma chamber. As shown in section 6.1, this requirement also does not fit field and laboratory data from the Oman ophiolite.

The sill intrusion model also requires that magma can crystallise to form the lower crust throughout the thickness of the crust at the spreading axis, and that therefore the latent heat of crystallisation can be removed from the crust from beneath the axial magma chamber. This question of heat removal is discussed in section 6.2.

6.1. Cumulus Chemistry of Gabbroic Sections of the Oman Ophiolite

To test the prediction of the sill-intrusion model that magma composition should change gradually with height in the lower

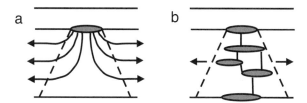

Figure 14. Schematic cross sections showing the two contrasted models for the formation of the lower crust at a fast-spreading ridge. (a) The gabbro glacier model, (b) The sill-intrusion model. Vertical dimension of sketches: ~6 km; horizontal dimension: ~8 km.

crust, we assembled data on olivine compositions from several extended crustal sections in the Oman ophiolite [*Pallister and Hopson*, 1981; *Lippard et al.*, 1986]. Data from [*MacLeod and Yaouancq*, 2000] for another section through the Oman gabbros lead to essentially the same picture.

Close to the junction between the gabbroic cumulates and the underlying harzburgite-dunite unit, corresponding to the Moho, the harzburgite is cut by sills that have a mineral chemistry close to equilibrium with primary magma [*Boudier and Nicolas*, 1995; *Kelemen et al.*, 1997]. Similar compositions are found in cumulates within 200 metres or so of the Moho in other sections. However, above that level, olivine compositions in cumulates show no trend with height in the cumulate section, even through sections of over 1000 m (Figure 15). The olivine compositions show a narrow spread in each section, though the mean composition of individual sections differs, and the olivine compositions throughout most of the sections can be calculated to be in equilibrium with evolved magma such as those of the sheeted dykes and lavas of the Oman ophiolite (Figure 16). Similar conclusions can be drawn

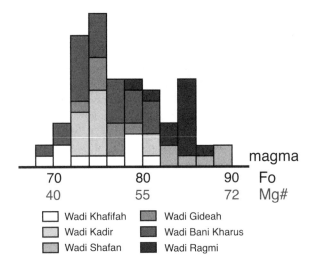

Figure 16. Histogram of olivine compositions from the lower crust of the Oman ophiolite from the same 6 sections as in Figure 15. The smallest blocks are 1 unit high. The Mg number of the magma in equilibrium with olivine has been calculated using a distribution coefficient of 0.28. Melt in equilibrium with mantle olivine has an Mg number of about 0.72. The sheeted dykes of the Oman ophiolite have Mg numbers ranging from about 0.65 to 0.40 [*Lippard et al.*, 1986], corresponding to the values calculated from the cumulus olivine compositions.

from the vertical profile of Cr in clinopyroxene in gabbros of Oman [*Coogan et al.*, 2002b].

These observations are in direct contradiction to the fundamental requirements of the sill intrusion model in its simplest form. They do not, however, rule out more complex sill intrusion models in which each sill would represent a mantle-derived magma, that partly crystallized and partly erupted, though such a hybrid model would require that fractionation in each sill proceeded to the same extent.

6.2. Cooling of the Lower Crust Beneath Fast-Spreading Ridges

The seismic structure of the lower crust close to the fast-spreading EPR axes is relatively well-defined as a result of several major experiments [*Harding et al.*, 1989; *Toomey et al.*, 1990; *Dunn et al.*, 2000]. Broadly speaking there is a trapezoidal low velocity zone in the lower crust immediately below the spreading axis, as wide as the axial magma chamber at the top of the lower crust and 4 to 6 km wide at the Moho [*Dunn et al.*, 2000]. Seismic velocities within this zone suggest that it consists in a crystal mush, with up to 18% magma [*Crawford and Webb*, 2002]. At the edges of the low-velocity zone is a narrow band in which the seismic velocity increases rapidly, which is interpreted as a transition from crystal mush to entirely solid gabbro.

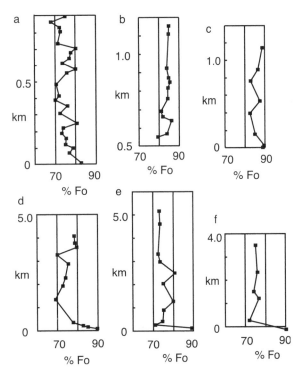

Figure 15. Olivine compositions as a function of height above the Moho in the lower crust of the Oman ophiolite. (a), (b) and (c) sections from the Wadi Bani Kharus, Wadi Ragmi and Wadi Shafan, adapted from *Lippard et al.* [1986]. (d), (e) and (f) sections from the Wadi Khafifah, Wadi Kadir and Wadi Gideah (east section) adapted from *Pallister and Hopson* [1981]. Note the very different scales of height.

If the lower crust is largely constructed by in-situ crystallisation of magma within a stack of sills at the spreading axis, then there must be a means of removing the latent heat of crystallisation from these sills, of a layer about 4 km thick of gabbro. Crystallisation on this scale can be achieved in two ways. One is by reaction of the magma with the enclosing rock, so that as crystals are precipitated, the more fusible components of the rock are melted and contributed to the rising magma. This process is probably important in the evolution of magmas rising through the mantle beneath a spreading axis, where harzburgite may be transformed to dunite by precipitation of olivine from the rising magma, accompanied by dissolution of pyroxene [*Kelemen et al.*, 1990 ; *Dick and Natland*, 1996]. However, the net addition of solid material in the process is very small. The country rock is transformed, but its mass is not significantly increased, while the volume of magma leaving the top of the system is much the same as that coming in at the base. The other way to crystallise magma is through the presence of a thermal gradient down which heat can be conducted away from the magma.

Numerical models by *Chen* [2001] suggest that it is impossible, if hydrothermal cooling is restricted to crust cooler than 600-700°C, to develop a thermal gradient steep enough to remove the latent heat from beneath the shallow axial melt lense. These simulations approximate hydrothermal cooling of the crust by assuming an enhanced thermal conductivity in regions thought to be affected by hydrothermal circulations. More recent simulations that explicitly include hydrothermal circulation [*Cherkaoui et al.*, 2003] suggest that when the permeability exceeds a threshold value, hydrothermal circulation can cool the entire crust very near the ridge axis, producing steep isotherms in the lower crust. These recent thermal simulations, however, still do not cool the lower crust right on axis, and therefore do not allow for in situ crystallisation of magma in the axial low velocity zone, to form the crystal mush that is imaged by seismic experiments [*Harding et al.*, 1989; *Toomey et al.*, 1990 ; *Dunn et al.*, 2000]. We develop a simple model here that supports this conclusion.

In this model, we ask what the temperature structure within the axial crystal mush would be if it is advected towards a hydrothermally-cooled surface, and cooled by conduction from that surface. This thermal structure is defined by the Peclet number of the system, calculated as:

$$\frac{u\rho c_{\mathrm{p}}L}{K} \qquad (5)$$

where u is the rate at which the crystal mush is being transported towards the cooled surface (the half-spreading rate in the configuration represented in Figure 17), ρ is the rock density, c_{p} the specific heat, L the distance over which the gabbro

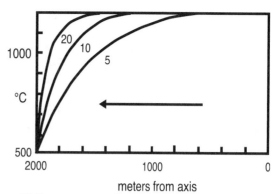

Figure 17. Temperature profiles in gabbroic lower crust containing 10% of melt, spreading away from a spreading axis at 50 mm/yr, calculated for a Peclet number of 10 (see text). Curves for Peclet numbers of 5 and 20 are shown for comparison purposes. Both of the latter curves lie well outside any uncertainties in calculation of the temperature profile.

is being advected (half of the width of the low-velocity zone) and K is the thermal conductivity of the rock. Following *Jaeger* [1964], we treat latent heat of the magma enclosed within the gabbro as specific heat from rock of enhanced temperature. Assuming a half spreading rate of 50 mm/yr, a thermal conductivity of 2 W m^{-1} K^{-1} , a 4 km-wide low velocity zone, and 10% of magma in the crystal mush, the Peclet number is ~10, and the appropriate temperature curve for that Peclet number [*Bredehoeft and Papadopoulos*, 1965] is shown in Figure 17. Curves for Peclet numbers of 5 and 20 show the sensitivity of the thermal profile to Peclet number.

The curve corresponding to a Peclet number of 10 in Figure 17 shows that under these conditions, melt cannot crystallise until it is more than a kilometre from the spreading axis. In situ crystallisation thus cannot contribute to the formation of a crystal mush in the lower crust at the spreading axis or within a kilometre on either side. If this model is correct, the axial crystal mush imaged beneath the EPR [*Harding et al.*, 1989; *Toomey et al.*, 1990; *Crawford et al.*, 1999; *Dunn et al.*, 2000; *Crawford and Webb*, 2002] would thus call for a component of crystal settling, from the roof or walls of the axial low velocity zone. An alternative may be offered by the model recently proposed by *Maclennan et al.* [2004], that allows for a depth-dependent melt composition and for melt intrusion over a finite width of 2 km within the axial low velocity zone. In this model, crystallisation does occur under the ridge axis.

One of the most interesting possibilities for constraining lower crustal cooling lies in determination of cooling rates of gabbros [*Garrido et al.*, 2001 ; *Coogan et al.*, 2002b]. Both gabbro glacier and sill intrusion models include implicitly very different cooling rates for the gabbros of the lower crust. Determination of cooling rates on both ophiolitic and oceanic materials thus allows a comparison between these two models. Cooling rates have already been calculated from Ca contents of olivines, and Ca profiles in olivine crystals in Oman gabbros [*Coogan et al.*, 2002b]. These calculations have been made difficult because of disagreement between different authors on the experimental determination of diffusion parameters of Ca in olivine, resulting cooling rates differing by several orders of magnitude. Better experimental data should reduce these uncertainties to a level at which the calculations are useful in discrimination. However, already the profiles of relative cooling rate detemined by these methods indicate that cooling rates decrease downward in the Oman gabbros [*Coogan et al.*, 2002b].

7. OCEANIC GABBROS AND OCEANIC MAGMA CHAMBERS AT FAST AND SLOW-SPREADING RIDGES

In the discussion of the construction of the lower crust, evidence from ophiolites and the oceans is closely intertwined. This is because there has not yet been a systematic sampling of a continuous, stratigraphically well-located section of lower ocean crust. But there are many samples of gabbros from the oceans, and these do show similar chemical characteristics to the gabbros from Oman. The most striking difference is the rarity of layered gabbros from the oceans, while they are common in the lower parts of the gabbros in Oman [*Pallister and Hopson*, 1981 ; *Nicolas et al.*, 1988].

Compilations of mineral chemistry from ODP Legs 118 and 176 (slow-spreading Southwest Indian Ridge; [*Bloomer et al.*, 1991; *Natland and Dick*, 2001]), 147 (Hess Deep, fast spread EPR crust ; [*Natland and Dick*, 1996; *Pedersen et al.*, 1996]), and 153 (slow spread MAR crust ; [*Cannat et al.*, 1997a; *Ross and Elthon*, 1997]) show that gabbros crystallised from melts in equilibrium with mantle olivine (melts with Mg numbers close to 72) are rare, though there are some in Hole 923A (ODP leg 153). Ferrogabbros are present in collections from slow and ultraslow-spreading ridges. They appear to be less common in collections from fast-spreading ridges, though this may be a sampling bias (far fewer gabbros have been collected from fast-spreading ridges than from elsewhere) or may reflect a real relationship to spreading rate [*Dick et al.*, 1991]. But most of the gabbro samples recovered from the oceans are intermediate in chemistry, are clearly cumulates, because of their poverty in incompatible elements, and must have formed as cumulates from magmas of intermediate degrees of evolution, such as most of those erupted at the mid-ocean ridges.

This predominance of gabbros of intermediate compositions can most simply be explained, both for ophiolites and for the ocean crust, if the magmatic systems at spreading centres involve the processing of magma through regularly replenished magma chambers [*O'Hara*, 1977]. This conclusion seems reasonable for fast-spreading ridges, where long-lived magma chambers are expected based on seismic evidence, but is unexpected for slow-spreading ridges, where magma chambers are considered, from geological and geophysical evidence, and from numerical modelling, to be short-lived (see Section 2.2). *Robson and Cann* [1982] simulated the growth of a magma chamber with random input of blobs of magma. Their simulation indicated that it would take a relatively short time, typically a few tens of thousands of years, to reach intermediate compositions, though perhaps not at steady state, even at relatively slow-spreading ridges. Magma chambers that are considered as short-lived based on geological and geophysical data, may thus reach intermediate compositions well before they become totally solid.

While gabbro compositions seem to be similar on first order (in detail there are indications for higher trapped melt fractions in gabbros from slow-spreading ridges [*Ross and Elthon*, 1997 ; *Coogan et al.*, 2001]), the spatial distribution of rock types and mineral compositions, and the deformation features in gabbro suites from slow spread crust are fundamentally different from those observed in Oman. In the Oman suite, very primitive cumulates are found at the very base of the section, and very evolved gabbros near the gabbro-dyke transition, while intermediate compositions dominate the large part of the crustal section in between [*Pallister and Hopson*, 1981; *Kelemen and Aharanov*, 1998; *MacLeod and Yaouancq*, 2000]. Oman gabbros are deformed at temperatures close to magmatic, but show little evidence for lower temperature ductile shearing [*Nicolas and Ildefonse*, 1996]. By contrast, gabbros from slow-spreading ridges commonly show extensive ductile shearing at a range of temperatures, and primitive and evolved rock types are juxtaposed at all scales, forming metre to millimetre thick intervals with very distinct mineral compositions [*Bloomer et al.*, 1991; *Dick et al.*, 1991; *Cannat et al.*, 1997a]. Contacts between these distinct rock types may be sharp, suggesting that they correspond to intrusions into solidified gabbros, or diffuse, suggesting intrusion into crystal mushes. This is best exemplified in the 1500 metre thick section of gabbros drilled at site ODP 735 near the Atlantis II Transform on the Southwest Indian Ridge. At this site, five plutons, 200 m to 500 m thick, appear to have been separately intruded as numerous small injections of magma [*Natland and Dick*, 2001], and variably deformed in dominantly normal ductile shear zones [*Cannat et al.*, 1991; *Dick et al.*, 2000]. The picture emerging for slow-spreading ridge gabbros thus involves crystallisation in plutons formed by repeated melt injections in a deforming environment.

8. AN ESTIMATE OF THE RATIO OF CUMULATES TO LAVAS PLUS DYKES AT THE MID-ATLANTIC RIDGE

To constrain the supply of heat to the crust, is important to obtain an estimate of the proportion of the crust made of lavas and dykes (liquid samples of the magma chamber), and the proportion made of cumulates. The ophiolite model gives an approximation to the structure at fast-spreading ridges, indicating a gabbro thickness of about 4 km, similar to the measured thickness of seismic layer 3. The thickness of gabbro at the MAR cannot be estimated in the same way, since, as pointed out above, there are good arguments that a proportion of the crust is made of wholly or partly serpentinised peridotite. However, it is possible to estimate the ratio of lavas plus dykes (samples of magma available in the crust) to that of gabbros (cumulates from the magma) indirectly, from the geochemistry of the lavas and dykes.

In the simple model proposed by *Cann* [1981], lavas and dykes are assumed to be samples of liquid from a steady-state magma chamber, and a mass balance calculation is applied to balance the supply of elements into the magma chamber through primary magma, with their removal from the magma chamber as both liquids and cumulates. Figure 18 shows how the model works. The mass balance yields the relation:

$$C_I = C_S(1-f) + C_L f \qquad (6)$$

where C_I is the concentration of an element in melt from the mantle, C_L is the mean concentration of that element in the lavas and dykes, and f is the fraction of the liquid coming from the mantle that ends up as lavas and dykes. Since

$$\frac{C_S}{C_L} = D \qquad (7)$$

with D the bulk solid/liquid distribution coefficient for that element,

$$C_I = DC_L(1-f) + C_L f \qquad (8)$$

and this yields the relation:

$$\frac{C_I}{C_L} = D - Df + f \qquad (9)$$

The mass balance is best struck with a moderately incompatible element, for which there is a marked difference between the initial concentration in melts from the mantle (C_I) and the mean composition of the lavas and dykes (C_L), but not too much variation from place to place in the oceans. For such elements, the precise value of D is not critical for the calcu-

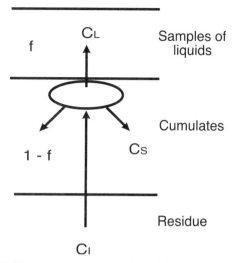

Figure 18. Diagrammatic representation of the mass-balance model relating magma geochemistry to the ratio of cumulates to lavas plus dykes (see text).

lation, as it is for compatible elements such as Ni. Ti is an appropriate element to use, and is convenient to use too since it is routinely analysed in glasses by electron probe.

A search through the RIDGE Petrological Database at Lamont [*Lehnert et al.*, 2000] shows that there are a number of glass analyses from the MAR (we used the MAR 52°N-52°S dataset compiled in the database) with Mg numbers (calculated assuming that 0.9 of the Fe is present as Fe^{II}) close to 0.72, the value in equilibrium with olivine in the upper mantle. The mean TiO_2 content of these samples is 0.7%. The mean TiO_2 content of all the basalts from the MAR dataset is 1.42%, giving an overall C_I/C_L of 0.49. Assuming a solid/liquid distribution coefficient for Ti of 0.1 (a value that is not critical for the calculation), the fraction of lavas and dykes (f) calculates as 0.43.

This fraction is greater than that of the classic ophiolite model of fast-spreading crust, which, with about 2 km of lavas and dykes, and a totally magmatic crust, would yield f about 0.33. This lower value is consistent with the higher mean TiO_2 content of EPR basalts (1.64%) in the compilation by *Lehnert et al.* [2000] in the RIDGE Petrological Database at Lamont.

9. EPISODICITY OF MAGMA INPUT

Up until now, our discussion has assumed that magma is supplied continuously to a spreading segment, but there is evidence that magma (and heat) is in fact supplied episodically at a range of time scales. Evidence for episodic magma supply mainly comes from the morphology of the axial zone, from the size of mid-ocean ridge eruptions, and from the heat output of major hydrothermal systems.

At fast-spreading ridges the bathymetric morphology of the axial zone allows the distinction of inflated and deflated segments [*Macdonald*, 1998]. This inflation and deflation is correlated with the seismic character of the axial magma chamber and/or the presence or absence of a reflector from the magma chamber [*Hooft et al.*, 1997], and hence with the amount of magma in the chamber. Inflated segments are considered to have been replenished recently, and deflated segments to be awaiting replenishment. On a segment scale, the cross sectional area of the crest of the EPR varies between about 5 km^2 down to about 2 km^2 [*Macdonald*, 1998]. For a segment length of 50 km, this corresponds in a difference in volume of 150 km^3. At a spreading rate of 100 mm/yr, at steady state, a 50 km segment would require 0.03 km^3/yr of magma to generate the whole crust. The difference between inflated and deflated volumes of the ridge crest would thus represent 4500 years of construction of the whole crust, or longer if the inflation is related only to construction of part of the crust. This is an indication that magma supply at the EPR fluctuates on a thousand year timescale.

Fluctuation at shorter timescales at the fast-spreading EPR is indicated by the occasional eruption of very large lava flows. *Macdonald et al.* [1989] recognised a single flow unit of plan area 220 km^2. If this flow is 10 metres thick, then its volume would be about 2 km^3, which is a 60 year supply of magma for the whole crustal thickness of a 50 km long segment spreading at 100 mm/yr, or a 600 year supply for the extrusive section alone. Alternatively it would represent the complete emptying of an axial magma chamber 1 km wide, 50 km long and 40 metres thick, comparable in size with the estimates from seismic surveys of *Singh et al.* [1998].

Large axial volcanic ridges at slow-spreading centres are typically 500 m high and 5 km wide, giving a cross-sectional area of 1.3 km^2. Such a ridge in a 50 km long segment would have a volume of about 50 km^3, allowing for tapering of the ridge towards its ends. For a spreading rate of 25 mm/yr, the steady state total magma supply for a 50 km long segment would be about 7 km^3 per thousand years (assuming an average melt thickness of 6 km), and the steady state supply of extrusives about 0.5 km^3 per thousand years (assuming an average lava thickness of 500 m). The volume of extrusives in a large axial volcanic ridge thus represents at least tens and probably hundreds of thousands of years supply.

There is also limited seismic evidence (mentioned in section 4.2) for an episodicity of magma injections from the mantle at the MAR south of the Kane FZ, with periods of 400 to 800 kyrs [*Canales et al.*, 2000a].

Shorter scales of episodicity of magma supply are indicated by the activity of the TAG hydrothermal system in the MAR. Radiometric dating of sulphides from the TAG mound has shown that the system there has been active for short periods every few thousand years [*Lalou et al.*, 1995]. The heat output estimated for the TAG hydrothermal system (~1000 MW; [*Humphris and Cann*, 2000]) is far greater than can be supplied in steady state over a reasonable ridge axis length (~16 MW/km of ridge axis ; section 3), and the conclusion must be that a pulse of magma, with high rates of magma supply (similar to those observed in volcanoes in Hawaii and Iceland), arrives in the vicinity of the TAG mound every few thousand years to reawaken the hydrothermal system [*Humphris and Cann*, 2000].

A similar scale of episodicity is shown by the larger eruptive features that form part of axial volcanic ridges at slow-spreading centres. Single eruptive features such as lava terraces and major lava flows have volumes of 1 km^3 or more [*Smith and Cann*, 1992]. Such features must have formed in a single prolonged eruption, since they are positive topographic features than would be by-passed by later flows, and their volume represents a few thousand years of eruptive supply at a slow-spreading ridge. Similar features, of similar size and shape, are found on the Puna Ridge, the offshore extension of the Kilauea East Rift [*Smith and Cann*, 1999], and such volumes of lava have been seen to be erupted from Kilauea in the space of a few years. By analogy with the eruptions in Kilauea, the terraces and major lava flows on slow-spreading centres would have been constructed in a similar space of time.

There is thus broad concurrence that magma supply to a single segment of a mid-ocean ridge fluctuates both at fast and slow-spreading ridges on timescales of a few thousand years. In addition, superimposed on this timescale, there are suggestions from slow-spreading ridges for episodicity on longer timescales (hundreds of thousand years to 1 million years).

10. TECTONIC ADVECTION OF HEAT AT SLOW-SPREADING RIDGES

Mantle-derived peridotites and gabbros are tectonically uplifted throughout the axial lithosphere of the magma-poor ends of slow-spreading segments (Figure 10), and advect heat as they rise. At the centres of slow-spreading segments, the larger magma input to the crust should reduce this tectonic uplift component, but there remains the possibility for peridotites and for gabbros that crystallised in the mantle part of the axial lithosphere, to be uplifted at least to form part of the lower crust. Axial tectonics may therefore affect the thermal regime of slow-spreading ridges in two ways: one is to favour seawater penetration through open cracks and permeable fault zones, and the other is to advect the heat carried by deeply-derived rocks as they are tectonically uplifted through the axial lithosphere. A critical factor in modelling this tectonic component of heat advection is the extent to which tectonic strain associated with this uplift is localized. If tectonic strain is strongly localized, uplift rates may be high, allowing for more heat to be tectonically advected to the crust. By contrast, moderate strain localization would lead to moderate rates of uplift and heat advection.

Most peridotites from slow-spreading ridges (away from fracture zones) preserve coarse grained ductile deformation textures acquired under low deviatoric stress in near solidus temperature conditions [*Cannat et al.*, 1990; *Cannat and Casey*, 1995; *Ceuleneer and Cannat*, 1997]. Preservation of these textures indicates that deformation under higher deviatoric stress in the ductile part of the axial lithosphere was not pervasive, but localized into shear zones. The corrugated surfaces identified in the off-axis domains of slow-spreading ridges [*Cann et al.*, 1997; *Tucholke et al.*, 1998] are larger scale indicators of significant localization of strain in the axial lithosphere. These corrugated surfaces are interpreted as the fossil fault planes of major axial detachment faults [*Cann et al.*, 1997; *Blackman et al.*, 1998; *Tucholke et al.*, 1998; *Lavier et al.*, 2000; *Buck et al.*, 2003], and are therefore inferred to form during periods of extreme axial strain localization. Corrugated surfaces, although common, cover only a small proportion of the off-axis Mid-Atlantic seafloor [*Tucholke et al.*, 1998]. Periods when tectonic strain at the MAR is accommodated by very large offset detachments therefore probably alternate with longer periods when tectonic strain is accommodated by smaller offset axial faults [*Buck and Poliakov*, 1998; *Lavier et al.*, 2000; *Buck et al.*, 2003], possibly leading to more moderate rates of uplift and tectonic heat advection.

11. CONCLUSIONS

In this paper, we have examined many aspects of how heat from melts and from the mantle is transferred to the oceanic crust at mid-ocean ridges. In this final section, we emphasize our principal conclusions. Many of these conclusions concern the tectonically and lithologically more complex slow-spreading ridges and will, we hope, be of help to develop improved thermal models of these complex settings.

1. The specific heat of peridotites and gabbros is only marginally dependent on modal composition, and varies between ~ 1300 J kg^{-1} °K^{-1} at temperatures > 1000°C, and ~ 800 J kg^{-1} K^{-1} at 0°C (Figure 4). The latent heat of crystallisation of basaltic melts in the crust is ~ 0.5 MJ kg^{-1}. Heat released by serpentinisation of peridotite is ~ 0.25 MJ kg^{-1} [*Fyfe and Lonsdale*, 1981].

2. In thermal terms, it is useful to discuss slow-spreading ridges under two crustal end-members : a magmatically-dominated, segment centre type, end-member, and a magma-poor, segment end type, end-member, keeping in mind that many sites may be intermediate between these two end-members.

The crustal structure for the magmatically-dominated end-member should be comparable, although probably not as systematically layered, to that of fast-spreading ridges or large ophiolites, with units of extrusives, dykes and gabbroic plutonics. However, many dykes at slow-spreading ridges probably propagate laterally from isolated central magma chambers. And these magma chambers are expected to have a limited life. These are important differences in terms of heat distribution with fast-spreading ridge settings.

The crustal structure for the magma-poor end-member would be heterogeneous, with a tectonically uplifted component of mantle-derived peridotites and short-lived gabbro plutons, as well as dykes feeding lava flows. Dyke swarms propagating from the segment centre would also feed lava flows at segment ends, perhaps associated with underlying diabase bodies.

3. The evidence reviewed in section 5 gives strong support for partial crystallisation of magma in the form of gabbros at sub-crustal depth below slow-spreading ridges. Most of this evidence comes from basalts, many of which come from segment centre settings. Significant sub-crustal melt crystallisation is therefore not a characteristic only of the magma-poor, segment-end type settings. The depths calculated range at least to 12 km (equivalent to a pressure of 0.4 GPa). Solidus temperatures for gabbros of intermediate compositions are therefore probably reached at these depths, even at segment centres.

Such cold thermal conditions on axis would also favor the development of deep-reaching normal faults. Gabbros that have crystallised at sub-crustal depths may therefore subsequently be tectonically incorporated into the crust (Figure 10). This is a strong argument to propose that slow-spread crust, even at segment centres, is not as systematically layered as that of fast-spreading ridges or large ophiolites. Instead, it probably comprises a mix of tectonised deeply-derived material hosting crustal magmatic plutons and dykes [*Cannat*, 1996]. In section 3, we calculated that the heat supplied by magma to near-axis hydrothermal systems at a slow (25 mm/yr) spreading segment centre should be ~18 to 25 MW per km of ridge crest. In this calculation, we assumed that enough melt was provided to build a 6 to 8 km-thick crust, and that all melts crystallised at crustal levels. Crystallising part of these melts below the crust will release latent heat, and some specific heat, at greater depths. While not changing the overall heat supply to the ridge axis, this could make that heat less available to high heat flux, black smokers type axial hydrothermal systems.

4. In the slow-spreading, magma-poor end-member, the rising peridotite would be a significant source of heat. Tectonic heat advection would occur as deeply-derived rocks (peridotites and gabbros) are uplifted through the axial lithosphere. In section 10, we discussed how strain localization along large-offset detachment faults could lead to higher uplift rates and therefore enhance this tectonic heat advection component. In section 3, we calculated that in the extreme case of completely amagmatic accretion, rising peridotites would contribute about 16 MW km^{-1} at a ridge spreading at 25 mm/yr. This is not much less than the heat supply calculated for magmatically-dominated segment centres at the same spreading rate. Most of this would be specific heat from cooling the peridotite into the 8 km-thick brittle lithosphere inferred for slow-spreading ridge segment ends (based on passive seismic studies [*Toomey et al.*, 1988; *Kong et al.*, 1992; *Wolfe et al.*, 1995]). As noted in section 3, it is unlikely that a thin thermal boundary layer would develop in this amagmatic end-member setting as it presumably does when magmatic bodies are present [*Lister*, 1995]. It is also possible that permeability in serpentinising peridotites would be less due to volume increase associated with serpentinisation [*Wilcock and Delaney*, 1996]. This could restrict efficient extraction of heat by high temperature hydrothermal systems, and pervasive serpentinisation of large volumes of peridotite in end of segment-type settings, to the vicinity of frequently reactivated faults, and of cooling gabbro or diabase intrusions.

5. In section 8, we use the ratio of incompatible elements concentration in primary MORB from the MAR to that in mean MORB to propose that the ratio of cumulates to lavas plus dykes at this slow-spreading ridge is about 0.43. This fraction is greater than that of the classic ophiolite model of fast-spreading crust, which, with about 2 km of lavas and dykes, and a totally magmatic crust, would yield a ratio of about 0.33. From a thermal point of view, this suggests that a greater proportion of the heat supplied by melts to the slow-spreading MAR would go into rapidly cooled magmatic rocks (diabase and basalt as opposed to gabbros).

6. The two current end-member models (gabbro glacier model and sill intrusion model) for construction of the lower crust at fast-spreading ridges have very distinct implications in terms of near axis heat supply. In section 6, we discussed these two end-member models, using the Oman ophiolite as an analogue for fast-spread oceanic crust. We find that neither of the two models fits the textural and petrological characteristics of the Oman gabbros. In addition, the sill intrusion model in its simplest form does not satisfactorily explain the formation of the axial crystal mush imaged beneath the EPR [*Harding et al.*, 1989; *Toomey et al.*, 1990 ; *Crawford et al.*,

1999; *Dunn et al.*, 2000; *Crawford and Webb*, 2002]. New hybrid models, that would for example combine in situ crystallisation near the walls of the axial low velocity zone, a finite width of magma injection in the lower crust, and a component of crystal settling from the roof and walls of this low velocity zone, may prove more successful. To develop such new models, a better understanding of near axis hydrothermal cooling processes in fast-spread lower crust appears critical. Ultimately, however, answering the question will require adequate sampling of the lower crustal rocks formed at fast-spreading ridges.

7. The volume of melt supplied to the ridge for each increment of plate spreading is a critical parameter of ridge thermal models. It is commonly assumed that it is enough to build a magmatic layer equivalent to the seismically defined crustal layer. This is supported by the global agreement between seismic crustal thicknesses and melt thicknesses calculated from MORB compositions and axial depths [*Klein and Langmuir*, 1987; *White et al.*, 1992a]. These indirect approaches yield melt thickness estimates that are independent of how melt is distributed between the crust and upper mantle, but strongly dependent on mantle composition, temperature, degree of melting, and dynamics. The global agreement between seismic crustal thicknesses and melt thicknesses calculated from MORB compositions and axial depths is therefore also an indirect argument to say that the simple passive sub-axial mantle flow assumed in the calculation is applicable, at least at the regional, multi-segment scale [*Klein and Langmuir*, 1987; *Langmuir et al.*, 1992].

The hypothesis of melt thickness equals seismic crustal thickness is, however, challenged by geological evidence for the emplacement of variably serpentinised peridotites in slow-spread crust [*Dick*, 1989; *Cannat*, 1993]. It is also questioned by petrological data suggesting sub-crustal melt crystallisation (see section 5), although we have seen in section 5.3 that sub-crustal melt crystallisation does not necessary mean less melt products in the crust as it leaves the axial region (Figure 10). In section 4, we used MORB composition and regional axial depths to model melt thickness at nine well-studied mid-ocean ridge regions, and compared the results to seismically determined crustal thickness in each region. This approach differs from the global approach of [*Klein and Langmuir*, 1987; *White et al.*, 1992a] in that it focuses on the detail of how the different melt thickness indicators agree or disagree at the regional scale. We got consistent results for the EPR regions and for two of the six MAR regions, indicating that seismic crustal thickness in these regions does provide a good estimate of the melt supply per increment of plate separation. The other four MAR regions yielded inconsistent results. In the case of two near

Table A.1. Correction factors to the LDEO point beam microprobe calibration [*Reynolds*, 1995].

	SiO2	TiO2	Al2O3	FeO	MgO	CaO	Na2O	K2O
Smithonian	0.998	-	0.988	0.987	1.047	-	1.019	-
Hawaii	0.998	-	0.988	0.987	1.047	-	1.019	-
Tulsa	-	-	0.996	-	-	1.012	1.029	1.067
LDEO broad beam	-	-	-	-	-	-	0.962	-

hot spot regions, melt thickness estimated from MORB composition was twice the seismic crustal thickness, and greater than predicted from regional axial depths. In the other two MAR regions, melt thickness estimated from MORB composition was less than the measured seismic crustal thickness. We discussed these discrepancies, and proposed that the presence of variably serpentinised peridotite in the crust may affect melt supply estimates from seismic crustal thickness and axial depth, while complexities in melt and mantle dynamics beneath the ridge may affect melt supply estimates from MORB chemistry. Our conclusion is that determination of the melt (and heat) supply to the crust of slow-spreading ridges is not straightforward and requires a good integration of local scale geological, geophysical, and geochemical data.

8. The evidence discussed in section 8 suggests that magma supply to the crust of a single segment of a mid-ocean ridge fluctuates both at fast and slow-spreading ridges on timescales of a few thousand years. Such episodicity is probably critical for thermal models of axial hydrothermal systems, particularly at slow-spreading ridges [*Wilcock and Delaney*, 1996]. Heat fluxes estimated for large hydrothermal vent fields at both fast and slow ridges are of the order of 1000 MW or greater [*Rosenberg et al.*, 1988; *Baker*, 1994; *Baker et al.*, 1996; *Humphris and Cann*, 2000]. With steady state magma supply to the crust (estimated heat flux of ~18 MW/km ; section 3), this would mean tapping heat from > 50 km along-axis of a slow (25 mm/yr) spreading ridge.

The alternative is to derive heat for high temperature black smokers type hydrothermal systems from episodic melt injection events, with rates of magma supply higher than in the steady state hypothesis [*Wilcock and Delaney*, 1996; *Humphris and Cann*, 2000]. Such high magma supply events may occur on timescales of the order of a few thousand years. This magmatic episodicity should not substantially affect the thermal state of the whole axial region of a slow-spreading ridge (the MAR axial valley typically corresponds to crustal ages of zero up to 1 myr). However, there are also suggestions from slow-spreading ridges for episodicity on longer timescales (hundreds of thousand years to 1 million years). If confirmed, these ought to be taken into account in thermal models of the axial region.

APPENDIX A. MORB NA8.0 IN NINE EPR AND MAR REGIONS

Basalt glass major elements data used in our compilation are stored in the LDEO RIDGE Geochemical Database [*Lehnert et al.*, 2000]. All but the MAR16N data are microprobe data and were corrected back to the LDEO microprobe point beam calibration using the correction factors of *Reynolds* [1995] (Table A1). Average $Na_{8.0}$ values were calculated as follows for samples with K_2O/TiO_2 = 0.2, and 6.5 = MgO = 8.5% using the formula proposed by *Klein and Langmuir* [1987] : $Na_{8.0}$= $Na_2O + 0.373.MgO - 2.98$; or the best linear fit for Na_2O vs. MgO trends (Table A.2). $Na_{8.0}$ values used in the paper are the "best fit" ones. Table A.2 also shows average K_2O/TiO_2 ratios for the whole set of data, and for samples with K_2O/TiO_2 = 0.2 only. Four regions of the MAR have average K_2O/TiO_2 ratios = 0.14 (MAR37N, MAR34N, MAR16N, and MAR33S). Only one region (MAR34N) has a significant number of samples with K_2O/TiO_2 > 0.2. The average $Na_{8.0}$ for these samples (2.334 ± 0.189) is similar to that of MAR34N samples with lower K_2O/TiO_2 ratio (Table A2).

APPENDIX B. ESTIMATING MELT THICKNESS FROM MORB REE DATA

MORB REE concentration data used in our compilation are stored in the LDEO Ridge Geochemical Database and were used to predict oceanic crustal thicknesses using the inversion technique method described by *McKenzie and O'Nions* [1991] and *White et al.* [1992b]. We used the depleted MORB mantle source of *White et al.* [1992b] for all of these calculations. The use of this method involves a number of assumptions about magmatic behaviour at ridges, many of which are also found in the $Na_{8.0}$ method of *Klein and Langmuir* [1987]. The first assumption is that mantle melting is driven by adiabatic decompression of mantle which has horizontal isotherms beneath the base of the melting region. This adiabatic decompression is caused by passive upwelling of isoviscous asthenosphere in reponse to plate separation, and this upwelling will take the form of a corner flow. It can be shown that the melt production caused by corner flow upwelling is equivalent to that of mantle upwelling with a constant velocity in columns within a triangular melting

Table A.2. Average K_2O/TiO_2 ratios and $Na_{8.0}$ values for nine EPR and MAR regions. See Table 3 for references. N: total number of samples; n: number of samples with $K_2O/TiO_2 = 0.2$. $Na_{8.0}$ (K+L): calculated with *Klein and Langmuir* [1987] formula. $Na_{8.0}$ (best fit) : calculated using the best linear fit for Na_2O vs. MgO trends.

RIDGE REGION	full dataset			$K_2O/TiO_2 \leq 0.2$						
	N	K2O/TiO2		n	K2O/TiO2		Na8.0 (K+L)		Na8.0 (best fit)	
		av.	std		av.	std	av.	std	av.	std
EPR 12-12.7°N	194	0.08	0.05	185	0.07	0.03	2.60	0.20	2.60	0.18
EPR 9.5-9.7°N	111	0.09	0.06	96	0.08	0.03	2.75	0.09	2.63	0.07
EPR 14-18.5°S	92	0.07	0.04	86	0.07	0.03	2.70	0.22	2.66	0.14
MAR 36.3-37.3°N (FAMOUS)	205	0.15	0.13	203	0.13	0.03	2.16	0.10	2.17	0.08
MAR 33.5-35.5°N (OH1-OH3)	81	0.14	0.10	49	0.11	0.05	2.30	0.16	2.37	0.12
MAR 23.8-24.7° (MARNOK)	11	0.08	0.01	11	0.08	0.01	2.97	0.14	2.97	0.04
MAR 22.8-23.7°N (MARK)	56	0.08	0.01	56	0.08	0.01	3.08	0.14	2.98	0.12
MAR 15.4-16.4°N	10	0.14	0.06	10	0.14	0.06	3.18	0.07	3.05	0.04
MAR 32-33.5°S	22	0.14	0.05	21	0.13	0.03	2.68	0.14	2.79	0.07

region, and that the melt production is independent of spreading rate or ridge angle [*Plank and Langmuir*, 1992]. This assumption does not hold at very slow-spreading rates (<15 mm yr^{-1} full spreading rate; [*Bown and White*, 1994]) when conductive cooling of the upwelling mantle is important. It is then assumed that mantle passing through the melting region melts according to incremental fractional melting equations, so that the composition of the instantaneous fractional melts and the co-existing residual mantle can be calculated throughout the melting region. All of the melts generated are assumed to be instantaneously extracted from the melting region, and are fully mixed before eruption. If these assumptions hold and no crystallisation takes place, then the average composition of a suite of basaltic samples will be the same as that of the average composition of the melt generated within the melting region.

The REE composition of the average melt is sensitive both to the depth and degree of melting. Melts generated at small extents of melting have high REE concentrations. Further melting will dilute these concentrations. Melts generated within the garnet stability field (>60 km depth in our models) have low heavy REE concentrations (e.g., Yb, Lu) because these elements are compatible in garnet. Heavy REEs are, however, more incompatible within the spinel stability field (<80 km depth). If the melt fraction against depth relationship is known, it is possible, using the assumptions outlined above, to calculate the average REE concentration of the mantle melts. However, in the melting problem the REE compositions are known while the melt fraction against depth relationship is unknown. The inversion approach used by *McKenzie and O'Nions* [1991] and *White et al.* [1992b] performs many forward calculations with varying melt fraction against depth

relationships until it finds a suitable fit to the REE concentrations. This best fitting melt fraction against depth relationship can then be used to calculate the crustal thickness, using the expressions given by *White et al.* [1992b]. Crystallisation of MORB raises its REE concentrations, and must be taken account of in the modelling of REEs and calculation of crustal thicknesses. *McKenzie and O'Nions* [1991] used the Mg-Fe contents of the observed MORB and calculated primary melts to estimate the degree of crystallisation, F, of mafic phases (Table 3). The crustal thickness estimates are then increased by a factor of $1/(1-F)$. The extent of crystallisation, and hence the crustal thickness, is underestimated if plagioclase or other non-mafic phases form an important part of the crystallising assemblage (see discussion in *Maclennan et al.* [2001a]).

APPENDIX C. MODELLING MANTLE MELTING, MELT THICKNESS, MELT SODIUM CONTENT, AND AXIAL DEPTH

Figures 5, 6, 7, 8 and 9 show melt thicknesses, melt sodium contents, and average regional axial depths predicted for a range of mantle temperatures at 100 km depth (Tzzero), and, in the case of Figure 9, for a range of final depths of melting (zf), using a simple mantle melting model. This model is closely derived from the *Klein and Langmuir* [1987] model and is based on many assumptions (tent-shaped region of mantle melting, passive mantle flow, complete melt extraction, perfect focusing of melts toward the ridge axis, and a uniform sodium content in the mantle at the onset of melting; see Appendix B). We use the mantle solidus parameterisation proposed by *Hirschmann* [2000] : $T = -5.904.P^2 + 139.44.P$

+1108.08 (P in Gpa), and we allow for 1D conductive cooling of the uprising mantle from above, with the possibility (not used in this paper) to vary mantle vertical velocity, and the vertical penetration of enhanced hydrothermal cooling. Melt Na, K, Ti, Mg, and Fe contents are calculated for pooled fractional melts as indicated in Appendix A of *Langmuir et al.* [1992], for the following initial mantle composition (weight %): Si=45.57; Ti=0.170; Al=2.70; Fe=8.01; Mn=0.13; Mg=40.80; Ca=2.41; Na=0.27; K=0.002; Cr=0.44. Melt $Na_{8.0}$ contents are calculated using 1 atm, closed-system liquid lines of descent as indicated in Appendix C of *Langmuir et al.* [1992]. Predicted regional axial depths are calculated as described in *Klein and Langmuir* [1987], except that we did not take into account the possible change of density of mantle rocks at the spinel to plagioclase facies transition. Spinel is an ubiquitous phase in abyssal peridotites, indicating that full reequilibration of sub-axial mantle rocks in the plagioclase facies is the exception rather than the rule. Parameters of our melting model, and the values used in this paper, are listed in Table C1. A Matlab script of this melting model is available upon request to the first author of this paper (cannat@ipgp.jussieu.fr).

Acknowledgments. The authors would like to thank Jian Lin, Chris German and Lindsay Parson for their work as editors, Peter Michael, Bramley Murton, and Laurence Coogan for their thorough and extremely helpful reviews, the US RIDGE program and the team in charge of creating and maintaining the RIDGE Geochemical Database at Lamont, and Javier Escartin for an early informal review. This work was supported by an EU Marie Curie Individual Research Fellowship to J.M. This is IPGP contribution # 1968.

REFERENCES

Adamson, A. C., Basement stratigraphy, Deep Sea Drilling Project Hole 504B, in *Initial Reports Deep Sea Drilling Project*, pp. 121–127, Governement Printing Office, Washington D.C., 1985.

Alt, J. C., D. A. H. Teagle, C. Laverne, D. A. Vanko, W. Bach, J. Honorez, K. Becker, M. Ayadi, and P. Pezard, Ridge-flank alteration of upper ocean crust in the eastern Pacific: synthesis of results for volcanic rocks of holes 504B and 896A, *Proc. Ocean Drill. Pro-*

Table C.1. Parameters used in mantle melting model.

NAME AND UNIT OF PARAMETER	abbreviation and value used in melting model
mantle upwelling velocity (cm/yr)	u=5
mantle adiabatic temperature gradient (°/km)	adgrad=0.3
latent heat of melting in mantle (cal/g.°)	Hf=150
mantle heat capacity (cal/g)	cp=0.3
dT/dF (°/%)	dTdF=3.5
dT/dF for mantle that has already melted by 22%	dTdF22=6.8
thermal diffusivity of mantle rocks (m2/s)	$k=8.04.10^{-7}$
upper boundary for conductive cooling model (km)[a]	za=0
depth (km) at which conductive cooling is assumed to be negligable	zzero=100
max. depth of model (km)	zz=200
temperature at za (°C)[a]	Tza=1
maximum depth of hydrothermal cooling (km)[a]	zhydr=0
enhanced conductivity in domain of hydrothermal circulation	coeff=1
compensation depth (km) for regional isostatic compensation	zcomp=zz
axial depth (km) for reference column	zref=3
melt thickness (km) for reference column	mref=6.06
mean mantle density (kg/m3) for reference column	dmref=3313.9
mantle thermal expansion coeff. (/°)	$exptherm=3.10^{-5}$
mantle compressibility modulus (/kbar)	$modcomp=10^{-3}$
density (kg/m3) of fertile mantle in spinel-garnet facies (25°C, 1atm)	dmsg=3340
density (kg/m3) of residual mantle (F:30%) in spinel-garnet facies (25°C, 1atm)	dmd=3295
density of crust (kg/m3)	dc=2700
density of water (kg/m3)	de=1010
percent serpentinized peridotite in crust[b]	serp= 0

[a] In Figure 9 variations in zf are imposed by arbitrarily setting Tza to 600°C and varying za = zhydr between 0 and 46 km. This configuration has no geological relevance and is not used to predict regional axial depths.
[b] In Figure 8 serp varies between 0 and 40%.

gram., Sci. Results, 148, 435–452, 1996.

Anderson, R. N., J. Honnorez, and K. Becker, DSDP Hole 504B, the first reference section over 1 km throught layer 2 of the oceanic crust, *Nature, 300*, 589–594, 1982.

Aumento, F., W. G. Melson, J. M. Hall, H. Bougault, and et al., Site 335, in *Initial Reports of the Deep Sea Drilling Project vol. 37*, edited by P. T. Robinson, and R. C. Howe, Texas A & M University, Ocean Drilling Program, College Station, TX, United States, Washington D.C., 1977.

Auzende, J.-M., D. Bideau, E. Bonatti, M. Cannat, J. Honnorez, Y. Lagabrielle, J. Malavieille, V. Mammaloukas-Frangoulis, and C. Mevel, Direct observations of a section through slow-spreading oceanic crust, *Nature, 337*, 726–729, 1989.

Babcock, J. M., A. J. Harding, G. M. Kent, and J. A. Orcutt, An examination of along-axis variations of magma chamber width and crustal structure on the East Pacific Rise between 13°30'N and 12°20'N, *Journal of Geophysical Research, 103* (B12), 30451–30467, 1998.

Baker, E. T., A 6-year time series of hydrothermal plumes over the Cleft segment of the Juan de Fuca Ridge, *Journal of Geophysical Research, 99*, 4889–4904, 1994.

Baker, E. T., Y. J. Chen, and J. Phipps Morgan, The relationship between near-axis hydrothermal cooling and the spreading rate of mid-ocean ridges, *Earth and Planetary Science Letters, 142*, 137–145, 1996.

Barth, G. A., and J. C. Mutter, Variability in crustal thickness and structure : multichannel seismic reflection results from the northern East Pacific Rise, *J. Geophys. Res., 101*, 17951–17976, 1996.

Batiza, R., and Y. Niu, Petrology and magma chamber processes at the East Pacific Rise 9°30'N., *Journal of Geophysical Research, 97* (B5), 6779–6797, 1992.

Becker, K., and et al., Drilling deep into young oceanic crust, Hole 504B, Costa Rica Rift, *Reviews of Geophysics, 27*, 79–102, 1989.

Bell, R. E., and W. R. Buck, Crustal control of ridge segmentation inferred from observations of the Reykjanes Ridge, *Nature, 357*, 583–586, 1992.

Bender, J. F., F. N. Hodges, and A. E. Bence, Petrogenesis of basalts from the project FAMOUS area: experimental study from 0 to 15 kbars, *Earth and Planetary Science Letters, 41*, 277–302, 1978.

Berman, R. G., Internally consistent thermodynamic data for minerals in the system Na2O-K2O-CaO-Mg0-FeO-Fe2O3-Al2O3-SiO2-TiO2-H2O-CO2, *Journal of Petrology, 29*, 445–522, 1988.

Blackman, D. K., J. R. Cann, B. Janssen, and D. K. Smith, Origin of extensional core complexes: Evidence from the Mid-Atlantic Ridge at Atlantis Fracture Zone, *Journal of Geophysical Research, 103* (B9), 21315–21333, 1998.

Bloomer, S. H., J. H. Natland, and R. L. Fisher, Mineral relationships in gabbroic rocks from fracture zones of Indian Ocean ridges: evidence for extensive fractionation, parental diversity and boundary-layer recrystallization, in *Magmatism in the Ocean Basins*, edited by A. D. Saunders, and M. J. Norry, pp. 107–124, Geological Society Special Publication, 1989.

Bloomer, S. H., P. S. Meyer, H. J. B. Dick, K. Ozawa, and J. H. Natland, Textural and mineralogical variations in gabbroic rocks from Hole 735B, in *Proceedings of the Ocean Drilling Program, Scientific Results*, edited by R. P. Von Herzen, C. J. Robinson, and A.

C. Adamson, pp. 21–40, Ocean Drilling Program, College Station, 1991.

Boudier, F., and A. Nicolas, Nature of the Moho transition zone in the Oman ophiolite, *Journal of Petrology, 36* (3), 777–796, 1995.

Boudier, F., A. Nicolas, and B. Ildefonse, Magma chambers in the Oman ophiolite: fed from the top and the bottom, *Earth and Planetary Science Letters, 144*, 239–250, 1996.

Bougault, H., L. Dmitriev, J. G. Schilling, A. Sobolev, J. L. Joron, and H. D. Needham, Mantle heterogeneity from trace elements: MAR triple junction near 14°N, *Earth and Planetary Science Letters, 88*, 27–36, 1988.

Bown, J. W., and R. S. White, Variation with spreading rate of oceanic crustal thickness and geochemistry, *Earth and Planetary Science Letters, 121*, 435–449, 1994.

Bredehoeft, J. D., and I. S. Papadopoulos, Rates of vertical groundwater movement estimated from the earth's thermal profile, *Water Resources Research, 11*, 325–328, 1965.

Bryan, W. B., G. Thompson, and J. Ludden, Compositional variation in normal MORB from 22°–25°N: Mid-Atlantic Ridge and Kane Fracture Zone., *Journal of Geophysical Research, 86* (B12), 11815–11836, 1981.

Buck, W. R., and A. N. B. Poliakov, Abyssal hills formed by stretching oceanic lithosphere, *Nature, 392*, 272–275, 1998.

Buck, W. R., L. Lavier, and A. Poliakov, Modes of faulting at mid-ocean ridges, *Geophysical Research Abstracts,, 5*, 13333, 2003.

Canales, J. P., R. S. Detrick, S. Bazin, A. J. Harding, and J. A. Orcutt, Off-axis crustal thickness across and along the East Pacific Rise within the MELT area, *Science, 280*, 1218–1221, 1998.

Canales, J. P., J. A. Collins, J. Escartin, and R. S. Detrick, Seismic structure across the rift valley of the Mid-Atlantic Ridge at 23°20'N (MARK area): Implications for crustal accretion processes at slow-spreading ridges, *Journal of Geophysical Research, 105* (B12), 28411–28425, 2000a.

Canales, J. P., R. S. Detrick, J. Lin, J. A. Collins, and D. R. Toomey, Crustal and upper mantle seismic structure beneath the rift mountains and across a non-transform offset at the Mid-Atlantic Ridge (35°N), *Journal of Geophysical Research, 105* (B2), 2699–2719, 2000b.

Canales, J. P., R. S. Detrick, D. R. Toomey, and S. D. Wilcock, Segment-scale variations in crustal structure of 150 to 300 ky-old fast-spreading oceanic crust (East Pacific Rise 8°15'N - 10°15'N) from wide angle seismic refraction profiles, *Geophys. J. Intern., 152*, 766–794, 2003.

Cann, J. R., A model for oceanic crustal structure developed, *Geophys. J. R. Astr. Soc., 39*, 169–187, 1974.

Cann, J. R., Basalts from the ocean floor, in *The Sea—The Oceanic Lithosphere*, edited by C. Emiliani, pp. 363–390, Wiley, New York, 1981.

Cann, J. R., D. K. Blackman, D. K. Smith, E. McAllister, B. Janssen, S. Mello, E. Avgerinos, A. R. Pascoe, and J. Escartín, Corrugated slip surfaces formed at North Atlantic ridge-transform intersections, *Nature, 385*, 329–332, 1997.

Cannat, M., Emplacement of mantle rocks in the seafloor at mid-ocean ridges, *Journal of Geophysical Research, 98* (B3), 4163–4172, 1993.

Cannat, M., How thick is the magmatic crust at slow-spreading oceanic ridges?, *Journal of Geophysical Research*, *101* (B2), 2847–2857, 1996.

Cannat, M., T. Juteau, and E. Berger, Petrostructural analysis of the Leg 109 serpentinized peridotites, in *Proceedings of the Ocean Drilling Program, Scientific Results*, edited by R. Detrick, J. Honnorez, W.B. Bryan, T. Juteau, and e. al., pp. 47–56, Ocean Drilling Program, College Station, TX, 1990.

Cannat, M., M. Mével, and D. Stakes, Normal ductile shear zones at an oceanic spreading ridge: Tectonic evolution of site 735 gabbros (Southwest Indian Ocean), in *Proceedings of the Ocean Drilling Program, Scientific Results*, edited by R. P. Von Herzen, C. J. Robinson, and A. C. Adamson, pp. 415–429, Ocean Drilling Program, College Station, 1991.

Cannat, M., D. Bideau, and H. Bougault, Serpentinized peridotites and gabbros in the Mid-Atlantic Ridge axial valley at 15°37'N and 16°52'N, *Earth and Planetary Science Letters*, *109*, 87–106, 1992.

Cannat, M., and J. F. Casey, An ultramafic lift at the Mid-Atlantic Ridge: Successive stages of magmatism in serpentinized peridotites from the 15°N region, in *Mantle and lower crust exposed in oceanic ridges and in ophiolites*, edited by R. L. M. Vissers, and A. Nicolas, pp. 5–34, Kluwer Academic publishers, Netherlands, 1995.

Cannat, M., C. Mével, M. Maia, C. Deplus, C. Durand, P. Gente, P. Agrinier, A. Belarouchi, G. Dubuisson, E. Humler, and J. R. Reynolds, Thin crust, ultramafic exposure, and rugged faulting patterns at the Mid-Atlantic Ridge (22°–24°N), *Geology*, *23*, 49–52, 1995.

Cannat, M., G. Ceuleneer, and J. Fletcher, Localization of ductile strain and the magmatic evolution of gabbroic rocks drilled at the Mid-Atlantic Ridge (Lat. 23°N), in *Proceedings of the Ocean Drilling Program, Scientific Results*, edited by J. A. Karson, M. Cannat, D. J. Miller, and D. Elthon, pp. 77–98, Ocean Drilling Program, College Station, 1997a.

Cannat, M., F. Chatin, H. Whitechurch, and G. Ceuleneer, Gabbroic rocks trapped in the upper mantle at the Mid-Atlantic Ridge, in *Proceedings of the Ocean Drilling Program, Scientific Results, 153*, edited by J. A. Karson, M. Cannat, D. J. Miller, and D. Elthon, pp. 243–264, ODP, 1997b.

Cannat, M., Y. Lagabrielle, H. Bougault, J. Casey, N. de Coutures, L. Dmitriev, and Y. Fouquet, Ultramafic and gabbroic exposures at the Mid-Atlantic Ridge: Geological mapping in the 15°N region, *Tectonophysics*, *279*, 193–213, 1997c.

Cannat, M., C. Rommevaux-Jestin, D. Sauter, C. Deplus, and V. Mendel, Formation of the axial relief at the very slow-spreading Southwest Indian Ridge (49°–69°E), *Journal of Geophysical Research*, *104* (B10), 22825–22843, 1999.

Carbotte, S. M., J. C. Mutter, and L. Xu, Contribution of volcanism and tectonism to axial and flank morphology of the southern East Pacific Rise, 17°10'–17°40'S, from a study of layer 2A geometry, *Journal of Geophysical Research*, *102*, 10165–10184, 1997.

Castillo, P. R., E. Klein, C. Langmuir, S. Shirey, R. Batiza, and W. White, Petrology and Sr, Nc, Pb isotope geochemistry of midocean ridge basalt glasses from the 11°45'N to 15°00'N segment of the East Pacific Rise, *Geochemistry, Geophysics, Geosystems*, *1*, 1999GC000024, 2000.

Ceuleneer, G., and M. Cannat, High-temperature ductile deformation of Site 920 peridotites, in *Proceedings of the Ocean Drilling Program, Scientific Results, 153*, edited by J. A. Karson, M. Cannat, D. J. Miller, and D. Elthon, pp. 23–34, ODP, 1997.

Chen, Y. J., Thermal effect of gabbro accretion from a deeper second melt lens at the fast-spreading East Pacific Rise, *Journal of Geophysical Research*, *106* (B5), 8581–8588, 2001.

Chen, Y., and W. J. Morgan, Rift valley/no rift valley transition at mid-ocean ridges, *J. Geophys. Res.*, *95*, 17571–17581, 1990.

Chen, Y. J., and J. Phipps Morgan, The effects of spreading rate, the magma budget, and the geometry of magma emplacement on the axial heat flux at mid-ocean ridges, *Journal of Geophysical Research*, *101* (B5), 11475–11482, 1996.

Cherkaoui, A. S. M., W. S. D. Wilcock, R. A. Dunn, and D. R. Toomey, A numerical model of hydrothermal cooling and crustal accretion at a fast-spreading mid-ocean ridge, *Geochemistry Geophysics Geosystems*, *4* (9), 2001GC000215, 2003.

Cochran, J. R., G. J. Kurras, M. H. Edwards, and B. J. Coakley, The Gakkel Ridge: bathymetry, gravity anomalies, and crustal accretion at extremely slow-spreading rates, *Journal of Geophysical Research*, *108* (B2), 2002JB001830, 2003.

Constantin, M., R. Hékinian, D. Ackerman, and P. Stoffers, Mafic and ultramafic intrusions into upper mantle peridotites from fast-spreading centers of the Easter microplate (Southeast Pacific), in *Mantle and lower crust exposed in oceanic ridges and in ophiolites*, edited by R. L. M. Vissers, and A. Nicolas, pp. 71–120, Kluwer, Rotterdam, 1995.

Coogan, L. A., C. J. MacLeod, H. J. B. Dick, S. J. Edwards, A. Kvassnes, J. H. Natland, P. T. Robinson, G. Thompson, and M. J. O'Hara, Whole-rock geochemistry of gabbros from the Southwest Indian Ridge: constraints on geochemical fractionations between the upper and lower oceanic crust and magma chamber processes at (very) slow-spreading ridges, *Chemical Geology*, *178*, 1–22, 2001.

Coogan, L. A., K. M. Gillis, C. J. MacLeod, G. M. Thompson, and R. Hékinian, Petrology and geochemistry of the lower ocean crust formed at the east Pacific Rise and exposed at Hess Deep: a synthesis and new results, *Geochemistry Geophysics Geosystems*, *3* (11), 2001GC000230, 2002a.

Coogan, L. A., G. R. T. Jenkin, and R. N. Wilson, Constraining the cooling rate of the lower oceanic crust: a new approach applied to the Oman ophiolite, *Earth and Planetary Science letters*, *199*, 127–146, 2002b.

Crane, K., Structural evolution of the East Pacific Rise axis from 13°10'N to 10° 35'N: interpretations from SeaMARC I data, *Tectonophysics*, *136*, 65–124, 1987.

Crawford, W., S. C. Webb, and J. A. Hildebrand, Constraints on melt in the lower crust and Moho at the East Pacific Rise, 9°48'N, using seafloor compliance measurements, *Journal of Geophysical Research*, *104* (B2), 2923–2939, 1999.

Crawford, W. C., and S. C. Webb, Variations in the distribution of magma in the lower crust and at the Moho beneath the East Pacific Rise at 9°–10°N, *Earth and Planetary Science Letters*, *202*, 117–130, 2002.

Danyushevsky, L. V., The effect of small amounts of H2O on crystallisation of mid-ocean ridge and backarc basin magmas, *Journal of Volcanology and Geothermal Research*, *110*, 265–280, 2001.

Danyushevsky, L. V., A. V. Sobolev, and L. V. Dmitriev, Estimation of the pressure of crystallisation and H2O content of MORB and BABB glasses: calibration of an empirical technique, *Mineralogy and Petrology*, *57*, 185–204, 1996.

Detrick, R. S., P. Buhl, E. Vera, J. Mutter, J. Orcutt, J. Madsen, and T. Brocher, Multichannel seismic imaging of a crustal magma chamber along the East Pacific Rise between 9°N and 13°N, *Nature*, *326*, 35–41, 1987.

Detrick, R. S., J. C. Mutter, P. Buhl, and I. I. Kim, No evidence from multichannel reflection data for a crustal magma chamber in the MARK area on the Mid-Atlantic Ridge, *Nature*, *347*, 61–64, 1990.

Detrick, R. S., H. A. J., G. M. Kent, J. A. Orcutt, J. C. Mutter, and P. Buhl, Seismic structure of the southern East Pacific Rise, *Science*, *259*, 499–503, 1993a.

Detrick, R. S., R. S. White, and G. M. Purdy, Crustal structure of North-Atlantic fracture zones, *Reviews of Geophysics*, *31* (4), 439–459, 1993b.

Detrick, R. S., H. D. Needham, and V. Renard, Gravity anomalies and crustal thickness variations along the Mid-Atlantic Ridge between 33°N and 40°N, *Journal of Geophysical Research*, *100* (B3), 3767–3787, 1995.

Dewey, J. F., and W. S. F. Kidd, Geometry of plate accretion, *Geological Society of America Bulletin*, *79*, 411–423, 1977.

Dick, H. J. B., Abyssal peridotites, very slow-spreading ridges and ocean ridge magmatism, in *Magmatism in the Ocean Basins*, edited by A. D. Saunders, and M. J. Norry, pp. 71–105, Geological Society Special Publication, 1989.

Dick, H. J. B., R. L. Fisher, and W. B. Bryan, Mineralogic variability of the uppermost mantle along mid-ocean ridges, *E. Planet. Sci. Lett.*, *69*, 88–106, 1984.

Dick, H. J. B., P. S. Meyer, S. Bloomer, S. Kirby, D. Stakes, and C. Mawer, Lithostratigraphic evolution of an *in-situ* section of oceanic layer 3, *Proceedings of the Ocean Drilling Program, Scientific Results*, *118*, 439–538, 1991.

Dick, H. J. B., and J. H. Natland, Late-stage melt evolution and transport in the shallow mantle beneath the East Pacific Rise, in *Proc. Ocean Drill. Program., Sci. Results leg 147*, edited by C. Mevel, K.M. Gillis, J.F. Allan, and P.S. Meyer, pp. 103–134, 1996.

Dick, H. J. B., J. H. Natland, J. C. Alt, W. Bach, D. Bideau, J. S. Gee, S. Haggas, J. G. H. Hertogen, G. Hirth, P. M. Holm, B. Ildefonse, G. J. Iturrino, B. E. John, D. S. Kelley, E. Kikawa, A. Kindom, P. J. LeRoux, J. Maeda, P. S. Meyer, D. J. Miller, H. R. Naslund, Y.-L. Niu, P. T. Robinson, J. Snow, R. A. Stephen, P. W. Trimby, H.-U. Worm, and A. S. Yoshinobu, A long in situ section of the lower oceanic crust: results of ODP leg 176 drilling at the Southwest Indian Ridge, *Earth and Planetary Science Letters*, *179*, 31–51, 2000.

Dick, H. J. B., J. Lin, and H. Schouten, An ultraslow-spreading class of ocean ridge, *Nature*, *426*, 405–412, 2003.

Dosso, L., H. Bougault, C. Langmuir, C. Bollinger, O. Bonnier, and J. Etoubleau, The age and distribution of mantle heterogeneity along the Mid-Atlantic Ridge (31–41°N), *Earth and Planetary Science Letters*, *170*, 269–286, 1999.

Dunn, R. A., D. R. Toomey, R. S. Detrick, and W. S. Wilcock, Continuous mantle melt supply beneath an overlapping spreading cen-

ter on the East Pacific Rise, *Science*, *291*, 1955–1958, 2001.

Dunn, R. A., D. R. Toomey, and S. C. Solomon, Three-dimensional seismic structure and physical properties of the crust and shallow mantle beneath the East Pacific Rise at 9°30'N., *Journal of Geophysical Research*, *105* (B10), 23537–23555, 2000.

Escartín, J., and M. Cannat, Ultramafic exposures and the gravity signature of the lithosphere near the Fifteen-Twenty Fracture Zone (Mid-Atlantic Ridge, 14°–16.5°N), *Earth and Planetary Science Letters*, *171* (3), 411–424, 1999.

Escartin, J., M. Cannat, G. Pouliquen, and A. Rabain, Crustal thickness of V-shaped ridges south of the Azores: interaction of the Mid-Atlantic Ridge (36°–39°N) and the Azores hot spot, *Journal of Geophysical Research*, *106* (B10), 21719–21735, 2001.

Fialko, Y. A., and A. M. Rubin, Thermodynamics of lateral dyke propagation: implications for crustal accretion at slow-spreading ridges, *Journal of Geophysical Research*, *103*, 2501–2514, 1998.

Flower, M. F. J., Thermal and kinematic control on ocean-ridge magma fractionation : contrasts between Atlantic and pacific spreading axes., *J. Geol. Soc. London*, *138*, 695–712, 1981.

Forsyth, D. W., Geophysical constraints on mantle flow and melt generation at mid-ocean ridges, in *Mantle flow and melt generation at mid-ocean ridges*, edited by J. Phipps Morgan, D. K. Blackman, and J. Sinton, pp. 1–65, American Geophysical Union, Washington, 1992.

Fowler, C. M. R., Crustal structure of the Mid-Atlantic Ridge crest at 37°N, *Geophysical Journal of Royal Astronomical Society*, *47* (3), 459–491, 1976.

Francheteau, J., R. Armijo, J. L. Cheminee, R. Hekinian, P. F. Lonsdale, and N. Blum, I Ma East Pacific Rise ocean crust and upper mantle exposed by rifting in Hess Deep (equatorial Pacific Ocean), *Earth and Planetary Science Letters*, *101*, 281–295, 1990.

Francheteau, J., R. Armijo, J. L. Cheminee, R. Hekinian, P. F. Lonsdale, and N. Blum, Dyke complex of the East Pacific Rise exposed in the walls of Hess Deep and the structure of the upper oceanic crust, *Earth and Planetary Science Letters*, *111* (1), 109–121, 1992.

Francis, T. J. G., and I. T. Porter, Median valley seismology: the Mid-Atlantic Ridge near 45°N, *Geophys. J. R. Astr. Soc.*, *34*, 279–311, 1973.

Fyfe, W. S., and P. Lonsdale, Ocean floor hydrothermal activity, in *The Sea, vol.7 : The Oceanic Lithosphere*, edited by C. Emiliani, pp. 589–638, Wiley, New York, 1981.

Garrido, C. J., P. B. Kelemen, and G. Hirth, Variation of cooling rate with depth in lower crust formed at an oceanic spreading ridge: Plagioclase crystal size distributions in gabbros from the Oman ophiolite, *Geochemistry, Geophysics, Geosystems*, *2*, 2000GC000136, 2001.

Gillis, K. M., The rootzone of an ancient hydrothermal system exposed in Troodos ophiolite, Cyprus, *Journal of Geology*, *110*, 57–74, 2002.

Grove, T. L., R. J. Kinzler, and W. B. Bryan, Fractionation of Mid-Ocean Ridge Basalt (MORB), in *Mantle flow and melt generation at mid-ocean ridges*, edited by J. Phipps Morgan, D. K. Blackman, and J. Sinton, pp. 281–310, American Geophysical Union, Washington, 1992.

Harding, A. J., J. A. Orcutt, M. E. Kappus, E. E. vera, J. C. Mutter,

P. Buhl, R. S. Detrick, and T. M. Brocher, Structure of young oceanic crust at 13°N on the East Pacific Rise from expanding spread profiles, *Journal of Geophysical Research*, *94* (B9), 12163–12196, 1989.

Hekinian, R., D. Bideau, J. Francheteau, J. L. Cheminee, R. Armijo, P. F. Lonsdale, and N. Blum, Petrology of the East Pacific Rise crust and upper mantle exposed in Hess Deep (Eastern Equatorial Pacific), *Journal of Geophysical Research*, *98*, 8069–8094, 1993.

Hellebrand, E., J. E. Snow, and R. Muehe, Mantle melting beneath Gakkel Ridge (Arctic Ocean); abyssal peridotite spinel compositions, *Chemical Geology*, *182* (2–4), 227–235, 2002.

Hess, P. C., Phase equilibria constraints on the origin of ocean floor basalts, in *Mantle flow and melt generation at mid-ocean ridges*, edited by J. Phipps Morgan, D. K. Blackman, and J. Sinton, pp. 67–102, American Geophysical Union, Washington, 1992.

Hirschmann, M. M., Mantle solidus : Experimental constraints and the effects of peridotite composition., *Geochem. Geophys. Geosyst.*, *1*, 2000.

Hooft, E. E. E., H. Schouten, and R. S. Detrick, Constraining crustal emplacement processes from the variation in Layer 2A thickness at the East Pacific Rise, *Earth and Planetary Science Letters*, *142*, 289–309, 1996.

Hooft, E. E. E., R. S. Detrick, and G. Kent, Seismic structure and indicators of magma budget along the southern East Pacific Rise, *Journal of Geophysical Research*, *102* (B12), 27319–27340, 1997.

Hooft, E. E. E., R. S. Detrick, D. R. Toomey, J. A. Collins, and J. Lin, Crustal thickness and structure along three contrasting segments of the Mid-Atlantic Ridge, 33.5°–35°N, *J. Geophys. Res.*, *105*, 8205–8226, 2000.

Humphris, S. E., W. B. Bryan, G. Thompson, and L. K. Autio, Morphology, geochemistry, and evolution of Serocki volcano, in *Proc. ODP, Scientific Results, 106/109*, edited by R. S. Detrick, J. Honnorez, W. B. Bryan, and T. Juteau, pp. 67–84, 1990.

Humphris, S. E., and J. R. Cann, Constraints on the energy and chemical balances of the modern TAG and ancien Cyprus seafloor sulfide deposits, *Journal of Geophysical Research*, *105* (B12), 28477–28488, 2000.

Iturrino, G. J., D. J. Miller, and N. I. Christensen, Velocity behavior of lower crustal and upper mantle rocks from a fast-spreading ridge at Hess Deep, *Proceedings of the Ocean Drilling Program, Scientific Results*, *147*, 417–440, 1996.

Jaeger, J. C., The temperature in the neighbourhood of a cooling intrusive sheet, *American Journal of Science*, *255*, 306–318, 1957.

Jaeger, J. C., Thermal effects of intrusions, *Review of Geophysics*, *2*, 443–466, 1964.

Karson, J. A., Structure of uppermost oceanic crust created at fast to intermediate-rate spreading centers, *Annual Reviews of Earth and Planetary Sciences*, *30* (347-384), 10.1146 / annurev.earth. 30.091201.141132, 2002.

Karson, J. A., G. Thompson, S. E. Humphris, J. M. Edmon, W. B. Bryan, J. B. Brown, A. T. Winters, R. A. Pockalny, J. F. Casey, A. C. Campbell, G. P. Klinkhammer, M. R. Palmer, R. J. Kinzler, and M. M. Sulanowska, Along-axis variations in seafloor spreading in the MARK Area, *Nature*, *328*, 681–685, 1987.

Karson, J. A., S. D. Hurst, and P. Londsale, Tectonic rotations of dykes in fast-spreading oceanic crust exposed near Hess Deep,

Geology, *20*, 685–688, 1992.

Kelemen, P. B., Reaction between ultramafic rock and fractionating basaltic magma I. Phase reactions, the origin of calc-alkaline magma series, and the formation of discordant dunite, *Journal of Geology*, *31* (1), 51–98, 1990.

Kelemen, P. B., D. B. Joyce, J. D. Webster, and J. R. Holloway, Reaction between ultramafic rock and fractionating basaltic magma II. Experimental investigation of reaction between olivine tholeiite and harzburgite at 1150–1050°C and 5 kb, *Journal of Geology*, *31* (1), 99–134, 1990.

Kelemen, P. B., K. Koga, and N. Shimizu, Geochemistry of gabbro sills in the crust/mantle transition zone of the Oman ophiolite: Implications for the origin of the oceanic lower crust, *Earth and Planetary Science Letters*, *146*, 475–488, 1997.

Kelemen, P. B., and E. Aharanov, Periodic formation of magma fractures and generation of layered gabbros in the lower crust beneath oceanic spreading ridges, in *Faulting and magmatism at mid-ocean ridges*, edited by W. R. Buck, P. T. Delaney, J. A. Karson, and Y. Lagabrielle, pp. 153–176, AGU, Washington D. C., 1998.

Kent, G. M., A. J. Harding, and J. Orcutt, Distribution of magma beneath the East Pacific Rise between the Clipperton transform and the 9°17'N deval from forward modeling of common depth point data, *Journal of Geophysical Research*, *98* (B8), 13945–13969, 1993.

Kidd, R. G. W., A model for the process of formation of the upper oceanic crust, *Geophys. J. R. Astr. Soc.*, *50*, 149`ae`–183, 1977.

Kinzler, R. J., and T. L. Grove, Primary magmas of mid-ocean ridge basalts 1. Experiments and methods, *J. Geophys. Res.*, *97*, 6885–6906, 1992.

Klein, E. M., and C. H. Langmuir, Global correlations of ocean ridge basalt chemistry with axial depth and crustal chemistry, *Journal of Geophysical Research*, *92*, 8089–8115, 1987.

Kojitani, H., and M. Akaogi, Melting enthalpies of mantle peridotite: calorimetric determinations in the system CaO-MgO-Al2O3-SiO2 and application to magma generation., *Earth and Planetary Science Letters*, *153*, 209–222, 1997.

Kong, L. S. L., S. C. Solomon, and G. M. Purdy, Microearthquake characteristics of a mid-ocean ridge along-axis high, *J. Geophys. Res.*, *97* (B2), 1659–1685, 1992.

Korenaga, J., and P. B. Kelemen, Origin of gabbro sills in the Moho transition zone of the Oman ophiolite: Implications for magma transport in the oceanic lower crust, *Journal of Geophysical Research*, *102* (B12), 27729–27749, 1997.

Kuo, B. Y., and D. W. Forsyth, Gravity anomalies of the ridge-transform system in the South Atlantic between 31° and 34.5°S: upwelling centers and variation in crustal thickness, *Mar. Geophy. Res.*, *10*, 205–232, 1988.

Lalou, C., J. L. Reyss, E. Brichet, P. A. Rona, and G. Thompson, Hydrothermal activity on a 105-year scale at a slow-spreading ridge, TAG hydrothermal field, *Journal of Geophysical Research*, *100*, 17855–17862, 1995.

Lange, R. A., and S. E. Carmichael, Densities of Na2O-nK2O-CaO-MgO-FeO-Fe2O3-Al2O3-TiO2-SiO2 liquids: new measurements and derived partial molar properties, *Geoch. Cosm. Acta*, *51*, 2931–2946, 1987.

Lange, R. A., and A. Navrotsky, Heat capacities of Fe2O3-bearing silicate liquids, *Contrib. Mineral. Petrol.*, *110*, 311–320, 1992.

Langmuir, C. H., E. M. Klein, and T. Plank, Petrological systematics of mid-ocean ridge basalts: constraints on melt generation beneath ocean ridges, in *Mantle flow and melt generation at Mid-Ocean Ridges*, edited by J. Phipps Morgan, D. K. Blackman, and J. M. Sinton, pp. 183–280, Am. Geophys. Union Geophys. Monogr., 1992.

Lavier, L., W. R. Buck, and A. N. B. Poliakov, Factors controlling normal fault offset in an ideal brittle layer, *Journal of Geophysical Research*, *105* (B10), 23431–23442, 2000.

Lawson, K., R. C. Searle, J. A. Pearce, P. Browning, and P. Kempton, Detailed volcanic geology of the MARNOK area, Mid-Atlantic Ridge north of Kane transform, in *Tectonic, magmatic, hydrothermal and biological segmentation of mid-ocean ridges*, edited by C. J. MacLeod, P. A. Tyler, and C. L. Walker, pp. 61–102, Geological Society, London, 1996.

Lehnert, K., Y. Su, C. H. Langmuir, B. Sarbas, and U. Nohl, A global geochemical database structure for rocks, *Geochemistry, Geophysics, Geosystems*, *1*, 1999GC000026, 2000.

Lin, J., and E. M. Parmentier, Mechanisms of lithospheric extension at mid-ocean ridges, *Jour. Geophys. Res.*, *96*, 1–22, 1989.

Lin, J., G. M. Purdy, H. Schouten, J. C. Sempere, and C. Zervas, Evidence from gravity data for focused magmatic accretion along the Mid-Atlantic Ridge, *Nature*, *344*, 627–632, 1990.

Lin, J., and J. Phipps Morgan, The spreading rate dependence of three-dimensional mid-ocean ridge gravity structure, *Geophysical Research Letters*, *19* (1), 13–15, 1992.

Lippard, S. J., A. W. Shelton, and I. G. Gass, The ophiolite of Northern Oman, *Geological Society of London Memoir*, *11*, 1986.

Lister, C. R. B., On the penetration of water into hot rock, *Geophys. J. R. Astr. Soc.*, *39*, 465–509, 1974.

Lister, C. R. B., Heat transfer between magmas and hydrothermal systems, or, six lemmas in search of a theorem, *Geophysical Journal International*, *120*, 45–59, 1995.

Longhi, J. E., and V. Pan, A reconnaissance study of phase boundaries in low-alkali basaltic liquids, *Journal of Petrology*, *29*, 115–149, 1988.

Lowell, R. P., and P. A. Rona, Seafloor hydrothermal systems driven by the serpentinization of peridotite, *Geophysical Research Letters*, *29* (11), 10.1029/2001GL04411, 2002.

Macdonald, K. C., Linkages between faulting, volcanism, hydrothermal activity and segmentation at fast-spreading centers, in *Faulting and magmatism at mid-ocean ridges*, edited by W. R. Buck, P. T. Delaney, J. A. Karson, and Y. Lagabrielle, pp. 27–58, AGU, Washington D. C., 1998.

Macdonald, K. C., R. M. Haymon, and A. Shor, A 220 km2 recently erupted lava field on the East Pacific Rise near lat. 8°S, *Geology*, *17*, 212–216, 1989.

Maclennan, J., D. McKenzie, and K. Gronvold, Plume-driven upwelling under central Iceland, *Earth and Planetary Science Letters*, *194* (1–2), 67–82, 2001a.

Maclennan, J., D. McKenzie, K. Gronvöld, and L. Slater, Crustal accretion under northern Iceland, *Earth and Planetary Science Letters*, *191*, 295–310, 2001b.

Maclennan, J., T. Hulme, and S. C. Singh, Thermal models of oceanic crustal accretion: linking geophysical, geological and petrological

observations, *Geochemistry Geophysics Geosystems*, *5*, 2003GC000605, 2004.

MacLeod, C. J., and G. Yaouancq, A fossil melt lens in the Oman ophiolite: implications for magma chamber processes at fast-spreading ridges, *Earth and Planetary Science letters*, *176*, 357–373, 2000.

Madsen, J. A., R. S. Detrick, J. C. Mutter, P. Buhl, and J. A. Orcutt, A two- and three-dimensional analysis of gravity anomalies associated with the East Pacific Rise at 9°N and 13°N, *Journal of Geophysical Research*, *95* (B4), 4967–4987, 1990.

Magde, L. S., D. W. Sparks, and R. S. Detrick, The relationship beween buoyant mantle flow, melt migration and bull's eyes at the Mid-Atlantic Ridge between 33°N and 35°N, *Earth and Planetary Science Letters*, *148* (1–2), 59–67, 1997.

Magde, L. S., A. H. Barclay, D. R. Toomey, R. S. Detrick, and J. A. Collins, Crustal magma plumbing within a segment of the Mid-Atlantic Ridge, 35°N, *Earth and Planetary Science letters*, *175*, 55–67, 2000.

Maia, M., and P. Gente, Three-dimensional gravity and bathymetry analysis of the Mid-Atlantic Ridge between 20°N and 24°N: Flow geometry and temporal evolution of the segmentation, *Journal of Geophysical Research*, *103* (B1), 951–974, 1998.

Martin, B., and W. S. Fyfe, Some experimental and theoritical observations on the kinetics of hydration reactions with particular reference to serpentinization, *Chemical Geology*, *6*, 185–202, 1970.

McKenzie, D., and M. J. Bickle, The volume and composition of melt generated by extension of the lithosphere, *J. Petrology*, *29*, 625–679, 1988.

McKenzie, D., and R. K. O'Nions, Partial melt distributions from inversion of Rare Earth Element concentrations, *Journal of Petrology*, *32* (5), 1021–1091, 1991.

Meyer, P. S., and W. B. Bryan, Petrology of basaltic glasses from the TAG segment: Implications for a deep hydrothermal heat source, *Geophysical Research Letters*, *23* (23), 3425–3438, 1996.

Michael, P. J., and R. L. Chase, The influence of primary magma composition, H_2O and pressure on mid-ocean ridge basalt differentiation, *Contrib. Mineral. Petrol.*, *96*, 245–263, 1987.

Michael, P. J., and W. C. Cornell, Influence of spreading rate and magma supply on crystallization and assimilation beneath mid-ocean ridges: Evidence from chlorine and major chemistry of mid-ocean ridge basalts, *Journal of Geophysical Research*, *103* (B8), 18325–18356, 1998.

Michael, P. J., D. W. Forsyth, D. K. Blackman, P. J. Fox, B. B. Hanan, A. J. Harding, K. C. Macdonald, G. A. Neumann, J. A. Orcutt, M. Tolstoy, and C. M. Weiland, Mantle control of a dynamically evolving spreading center: Mid-Atlantic Ridge 31–34°S, *Earth and Planetary Science Letters*, *121*, 451–468, 1994.

Miller, D. J., and N. I. Christensen, Seismic velocities of lower crustal and upper mantle rocks from the slow-spreading Mid-Atlantic Ridge, south of the Kane Transform zone (MARK), *Proceedings of the Ocean Drilling Program, Scientific Results*, *153*, 437–454, 1997.

Minshull, T. A., R. S. White, J. C. Mutter, P. Buhl, R. S. Detrick, C. A. Williams, and E. Morris, Crustal structure at the Blake Spur fracture zone from expanding spread profiles, *Journal of Geophysical Research*, *96* (B6), 9955–9984, 1991.

Minshull, T. A., M. R. Muller, C. J. Robinson, R. S. White, and M. J. Bickle, Is the oceanic Moho a serpentinization front?, in *Modern Ocean Floor Processes and the Geological Record*, edited by R. A. Mills, and K. Harrison, pp. 71–80, Geol. Soc. London, London, 1998.

Morton, J. L., and N. H. Sleep, A mid-ocean ridge thermal model: Constraints on the volume of axial hydrothermal heat flux, *Journal of Geophysical Research*, *90* (B13), 11345–11353, 1985.

Mottl, M. J., Partitioning of energy and mass fluxes between mid-ocean ridge axes and flanks at high and low temperature, in *Energy and Mass Transfer in Marine Hydrothermal Systems*, edited by P. E. Halbach, V. Tunnicliffe, and J. R. Hein, pp. 271–286, Dahlem University Press, 2003.

Muller, M. R., T. A. Minshull, and R. S. White, Segmentation and melt supply at the Southwest Indian Ridge, *Geology*, *27* (10), 867–870, 1999.

Natland, J. H., and H. J. B. Dick, Melt migration through high-level gabbroic cumulates of the East Pacific Rise at Hess Deep: the origin of magma lenses and the deep crustal structure of fast-spreading ridges, in *Proc. ODP, Sci. Results, 147*, edited by C. Mével, K. M. Gillis, J. F. Allan, and P. S. Meyer, pp. 21–58, 1996.

Natland, J. H., and H. J. B. Dick, Formation of the lower oceanic crust and the crystallization of gabbroic cumulates at a very slowly spreading ridge, *Journal of Volcanology and Geothermal Research*, *110*, 191–233, 2001.

Navin, D. A., C. Pierce, and M. C. Sinha, The RAMSESSES experiment – II. Evidence for accumulated melt beneath a slow-spreading ridge from wide-angle refraction and multichannel reflection seismic profiles, *Geophysical Journal International*, *135*, 746–772, 1998.

Navrotsky, A., D. Ziegler, R. Oestrike, and P. Maniar, Calorimetry of silicate melts at 1773°K : measurement of enthalpies of fusion and of mixing in the systems diopside-anorthite-albite and anorthite-forsterite., *Contrib. Mineral. Petrol.*, *101*, 122–130, 1989.

Nehlig, P., Interactions between magma chambers and hydrothermal systems: oceanic and ophiolitic constraints, *Journal of Geophysical Research*, *98* (B11), 19621–19633, 1993.

Nicolas, A., and B. Ildefonse, Flow mechanism and viscosity in basaltic magma chambers, *Geophysical Research Letters*, *23* (16), 2013–2016, 1996.

Nicolas, A., I. Reuber, and K. Benn, A new magma chamber model based on structural studies in the Oman Ophiolite, *Tectonophysics*, *151*, 87–105, 1988.

Nicolas, A., D. Mainprice, and F. Boudier, High temperature seawater circulation troughout crust of oceanic ridges. A model derived from the Oman ophiolite, *Geophysical Research Abstracts,*, *5*, 09476, 2003.

Niida, K., Mineralogy of MARK peridotites: replacement through magma channeling examined from Hole 920D, MARK area, in *Proceedings of the Ocean Drilling Program, Scientific Results*, edited by J. A. Karson, M. Cannat, D. J. Miller, and D. L. Elthon, pp. 265–275, Ocean Drilling Program, College Station, 1997.

Nimis, P., and P. Ulmer, Clinopyroxene geobarometry of magmatic rocks Part 1 : An expanded structural geobarometer for anhydrous and hydrous, basic and ultrabasic systems., *Contrib. Mineral. Petrol.*, *133*, 122–135, 1998.

Niu, Y., D. Bideau, R. Hékinian, and R. Batiza, Mantle compositional control on the extent of mantle melting, crust production, gravity anomaly, ridge morphology, and ridge segmentation: a case study at the Mid-Atlantic Ridge, 33–35°N, *Earth and Planetary Science letters*, *186*, 383–399, 2001.

O'Hara, M. J., Geochemical evolution during fractional crystallisation of a periodically refilled magma chamber, *Nature*, *266* (5602), 503–507, 1977.

Orr, R. L., High-temperature heat contents of magnesium orthosilicate and ferrous orthosilicate, *J. Am. Chem. Soc.*, *75*, 528–529, 1953.

Ozawa, K., P. S. Meyer, and S. H. Bloomer, Mineralogy and textures of iron-titanium oxide gabbros from Hole 735B, in *Proceedings of the Ocean Drilling Program, Scientific Results*, edited by R. P. Von Herzen, C. J. Robinson, and A. C. Adamson, pp. 41–74, Ocean Drilling Program, College Station, 1991.

Pallister, J. S., Structure of the sheeted dyke complex of the Samail ophiolite near Ibra, Oman, *Journal of Geophysical Research*, *86*, 2661–2672, 1981.

Pallister, J. S., and C. Hopson, Samail Ophiolite plutonic suite: Field relations, phase variation, cryptic variation and layering, and a model of a spreading ridge magma chamber, *Journal of Geophysical Research*, *86* (B4), 2593–2644, 1981.

Pedersen, R. B., J. Malpas, and T. J. Falloon, Petrology and geochemistry of gabbroic and related rocks from Site 894, HessDeep, in *Proc. ODP, Sci. Results, 147*, edited by C. Mével, K.M. Gillis, J. F. Allan, and P. S. Meyer, pp. 3–19, 1996.

Penrose Conference on Ophiolites, *Geotimes*, 17, 24–25, 1972.

Phipps Morgan, J., E. M. Parmentier, and J. Lin, Mechanisms for the origin of Mid-Ocean Ridge axial topography: implications for the thermal and mechanical structure of accreting plate boundaries, *J. Geophys. Res.*, *92*, 12823–12836, 1987.

Phipps Morgan, J., and D. W. Forsyth, Three-dimensional flow and temperature perturbations due to a transform offset: Effects on oceanic crustal and upper mantle structure, *Journal of Geophysical Research*, *93*, 2955–2966, 1988.

Phipps Morgan, J., and Y. J. Chen, The genesis of oceanic crust: magma injection, hydrothermal circulation, and crustal flow, *Journal of Geophysical Research*, *98* (B4), 6283–6297, 1993.

Plank, T., and C. H. Langmuir, Effects of the melting regime on the composition of the oceanic crust, *Journal of Geophysical Research*, *97*, 19749–19770, 1992.

Presnall, D. C., T. H. Dixon, T. H. O'Donnel, and S. A. Dixon, Generation of mid-ocean ridge tholeiites, *Journal of Petrology*, *20*, 3–35, 1979.

Purdy, G. M., and R. S. Detrick, Crustal structure of the Mid-Atlantic Ridge at 23°N from seismic refraction studies, *Journal of Geophysical Research*, *91* (B3), 3739–3762, 1986.

Putirka, K., M. Johnson, R. Kinzler, J. E. Longhi, and D. Walker, Thermobarometry of mafic igneous rocks based on clinopyroxene-liquid equilibria, O-30 kbar., *Contrib. Mineral. Petrol.*, *123*, 92–108, 1996.

Reynolds, J. R., Segment-scale systematics of mid-ocean ridge magmatism and geochemistry, Ph.D. thesis thesis, Columbia University, New York, 1995.

Reynolds, J. R., and C. Langmuir, Petrological systematics of the

Mid-Atlantic Ridge south of Kane: Implications for ocean crust formation., *Journal of Geophysical Research*, *102* (B7), 14915–14946, 1997.

Richet, P., and Y. Bottinga, Anorthite, andesine, wollastonite, diopside, cordierite and pyrope: Glass transitions, thermodynamics of melting, and properties of the amorphous phases, *Earth and Planetary Science Letters*, *67*, 1984.

Richet, P., and Y. Bottinga, Thermochemical properties of silicate glasses and liquids: A review., *Review of Geophysics*, *24* (1), 1–25, 1986.

Richet, P., and G. Fiquet, High-temperature heat capacity and premelting of minerals in the system MgO-CaO-Al2O3-SiO2, *Journal of Geophysical Research*, *96* (B1), 445–456, 1991.

Richet, P., F. Leclerc, and L. Benoist, Melting of forsterite and spinel, with implications for the glass transition of Mg2SiO4 liquid, *Geophysical Research Letters*, *20*, 1675–1678, 1993.

Robie, R. A., B. S. Hemingway, and J. R. Fisher, Thermodynamic properties of minerals and related substances at 298.15K and 1 bar and at higher temperatures, *U. S. Geological Survey Bulletin*, *1452*, 456 pp., 1978.

Robson, D., and J. R. Cann, A geochemical model of mid-ocean ridge magma chambers, *Earth and Planetary Science Letters*, *60*, 93–104, 1982.

Rosenberg, N. D., J. E. Lupton, D. Kadko, R. Collier, and M. D. Lilley, Estimation of heat and chemical fluxes from a seafloor hydrothermal vent using radon measurements, *Nature*, *334*, 604–607, 1988.

Ross, K., and D. L. Elthon, Cumulus and post-cumulus crystallization in the oceanic crust: major and trace element geochemistry of Leg 153 gabbroic rocks, in *Proceedings of the Ocean Drilling Program, Scientific Results*, edited by J. A. Karson, M. Cannat, D. J. Miller, and D. L. Elthon, pp. 333–350, Ocean Drilling Program, College Station, 1997.

Schilling, J. G., Azores mantle blob: Rare-earth evidence, *Earth and Planetary Science Letters*, *25*, 103–115, 1975.

Seyfried, W. E., D. I. Foustoukos, and D. E. Allen. "Ultramafic-hosted hydrothermal systems at mid-ocean ridges: chemical and physical controls on pH, redox and carbon reduction reactions." *this volume*.

Seyler, M., M. J. Toplis, J.-P. Lorand, A. Luguet, and M. Cannat, Clinopyroxene microtextures reveal incompletely extracted melts in abyssal peridotites, *Geology*, *29* (2), 155–158, 2001.

Seyler, M., M. Cannat, and C. Mével, Evidence for major element heterogeneity in the mantle source of abyssal peridotites from the southwest Indian Ridge (52 to 68° East), *Geochemistry, Geophysics, Geosystems*, *4* (2), 2002GC000305, 2003.

Shen, Y., and D. W. Forsyth, Geochemical constraints on initial and final depths of melting beneath mid-ocean ridges, *Journal of Geophysical Research*, *100* (B2), 2211–2237, 1995.

Shipboard Scientific Party, Site 920, in *Proc. ODP Initial report*, edited by M. Cannat, J. A. Karson, and D. J. Miller, pp. 45–119, Ocean Drilling Program, College Station, 1995.

Singh, S. C., G. M. Kent, J. S. Collier, A. J. Harding, and J. O. Orcutt, Melt to mush variations in crustal magma properties along the ridge crest at the southern East Pacific Rise, *Nature*, *394*, 874–878, 1998.

Sinha, M. C., D. A. Navin, L. M. MacGregor, S. Constable, C. Pierce, A. White, G. Heinson, and M. A. Inglis, Evidence for accumulated melt beneath the slow-spreading Mid-Atlantic Ridge, *Philosophical Transactions of the Royal Society of London*, *355*, 233–253, 1997.

Sinton, J. M., and R. S. Detrick, Mid-ocean ridge magma chambers, *J. Geophys. Res.*, *97* (B1), 197–216, 1992.

Sinton, J. M., S. M. Smaglik, J. J. Mahoney, and K. C. Macdonald, Magmatic processes at superfast-spreading Mid-Ocean Ridges: Glass compositional variations along the East Pacific Rise 13°–23°S, *Journal of Geophysical Research*, *96* (B4), 6133–6155, 1991.

Sleep, N. H., Formation of oceanic crust: Some thermal constraints, *Journal of Geophysical Research*, *80* (29), 4037–4042, 1975.

Sleep, N., and G. A. Barth, The nature of the lower crust and shallow mantle emplacement at low spreading rates, *Tectonophysics*, *279*, 181–191, 1997.

Smith, D. K., and J. R. Cann, The role of seamount volcanism in crustal construction at the Mid-Atlantic Ridge (24°-30°N), *J. Geophys. Res.*, *97* (B2), 1645-1658, 1992.

Smith, D.K., and J.R. Cann, Constructing the upper crust of the Mid-Atlantic Ridge; a reinterpretation based on the Puna Ridge, Kilauea volcano, *Journal of Geophysical Research*, *104* (B11), 25379–25399, 1999.

Stakes, D. S., J. W. Shervais, and C. A. Hopson, The volcanic-tectonic cycle of the FAMOUS and AMAR valleys, Mid-Atlantic Ridge (36°47'N): Evidence from basalt glass and phenocryst compositional variations for a steady state magma chamber beneath the valley midsections, AMAR3, *Journal of Geophysical Research*, *89* (B8), 6995–7028, 1984.

Stebbins, J. F., S. E. Carmichael, and D. F. Weill, The high-temperature liquid and glass heat contents and the heats of fusion of diopside, albite, sanidine, and nepheline, *Am. Mineral.*, *68*, 717–730, 1983.

Stebbins, J. F., S. E. Carmichael, and L. K. Moret, Heat capacities and entropies of silicate liquids and glasses, *Contrib. Mineral. Petrol.*, *86*, 131–148, 1984.

Tartarotti, P., S. Susini, P. Nimis, and L. Ottolini, Melt migration in the upper mantle along the Romanche Fracture Zone (Equatorial Atlantic), *Lithos*, *63*, 125–149, 2002.

Tolstoy, M., A. J. Harding, and J. A. Orcutt, Crustal thickness on the Mid-Atlantic Ridge: Bull's eye gravity anomalies and focused accretion, *Science*, *262* (726–729), 726–729, 1993.

Tolstoy, M., A. J. Harding, and J. A. Orcutt, Deepening of the axial magma chamber on the southern East Pacific Riste toward the Garrett Fracture Zone, *Journal of Geophysical Research*, *102* (B2), 3097–3108, 1997.

Toomey, D. R., G. M. Purdy, and S. C. Solomon, Microearthquakes beneath the median valley of the Mid-Atlantic Ridge near 23°N: tomography and tectonics, *Jour. Geophys. Res.*, *93*, 9093–9112, 1988.

Toomey, D. R., G. M. Purdy, S. C. Solomon, and W. S. D. Wilcock, The three-dimensional seismic velocity structure of the East Pacific Rise near latitude 9°30' N, *Nature*, *347*, p. 639–645, 1990.

Tormey, D. R., T. L. Grove, and W. B. Bryan, Experimental petrology of normal MORB near the Kane fracture zone: 22°–25°N,

Mid-Atlantic Ridge., *Contrib. Mineral. Petrol.*, *96*, 121–139, 1987.

Tryggvason, E., Subsidence events in the Krafla area, north Iceland, 1975–1979, *Journal of Geophysics*, *47*, 141–153, 1980.

Tucholke, B. E., J. Lin, and M. C. Kleinrock, Megamullions and mullion structure defining oceanic metamorphic core complexes on the Mid-Atlantic Ridge, *Journal of Geophysical Research*, *103* (B5), 9857–9866, 1998.

Vera, E. E., C. J. Mutter, P. Buhl, J. A. Orcutt, A. J. Harding, M. E. Kappus, R. S. Detrick, and T. M. Brocher, The structure from 0- to 0.2-m.y.-old oceanic crust at 9°N on the East Pacific Rise from expanded spread profiles, *Journal of Geophysical Research*, *95* (B10), 15529–15556, 1990.

Walker, G. P. L., The dyke complex of Koolau volcano, Oahu: internal structure of a Hawaiian rift zone, in *Volcanism in Hawaii*, edited by R. W. Decker, T. L. Wright, and P. H. Stauffer, pp. 961–993, U.S. Geol. Surv. Prof. Paper 1350, 1987.

Wang, X., J. R. Cochran, and J. A. Barth, Gravity anomalies, crustal thickness and the pattern of mantle flow at the fast-spreading East Pacific Rise, 9°–10°N : evidence for three-dimensional upwelling, *J. Geophys. Res.*, *101*, 17927–17940, 1996.

Weaver, J., and C. Langmuir, Calculation of phase equilibrium in mineral-melt systems, *Computers and geosciences*, *16*, 1–19, 1990.

White, W. M., and J.-G. Schilling, The nature and origin of geochemical variations in Mid-Atlantic Ridge basalts from the Central North Atlantic, *Geochem. Cosmoch. Acta*, *42*, 1501–1516, 1978.

White, R. S., D. McKenzie, and K. O'Nions, Oceanic crustal thickness from seismic measurements and rare earth element inversions, *Journal of Geophysical Research*, *97* (B13), 19683–19715, 1992a.

White, R. S., D. McKenzie, and R. K. O'Nions, Oceanic crustal thickness from seismic measurements and rare earth element inversions, *Journal of Geophysical Research*, *97* (B13), 19683–19715, 1992b.

Wilcock, W. S. D., and J. R. Delaney, Mid-ocean ridge sulfide deposits: Evidence for heat extraction from magma chambers or cracking fronts?, *Earth and Planetary Science Letters*, *145*, 49–64, 1996.

Wolfe, C., G. M. Purdy, D. R. Toomey, and S. C. Solomon, Microearthquake characteristics and crustal velocity structure at 29°N of the Mid-Atlantic Ridge: The architecture of a slow-spreading segment, *Journal of Geophysical Research*, *100* (B12), 24449–24472, 1995.

Wright, D. J., R. M. Haymon, and D. J. Fornari, Crustal fissuring and its relationship to magmatic and hydrothermal processes on the East Pacific Rise crest (9°12' to 54°N), *Journal of Geophysical Research*, *100* (B4), 6097–6120, 1995.

Yang, H. J., R. J. Kinzler, and T. L. Grove, Experiments and models of anhydrous , basaltic olivine-plagioclase-augite saturated melts from 0.01 to 10 kbar., *Contrib. Mineral. Petrol.*, *124*, 1–18, 1996.

J. Cann, School of Earth Sciences, University of Leeds, Leeds LS2 9JT, UK

M. Cannat, Laboratoire de Géosciences Marines, CNRS-Institut de Physique du Globe de Paris, 4 place Jussieu 75252 Paris cedex 05, France.

J. Maclennan, School of GeoSciences, University of Edinburgh, Edinburgh EH9 3JW, UK

Effects of Hydrothermal Cooling and Magma Injection on Mid-Ocean Ridge Temperature Structure, Deformation, and Axial Morphology

Mark D. Behn[1], Jian Lin[2], and Maria T. Zuber[3]

Fault development at mid-ocean ridge spreading centers is strongly dependent on the thermal state of the axial lithosphere. Thermal conditions at a ridge axis are a combined function of spreading rate, mantle temperature, magma injection, and hydrothermal circulation. In this study, we test the sensitivity of fault development in slow-spreading environments to the efficiency of hydrothermal cooling and the depth extent of magma injection near the ridge axis. A 3-D finite difference scheme is first used to calculate axial temperature structure, and deformation is then modeled in 2-D vertical sections of lithosphere using a visco-plastic finite element model. Strain-rate softening in the brittle regime is used to simulate the rate-dependence of frictional strength observed in laboratory studies. This formulation results in the formation of localized zones of high strain rate (analogous to faults) that develop in response to the rheology and boundary conditions and are not imposed a priori. Comparing our numerical experiments with observed faulting at the center and ends of several segments along the slow-spreading Mid-Atlantic Ridge, we find that temperatures near the segment end must be warmer than predicted by previous models. These predicted high temperatures can be explained by either inefficient hydrothermal cooling in the shallow crust or heating of the upper mantle through magmatic accretion below the Moho. Because geophysical and geochemical evidence support efficient hydrothermal cooling in young oceanic lithosphere, we favor a model in which heat is supplied to the upper mantle beneath the ends of slow-spreading segments by either crystallization of rising asthenospheric melts or episodic lateral dike propagation from the segment center.

[1]Department of Terrestrial Magnetism, Carnegie Institution of Washington, Washington, DC.

[2]Department of Geology and Geophysics, Woods Hole Oceanographic Institution, Woods Hole, MA.

[3]Department of Earth, Atmospheric, and Planetary Sciences, Massachusetts Institute of Technology, Cambridge, MA, also at Laboratory for Terrestrial Physics, NASA Goddard Space Flight Center, Greenbelt, Maryland.

Mid-Ocean Ridges: Hydrothermal Interactions Between the Lithosphere and Oceans
Geophysical Monograph Series 148

10.1029/148GM06

1. INTRODUCTION

The thermal state of young oceanic lithosphere is a complex function of spreading-rate, mantle temperature, magma injection, and hydrothermal cooling. Because of the difficulties in directly measuring subsurface thermal structure, researchers frequently rely on observables such as axial topography and abyssal hill fabric to infer thermal conditions near oceanic spreading centers. Fault evolution is highly sensitive to rheological heterogeneities within the brittle lithosphere and the underlying ductile asthenosphere [*Davy et al.*, 1995; *Heimpel and Olsen*, 1996; *Cowie*, 1998]. Thus, by incorporating experimentally determined rheological laws in thermo-mechanical models of lithospheric stretching, the observed fault patterns can be used to place important constraints on axial thermal structure [e.g., *Sleep and Rosendahl*, 1979; *Phipps Morgan et al.*, 1987; *Lin and Parmentier*, 1989; *Chen and Morgan*, 1990a; *Shaw and Lin*, 1996; *Poliakov and Buck*, 1998].

Using this approach *Malinverno and Cowie* [1993] and *Shaw and Lin* [1996] attributed the first-order dependence of fault style on spreading rate to changes in the mechanical strength of the lithosphere caused by the difference in thermal regime between fast- and slow-spreading ridges. Systematic variations in rift morphology are also observed along individual segments of the slow-spreading Mid-Atlantic Ridge (MAR). For example, MAR segments are often characterized by hour-glass shaped rift valleys that become wider and deeper toward the segment ends [e.g., *Sempéré et al.*, 1993]. Coinciding with this change in rift morphology is an observed increase in the size and spacing of normal faults and abyssal hills from the segment center to the segment ends [e.g., *Shaw*, 1992; *Shaw and Lin*, 1993]. These changes in axial topography and fault style have been attributed to segment-scale variations in axial thermal structure [*Shaw and Lin*, 1993; 1996].

Along-axis thermal gradients can develop due to 3-D mantle upwelling [*Phipps Morgan and Forsyth*, 1988], the juxtaposition of old and therefore cold lithosphere across transform and non-transform offsets [*Macdonald and Luyendyk*, 1977; *Forsyth and Wilson*, 1984], and enhanced magmatic accretion near segment centers [*Kuo and Forsyth*, 1988; *Lin et al.*, 1990; *Lin and Phipps Morgan*, 1992; *Tolstoy et al.*, 1993; *Detrick et al.*, 1995; *Hooft et al.*, 2000]. However, several lines of evidence suggest that the along-axis thermal gradients at many slow-spreading segments may be smaller than previously proposed. *Escartín et al.* [1999] used a combination of deep-towed side-scan sonar and high resolution bathymetry data to show that fault spacing at the 29°N segment of the MAR does not necessarily increase systematically toward the segment end. Instead, fault spacing remains relatively constant from the segment center to the outside-corner crust near the ridge-transform intersection, increasing only on the inside-corner crust. The asymmetry across the segment end suggests faulting in this region may be affected more by mechanical interactions with the active segment offset than by an along-axis temperature gradient. The lack of a significant increase in the maximum depth of seismicity along the 29°N segment is also indicative of a small along-axis thermal gradient [*Wolfe et al.*, 1995].

Chen and Lin [1999] showed that much of the hour-glass shape of the rift valley south of the Atlantis transform (MAR 28°–30°N) is a result of isostatic compensation of along-axis crustal thickness variations. When this crustal signal is removed

the residual (or tectonic) rift-valley width remains relatively constant along-axis. A similar difference between the topographically-defined and the tectonically-defined rift-valley is observed at ends of the OH-1 segment of the MAR (see Figure 1). The OH-1 segment is the northernmost of three major segments bounded by the Oceanographer and Hayes Fracture Zones, and is characterized by a prominent hour-glass shaped morphology [*Gràcia et al.*, 1999; *Rabain et al.*, 2001]. Defining the rift valley based on the peak topography on either side of the ridge axis, the rift half-width is observed to be 10–20 km near the ends of the OH-1 segment. In contrast, if rift width is measured as the distance between the first large fault scarps (>300–500 m) on either side of the ridge axis, the segment ends are characterized by rift half-widths of only 5–7 km. The location of these large fault scarps indicates that significant brittle strain has accumulated near the ridge axis and implies a relatively thin brittle lithosphere even at the ends of the OH-1 segment. The narrowness of the tectonic rift valley is particularly surprising at the northern end of the segment, where the large age discontinuity across the Oceanographer Transform should result in significant conductive cooling and thicker lithosphere at the segment end.

Similar differences in the topographic and tectonic rift-valley width are observed at many other segments along the MAR. Figure 2 illustrates tectonic rift half-width (defined by half of the distance between the inner-most pair of large fault scarps) as a function of crustal thickness for a number of MAR segments where crustal thickness has been measured seismically. In all but one of the segments examined, the tectonic half-width near the segment end is less than 8 km and tends to be 4–8 km narrower than the topographically-defined rift-valley. We hypothesize that the tectonic rift valley is a direct reflection of the zone of active faulting at a ridge axis, while the topographic rift valley is the combined expression of several factors, including isostatic uplift associated with along-axis variations in crustal thickness [e.g., *Chen and Lin*, 1999] and flexural uplift associated with faulting. In many locations several large fault scarps are located between the rift axis and the location of maximum topography (e.g., Figure 1b). In these situations the overall morphology of the rift valley likely reflects flexural uplift associated with multiple generations of fault formation.

In this study, we examine the implications of tectonic rift width for axial thermal structure in slow spreading environments. Specifically, we test the sensitivity of fault development to the efficiency of hydrothermal cooling and the depth extent of magma injection near a ridge axis using a visco-plastic finite element model. Our modeling approach incorporates the rate-dependence of frictional strength observed in laboratory studies and results in the formation of localized zones of high strain rate (analogous to faults), which develop in

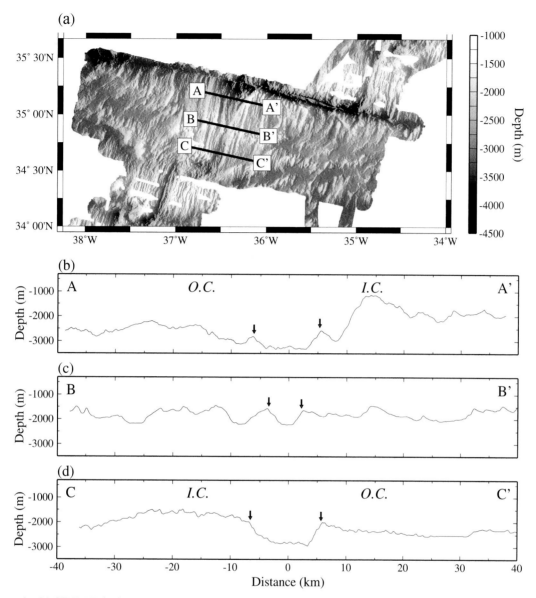

Figure 1. (a) SIMRAD bathymetry map of the OH-1 and OH-2 segments south of the Oceanographer Fracture Zone (100-m grid spacing) at the MAR [*Rabain et al.*, 2001]. Illumination is from N285°. Solid lines illustrate profiles shown in (b) A–A', (c) B–B' and (d) C–C'. Small arrows denote the locations of the innermost major fault scarps (> 500 m). The observed rift half-width increases from 3–4 km at the segment center to 4–7 km at the segment ends.

response to the rheology and boundary conditions. This formulation is advantageous because it allows us to investigate the style of faulting near a ridge axis over a wide range of thermal conditions without imposing the location of the faults a priori. Comparing the results of our numerical experiments to the tectonic rift width at the center and ends of segments along the MAR, we find that crustal and upper mantle temperatures near segment ends must be warmer than predicted by previous models [e.g., *Shaw and Lin*, 1996]. To generate these high temperatures, we show that either hydrothermal

cooling must be very inefficient in the shallow crust or heat must be emplaced in the upper mantle due to magmatic accretion below the Moho.

2. MODEL DESCRIPTION

We use a two-step approach to model the zone of active faulting as a function of axial thermal structure in slow-spreading environments. A 3-D temperature model is first calculated incorporating the mantle flow field, magma injection, and

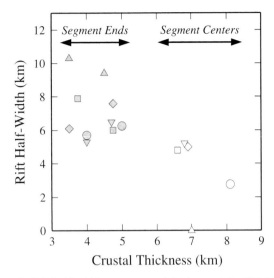

Figure 2. Rift half-width versus crustal thickness at the OH-1 (circles), OH-2 (diamonds), OH-3 (squares), 29°N (inverted triangles), and 33°S (triangles) segments of the Mid-Atlantic Ridge. Segment centers are denoted with open symbols and segment ends with filled symbols. Rift half-width was determined from the distance between the first major fault scarp (> 300–500 m) on either side of the spreading axis. Crustal thickness estimates are taken from a combination of seismic refraction and gravity data: OH-1, OH-2, and OH-3 [*Hooft et al.*, 2000], 29°N [*Lin et al.*, 1990; *Wolfe et al.*, 1995], and 33°S [*Kuo and Forsyth*, 1988; *Tolstoy et al.*, 1993]. Note that rift-half width generally decreases with increasing crustal thickness.

hydrothermal cooling. The resulting temperature field is then used to calculate the location of fault formation in 2-D vertical sections of lithosphere across the center and end of a typical ridge segment.

2.1. 3-D Thermal Structure

The 3-D temperature structure is calculated using the technique of *Shaw and Lin* [1996], which incorporates conductive and advective heat transport in mantle flow driven by the separation of surface plates. The predicted mantle flow field is strongly sensitive to the plate boundary geometry. For example, longer offsets between spreading segments tend to generate a more 3-D pattern of mantle upwelling, resulting in enhanced cooling near the segment ends [*Phipps Morgan and Forsyth*, 1988]. The plate geometry used in this study consists of a 100-km spreading segment bounded by a 150-km offset to the north and a 35-km offset to the south, simulating the effects of a transform and non-transform offset, respectively (Figure 3a). This spreading geometry is similar to that observed at the OH-1 segment of the MAR and is ideal for examining end-member models of along-axis thermal variability at slow-spreading ridge segments.

Hydrothermal cooling is modeled by enhancing the thermal conductivity by a factor Nu above a threshold depth of 6 km and in regions where the temperature is below 700°C. This parameterization implies that in situations where shallow crustal temperatures are high, hydrothermal cooling will only extend down to the 700°C isotherm. The threshold depth represents the maximum pressure at which cracks are predicted to remain open for hydrothermal circulation, while the cut-off temperature is consistent with the maximum alteration temperature observed in lower crustal rocks [*Kelley and Gillis*, 2002].

Magma injection is simulated by including a heat source at the ridge axis that includes both the injection temperature and the latent heat of the crystallizing magma. The injection zone has a width of 0.5 km and a height of h_a (Figure 3b). *Shaw and Lin* [1996] considered only cases in which the height of the injection zone was equal to the crustal thickness, t_c. In this study, however, we relax that condition and allow h_a to vary independently of t_c. The implications of this assumption will be discussed in detail later in the text.

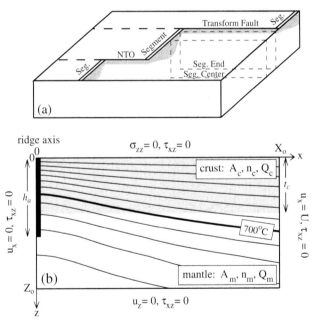

Figure 3. (a) 3-D model geometry used to calculate temperature structure along a ridge segment. Dashed lines illustrate the location of the 2-D vertical sections used to model fault formation at the center and end of the segment. (b) Model setup for mechanical models of lithospheric stretching. The model space is symmetric about the ridge axis, with dimensions X_o = 45 km and Z_o = 40 km. A uniform horizontal velocity of 1 cm/yr is applied to the right-hand side of the model space, and extension is continued until 1% total strain is achieved. Crustal thickness, t_c, is constant across-axis, with the crust and mantle rheological parameters given by n_c, A_c, Q_c and n_m, A_m, Q_n, respectively (Table 1). The magma injection zone (solid black region) has a width of 0.5 km and a height of h_a.

2.2. Mechanical Model of Fault Formation

The calculated 3-D temperature models are used as the input thermal conditions to a Lagrangian visco-plastic strain-rate-dependent finite element model. Strain-rate softening in the brittle layer is used to simulate the rate-dependence of frictional strength observed in laboratory studies [e.g., *Dieterich*, 1979; *Ruina*, 1983]. Both analytical [*Montési and Zuber*, 2002] and numerical [*Neumann and Zuber*, 1995; *Behn et al.*, 2002; *Montési and Zuber*, 2003] studies have shown that strain-rate softening in the brittle regime can result in the formation of localized zones of high strain rate, analogous to faults. Because these zones develop in response to the rheology and boundary conditions, the calculated deformation field can be used to predict the preferred location of fault initiation over a range of thermal and rheologic conditions.

For viscous flow we assume a non-Newtonian temperature-dependent rheology [*Kirby*, 1983; *Kohlstedt et al.*, 1995]

$$\dot{\varepsilon} = A(\sigma_1 - \sigma_3)^n \exp(-Q/RT) \quad (1)$$

where $\dot{\varepsilon}$ is the uniaxial strain rate, σ_1 and σ_3 are the maximum and minimum principle stresses, n is the power law exponent, Q is the molar activation energy, A is a material strength constant, T is the temperature, and R is the gas constant. Although the relationship between stress and strain rate is nonlinear, we can define a linearized viscosity law [e.g., *Chen and Morgan*, 1990b; *Boutilier and Keen*, 1994] by

$$\tau_{ij} = \sqrt{2}\eta\dot{\varepsilon}_{ij} \quad (2)$$

where τ_{ij} is the stress tensor, η is the effective viscosity, and $\dot{\varepsilon}_{ij}$ is the strain-rate tensor. This linearization leads to an expression for the apparent effective Newtonian viscosity

$$\eta = B\dot{\varepsilon}_{II}^{(1-n)/n} \exp(Q/nRT) \quad (3)$$

where $\dot{\varepsilon}_{II}$ is the second invariant of the stain rate tensor, and B is a material constant related to A.

In the brittle regime, strength is assumed to be controlled by a frictional resistance law [e.g., *Byerlee*, 1978; *Scholz*, 1990]

$$\tau_{max} = \sigma_o - \mu\sigma_n \quad (4)$$

where σ_o is the cohesive strength, μ is the coefficient of friction, and σ_n is approximately equal to the lithostatic stress. The rate dependence of frictional strength is simulated by defining an apparent friction coefficient, μ', as

$$\mu' = \mu_0[1 - \gamma\log_{10}(\dot{\varepsilon}_{II}/\dot{\varepsilon}_0)] \quad (5)$$

where μ_o is the reference coefficient of friction, γ is the strain-rate softening coefficient, and $\dot{\varepsilon}_o$ is the reference strain-rate. This formulation not only simulates strain-rate weakening for $\dot{\varepsilon}_{II} > \dot{\varepsilon}_o$ but also generates strengthening in regions where $\dot{\varepsilon}_{II} < \dot{\varepsilon}_o$. *Behn et al.* [2002] showed that $\gamma \geq 0.10$ results in efficient strain localization in models of lithospheric deformation for plausible rheological structures, and we choose $\gamma = 0.15$ for the numerical experiments presented in this study. Acknowledging that this approach neglects many of the complexities of the earthquake process, we interpret these regions of high strain-rate to be analogous to fault zones. Note that in the visco-plastic formulation implemented here, the pattern of deformation is found to be relatively insensitive to the values of μ_o and σ_o.

Following the procedures described in *Neumann and Zuber* [1995] and *Behn et al.* [2002] we calculate deformation in two 2-D vertical sections of lithosphere near the center and end of the segment. At each time-step the element viscosities are calculated from the temperature and evolving strain-rate fields. If the resulting maximum principle shear stress calculated from Equation 2 is greater than the frictional failure criterion, τ_{max}, the effective viscosity of the element is reset to $\eta = \tau_{max}/\sqrt{2}\dot{\varepsilon}_{II}$. The initial element viscosities are calculated assuming a uniform background strain-rate of $10^{-14}s^{-1}$.

We stress that the numerical experiments presented here should be treated only as a proxy for the initial pattern of faulting that develops for a given set of thermal conditions, rather than as a method to study the evolution of individual faults over geologic time. The rotation of fault blocks in highly extended terrains generates large flexural stress [e.g., *Forsyth*, 1992; *Buck*, 1993] that are not accounted for in our visco-plastic formulation. By limiting our calculations to 1% total strain, we can safely ignore these elastic stresses and also eliminate numerical inaccuracies associated with the distortion of model elements.

Crustal thicknesses of 8 km and 3 km are used for the center and end of the segment, respectively. These values are based on end-member segment-scale variations in crustal thickness observed in seismic refraction experiments at the MAR [e.g., *Tolstoy et al.*, 1993; *Hooft et al.*, 2000]. We assume the dry diabase flow law of *Mackwell et al.* [1998] for the crust ($n = 4.7$, $A = 1.9 \times 10^2$ MPa^{-n} s^{-1}, $Q = 485$ kJ mol^{-1}) and the dry dunite flow law of *Chopra and Paterson* [1984] for the mantle ($n = 3$, $A = 1.0 \times 10^3$ MPa^{-n} s^{-1}, $Q = 520$ kJ mol^{-1}). These dry rheologies are appropriate for young oceanic lithosphere that has been dewatered due to melting associated with crustal formation [*Hirth et al.*, 1998].

Deformation is driven by applying a uniform horizontal velocity of 1 cm/yr to the right-hand side of the model space ($x = X_o$). For numerical efficiency, a symmetry condition is imposed on the left-hand side of model ($x = 0$) by setting the horizontal velocity, u_x and the shear stress, τ_{xz}, equal to zero. The model dimensions ($X_0 = 45$ km, $Z_0 = 40$ km) are specified to ensure that the boundaries do not influence the final solution and the finite element grid is adjusted to give maximum resolution (grid size of 300 m × 300 m) near the ridge axis. The boundary conditions and driving forces are illustrated in Figure 3b and a list of all model parameters is given in Table 1.

3. NUMERICAL RESULTS

Deformation is calculated over a wide range of thermal conditions, varying both the height of the magma injection zone, h_a, and the efficiency of hydrothermal cooling, Nu. Plate 1 shows the predicted pattern of deformation at the segment center for $Nu = 4$ and $h_a = 6$ km. Strain is calculated to be focused within two inward-dipping shear zones that intersect the surface at distances of ~4 km and ~12 km from the ridge axis, respectively, and an outward-dipping shear zone connecting the two inward-dipping zones at depth of 5–10 km. In general, stress increases with depth to a maximum value of ~140 MPa at the brittle-ductile transition and then decreases abruptly in the underlying ductile asthenosphere (Plate 1, top-left panel). Corresponding effective viscosities are shown in Plate 1 (lower-left panel), with elements undergoing brittle failure shaded in grey. Within the shear zones stresses

are low due to the efficient strain-rate weakening in these regions. Also, note that the brittle-ductile transition has been deformed beneath the axis, illustrating the coupling between the evolving stress and strain fields.

The flow-field shown in Plate 1 (top-right panel) is very similar to that observed in analogue experiments of extension in clay layers [e.g., *Cloos*, 1968], as well as earlier studies based on slip-line analysis [*Lin and Parmentier*, 1990]. These similarities include both the central graben development, and also the formation of weak antithetic shear zones near the surface intercept of the outer fault. The development of multiple active shear zones, both synthetic and antithetic, illustrates the importance of the interaction between stress and strain in determining the preferred style of deformation. Similar styles of faulting have been observed in modeling studies that incorporate cohesion loss on fault surfaces through a strain-weakening rheology in the brittle crust [*Poliakov and Buck*, 1998; *Lavier et al.*, 2000; *Lavier and Buck*, 2002]. Both approaches incorporate the coupling between stress and strain through non-linear frictional resistance laws, and represent an important advance over models that infer fault development using an elastic plate formulation [e.g., *Forsyth*, 1992; *Shaw and Lin*, 1996; *Escartín et al.*, 1997]. In these elastic formulations fault location and dip must be specified, rather than developing in response to the evolving stress field. Thus, this

Table 1. Model Parameters

	Definition	Value	Units
u	spreading half-rate	1.0	cm yr^{-1}
T_m	mantle temperature	1350	°C
Nu	efficiency of hydrothermal cooling	1–12	
T_{cut}	maximum temperature of hydrothermal cooling	700	°C
z_{cut}	maximum depth of hydrothermal cooling	6	km
h_a	magma injection zone height	2–10	km
T_i	magma injection temperature	1150	°C
l_h	latent heat of magma	7.5×10^5	J kg^{-1}
c_p	specific heat of magma	1250	J kg^{-1} °C^{-1}
t_c	crustal thickness (seg. center, seg. end)	8, 3	km
W	width of the emplacement zone	0.5	km
κ	thermal diffusivity	10^{-6}	m^2 s^{-1}
R	gas constant	8.3144	J mol^{-1} K^{-1}
n_c, n_m	power law exponent (crust, mantle)	4.7, 3	
A_c, A_m	material strength constant (crust, mantle)	1.9×10^2, 1.0×10^3	MPa^{-n} s^{-1}
Q_c, Q_m	molar activation energy (crust, mantle)	485, 520	kJ mol^{-1}
ρ_c, ρ_m	density (crust, mantle)	2700, 3300	kg m^{-3}
η_{eff}	effective viscosity		Pa s
$\tau_{\max}$	maximum differential stress		MPa
μ'	effective coefficient of friction		
μ_0	reference coefficient of friction	0.6	
γ	strain rate softening coefficient	0.15	
$\dot{\varepsilon}_{II}$	second invariant of the strain rate tensor		s^{-1}
$\dot{\varepsilon}_0$	reference strain rate	10^{-14}	s^{-1}
w_r	rift half-width		km
x, z	horizontal and vertical coordinates		km
X_0, Z_0	horizontal and vertical dimensions of model space	45, 40	km

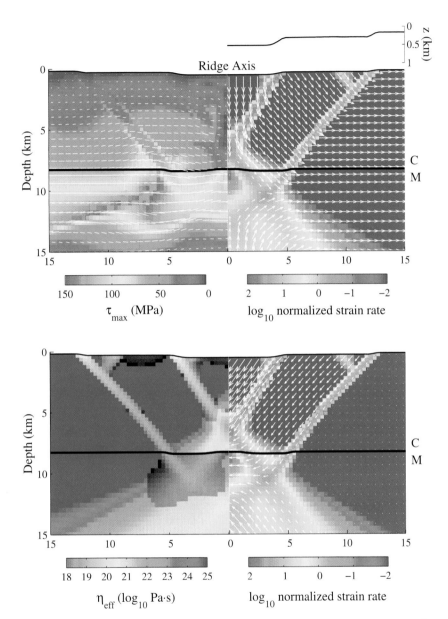

Plate 1. Numerical simulation illustrating the predicted pattern of deformation at the segment center for $Nu = 4$ and $h_a =$ 6 km. (Top-right) Calculated strain-rates normalized to $\dot{\varepsilon}_o$ after 1% total strain. White arrows show flow field relative to horizontal motion of the off-axis lithosphere. Black line illustrates boundary between the crust (C) and mantle (M). (Top-left) Maximum differential stress, τ_{max}. White lines indicate direction of maximum tensile stress with length scaled to the magnitude of τ_{max}. (Bottom-right) Normalized strain-rates with white arrows showing flow field relative to no motion of the off-axis lithosphere. (Bottom-left) Effective viscosity, η_{eff}. Grey shading illustrates elements undergoing brittle failure. Strain is calculated to be focused in two inward-dipping shear zones that intersect the surface with rift-half widths of ~4 and ~12 km from the ridge axis, respectively.

approach is not suitable for accurately modeling the interaction and initiation of multiple sets of faults.

Plates 2 and 3 show the calculated strain-rate fields as a function of thermal structure for the center and northern end of the segment, respectively. We examine only the northern end of the segment because conductive cooling across the transform will generate a larger gradient in thermal conditions relative to the segment center than will the shorter non-transform offset to the south. By comparing the northern end of the segment with the segment center, we can investigate the effects of end-member thermal conditions on fault formation in slow-spreading environments.

Deformation is calculated for $Nu = 1-12$ and $h_a = 2-10$ km. The wide range of thermal parameters was chosen to test the sensitivity of the predicted deformation field to a broad spectrum of possible temperature structures. In general, rift half-width, w_r, is calculated to increase with decreasing values of h_a and increasing values of Nu (Figure 4). We define w_r as the distance from the ridge axis ($x = 0$ km) to the surface intercept of the inward-dipping shear zone. (Note that for thermal conditions in which two shear zones form, both an inner and outer rift half-width are predicted). We also find that for a given set of thermal parameters, w_r is greater at the segment end than at the segment center. These first-order predictions are consistent with the results of *Shaw and Lin* [1996].

However, the numerical experiments conducted in this study also show interesting features not observed in previous studies. In particular, the formation of multiple inward-dipping faults is calculated to occur as a transitional regime between cold, wide single-fault rift zones and warm, narrow single-fault rift zones. For example, when $Nu = 8$, the calculated strain-rates at the segment center show wide rift zones ($w_r \approx$ 15 km) defined by a single fault for $h_a = 2-6$ km, and a narrow single-fault rift ($w_r \approx 6$ km) for $h_a = 10$ km (Plate 2). However, when $h_a = 8$ km, the deformation field is characterized by two inward-dipping shear zones ($w_r \approx 6$ & 13 km). A similar style of transition from wide to narrow single-fault rifts is observed for $h_a = 6$ km and decreasing values of Nu (Plate 2). In reality, topographic stresses associated with continued slip on multiple faults may cause one fault to become preferred over the other. However, these results show the potential for slip in multiple fault zones near the ridge axis.

We also find that when high temperatures are present in the shallow crust beneath the ridge axis, a narrow (≤ 1 km) collapse feature is often predicted to form (red arrows in Plates 2 and 3). These collapse structures are typically bounded by nearly vertical shear zones. Similar features, often referred to as axial caldera troughs, are observed at many fast spreading ridges and are strongly correlated to the presence of a

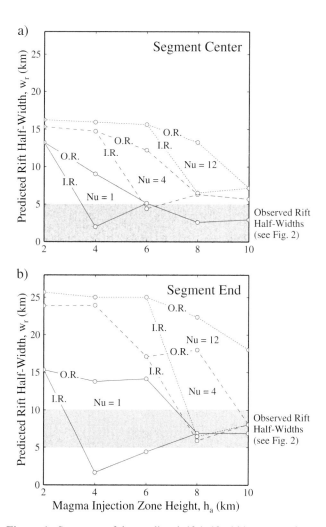

Figure 4. Summary of the predicted rift half-width, w_r, as a function of the magma injection zone height, h_a, at the (a) segment center and (b) segment end. Each symbol illustrates the results of one numerical experiment (see Plates 2 and 3). Calculated rift half-widths are shown for $Nu = 1$ (solid), $Nu = 4$ (dashed), and $Nu = 12$ (dotted). Note that in some cases two values of w_r are predicted, corresponding to two sets of calculated faults. I.R. and O.R. refer to the inner and outer faults, respectively. In general, rift half-width is calculated to increase with decreasing values of h_a and increasing values of Nu.

shallow axial mamga chamber [e.g., *Macdonald and Fox*, 1988].

4. COMPARISON OF PREDICTED AND OBSERVED RIFT WIDTHS

The predicted deformation patterns were first compared to the axial bathymetry at OH-1 segment to determine the thermal conditions that are most consistent with the observed tectonic rift width. The OH-1 segment represents an ideal location to constrain axial thermal structure, because the wealth of geophysical data that have been collected in the region. Specifically, the along-axis variability in tectonic rift width is well constrained by seafloor mapping of the axial valley floor and adjacent ridge flanks [Gràcia et al., 1999; Rabain et al., 2001]. In addition, seismic data constrain the along-axis variations in crustal thickness from a maximum of 8 km near the segment center to 4 km and 5 km at the northern and southern segment ends, respectively [Hooft et al., 2000; Hosford et al., 2001]. It is possible that multiple sets of faults are active at the OH-1 segment. A microearthquake study [Barclay et al., 2001] and hydroacoustic monitoring [Smith et al., 2002] show a tendency for seismicity to cluster at the inside corners of the OH-1 segment. However, the limited number of events observed during these experiments are not sufficient to determine all faults that are currently active. Thus, because the off-axis extent of active deformation cannot be determined, we compare the rift half-width defined by the innermost calculated shear zone to the innermost observed fault scarp.

At the center of the OH-1 segment, the tectonic rift half-width is observed to be 3–4 km (Figure 1c). Assuming the height of the magma injection zone is not less than the crustal thickness (i.e., $h_a \geq t_c$ the observed rift half-width is most consistent with $Nu = 1–2$ (Plate 2, $h_a = 8–10$ km). These thermal structures predict active faulting to depths of ~5 km, consistent with the observed depth of microseismicity at the segment center [Barclay et al., 2001]. In this case, the calculated location of the innermost shear zone is not highly sensitive to the efficiency of hydrothermal cooling for $h_a \geq 8$ km (Figure 4). Thus, it is difficult to rule out cases with $Nu \geq 4$.

Near the northern end of the OH-1 segment, prominent inside-corner high topography is located 7–14 km from the ridge axis (Figure 1b). However the observed half-width of the inner rift valley appears to be only 4–7 km (Figures 1b and 1d). Such a narrow rift half-width is consistent with $h_a \geq 4$ km for $Nu = 1$; $h_a \geq 6$ km for $Nu = 2$; and $h_a \geq 8$ km for $Nu \geq 4$ (Plate 3). This implies that with the exception of the numerical experiments with no hydrothermal cooling ($Nu = 1$), the observed rift half-width requires a magma injection zone that extends 2–6 km below the base of the seismically-determined crust at northern end of the OH-1 segment. It is interesting that in many of these cases deformation is predicted to localize in two shear zones, with the outer fault defining a rift half-width ≥ 14 km.

We also compared the results of our numerical experiments to the observed tectonic rift width at several other MAR segments where crustal thickness has been determined seismically (Figure 2). Rift half-width at the center of these segments ranges from 0 to 5 km, while the seismically-determined crustal thickness varies from 6 to 8 km. We find that the predicted inner rift half-width is consistent with the observed

half-widths for $h_a \approx t_c$ and a range of Nusselt numbers from 1–8 (Plate 2 and Figure 4a). Near the segment ends, observed rift half-width varies from 5 to 10 km with crustal thicknesses of 3–5 km. These narrow rift half-widths are predicted only by models with either $h_a \approx t_c$ and $Nu = 1$, or $h_a > t_c$ and $Nu > 1$ (Plate 3 and Figure 4b). Thus, except in cases with no enhanced hydrothermal cooling, the magma injection zone must extend below the seismically-determined Moho near the segment ends to match the observed topography. In future studies, it will be interesting to test models that include either an isolated magma injection zone in the upper mantle or in multiple discrete zones throughout the crust and mantle to determine if these accretion zone geometries also generate faulting consistent with the observed tectonic rift widths at the ends of slow-spreading segments.

5. DISCUSSION

Previous studies of axial thermal structure [e.g., *Phipps Morgan and Chen*, 1993; *Shaw and Lin*, 1996] typically assume efficient hydrothermal cooling in the shallow lithosphere ($Nu = 8$) and a magmatic accretion zone isolated to the crust ($h_a \leq t_c$). In this formulation, heat supply at the center of a slow-spreading segment is greater than near the segment ends due to the along-axis gradient in crustal thickness. However, using these thermal parameters we predict rift widths that are significantly wider than observed at the ends of many MAR segments. For example, deformation at the northern end of the OH-1 segment is calculated to localize in a single fault zone with $w_r \approx 20$ km. Such a wide rift is not consistent with the observed faulting (Figure 1) and indicates that the thermal parameters used in these earlier studies may significantly underestimate temperatures near the end of the OH-1 segment. We propose two alternative explanations for this discrepancy and discuss their implications for magmatic and hydrothermal processes at slow-spreading ridges.

One explanation for the elevated temperatures at the ends of many MAR segments is that hydrothermal circulation does not significantly enhance thermal conductivity in the shallow crust. Assuming $Nu = 1$ and $h_a = 8$ km, the predicted deformation at the segment center matches the observed rift half-widths shown in Figure 2. A narrow inner rift half-width consistent with the data can also be generated at the segment end with $Nu = 1$ and $h_a = 4$ km (Plate 3). A mechanism that could potentially limit the efficiency of hydrothermal cooling is the precipitation of hydrothermal minerals blocking fluid pathways [e.g., *Sleep*, 1991]. The sealing time of a porous medium is inversely proportional to the mass flux squared [*Lowell and Yao*, 2002]. This results in a negative feedback, in which the regions of most rapid fluid flow, and thus most effi-

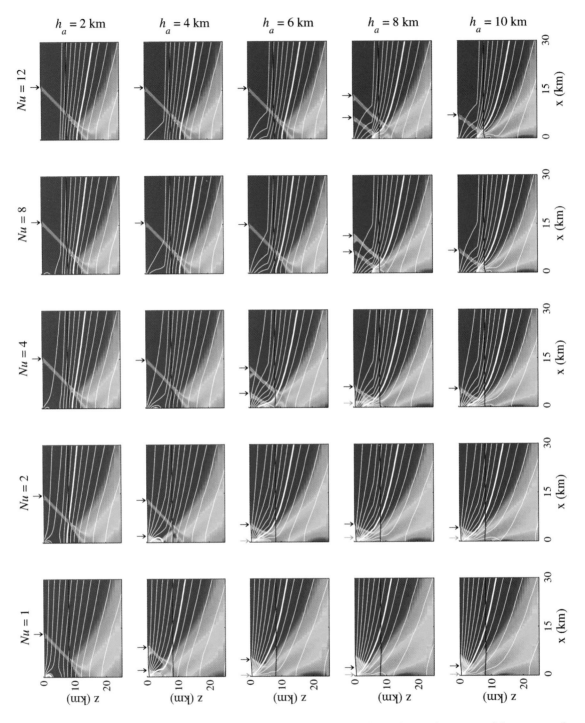

Plate 2. Numerical simulations illustrating the predicted pattern of deformation at the center of the segment for $Nu = 1$–12 and $h_a = 2$–10 km. Each panel shows the results of one numerical experiment, with Nu increasing from left to right and h_a increasing from top to bottom. Colors indicate calculated strain-rates normalized to $\dot{\varepsilon}_o$ after 1% total strain (color scale shown in Plate 1). White lines show initial temperature field prior to deformation with contour interval of 100°C. Black arrows indicate surface intercept of major inward-dipping fault zones, while red arrows denote vertical collapse features. A crustal thickness of 8 km is used for all numerical experiments at the segment center. Rheological parameters used for the crust and mantle are given in Table 1.

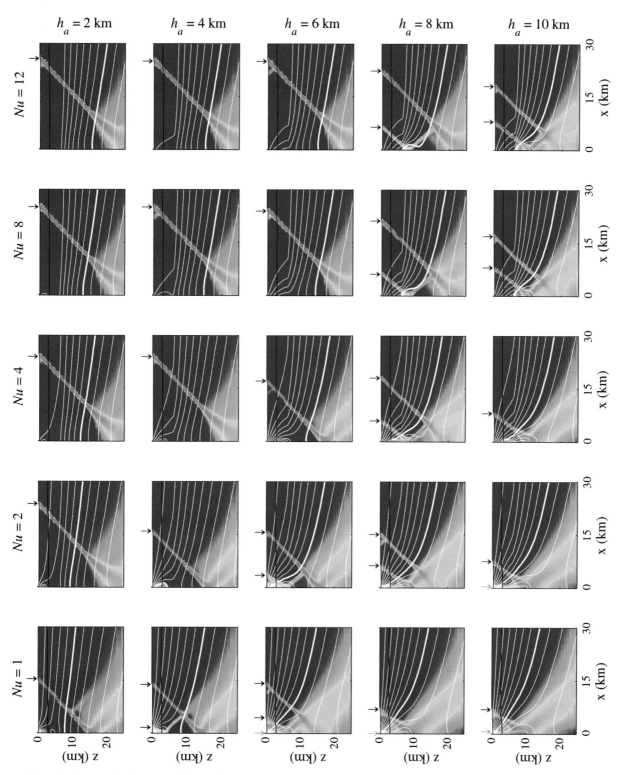

Plate 3. Numerical simulations illustrating the predicted pattern of deformation at the northern end of the segment for $Nu = 1$–12 and $h_a = 2$–10 km. A crustal thickness of 3 km is used for all numerical experiments at the segment end. All other model parameters are the same as those used at the center of the segment (Plate 2).

cient hydrothermal cooling, are also the regions that experience the most rapid decrease in permeability.

However, rapid sealing of fluid pathways is not consistent with geophysical and geochemical evidence for efficient hydrothermal cooling in young oceanic lithosphere. Crustal and upper-mantle seismic structure at 9°30'N on the East Pacific Rise indicate rapid cooling in the shallow crust near the ridge axis [*Dunn et al.*, 2000]. This is supported by the systematic overprediction of heat flow in young oceanic crust by plate cooling models [*Stein and Stein*, 1992]. In addition, thermal modeling by *Phipps Morgan and Chen* [1993] showed that values of $Nu \geq 8$ are required to accurately predict the depth and presence of an axial magma chamber over a wide range of spreading rates. Finally, altered primary mineral phases, vein assemblages, fluid inclusions, and oxygen isotopic measurements all indicate that hydrothermal flow occurs in the lower crust at temperatures up to 700°C [*Kelley and Gillis*, 2002]. Therefore, we favor an alternative model in which heat is supplied to the upper mantle at the ends of the slow-spreading segments by magmatic accretion beneath the Moho.

Cannat [1996] proposed that when the axial lithosphere is thicker than the seismically-determined crust, melt extracted from the asthenosphere may crystallize in the upper lithospheric mantle (Model 1 in Figure 5). Because crystallization only requires that temperatures in the surrounding rock be below the melt solidus (1000°–1200°C [*Sinton and Detrick*, 1992]), significant latent heat may be released if rising asthenospheric melts begin crystallization below the Moho. This model is supported by petrogenetic studies of gabbros from slow-spreading ridges that showed fractional crystallization begins at pressures exceeding the ambient pressure at the base of the crust [*Flower*, 1981; *Elthon et al.*, 1992; *Ross and Elthon*, 1997]. Furthermore, melt/rock reactions in rocks from ODP Site 920 show evidence for gabbroic melts crystallizing in peridotites at temperatures only slightly below the melt solidus [*Cannat et al.*, 1997].

Magma could also be supplied to the upper mantle at the ends of the slow-spreading segments by lateral dike propagation from the segment center (Model 2 in Figure 5). Dike propagation over distances > 50 km has been documented at Krafla volcano in Iceland [*Einarsson and Brandsdottir*, 1980] and Axial volcano on the Juan de Fuca Ridge [*Dziak and Fox*, 1999; *Embley et al.*, 2000]. Dikes tend to propagate laterally at the level of effective neutral bouyancy [e.g., *Ryan*, 1987]. In magma-starved regions this level is controlled by a combination of the rock and magma density, local stress state and magma pressure, and broadly corresponds to the depth of the brittle-ductile transition [*Rubin*, 1990]. Thus, if the brittle-ductile transition lies below the Moho at the segment end, dike propagation from the segment center could transport significant amounts of heat and magma to the upper mantle near segment ends [*Fialko and Rubin*, 1998].

Earthquake swarms are often associated with diking events and volcanic eruptions at mid-ocean ridges [e.g., *Embley et al.*, 1995]. Since the installation of a NOAA/PMEL hydroacoustic network in the North Atlantic in 1999 [*Smith et al.*, 2002] no earthquake swarms have been observed at any of the segments shown in Figure 2. However, at a spreading half-rate of 1–1.2 cm/yr a 1-m dike would be predicted only once every 40–50 years. Thus, the lack of an observed event over the last 3 years does not preclude the lateral propagation of dikes as a viable mechanism for supplying heat to the ends of slow-spreading segments.

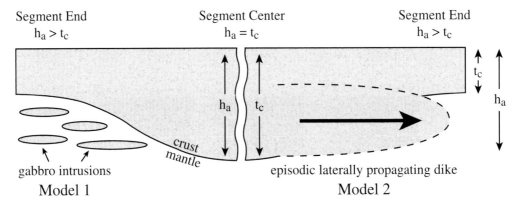

Figure 5. Schematic diagram of two possible mechanisms for magma injection in the upper mantle near the ends of slow-spreading segments. (Left) Model 1: Rising asthenospheric melts crystallize below the Moho when temperatures in upper lithospheric mantle are below the melt solidus [e.g., *Cannat*, 1996]. (Right) Model 2: Magma is supplied to the segment ends via lateral dike propagation from the segment center [e.g., *Fialko and Rubin*, 1998]. Either of these mechanisms can result in an episodic magma injection zone at the segment end that is thicker than the seismically determined crustal thickness (i.e., $h_a > t_c$).

The results of this study do not imply that sub-crustal heat input is required at the ends of all slow-spreading ridge segments. There is considerable evidence that slow-spreading ridges experience temporal variations in magma supply on time scales of 1–5 Myrs [e.g., *Canales et al.*, 2000; *Bonatti et al.*, 2003], which could strongly affect crustal and upper mantle thermal structure. In addition, transient magmatic events on shorter time scales may result in axial temperatures that are not in steady-state. However, the results presented here do illustrate the possible importance of sub-crustal accretion and the complexity in determining axial thermal structure in slow-spreading environments.

6. CONCLUSIONS

We have shown the sensitivity of fault development in slow-spreading environments to the efficiency of hydrothermal cooling and the magma injection zone near the ridge axis. The width of active deformation is calculated to increase with increasing efficiency of hydrothermal cooling and decreasing depth extent of the magma injection zone. We find that for a given set of thermal parameters, rift width is greater at the segment end than at the segment center, consistent with the results of earlier modeling studies [e.g., *Shaw and Lin*, 1993; 1996]. However, unlike these earlier studies, which assumed slip occurred on only one fault, our results predict that under certain circumstances multiple sets of inward-dipping faults may be active simultaneously.

Comparing our numerical results with the observed style of faulting at the center and ends of several segments along the MAR, we find that temperatures in the crust and upper mantle do not vary significantly along-axis. Specifically, the thermal regime at segment ends must be warmer than predicted by previous models in order to generate tectonic rift widths similar to those observed along the MAR. To produce these warm temperatures, we show that either hydrothermal cooling must be very inefficient in the shallow crust or that heat must be emplaced in the upper mantle due to magmatic accretion below the Moho. Geophysical and geochemical evidence supports efficient hydrothermal circulation in the shallow crust. Therefore, we favor a model in which heat is supplied to the upper mantle at the segment end by either crystallization of rising asthenospheric melts or episodic lateral dike propagation from the segment center.

Acknowledgments. This study benefited from constructive conversations with Roger Buck, Mathilde Cannat, Laurent Montési and Brian Tucholke. We thank Roger Searle, Lindsay Parsons, and an anonymous reviewer for their constructive comments that significantly improved an earlier version of this manuscript. This work was supported by a Carnegie Institution of Washington postdoctoral fellowship (M.B.); NSF grant EAR-0003888 and Andrew W. Mellon Foundation Endowed Fund for Innovative Research at WHOI (J.L.); and NASA Grant NAG5-4555 and a fellowship from the Radcliffe Institute for Advanced Study at Harvard University (M.Z.). WHOI contribution number 11090.

REFERENCES

Barclay, A. H., D. R. Toomey, and S. C. Solomon, Microearthquake characteristics and crustal Vp/Vs structure at the Mid-Atlantic Ridge, 35ºN, *J. Geophys. Res., 106*, 2017–2034, 2001.

Behn, M. D., J. Lin, and M. T. Zuber, A continuum mechanics model for normal faulting using a strain-rate softening rheology: Implications for thermal and rheological controls on continental and oceanic rifting, *Earth Planet. Sci. Lett., 202*, 725–740, 2002.

Bonatti, E., M. Ligi, D. Brunelli, A. Cipriani, P. Fabretti, V. Ferrante, L. Gasperini, and L. Ottolini, Mantle thermal pulses below the Mid-Atlantic Ridge and temporal variations in the formation of oceanic lithosphere, *Nature, 423*, 499-505, 2003.

Boutilier, R. R., and C. E. Keen, Geodynamic models of fault-controlled extension, *Tectonics, 13*, 439–454, 1994.

Buck, W. R., Effect of lithospheric thickness on the formation of high- and low-angle normal faults, *Geology, 21*, 933-936, 1993.

Byerlee, J., Friction of rocks, *Pure Appl. Geophys., 116*, 615–626, 1978.

Canales, J. P., J. A. Collings, J. Escartín, and R. S. Detrick, Seismic structure across the rift valley of the Mid-Atlantic Ridge at 23º20'N (MARK Area): Implications for crustal accretion processes at slow-spreading ridges, *J. Geophys. Res., 105*, 28, 411–28, 425, 2000.

Cannat, M., F. Chatin, H. Whitechurch, and G. Ceuleneer, Gabbroic rocks trapped in the upper mantle at the Mid-Atlantic Ridge, in *Proc. Ocean Drill. Program, Sci. Res.*, edited by Karson, J. A., M. Cannat, D. J. Miller, and D. Elthon, *153*, pp. 243–264, College Station, TX, 1997.

Cannat, M., How thick is the magmatic crust at slow spreading oceanic ridges?, *J. Geophys. Res., 101*, 2847–2857, 1996.

Chen, Y. J., and J. Lin, Mechanisms for the formation of ridge-axis topography at slow-spreading ridges: A lithospheric plate flexure model, *Geophys. J. Int., 136*, 8–18, 1999.

Chen, Y. and W. J. Morgan, Rift valley/No rift valley transition at mid-ocean ridges, *J. Geophys. Res., 95*, 17,571–17,581, 1990a.

Chen, Y. and W. J. Morgan, A nonlinear rheology model for mid-ocean ridge axis topography, *J. Geophys. Res., 95*, 17,583–17,604, 1990b.

Chopra, P. N., and M. S. Paterson, The role of water in the deformation of dunite, *J. Geophys. Res., 89*, 7861–7876, 1984.

Cloos, E., Experimental analysis of gulf coast fracture patterns, *Am. Assoc. Petr. Geol. Bull., 52*, 420-444, 1968.

Cowie, P. A., Normal fault growth in three-dimensions in continental and oceanic crust, in *Faulting and Magmatism at Mid-Ocean Ridges*, edited by Buck, W. R. and P. T. Delaney, and J. A. Karson, and Y. Lagabrielle, AGU Geophys. Monograph, *106*, 325–348, 1998.

Davy, P., A. Hansen, E. Bonnet, and S.-Z. Zhang, Localisation and fault growth in layered brittle-ductile systems: implications for deformation of the continental lithosphere, *J. Geophys. Res., 100*, 6281–6289, 1995.

Detrick, R. S., H. D. Needham, and V. Renard, Gravity anomalies and crustal thickness variations along the Mid-Atlantic Ridge between 33°N and 40°N, *J. Geophys. Res., 100,* 3767–3787, 1995.

Dieterich, J. H., Modeling of rock friction 1. Experimental results and constitutive equations, *J. Geophys. Res., 84,* 2161–2168, 1979.

Dunn, R. A., D. R. Toomey, and S. C. Solomon, Three-dimensional seismic structure and physical properties of the crust and shallow mantle beneath the East Pacific Rise at 9°30′N, *J. Geophys. Res., 105,* 23,537–23,555, 2000.

Dziak, R. P. and C. G. Fox, The January 1998 earthquake swarm at Axial volcano, Juan de Fuca Ridge: Hydroacoustic evidence of seafloor volcanic activity, *Geophys. Res. Lett., 26,* 3429–3432, 1999.

Einarsson, P. and B. Brandsdottir, Seismological evidence for lateral magma intrusion during the July 1978 deflation of the Krafla volcano in NE-Iceland, *J. Geophys., 47,* 160–165, 1980.

Elthon, D., M. Stewart, and D. K. Ross, Compositional trends of minerals in oceanic cumulates, *J. Geophys. Res., 97,* 15189–15199, 1992.

Embley, R. W., W. W. Chadwick, I. R. Jonasson, D. A. Butterfield, and E. T. Baker, Initial results of the rapid response to the 1993 CoAxial event: Relationships between hydrothermal and volcanic processes, *Geophys. Res. Lett., 22,* 143–146, 1995.

Embley, R. W., W. W. Chadwick, M. R. Perfit, M. C. Smith, and J. R. Delaney, Recent eruptions on the CoAxial segment of the Juan de Fuca Ridge: Implications for mid-ocean ridge accretion processes, *J. Geophys. Res., 105,* 16,501–16,525, 2000.

Escartín, J., P. A. Cowie, R. C. Searle, S. Allerton, N. C. Mitchell, C.J. Macleod, and A. P. Slootweg, Quantifying tectonic strain and magmatic accretion at a slow spreading ridge segment, Mid-Atlantic Ridge, 29°N, *J. Geophys. Res., 104,* 10,421–10,437, 1999.

Escartín, J., G. Hirth, and B. Evans, Effects of serpentinization on the lithospheric strength and the style of normal faulting at slow-spreading ridges, *Earth Planet. Sci. Lett., 151,* 181–189, 1997.

Fialko, Y. A. and A. M. Rubin, Thermodynamics of lateral dike propagation: Implications for crustal accretion at slow spreading mid-ocean ridges, *J. Geophys. Res., 103,* 2501–2514, 1998.

Flower, M. F. J., Thermal and kinematic control on ocean-ridge magma fractionation: Contrasts between Atlantic and Pacific spreading axes, *J. Geol. Soc. London, 138,* 695–712, 1981.

Forsyth, D. W., Finite extension and low-angle normal faulting, *Geology, 20,* 27–30, 1992.

Forsyth, D. W. and B. Wilson, Three-dimensional temperature structure of a ridge-transform-ridge system, *Earth Planet. Sci. Lett., 70,* 355–362, 1984.

Gràcia, E., D. Bideau, R. Hekinian, and Y. Lagabrielle, Detailed geological mapping of two contrasting second-order segments of the Mid-Atlantic Ridge between Oceanographer and Hayes fracture zones (33°30′N–35°N), *J. Geophys. Res., 104,* 22,903–22,921, 1999.

Heimpel, M. and P. Olsen, A seismodynamical model of lithospheric deformation: Development of continental and oceanic rift networks, *J. Geophys. Res., 101,* 16,155–16,176, 1996.

Hirth, G., J. Escartin, and J. Lin, The rheology of the lower oceanic crust: Implications for lithospheric deformation at mid-ocean ridges, in *Faulting and Magmatism at Mid-Ocean Ridges,* edited

by Buck, W. R. and P. T. Delaney, and J. A. Karson, and Y. Lagabrielle, AGU Geophys. Monograph, *106,* 291–303, 1998.

Hooft, E. E. E., R. S. Detrick, D. R. Toomey, J. A. Collins, and J. Lin, Crustal thickness and structure along three contrasting spreading segments of the Mid-Atlantic Ridge, 33.5°N–35°N, *J. Geophys. Res., 105,* 8205–8226, 2000.

Hosford, A., J. Lin, and R. S. Detrick, Crustal evolution over the last 2 m.y. at the Mid-Atlantic Ridge OH-1 segment, 35°N, *J. Geophys. Res., 106,* 13,269–13,285, 2001.

Kelley, D. S. and K. M. Gillis, Petrologic constraints upon hydrothermal circulation, in *InterRidge Theoretical Institute: Thermal Regime of Ocean Ridges and Dynamics of Hydrothermal Circulation,* edited by C. German, J. Lin, R. Tribuzio, A. Fisher, M. Cannat, and A. Adamczewska, InterRidge, Pavia, 2002.

Kirby, S, Rheology of the lithosphere, *Rev. Geophys. Space Phys., 21,* 1458–1487, 1983.

Kohlstedt, D., B. Evans, and S. Mackwell, Strength of the lithosphere: Constraints imposed by laboratory experiments, *J. Geophys. Res., 100,* 17587–17602, 1995.

Kuo, B. and D. W. Forsyth, Gravity anomalies of the ridge-transform system in the South Atlantic between 31° and 34.5°S: Upwelling centers and variations in crustal thickness, *Mar. Geophys. Res., 10,* 205–232, 1988.

Lavier, L. L. and W. R. Buck, Half graben versus large-offset low-angle normal fault: Importance of keeping cool during normal faulting, *J. Geophys. Res., 107,* 10.1029/2001JB000513, 2002.

Lavier, L. L., W. R. Buck, and A. N. B. Poliakov, Factors controlling normal fault offset in an ideal brittle layer, *J. Geophys. Res., 105,* 23,431–23,442, 2000.

Lin, J. and E. M. Parmentier, A finite amplitude necking model of rifting in brittle lithosphere, *J. Geophys. Res., 95,* 4909–4923, 1990.

Lin, J. and E. M. Parmentier, Mechanisms of lithospheric extension at mid-ocean ridges, *Geophys. J. Int., 96,* 1–22, 1989.

Lin, J. and J. Phipps Morgan, The spreading rate dependence of three-dimensional mid-ocean ridge gravity structure, *Geophys. Res. Lett., 19,* 13–16, 1992.

Lin, J., G. M. Purdy, H. Schouten, J.-C. Sempéré, and C. Zervas, Evidence from gravity data for focused magmatic accretion along the Mid-Atlantic Ridge, *Nature, 344,* 627–632, 1990.

Lowell, R. P. and Y. Yao, Anhydrite precipitation and the extent of hydrothermal recharge zones at ocean ridge crests, *J. Geophys. Res., 107,* 2183, doi:10.1029/2001JB001289, 2002.

Macdonald, K. C., and P. J. Fox, The axial summit graben and cross-sectional shape of the East Pacific Rise as indicators of axial magma chambers and recent volcanic eruptions, *Earth Planet. Sci. Lett., 88,* 119–131, 1988.

Macdonald, K. C. and B. P. Luyendyk, Deep-tow studies of the structure of the Mid-Atlantic Ridge crest near lat 37°N, *GSA Bull., 88,* 621–636, 1977.

Mackwell, S. J., M. E. Zimmerman, and D. L. Kohlstedt, High-temperature deformation of dry diabase with application to tectonics on Venus, *J. Geophys. Res., 103,* 975–984, 1998.

Malinverno, A. and P. A. Cowie, Normal faulting and the topographic roughness of mid-ocean ridge flanks, *J. Geophys. Res., 98,* 17,921–17,939, 1993.

Montési, L. G. J., and M. Zuber, A unified description of localization for application to large-scale tectonics, *J. Geophys. Res., 107*, 10.1029/2001JB000465, 2002.

Montési, L. G. J., and M. Zuber, Clues to the lithospheric structure of Mars from wrinkle ridge sets and localization instability, *J. Geophys. Res., 108*, 10.1029/2002JE001974, 2003.

Neumann, G. A., and M. T. Zuber, A continuum approach to the development of normal faults, in *Proc. 35th US Symposium on Rock Mechanics*, edited by J. Daemen and R. Schultz, pp. 191–198, Balkema, Lake Tahoe, Nevada, 1995.

Phipps Morgan, J. and Y. J. Chen, The genesis of oceanic crust: Magma injection, hydrothermal circulation, and crustal flow, *J. Geophys. Res., 98*, 6283–6297, 1993.

Phipps Morgan, J. and D. W. Forsyth, Three-dimensional flow and temperature perturbations due to a transform offset: Effects on oceanic crust and upper mantle structure, *J. Geophys. Res., 93*, 2955–2966, 1988.

Phipps Morgan, J., E. M. Parmentier, and J. Lin, Mechanisms for the origin of mid-ocean ridge axial topography: Implications for the thermal and mechanical structure of accreting plate boundaries, *J. Geophys. Res., 92*, 12,823–12,836, 1987.

Poliakov, A. N. B. and W. R. Buck, Mechanics of strectching elastic-plastic-viscous layers: Applications to slow-spreading mid-ocean ridges, in *Faulting and Magmatism at Mid-Ocean Ridges*, edited by Buck, W. R. and P. T. Delaney, and J. A. Karson, and Y. Lagabrielle, AGU Geophys. Monograph, *106*, 305–323, 1998.

Rabain, A., M. Cannat, J. Escartín, G. Pouliquen, C. Deplus, and C. Rommevaux-Jestin, Focused volcanism and growth of a slow-spreading segment (Mid-Atlantic Ridge, 35ºN), *Earth Planet. Sci. Lett., 185*, 211–224, 2001.

Ross, K. and D. Elthon, Cumulus and postcumulus crystallization in the oceanic crust: Major- and minor-element geochemistry of Leg 153 gabbroic rocks, in *Proc. Ocean Drill. Program, Sci. Res.*, edited by Karson, J. A., M. Cannat, D. J. Miller, and D. Elthon, *153*, pp. 333–350, College Station, TX, 1997.

Rubin, A. M., A comparison of rift zone tectonics in Iceland and Hawaii, *Bull. Volcanol., 52*, 302–319, 1990.

Ruina, A. L., Slip instability and state variable friction laws, *J. Geophys. Res., 88*, 10,359–10,370, 1983.

Ryan, M. P., Neutral buoyancy and the mechanical evolution of magmatic systems, in *Magmatic Processes: Physiochemical Principles*, edited by B. O. Mysen, pp. 259–288, Geochem Soc., University Park, Pa., 1987.

Scholz, C., The Mechanics of Earthquakes and Faulting, Cambridge University Press, New York, 1990.

Sempéré, J.-C., J. Lin, H. S. Brown, H. Schouten, and G. M. Purdy, Segmentation and morphotectonic variations along a slow-spreading center: The Mid-Atlantic ridge (24º00'N–30º40'N), *Mar. Geophys. Res., 15*, 153–200, 1993.

Shaw, P. R., Ridge segmentation, faulting and crustal thickness in the Atlantic Ocean, *Nature, 358*, 490–493, 1992.

Shaw, P. R. and J. Lin, Causes and consequences of variations in faulting style at the Mid-Atlantic Ridge, *J. Geophys. Res., 98*, 21,839–21,851, 1993.

Shaw, W. J. and J. Lin, Models of ocean ridge lithospheric deformation: Dependence on crustal thickness, spreading rate, and segmentation, *J. Geophys. Res., 101*, 17,977–17,993, 1996.

Sinton, J. M. and R. S. Detrick, Mid-ocean ridge magma chambers, *J. Geophys. Res., 97*, 197–216, 1992.

Sleep, N. H., Hydrothermal circulation, anhydrite precipitation, and thermal structure at ridge axes, *J. Geophys. Res., 96*, 2375–2387, 1991.

Sleep, N. H. and B. R. Rosendahl, Topography and tectonics of mid-ocean ridge axes, *J. Geophys. Res., 84*, 6831–6839, 1979.

Smith, D. K., M. Tolstoy, C. G. Fox, D. R. Bohnenstiehl, H. Matsumoto, M.J. Fowler, Hydroacoustic monitoring of seismicity at the slow-spreading Mid-Atlantic Ridge, *Geophys. Res. Lett., 29*, doi:10:1029/2001GL013912, 2002.

Stein, C. A., and S. Stein, A model for the global variation in oceanic depth and heat flow with lithospheric age, *Nature, 359*, 123–129, 1992.

Tolstoy, M., A.J. Harding, and J.A. Orcutt, Crustal thickness on the Mid-Atlantic Ridge: Bull's-eye gravity anomalies and focused accretion, *Science, 262*, 726–729, 1993.

Wolfe, C. J., G. M. Purdy, D. R. Toomey, and S. C. Solomon, Microearthquake characteristics and crustal velocity structure at 29ºN on the Mid-Atlantic Ridge: The architecture of a slow spreading segment, *J. Geophys. Res., 100*, 24,449–24,472, 1995.

M. D. Behn, Department of Terrestial Magnetism, Carnegie Institution of Washington, 5241 Broad Branch Road NW, Washington, DC 20015. (behn@dtm.ciw.edu)

J. Lin, Department of Geology and Geophysics, Woods Hole Oceanographic Institution, Woods Hole, MA 02543. (jlin@whoi.edu)

M. T. Zuber, Department of Earth Atmospheric and Planetary Sciences, Massachusetts Institute of Technology, Cambridge, MA 02139. (zuber@mit.edu)

Experimental Constraints on Thermal Cracking of Peridotite at Oceanic Spreading Centers

Brian deMartin

MIT/WHOI Joint Program in Oceanography, Marine Geology and Geophysics, Woods Hole, MA

Greg Hirth

Woods Hole Oceanographic Institution, Department of Geology and Geophysics, Woods Hole, MA

Brian Evans

Massachusetts Institute of Technology, Department of Earth, Atmospheric, and Planetary Sciences, Cambridge, MA

We couple two-dimensional micromechanical models for thermal cracking with fracture mechanics data for olivine to constrain the conditions where diffuse fluid flow could occur in the oceanic lithosphere. In addition, we ran controlled cooling rate experiments on hot-pressed olivine aggregates to test the micromechanical models of thermal cracking in peridotite and study the evolution of permeability due to thermal cracking. In our experiments, impermeable olivine aggregates, formed at elevated temperatures and pressures, are subsequently cooled at constant rates. *In situ* permeability measurements, made as the aggregate cooled, allow us to determine when an interconnected microcrack network develops. By varying grain size and confining pressure during the experiments, we were able to either enhance or suppress thermal cracking within the olivine samples. When the results of our experiments are coupled with micromechanical models, we estimate a polycrystalline olivine fracture toughness of approximately 0.6 MPa m$^{1/2}$. We scaled the results of our experiments to the Earth using models that account for the influence of grain size, cooling rate, and confining pressure on the onset of thermal cracking in olivine aggregates. The depth extent of thermal cracking is estimated by coupling micromechanical models of stress intensity resulting from anisotropic thermal contraction with thermal models for upwelling mantle at oceanic spreading centers. Our analysis indicates that thermal cracking of peridotite is likely at depths less than 4 to 6 km beneath the seafloor. These predictions agree well with the depth of a transition from serpentinized to unaltered peridotite in the oceanic lithosphere determined from seismic observations.

Mid-Ocean Ridges: Hydrothermal Interactions Between the Lithosphere and Oceans
Geophysical Monograph Series 148
Copyright 2004 by the American Geophysical Union
10.1029/148GM07

1. INTRODUCTION

Fluid flow at mid-ocean ridges cools the oceanic lithosphere [*Sclater et al.*, 1980], drives the biological and chemical evolution of hydrothermal systems [*Murton et al.*, 1999], and controls the rheology and tectonics of the lithosphere [*Francis*, 1981]. Fluids present in the mantle alter ultramafic rocks to serpentine, which can strongly influence the strength of the oceanic lithosphere [e.g., *Escartín et al.*, 1997a; *Moore et al.*, 1997]. Faults provide conduits for focused flow at oceanic spreading centers and likely provide the high-permeability channels necessary to sustain high-temperature hydrothermal vents. By contrast, seismic [*Canales et al.*, 2000] and ODP [*Cannat et al.*, 1995] studies on rotated footwall blocks of detachment faults at slow spreading ridges provide evidence for distributed serpentinization and, therefore, fluid flow in the oceanic lithosphere. In this case, diffuse flow can be accommodated through a zone of thermally fractured rock producing distributed serpentinization of the lithosphere at slow spreading ridges.

Thermally induced microcracks form due to the accumulation of residual stresses at the grain-scale. Pioneering work on thermally cycled steel [*Boas and Honeycombe*, 1947], ceramics [*Kuszyk and Bradt*, 1973], and crystalline rocks [*Simmons and Richter*, 1976] demonstrated the influence of thermal expansion anisotropy and mismatch on the development of residual stresses and the formation of microcracks. The magnitude of residual stresses is governed by the thermoelastic properties (e.g., thermal expansivity and elastic compressibility) of the minerals comprising the aggregate and the magnitude of the temperature and pressure changes in the elastic regime. For conditions at oceanic spreading centers, thermal expansion anisotropy and mismatch between mineral grains generate larger residual stresses than those owing to differences in elastic expansion during depressurization. Therefore, thermal effects control the accumulation of internal stresses in these environments. Because olivine is the most abundant phase in the upper mantle, we hypothesize that anisotropic contraction of cooling olivine grains controls the accumulation of residual stresses and, ultimately, the formation of microcracks in the oceanic lithosphere. We illustrate an example of serpentinization occurring predominantly along grain boundary microcracks in a natural peridotite in Figure 1.

To help understand the role of thermal cracking in allowing the onset of diffuse fluid flow in the crust and upper mantle, we ran experiments on dense aggregates of hot-pressed olivine. Experiments on continental crustal rocks, e.g. granite, [*Bauer and Johnson*, 1979] and quartzite [*Siddiqi*, 1997], demonstrate that thermally induced cracking can create an interconnected permeability network. Micromechanical models

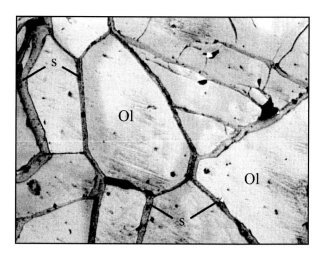

Figure 1. Reflected light micrograph of Balsam Gap dunite. Veins of lizardite and chrysotile serpentine (s) are distributed along grain boundaries between olivine (Ol) grains and along intergranular cracks, forming an interconnected network. Image provided by J. Escartín.

developed by studying the effects of thermal cracking on natural rocks provide important constraints on the mechanisms controlling microfracture [e.g., *Fredrich and Wong*, 1986; *Evans and Clarke*, 1980]. The stress intensity required for microcrack propagation can be estimated by coupling experimental observations with micromechanical models for thermal cracking [e.g., *Fredrich and Wong*, 1986]. In the following sections, we (1) review micromechanical models developed to constrain the growth of thermally induced fractures in polycrystalline, single-phase rocks, (2) describe experiments designed to constrain the thermal cracking process in olivine aggregates, and (3) couple the micromechanical models with results from thermal models of oceanic spreading centers to constrain conditions where thermal cracking occurs in the oceanic lithosphere.

2. MICROMECHANICAL MODELS

Theoretical studies of grain boundary cracking due to thermal expansion anisotropy have often focused on two-dimensional, elastically isotropic grain configurations [*Carlson et al.*, 1990; *Fredrich and Wong*, 1986; *Clarke*, 1980; *Evans*, 1978;]. In these models, stress singularities are predicted where three or more grains meet with identical, isotropic elastic moduli and unequal thermal expansion coefficients. Because of these stress singularities, grain boundary flaws present at grain boundary junctions are likely sites for microcrack nucleation. Using a closed form solution, *Evans* [1978] calculated the stress intensity factor along a grain boundary to estimate the critical flaw size required for growth of grain boundary microcracks during cooling. The predicted flaw size agreed well with data for magnesium titanate [*Kuszyk and Bradt*, 1973] if the grain boundary defects were about one tenth to three tenths the grain size. *Fredrich and Wong* [1986] used a similar model to calculate temperature changes where thermal cracks propagate in continental crustal rocks. Their calculations correlate well with stereological observations of grain boundary and intragranular crack densities in thermally cycled Westerly granite. *Carlson et al.* [1990] developed two-dimensional, plane stress, numerical simulations for four hexagonal grains embedded in an infinite, isotropic, elastic medium to constrain when grain boundary cracking would occur in a polyphase aggregate. The primary goal of *Carlson et al.* [1990] was to predict microcracking events as a function of pressure and temperature; their numerical simulations compare well with acoustic emission data on thermally cycled Westerly granite [*Wang et al.*, 1989].

The role of three-dimensional grain configurations in microcrack nucleation and development was analyzed by *Ghahremani and Hutchinson* [1990]. These authors compared the stress distribution near vertices, where four grains meet at a point, to the stress distribution around junctions, where three grains meet along a line. Qualitatively, the results of three-dimensional and two-dimensional models agree. When thermal expansion anisotropy is appreciable, both vertices and junctions are likely sites for microcrack nucleation. Quantitatively, vertices concentrate stress more than junctions, indicating that two-dimensional models underestimate stress intensities along grain boundaries. Nonetheless, the theoretical results of *Evans* [1978], *Fredrich and Wong* [1986], and *Carlson et al.* [1990] reproduce many important aspects of laboratory experiments on thermally cycled materials (e.g., critical flaw size, microcrack density, and microcracking rate as function of pressure and temperature). In this section, we discuss the general implications of the thermal cracking models. Particular emphasis is given to the *Fredrich and Wong* [1986] model, which presents a closed form solution to thermally induced stresses along a grain boundary, and therefore, provides a method to readily estimate the stress intensity factor in our cooling olivine aggregates.

The micromechanical models illustrate how the onset of thermal cracking depends on cooling rate, grain size, and confining pressure. At high temperatures, viscous creep processes dissipate residual stresses in polycrystalline aggregates. As a material cools, the rate of stress dissipation diminishes. At some temperature, which we refer to as the visco-elastic transition temperature (T'), creep processes are not rapid enough to effectively relax the thermally induced stresses. For faster cooling rates, T' is higher, and the temperature over which an aggregate accumulates residual stresses increases. Grain size affects T' if viscous relaxation

occurs via diffusion; larger grain sizes inhibit diffusion-controlled viscous processes and result in a higher T'. Grain size also controls the average flaw size in an aggregate [*Brace et al.*, 1972]. Because stress concentrations are greater at the tips of longer, narrower flaws, cracking is enhanced for larger grain sizes. Confining pressure decreases the stress concentration at the flaw tips. Thus, higher confining pressures suppress thermal cracking.

To predict the conditions where microcracks form in cooling aggregates, it is first necessary to quantify T'. Anisotropic thermal contraction in single-phase, polycrystalline aggregates generates tensile and compressive stresses along grain boundaries. At elevated temperatures, rapid diffusion facilitates the relaxation of residual stresses [e.g., *Evans and Clarke*, 1980]. Relaxation occurs via diffusive transport of material between adjacent grain faces, shown schematically in Figure 2a. This process is limited by the rate at which atoms are removed from boundaries under compression and deposited on boundaries in tension. If the relaxation rate is sufficiently rapid, atom removal and deposition occurs evenly along the grain boundary, resulting in a parabolic, steady-state stress distribution (Figure 2b). Positive stresses are tensional. Grain boundary stresses are at their minimum and maximum values where there is no net flux of atoms (e.g., $d\sigma/dt=0$). For the model shown in Figure 2a, there is no net flux at the grain junction (J) and the center of the grain face (O). Across the grain junction (A), the net flux of atoms is greatest, and stress is equal to half the mean stress along the boundary. The mean stress ($\langle\sigma\rangle$) in a cooling aggregate is a function of absolute temperature (T), grain size (L), effective grain boundary width (δ_b), diffusional displacement rate of the boundary ($\dot{\delta}_d$), Boltzmann's constant k_b, atomic volume of the diffusing species (Ω), and diffusion coefficient (D_b) [*Evans and Clarke*, 1980]:

$$\langle\sigma\rangle = \frac{k_b T L^2 \dot{\delta}_d}{6\Omega D_b \delta_b} \qquad (1)$$

To quantify T', the time derivative of stress along the boundary is related to the displacement rate of the grain boundary, $\dot{\delta}_e - \dot{\delta}_d$, where $\dot{\delta}_e$ is the elastic displacement rate of the grain boundary and $\dot{\delta}_d$ is the displacement rate induced by the diffusive transport of atoms along the grain boundary. At temperatures above T', $\dot{\delta}_d$ is much greater than $\dot{\delta}_e$, allowing viscous relaxation of residual stresses. At temperatures below T', $\dot{\delta}_d$ is much smaller than $\dot{\delta}_e$, and residual stresses accumulate rapidly.

Using expressions they derived for $\langle\sigma\rangle$, $\dot{\delta}_d$, and $\dot{\delta}_e$, *Evans and Clarke* [1980] formulated the following differential equation for the evolution of $\langle\sigma\rangle$ along a grain boundary in an aggregate cooling at a constant rate:

A

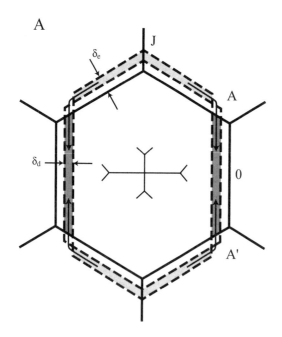

B

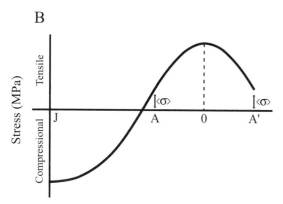

Figure 2. Schematic illustration and calculation of grain boundary stresses. (a) For an anisotropic grain cooling at high temperature, grain boundaries undergo an elastic displacement δ_e, owing to thermal contraction, and a diffusional displacement δ_d, owing to the diffusional transport of atoms. Shaded areas indicate the zones of mass transport. Arrows in the center of the grain illustrate the anisotropy in thermal expansion. (b) Plot of stress versus distance showing the superposition of elastic and diffusional stresses along grain boundaries. Compressional stresses along the grain boundary JA, lead to a net flux of atoms to the grain boundary AA'. Modified from *Evans and Clarke* [1980].

$$\frac{d\langle\sigma\rangle}{dT} = \frac{12\Omega\dot{\delta}_b D_0 E}{\sqrt{3}k_b L^3 \dot{T}}\frac{e^{-Q/RT}}{T}\langle\sigma\rangle - \frac{\beta E \Delta\alpha}{(1+\upsilon)}, \quad (2)$$

where $\dot{T}$ is cooling rate, R is the universal gas constant, Q is the activation enthalpy for grain-boundary diffusion, D_o is pre-exponential coefficient for grain-boundary diffusion of

the rate-limiting species, E is Young's modulus, υ is Poisson's ratio, and β is a coefficient that depends on the orientations of the adjacent grains. Thermal expansion anisotropy is characterized by $\Delta\alpha$, the deviation of the maximum thermal expansion coefficient from the mean value of the thermal expansion coefficients. Variable names and values are listed in Table 1. The first term on the right side of equation (2) accounts for diffusional relaxation of grain boundary stresses, while the second term describes the elastic contribution to the grain boundary displacement rate during cooling. A numerical solution to equation (2) is shown in Figure 3. T' is defined as the temperature obtained by extrapolating the elastic (linear) portion of the curve to a zero stress condition.

An approximate, analytic series expansion solution for T' in equation (2) is [*Evans and Clarke*, 1980]

$$T' \approx \frac{Q/R}{\ln\left[\dfrac{12\Omega\delta_b D_0 E}{\sqrt{3}nk_b L^3 \dot{T}}\right]}, \quad (3)$$

where n is a fitting parameter. For a given cooling rate and grain size, we determined T' from (2) and then solved (3) for n. Using a least-squares method, we calculated n for an olivine aggregate at a range of cooling rates and grain sizes appropriate for laboratory and natural conditions. With $n = 23$, equation (3) agrees with equation (2) within $\pm5°C$ for cooling rates

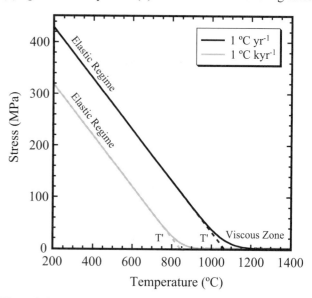

Figure 3. Stress accumulation in a cooling olivine aggregate for two different cooling rates calculated using equation (2). The visco-elastic transition temperature (T') is defined by extrapolating the elastic (linear) portion of the curve to zero stress. The grain size used in the calculation was 1 mm.

Table 1. Variable Symbols and Values Used

Variable	Variable Name	Value	Units	Reference
a	Flaw Size		m	
A	Cross Sectional Area Sample		m^2	
B_d	Storage Capacity Downstream Reservoir		$m^3\,Pa^{-1}$	
B_u	Storage Capacity Upstream Reservoir		$m^3\,Pa^{-1}$	
D_o	Grain Boundary Diffusion Coefficient	$1.5\ 10^{-08}\ \delta_b^{-1}$	$m\,s^{-2}$	[*Hirth and Kohlstedt*, 1995]
E	Young's Modulus	$197\ 10^9$	Pa	[*Hirth and Kohlstedt*, 1995]
h	Sample Height		m	
k	Permeability		m^2	
k_b	Boltzmann's Constant	$1.38\ 10^{-23}$	$J\,K^{-1}$	
K_I	Stress Intensity Factor		$Pa\,m^{1/2}$	
K_{IC}	Critical Stress Intensity Factor	0.6 ± 0.3	$Pa\,m^{1/2}$	[*this paper*]
L	1/2 Grain Size		m	
M_a	Mass in Air		kg	
M_w	Mass in Water		kg	
n	Fitting Parameter	23		[*this paper*]
P_c	Confining Pressure		Pa	
P_e	Effective Pressure		Pa	
P_f	Pore Fluid Pressure		Pa	
Q	Activation Enthalpy for Grain Boundary Diffusion	$3.75\ 10^5$	$J\,mol^{-1}$	[*Hirth and Kohlstedt*, 1995]
R	Universal Gas Constant	8.314	$J\,mol^{-1}\,kg^{-1}$	
τ	Period		s	
T	Temperature		K or °C	
$\dot{T}$	Cooling Rate		$K\,s^{-1}$	
T'	Visco-elastic Transition Temperature		K or °C	
x	Distance		m	
β	Boundary Angle	$\pi/6$	radians	
β_c	Storage Capacity Per Unit Volume		Pa^{-1}	
$\Delta\alpha$	Thermal Expansion Anisotropy	$3.1\ 10^{-6}$	K^{-1}	[*Bouhifd et al.*, 1996]
γ	Dimensionless Parameter			
δ_b	Effective Grain Boundary Width	$1.5\ 10^{-08}\ D_o^{-1}$	m	[*Hirth and Kohlstedt*, 1995]
$\dot{\delta}_d$	Diffusive Displacement Rate of Grain Boundary		$m\,s^{-1}$	
$\dot{\delta}_e$	Elastic Displacement Rate of Grain Boundary		$m\,s^{-1}$	
δP_f	Differential Pore Pressure		Pa	
η	Dynamic Viscosity of Pore Fluid		Pa s	
ρ_w	Density of Water	1000	$kg\,m^{-3}$	
ρ_s	Density of Sample		$kg\,m^{-3}$	
σ	Stress		Pa	
ψ	Dimensionless Parameter			
υ	Poisson's Ratio	0.246		[*Hirth and Kohlstedt*, 1995]
Ω	Atomic Volume	$1.23\ 10^{-29}$	m^3	[*Hirth and Kohlstedt*, 1995]

ranging from 1°C s^{-1} to 1°C kyr^{-1} and grain sizes of 1 mm to 0.1 m (Figure 4). In subsequent analyses, we use equation (3) to calculate T'.

Once an aggregate cools below T', thermally induced stresses accumulate owing to the anisotropy of thermal expansion of the mineral grains; the magnitude of the stress increase is proportional to the amount of cooling. Using a square inclusion model (Figure 5), *Fredrich and Wong* [1986] derive an expression for the normal stress acting on a two-dimensional grain boundary due to anisotropic thermal expansion,

$$\sigma_{yy}(x) = \frac{E\Delta\alpha(T'-T)}{2\pi(1-\upsilon^2)}\left(\frac{4L^2}{4L^2+(2L-x)^2}\right.$$
$$-\frac{4L^2}{4L^2+x^2}+Ln\left[\frac{2L-x}{x}\right] \qquad (4)$$
$$\left.-\frac{1}{2}Ln\left[\frac{4L^2+(2L-x)^2}{4L^2+x^2}\right]\right)$$

where x represents the distance from an arbitrary corner. Notice that the magnitude of σ_{yy} is proportional to $\Delta T = (T'-T)$, where T is temperature of the cooling aggregate. The stress along the grain boundary is asymmetric; one half of the boundary is under compression while the other half is under tension [*Fredrich and Wong*, 1986]. The stress field is singular at the grain corners. In reality, the singularity must be relaxed by inelastic deformation near the corner.

Microcracks in brittle materials nucleate along grain boundaries in response to stress concentrations caused by flaws, such as small cavities or precipitates. Observations in natural [e.g., *Sprunt and Brace*, 1974] and hot-pressed laboratory samples [e.g., *Olgaard and Evans*, 1988] reveal ubiquitous grain boundary flaws. Furthermore, mechanical observations indicate that the length of critical flaws that control deformation often scale with grain size [*Lawn*, 1993]. Thermally induced intergranular cracking occurs when the tensile stress intensity factor (K_I) at the tip of a grain boundary flaw exceeds the critical stress intensity (K_{IC}). The critical stress intensity, also called the fracture toughness, is a material property and is independent of cooling rate, grain size and confining pressure.

The stress intensity at the crack-tip depends on the geometry of the flaw, the external loading on the boundary, and the magnitude of the residual stresses [*Lawn*, 1993]. The magnitude of stress intensity at the crack tip is a superposition of the stress intensities produced by the residual stress along the boundary and the confining pressure. As residual stress along the boundary increases during cooling, stress intensity at the crack tip rises, promoting fracture. Conversely, as confining pressure increases, stress intensity at the crack tip decreases.

For a grain with flaw size a under confining pressure (P_c), the tensile stress intensity factor is given by [*Fredrich and Wong*, 1986],

$$K_{IC} = \sqrt{\frac{2}{\pi a}}\int_0^a \frac{\sigma_{yy}(x)\sqrt{x}}{\sqrt{a-x}}dx - P_c\sqrt{\pi a}. \qquad (5)$$

Intergranular crack propagation is, therefore, most likely for large changes in temperature and low confining pressures.

Figure 6 shows the dependence of the stress intensity factor on flaw size, temperature, and confining pressure for olivine aggregates calculated from equation (5) with a grain size of 1 mm. Crack growth occurs when $K_I > K_{IC}$. For K_{IC} in the range of 0.2 to 3.0 MPa m$^{1/2}$, typical values for crustal rocks [*Atkinson and Meredith*, 1987], thermal cracks can grow over a large range of flaw sizes. For all changes in temperature and pressure, the stress intensity decreases after reaching a peak value at a certain flaw size (Figure 6), indicating that grain boundary cracking induced by thermal expansion anisotropy is stable [*Fredrich and Wong*, 1986].

The confining pressure necessary to suppress cracking in a cooling rock can be calculated from equation (5) given a value of K_{IC}. Such a calculation can be used to evaluate the conditions where thermal fractures, and the concomitant development of a permeable crack network, may promote diffuse fluid flow in the lithosphere. Although olivine is the most abundant mineral in the upper mantle, relatively few fracture mechanics measurements have been conducted on olivine to constrain K_{IC}. *Atkinson* [1984] estimated K_{IC} for

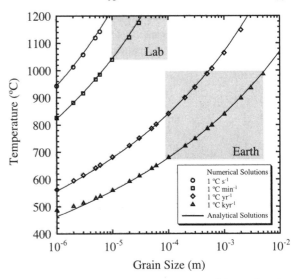

Figure 4. Visco-elastic transition temperature (T') as a function of grain size. Results from the numerical solution of equation (2), denoted by symbols, are compared to the series solution of equation (3), denoted by the solid lines. Regions indicating typical laboratory and Earth conditions are shaded.

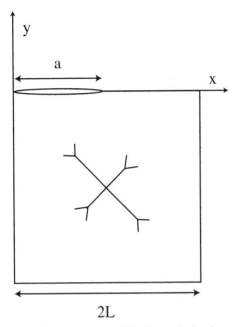

Figure 5. Coordinate system used for the analysis of grain boundary cracking due to thermal expansion anisotropy. Arrows inside the square grain indicate the orientation of the principal thermal expansion coefficients. Cracking occurs along the grain boundary due to the presence of a flaw with length *a*. Modified from *Fredrich and Wong* [1986].

olivine from *Swain and Atkinson's* [1978] fracture surface energy measurements on the [010] and [001] cleavage planes in single-crystal olivine. Using the Vickers indentation technique, *Swain and Atkinson* [1978] measured fracture surface energies of the preferred cleavage plane [010] and the [001] plane of 0.98 J m^{-2} and 1.26 J m^{-2}, corresponding to K_{IC} equal to 0.59 MPa m$^{1/2}$ and 0.73 MPa m$^{1/2}$, respectively [*Atkinson*, 1984]. Using the lower value of K_{IC} = 0.59 MPa m$^{1/2}$, we iteratively solved equation (5) to evaluate the influence of pressure, temperature change, and grain size on thermal cracking. The results of our analysis, illustrated in Figure 7, demonstrate the trade-offs among confining pressure, temperature change, and grain size on the initiation of thermal cracking. Based on the calculations illustrated in Figure 4, ΔT under geologic conditions at an oceanic spreading center may reach 600–900°C. Therefore, the results shown in Figure 7 indicate that as material rises and cools from high pressure and temperature beneath an oceanic ridge it will thermally crack when it reaches shallow depths.

There are at least two caveats for the application of the micromechanical models shown in Figure 7 to the analysis of thermal cracking beneath oceanic ridges. First, we used a two-dimensional model to calculate the stress intensity along grain boundaries. As discussed above, two-dimensional mod-

els predict smaller stress accumulations along grain boundaries than three-dimensional models, and therefore, may underestimate the actual stress intensity. Second, we used the fracture surface energy along the preferred cleavage plane of olivine to constrain K_{IC}. In an olivine aggregate, microcracks will form at flaws along grain boundaries, in addition to cleavage planes. The critical stress intensity factor for grain boundaries may be smaller than the K_{IC} for cleavage planes. To address these issues, we have run a series of thermal cracking experiments on cooling olivine aggregates. The conditions where microcracks form an interconnected network were determined by measuring permeability as the samples were cooled and depressurized. Coupling the results of the permeability measurements with the micromechanical models, we have a novel method for estimating K_{IC}, which can help us assess the uncertainty introduced by the limitations discussed above.

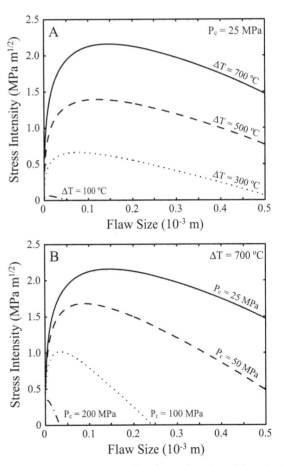

Figure 6. Stress intensity at the flaw tip as a function of flaw size for a grain size of 1 mm. (a) Effect of temperature change ($\Delta T = T'$-T). (b) Effect of confining pressure. Larger temperature changes or lower confining pressures increase the stress intensity values at flaw tips.

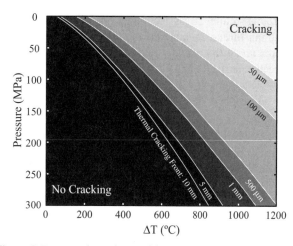

Figure 7. Pressure dependence of thermal cracking as a function of grain size and temperature change for $K_{IC} = 0.59$ MPa m$^{1/2}$. High confining pressures and small temperature changes suppress thermally induced grain boundary cracks. Larger grain sizes are more likely to crack than smaller grain sizes.

3. EXPERIMENTAL TECHNIQUES

The micromechanical models illustrate how we can trade-off cooling rate and grain size to investigate thermal cracking at laboratory time scales. At natural cooling rates of 1°C yr^{-1} to 1°C kyr^{-1}, T' for peridotite with a grain size of 1 mm ranges from 1070°C to 840°C, respectively (Figure 4). By contrast, we produced olivine aggregates with average grain sizes of 18 to 35 µm. At experimental cooling rates of 1°C s^{-1} to 1°C min^{-1}, T' is 1200°C. Thus, residual stresses in cooling mantle peridotite accumulate over a smaller temperature range than the laboratory aggregates. However, as discussed in the previous section, coarser-grained aggregates have larger flaws that generate greater residual stress concentrations at flaw tips. Therefore, while natural peridotites have lower T', the models indicate that by trading-off cooling rate and grain size similar values of K_I are produced under laboratory and natural conditions.

The experiments were designed to determine when thermal cracking creates an interconnected microfracture network in an originally uncracked peridotite. Samples were first hot-pressed to create impermeable, relatively uncracked olivine aggregates. After the hot-press, samples were cooled and depressurized at different rates to evaluate the influence of grain size, cooling rate, confining pressure, and temperature change on the development of a microfracture network. The formation of an interconnected fracture network was detected by measuring the permeability of the sample during cooling and depressurization. The experimental conditions at which the permeability network forms can then be coupled with the micromechanical models to estimate a critical stress intensity factor.

3.1. Sample Preparation and Experimental Procedure

Samples were prepared by hot, isostatic pressing of powders of San Carlos olivine (Fo$_{91}$). The powders, with grain sizes of 30–38µm and 38–63 µm, were first cold-pressed into Ni capsules (26 mm long, 11.7 mm O.D., 11.0 mm I.D.) with a uniaxial stress of approximately 100 MPa. Porous alumina discs (3 mm thick, 20% porosity) confined the olivine powder at both ends of the Ni capsule. After the cold-press, samples were dried in an evacuated oven for at least 12 hours at 75°C. Then they were inserted into an iron sleeve (Figure 8) and hot pressed at 1200°C and 300 MPa for 4 to 8 hours in an internally heated, argon-gas-medium apparatus [*Paterson*, 1970]. During the hot-press, thermally activated processes dissipate stresses, and produce a dense, low-porosity olivine aggregate. As with other high-temperature, high-pressure experiments on olivine [e.g., *Hirth and Kohlstedt*, 1995; *Karato*, 1989], we utilized a Ni cylinder to buffer oxygen fugacity at Ni-NiO.

Immediately following the hot-press, while the sample was still at elevated temperature and pressure, we conducted a sinusoidal permeability test [*Fischer and Paterson*, 1992] to confirm that the sample was sufficiently dense to close all

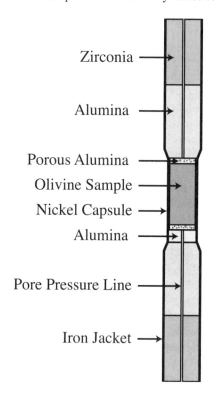

Figure 8. Experimental sample assembly. Olivine powder is first cold-pressed into the Ni capsule and porous alumina spacers are placed on both ends. The Ni capsule, containing the olivine sample, is then placed in an Fe jacket with ceramic pistons on both sides.

interconnected porosity. Once we had formed an impermeable aggregate, the sample was cooled at a constant rate between $0.1°C\ s^{-1}$ to $1°C\ s^{-1}$. During cooling, the pore pressure system was open to the atmosphere, and confining pressure was allowed to decrease with decreasing temperature. Samples were cooled to about 500°C, a temperature well below T', and a second sinusoidal permeability test was carried out. During this permeability test, the confining pressure was approximately 260 MPa and the pore pressure (P_f) was about 20 MPa. After measuring permeability at 500°C, pore pressure was dropped to atmospheric pressure, and temperature decreased to room temperature with an additional concomitant decrease in confining pressure to about 220 MPa. Permeability was then measured as a function of effective pressure ($P_e = P_c - P_f$) using the sinusoidal permeability technique.

At the end of experiments, the samples were extracted from the apparatus and their jackets dissolved with aqua regia. The samples were then ground into cylinders 10.5 mm long and 9 mm in diameter. Using a low-speed, diamond saw, 3 mm discs were cut from one end of each sample for microstructural analysis, and the remaining 7.5 mm × 9 mm cylindrical samples were jacketed with a 3 mm thick polyurethane rubber jacket and inserted into a wide-range permeameter [Bernabé, 1987]. The wide-range permeameter provided additional constraints on sample permeability as a function of effective pressure. We utilized long-period (3600 s) sinusoidal oscillations [Fischer and Paterson, 1992], as well as the transient pulse method [Brace et al., 1968] for one sample, to measure permeability. Once the samples were loaded into the wide-range permeameter, confining pressure was raised to 40 MPa. After confirming that there was no leak in confining pressure, the water pore fluid pressure was increased to 20 MPa. Twenty-four to forty-eight hours elapsed before the start of a permeability test. This hiatus allowed the system to equilibrate after the pressure adjustment. During the permeability measurements, pore pressure was held constant at 20 MPa, and confining pressure was increased incrementally from 40 MPa to 180 MPa.

3.2. Permeability Measurement Techniques

3.2.1. Sinusoidal Pore Pressure Method. The sinusoidal pore pressure technique is based on the measurement of the phase delay and attenuation of an oscillation of the pore fluid pressure as it propagates through the sample [Fischer, 1992]. A sinusoidal pressure oscillation is induced on the upstream side of the sample, and the pressure response is observed at the other end, the downstream side. The response of the downstream side is another sinusoid that is phase shifted and attenuated relative to the upstream oscillation.

The magnitudes of the phase shift and attenuation depend on the permeability (k) and the storage capacity per unit volume of the sample (β_c), as well as several experimental parameters: the height (h) and cross-sectional area (A) of the sample, the period of pore pressure oscillation (τ), the dynamic viscosity of the pore fluid (η), and the storage capacity of the downstream reservoir (B_d). Storage capacity is defined as the change in volume of pore fluid stored per unit change in pore fluid pressure ($B = \partial V/\partial P_f$). Permeability is calculated by using phase lag and attenuation measurements to numerically solve for two dimensionless parameters ψ and γ [Fischer and Paterson, 1992]. Storage capacity and permeability of the sample are then given in terms of ψ and γ by,

$$\beta_c = \frac{B_d}{hA\gamma}, \tag{6}$$

$$k = \frac{\pi\eta\beta_c}{\tau}\left(\frac{h}{\psi}\right)^2 = \frac{\pi\eta B_d h}{\tau A\psi^2\gamma}. \tag{7}$$

Phase delay and attenuation were determined using Fourier analysis. Transient effects at the initiation of the permeability test generally decayed within the first cycle, and only data from the second cycle onwards were used for processing. Eight to ten cycles were analyzed. An example of the upstream and downstream pore pressure recorded during a permeability measurement is shown in Figure 9a. To calculate phase delay and attenuation, we first removed small linear trends in the time-domain pressure data that arise from small leaks in the pore pressure system. Next, we converted data into the frequency domain, where the power and phase spectrum were calculated. After verifying that the maximum power occurred at the imposed frequency, the phase delay was calculated by subtracting the downstream phase value from the upstream phase value. Care was taken to avoid 2π offsets. The amplitude attenuation was found by setting the energy in all frequencies to zero, except that at the imposed value. The data were then inverted to the time domain, and the amplitude of the recovered signals was measured. Attenuation was defined as the ratio of the amplitude of the upstream side to the amplitude of the downstream side. The resolution of the sinusoidal pore pressure method was approximately $10^{-20}\ m^2$ during thermal cracking experiments and $5\times10^{-19}\ m^2$ the during wide-range permeameter experiments. The sensitivity during the thermal cracking experiments is higher because low-viscosity argon (as opposed to water) is used as the pore fluid.

3.2.2. Transient Pulse Method. In the transient pulse method, an approximately 0.1 MPa change of pore pressure is imposed on one side of a sample that was originally uniformly pressurized. The convention is to call this side the upstream side, regardless of the direction of fluid flow. The pore pressure is then allowed to return to equilibrium (Figure 9b). The time it takes for the differential pore pressure δP_f to return to equilibrium is a function of permeability. The decay curve of the differential pore pressure is approximately exponential and the decay time t is inversely proportional to the permeability [*Brace et al.*, 1968]

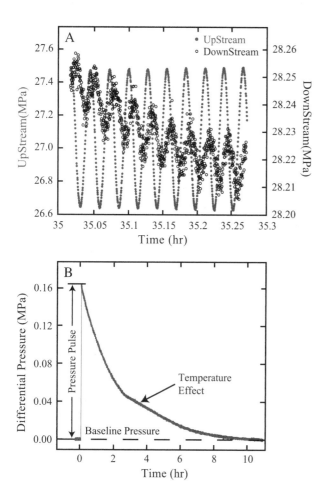

Figure 9. Unprocessed permeability data for OTC-3. (a) Sinusoidal permeability test. Upstream signal is denoted by solid, gray circles, while the downstream signal is represented by open, dark circles. The steady decrease in pore pressure on the downstream side is due to a small leak in the pore pressure system. Permeability is calculated by measuring the amplitude attenuation and phase lag between the two signals. (b) Transient pulse permeability test. The pressure difference between the upstream and downstream reservoirs (differential pressure) is plotted versus time. Permeability is calculated using the decay time of the differential pressure after a pressure pulse is generated in one reservoir. Small changes in the temperature of the laboratory generate perturbations in the differential pressure measurement.

$$\delta P_f(t) \propto \exp(-\xi t), \qquad (8)$$

where

$$\xi = \frac{Ak(B_u + B_d)}{\eta h B_u B_d}, \qquad (9)$$

and B_u is the storage capacity of the upstream reservoir. We calculated permeability by plotting the log of the differential pore pressure decay versus time. Using experimentally determined values for the apparatus storage capacities, as well as the cross sectional area and length of the sample, permeability is determined from the slope of the line. The resolution of the transient pulse method in the wide-range permeameter is approximately 5×10^{-20} m².

4. RESULTS OF THERMAL CRACKING EXPERIMENTS

We ran eight successful thermal cracking experiments designed to test the importance of cooling rate, hot press duration, initial grain size, and confining pressure on thermal cracking. The sample properties, experimental parameters, and the maximum pressure where permeability was observed for each experiment are summarized in Table 2. For all experiments, regardless of cooling rate, hot-press duration, or initial grain size the samples were impermeable after the hot-press and remained impermeable while at elevated temperatures. Permeability was observed for two coarser-grained samples as they depressurized at 20°C. The results of the permeability measurements on these samples, OTC-9, and OTC-12, are shown in Figure 10a. Several samples, OTC-1, OTC-10, OTC-11, and OTC-15, remained impermeable throughout the duration of the thermal cracking experiments and provide important constraints on K_{IC}. Sample OTC-3 was permeable at $P_e = 10$ MPa (Figure 10a). A confining pressure leak into the pore pressure system prevented us from making permeability measurements at higher pressures. We could not measure permeability during OTC-5 because there was a significant leak of pore pressure to the atmosphere.

We used a wide-range permeameter to further constrain the permeability of several samples after they were extracted from the high-temperature, gas-medium apparatus. Transient pulse measurements conducted on OTC-3 a month after the thermal cracking experiment are consistent with the *in situ* permeability measurement (Figure 10a). Long-period (1800–3600 s), sinusoidal permeability measurements were conducted on four samples (OTC-1, OTC-5, OTC-10, and OTC-11) 15–20 months after the thermal cracking experiments. While the finer-grained sample OTC-1 remained impermeable, coarser-grained samples OTC-10 and OTC-11 became permeable in the period since the thermal cracking experiment. In addition, we measured the permeability of sample OTC-5. We plot the results of the sinusoidal permeability measurements in the wide-range permeameter in Figure 10b. These results, performed months after fabrication of the samples, suggest that time is also an important variable in controlling the evolution of permeability.

Samples OTC-12 and OTC-15, run at identical conditions to OTC-3 and OTC-1, respectively, were designed to test the reproducibility of our experiments. As shown in Figure 10a, the permeabilities of OTC-12 and OTC-3 agree within a factor of two. Similarly, samples OTC-1 and OTC-15 both remained impermeable at all effective pressures.

4.1. Influence of Experimental Parameters on Permeability

The differences in the onset and magnitude of permeability in our samples reflect variations in the extent of microcracking that arise due to differences in cooling rate, hot-press duration, time between measurements, grain size, and confining pressure. Cooling rate had no discernable effect over the conditions we tested. As discussed below, the range in cooling rate was probably not large enough to significantly influence the visco-elastic transition temperature (T') for the grain sizes of our samples.

Longer hot presses inhibited the formation of an interconnected microcrack network during the thermal cracking experiments. The three coarser-grained samples with measurable permeability during the thermal cracking experiments were all hot-pressed for four hours prior to cooling and depressurization. By contrast, a coarse-grained sample hot-pressed for 8 hours (OTC-10) was impermeable during the thermal cracking experiment. The effect of hot-press duration is also constrained by OTC-11. This experiment was designed to determine if a mixed initial grain size enhanced grain growth

during hot pressing. During this experiment, the pore fluid system exhibited larger than normal noise. Thus, we extended the hot-press for an additional 2 hours to ensure that the sample was impermeable prior to cooling. After the eventual 6 hour hot-press, no permeability was observed over the entire range of effective pressures. Despite the high noise level for OTC-11, the spectral characteristics of the pore-fluid pressure data at low effective pressures indicated that the sample remained impermeable.

Measurements of permeability made after the samples were extracted from the high-temperature, gas-medium apparatus suggest that microcracking continues at ambient conditions. While OTC-10 and OTC-11 remained impermeable during the thermal cracking experiments, relatively high permeabilities for these same samples were measured in the wide-range permeameter 15 months after they were fabricated (Figure 10b). In addition, the permeability of OTC-5 measured 18 months after it was fabricated was almost an order of magnitude greater than that measured during thermal cracking experiments on samples with a similar grain size and hot press duration (compare Figure 10a and 10b).

Samples with the smallest initial powder size of 30–38µm (OTC-1 and OTC-15) remained impermeable throughout thermal cracking experiments. In contrast to the coarser-grained samples, the permeability of OTC-1 remained below the resolution of the wide-range permeameter during measurements made 20 months after the thermal cracking experiments.

During decompression of two samples (OTC-9 and OTC-12), permeability was observed at high effective pressure using the sinusoidal technique. These samples, hot-pressed under identical conditions, first exhibited measurable permeability at similar effective pressures. In the case of OTC-9, permeability was first measured after the confining pressure was lowered from 220 MPa to 170 MPa at 20°C. For OTC-12, no permeability was observed at 600°C and a confining pres-

Table 2. Summary of Thermal Cracking Experiments

Sample Number	Initial Grain Size (µm)	Hot Press Duration (hr)	Cooling Rate (K min^{-1})	Maximum P_e (MPa) Where Permeability Measured	Final Density (g cm^{-3})	Isolated Porosity (%)
OTC-1	30-38	4	25	n/a[a]	3.33	0.0
OTC-3	38-63	4	20	140[b]	3.27	1.8
OTC-5	38-63	4	6	140[b]	3.27	1.8
OTC-9	38-63	4	10	150	3.25	2.4
OTC-10	38-63	8	60	160[b]	3.24	2.7
OTC-11	mixed[c]	6	60	160[b]	3.30	0.9
OTC-12	38-63	4	20	190	3.28	1.5
OTC-15	30-38	4	25	n/a[a]	3.32	0.3

[a]Permeability never observed in OTC-1 and OTC-15.
[b]Permeability measurements in wide-range permeameter.
[c]OTC-11 had a mixed initial grain size one part 30-30 µm and 14 parts 38-63 µm.

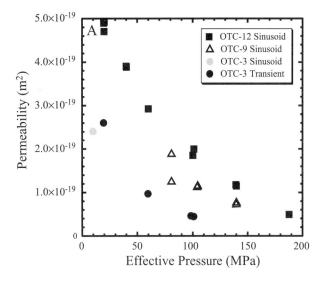

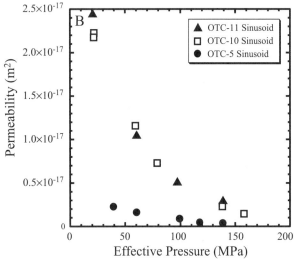

Figure 10. Summary of permeability measurements. (a) *In situ* and transient pulse permeability measurements. *In situ* permeability measurements on OTC-3, OTC-9, and OTC-12 were conducted using the sinusoidal oscillation method in the gas-medium apparatus with argon gas as the pore fluid. Transient pulse measurements on OTC-3 conducted a month after synthesis using water as the pore fluid agree with *in situ* permeability measurements at low-effective pressure. (b) Sinusoidal oscillation permeability measurements in the wide-range permeameter using water as the pore fluid 15–20 months after samples fabricated. The decrease in permeability with increasing effective pressure likely results from closure of low-aspect ratio cracks.

sure of 240 MPa. When temperature was lowered to 20°C with a concomitant decrease in confining pressure to 210 MPa the sample became permeable. The pore fluid pressure was 20 MPa during permeability measurements. Thus, permeability was first observed at effective pressures of 150 MPa

and 190 MPa, respectively, for these samples. After OTC-12 became permeable, permeability increased an order of magnitude (from 4×10^{-20} m^2 to 4×10^{-19} m^2) as the effective pressure decreased from 190 MPa to 20 MPa (Figure 10a). A similar influence of effective pressure on permeability is shown for the other permeable samples in Figure 10.

4.2. Microstructural Analysis

We used two methods to characterize the microstructure of the olivine aggregates following extraction from the gas-medium apparatus. First, we measured the density of the samples to quantify porosity. Second, we qualitatively analyzed the extent of thermal cracking and measured grain size using optical and scanning electron microscopy.

We measured the density (ρ) of thermally cracked samples in two ways. First, to estimate the total porosity we calculated density by measuring the height, radius, and mass of the cylin-

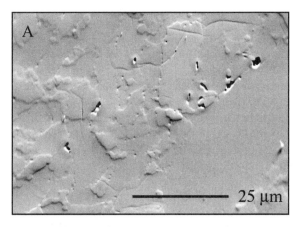

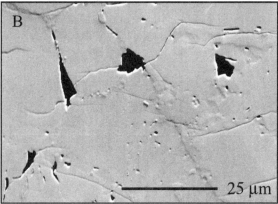

Figure 11. Orientation contrast images. The electron micrographs show two samples, cooled at similar rates, with different grain sizes. Images (1000x) reveal cracks in both samples, although cracking is less prevalent in the sample with the smaller grain size, OTC-1 (a), than the sample with the larger grain size, OTC-3 (b). Large pores, likely due to pluck-outs, are present in OTC-3.

drical samples. Total porosity was calculated assuming the difference in density between the samples and San Carlos olivine (ρ = 3330 kg m-3) was only attributable to the presence of air filled pores ($\rho \approx 0$). These rough density measurements yield a total porosity estimate of approximately 1% for the fine-grained samples and 3% for the coarse-grained samples. Error in measuring the height and radius of the cylindrical samples yields about 1% uncertainty in porosity. Second, to calculate the amount of isolated porosity we used Archimedes' method to measure the density of the samples (ρ_s),

$$\rho_s = \frac{M_a \rho_w}{(M_a - M_w)}, \qquad (10)$$

where M_a is the mass of the sample in air, M_w is the mass of the sample in water, and ρ_w is the density of water. Values for isolated porosity, listed in Table 2, range from 0.0% to 2.7% and are apparently independent of hot press duration, and cooling rate. Both methods used to determine density indicate that the samples fabricated using the finer grained powders have a higher density and lower porosity than coarser grained samples.

SEM micrographs reveal microcracks in all samples. However, the fracture density is greatest in the coarser-grained samples with measurable permeability. Orientation contrast images (1000x) of two samples, cooled at similar rates with different initial grain sizes, are shown in Figure 11. Both grain boundary and intragranular microcracks are more prevalent in the coarser grained sample (OTC-3). As discussed in the micromechanical model section, the observation of more prevalent cracking in the coarser grained sample is consistent with the influence of grain size on fracture toughness.

Pores on the order of the grain size are observed in the coarse grained sample (Figure 11b). The porosity measurements for the coarse grained sample indicate that these large pores are surface features produced by plucking out weakly attached grains during polishing and were not present during the thermal cracking experiments. The greater number of pluck-outs in this sample is consistent with the observation of more widespread microcracking in the coarse grained aggregates. The small pores (cross-sectional areas < 1 μm^2), observed in both the fine grained and coarse grained samples, were likely present and isolated following the hot-press step and could have been sites for crack initiation during cooling.

The average grain size of the samples shown in Figure 11 was determined by using electron backscatter diffraction (EBSD) to create grain orientation maps. We set the SEM to analyze electron backscatter patterns in 2 μm steps across a 90×10^3 mm^2 rectangular area of the sample surface. Reliable orientations were obtained for approximately 65% of the ana-

lyzed points. To determine the grain size, we first removed poorly determined grain orientations and then extrapolated areas of constant grain orientation to fill space between the well-analyzed areas. Individual grains were identified as regions of constant grain orientations, misoriented more than 10° with their neighbors. These analyses yield average grain areas of 110 μm^2 for the fine-grained sample and 380 μm^2 for the coarse-grained sample hot-pressed four hours. The grain size of these samples (18 μm and 33 μm) was calculated by multiplying the equivalent circle diameters of the average grain areas by 1.5 [*Underwood*, 1970].

5. DISCUSSION OF EXPERIMENTAL RESULTS

We observe evidence for the formation of thermally induced microcracks in olivine aggregates from permeability measurements and microstructural observations. Our observations indicate that the tendency for microcracks to form an interconnected network during thermal cracking experiments is influenced by grain size, hot-press duration, and effective pressure. However, for the range of grain sizes used during the experiments, cooling rates were too rapid to decrease the visco-elastic transition temperature significantly below 1200°C (Figure 4). Thus, the temperature change driving the increase in stress within the aggregates was approximately 1180°C for all samples. This conclusion is consistent with our observation that differences in cooling rate did not influence our results.

Permeability measurements reveal that longer hot press durations inhibited the formation of microfractures during thermal cracking experiments, although this conclusion is based on only two data points. This observation is contrary to what one would expect due to the kinetics of grain growth if the porosity of all the samples was the same [e.g., *Evans et al.*, 2001]. Longer hot press duration should yield larger grain sizes, which are more likely to fracture if the flaw size scales with grain size. However, the difference in hot-press duration of our experiments was not large enough to significantly change the grain size. Using data for grain growth of olivine [*Karato*, 1989], an increase in grain size of only 2 μm would be expected for a coarser-grained sample (i.e., grain size of approximately 35 μm) annealed for 8 versus 4 hours. We hypothesize that microcracking was apparently inhibited by longer hot-press durations because flaws (e.g. pores) were more effectively removed. Owing to the decreased number of flaws, crack initiation sites are apparently decreased, inhibiting the formation of an interconnected permeability network.

The permeability of three samples (OTC-5, OTC-10, OTC-11) measured 15–20 months after they were fabricated suggests that subcritical crack growth enhances the formation of an interconnected microcrack network. One caveat to this inter-

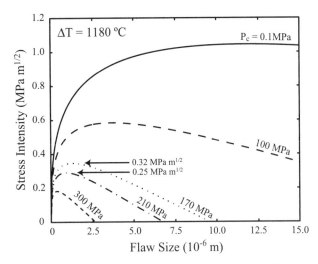

Figure 12. Stress intensity at the flaw tip as a function of flaw size calculated using equation (5) with a grain size of 33 μm and $\Delta T = 1180°C$. Permeability was first observed at 170 MPa for OTC-9 and 220 MPa for OTC-12. As described in the text, these observations yield a stress intensity necessary to induce an interconnected microcrack network of approximately 0.3 MPa m$^{1/2}$.

pretation is that the samples may be damaged during removal of the jacket, and the grinding and cutting done in preparation for the wide-range permeameter measurements. The interpretation of our microstructural observations may also be compromised by damage imparted during sample preparation. For this reason, in our analysis of the conditions where thermal cracking initiates in the oceanic lithosphere, we concentrate on the results of *in situ* permeability measurements made during the thermal cracking experiments.

5.1. Estimating K_{IC}

We use our experimental observations, in conjunction with micromechanical models to provide constraints on the critical stress intensity factor (K_{IC}) in olivine aggregates. As discussed below, our thermal cracking experiments yield three values for K_I necessary to induce an interconnected microcrack network. These values reflect differences in our experiments and provide a range of values for fracture toughness that may be applicable in the Earth.

The onset of permeability observed during two thermal cracking experiments at elevated pressure provides an estimate of K_I necessary to form an interconnected microcrack network. During these two experiments an interconnected microcrack network was first observed *in situ* after cooling the samples to room temperature at a confining pressure of 170 MPa (OTC-9) and 210 MPa (OTC-12). Values for K_I as a function of flaw size and a range of confining pressures are shown in Figure 12 for an aggregate cooled 1180°C with a

grain size of 33 μm. Maximum values of K_I for a confining pressure of 170 MPa and 210 MPa are 0.32 MPa m$^{1/2}$ and 0.25 MPa m$^{1/2}$, respectively. Averaging these values, we estimate a stress intensity of 0.29 MPa m$^{1/2}$ necessary to induce an interconnected microcrack network in a coarse-grained olivine aggregate hot-pressed 4 hours.

Samples that remained impermeable throughout the thermal cracking experiments provide additional values of K_I required to induce an interconnected microcrack network. For these samples, thermally induced stresses were insufficient to produce an interconnected microcrack network. However, using the same technique discussed above, we can estimate the maximum stress intensities attained in these samples. Samples OTC-1, with an average grain size of 18 μm, and OTC-15 were run at identical conditions using the same powders. Using equation (5) for a temperature change of 1180°C and a confining pressure of 0.1 MPa yields a maximum $K_I = 0.63$ MPa m$^{1/2}$. Similarly, a sample with a calculated grain size of 35 μm hot-pressed 8 hours (OTC-10) also remained impermeable over the course of the thermal cracking experiment. The maximum K_I calculated for this coarse-grained sample is 0.87 MPa m$^{1/2}$.

Stress intensities calculated from the coarse-grained thermal cracking experiments provide a range of estimates on K_I required to form an interconnected microcrack network. The lowest estimate is calculated from coarse-grained samples hot-pressed 4 hours. As described above, these samples may have had more crack initiation sites (e.g., flaws) owing to their larger initial grain sizes and shorter hot-press durations. Upon cooling and depressurization, the increased number of flaws enhanced the development of an interconnected microcrack network, and therefore, K_I was large enough to induce widespread cracking. Flaws are more effectively removed with longer hot-press durations, and thus, the coarse grain sample hot-pressed 8 hours yields a higher estimate of K_I necessary to form an interconnected microcrack network. However, the observation of permeability in this sample 15 months after the thermal cracking experiment implies that over time an interconnected microcrack network may form at stress intensities of approximately 0.87 MPa m$^{1/2}$.

The magnitude of stress intensity calculated from our thermal cracking experiments agrees well with previous measurements of the fracture toughness of olivine. The results of thermal cracking experiments yield three values of stress intensity that average to 0.6 MPa m$^{1/2}$, a value that corresponds well with the K_{IC} determined for the weak cleavage plane of olivine (0.59 MPa m$^{1/2}$) [*Atkinson*, 1984]. In addition, microstructural observations of some microcracks in the impermeable fine-grained sample (Figure 11a) indicate that locally K_I exceeded K_{IC}. This observation emphasizes that the permeability measurements do not directly correspond with the

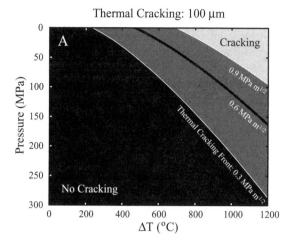

Thermal Cracking: 100 μm

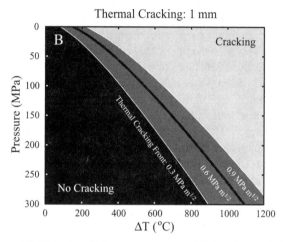

Thermal Cracking: 1 mm

Figure 13. Plot of confining pressure versus ΔT showing conditions where thermal fractures initiate for our experimentally estimated range of $K_{IC} = 0.6 \pm 0.3$ MPa m$^{1/2}$ and grain sizes of 100 μm (a) and 1 mm (b). Larger temperature changes and smaller grain sizes yield a larger pressure range for the thermal cracking front.

nucleation of individual grain boundary cracks. Rather, the permeability measurements provide a method of estimating the stress intensity necessary for the formation of an interconnected crack network. Given that K_{IC} is a material parameter (independent of pressure, temperature, cooling-rate, grain-size, etc.), the correlation of our calculated values of K_I with the olivine cleavage fracture toughness is encouraging. Using the bounds on K_I provided by the coarse-grained samples as uncertainties in our measurements, we conclude that K_{IC} is 0.6 ± 0.3 MPa m$^{1/2}$.

5.2. Modeling Thermal Cracking in the Oceanic Lithosphere

We use the constraints on K_{IC} from the thermal cracking experiments to estimate the range of conditions where ther-

mally induced microcracking occurs in the oceanic lithosphere. We first iteratively solved equation (5) to quantify the role of pressure, temperature change, and grain size on thermal cracking for the estimated range of K_{IC}. The results of our calculations, shown in Figure 13, yields a range of pressures where an interconnected microcrack network will develop for a given temperature change and grain size. Larger temperature changes and grain sizes fracture at higher pressures than smaller temperature changes and grain sizes. The uncertainty in pressure, arising from the uncertainty in K_{IC}, is greater for larger changes in temperature and smaller grain sizes. For the range of K_{IC} estimated from our experiments, thermal cracking would initiate at low confining pressures for moderate changes in temperature. For example, for $\Delta T > 400°C$ and a grain size of 1 mm, microcracks can nucleate over a range of confining pressures in the upper few kilometers of the oceanic lithosphere.

To calculate where thermal cracking may occur in slow spreading oceanic spreading system, we employed the mantle flow and temperature model of *Phipps Morgan and Forsyth* [1988] as modified by *Shaw and Lin* [1996] and *Behn et al.* [2002]. We model an oceanic ridge with a half-spreading rate of 10 mm yr^{-1} and a crustal thickness that varies from 8 km near the center of the segment to 3 km near the end of the segment [*Hooft et al.*, 2000]. Mantle temperature was set to 1320°C and the magma injection temperature at the spreading axis was set to 1150°C. Hydrothermal cooling was permitted where the lithosphere was cooler than the visco-elastic transition temperature. Presumably, in regions cooler than the visco-elastic temperature and hotter than where we predict thermal microfractures form, fluid flow will be limited to large-scale faults. In regions where thermal cracking is occurring, we would expect enhanced cooling to occur due to the greater flux of diffuse flow though the mantle. At present, the model does not incorporate these feedback processes.

The flow and temperature model implements a two step process to first calculate flow in the passively upwelling upper mantle and then determine the steady state temperature of this system. To calculate the visco-elastic transition temperature contour, the cooling rates along mantle streamlines were calculated. Cooling rates for the upwelling mantle were approximately 0.1°C ky^{-1}. Using equation (3) with a mantle grain size of 1 mm places the visco-elastic transition contour near the 800°C isotherm. The thermal cracking front can then be determined by calculating stress intensity using equation (5) along the streamline and comparing it with the critical stress intensity factor determined from experimental analysis. Because K_{IC} is a material parameter, it does not need to be scaled to natural conditions. However, the calculation does assume that the initial flaw size in the natural rock scales with grain size in the same way as the laboratory samples. We

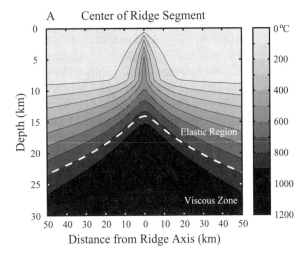

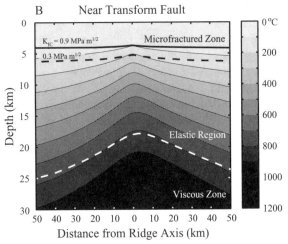

Figure 14. Thermal models for a slow spreading ridge (half-rate = 1 mm) showing where thermal cracking may occur at segment centers (a) and near transform faults (b). Cross-sections are perpendicular to the spreading axis. Thin black lines denote isotherms at 100°C intervals. The dashed white line denotes the visco-elastic transition temperature (T'). In (b), the black dashed line represents the thermal cracking front for $K_{IC} = 0.3$ MPa m$^{1/2}$, and the thick black line represents the thermal cracking front for $K_{IC} = 0.9$ MPa m$^{1/2}$. Pervasive cracking of the oceanic mantle is predicted to occur above these two lines. At the center of the ridge segment (a), thermal cracking is inhibited due to the presence of a thick (8 km) crustal layer. Near the segment ends (b) where the crust is thinner (3 km), thermal cracking is calculated to occur in mantle peridotite at depths less than 4–6 km.

show the uncertainty in our experiments by plotting the zone of cracking resulting from K_{IC} equal to 0.3 MPa m$^{1/2}$ through 0.9 MPa m$^{1/2}$.

The results of the flow and temperature model are shown in Figure 14. Each figure represents a slice though the ridge segment, perpendicular to the spreading axis. Figure 14a

shows the center of the segment, and Figure 14b shows the end of the segment, near the transform fault. Thermal fracturing of the upwelling mantle peridotite is limited to the segment end, where mantle material rises to shallow depths. At the center of the segment, where the oceanic crust is thicker, thermal cracking in the mantle peridotite is inhibited by a combination of a small decrease in temperature from the visco-elastic transition temperature and greater lithostatic pressure. Near the segment ends, where mantle peridotite rises to a depth of 3 km, lithostatic pressure is insufficient to inhibit the formation of thermal fractures owing to a larger temperature change. The thickness of the thermally cracked zone near the segment ends depends on K_{IC}. The smallest estimate of K_{IC} (0.3 MPa m$^{1/2}$) yields the thickest thermally cracked zone (6 km). Aggregates with higher K_{IC} must rise to shallower depths before cracking initiates.

Modeling results suggest that thermal cracking at the spreading axis should occur where upwelling, cooling mantle peridotite rises to a depth of 6 to 4 km at the ridge axis. If thermal cracking is pervasive, as inferred from our experimental results, an interconnected permeability network can form, allowing fluid to penetrate and alter the lithosphere. These measurements show that thermal expansion anisotropy of olivine is sufficient to create an interconnected permeability structure, at least near transform boundaries.

Several additional processes may effect the formation of microcracks in the oceanic lithosphere. First, thermal expansion mismatch between different mineral grains and anisotropic elastic properties may enhance microcrack formation. Second, the cooling effects of diffuse fluid flow may enhance the formation of microcracks in the oceanic lithosphere. Third, the slow growth of cracks resulting from subcritical crack growth may enhance cracking at the time scale of oceanic spreading. All three of these processes would enhance crack growth in the oceanic lithosphere and yield deeper zones of thermally fracture rocks. In contrast, differential stresses can generate recrystallized grain sizes smaller than 1 mm in the upper mantle [e.g., *Van der Wal et al.*, 1993; *Karato et al.*, 1980]. Smaller grain sizes will inhibit the formation of microcracks (e.g., notice the shallower zone of thermal cracking illustrated in Figure 13a) and may lead to zones of unaltered peridotite in the uppermost mantle.

Although the thermal models used here neglect several processes (e.g., multiple mineral phases, hydrothermal cooling, subcritical crack growth), they do illustrate the mutual dependence of pressure and temperature. Temperature decreases within the cooling mantle enhance the formation of microcracks, while pressure acts to suppress microcracking. Thus, thermal cracking is easiest at ridge segment ends where mantle peridotite uplifts to shallow depths. In zones where enhanced cooling occurs owing to hydrothermal circulation, the onset of thermal cracking will be deeper; in zones where the oceanic lithosphere is less effectively cooled, thermal cracking will only occur at shallower depths.

5.3. Serpentinization of Peridotite and Seismic Evidence for Thermal Cracking

The region of thermal cracking predicted by our analysis compares well with the conditions at which serpentinization is inferred to occur based on seismic surveys. Extensive exposures of serpentinized peridotites are present along oceanic fracture zones and the inside corner highs of slow spreading ridges [e.g., *Dick*, 1989; *Bonatti*, 1976; *Miyashiro et al.*, 1969]. The influx of volatiles into mantle peridotite along faults and subsequent diffuse flow of the fluid into the rock along thermal fractures at low temperatures results in serpentinization of mantle peridotite [*O'Hanley*, 1996]. Both the frictional strength [*Moore et al.*, 1997] and the fracture strength [*Escartín et al.*, 1997a] of serpentinized mantle peridotite are less than

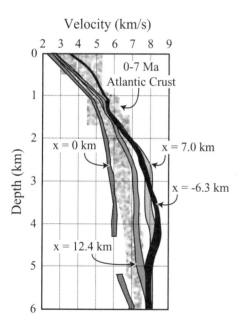

Figure 15. Velocity versus depth profiles for the MARK area of the Mid-Atlantic Ridge. The width of each profile represents two standard deviations after averaging 2D structures to form a 1D profile. Typical velocity profiles for 0–7 Ma oceanic crust are denoted by the patterned grey zone, while the black profile (*x* = -6.3 km) represents the velocity where serpentinized peridotite outcrops on the seafloor. The profile labeled *x* = 7.0 km represents the seismic structure for an area a similar distance from the axis on the opposite side of the ridge. The increase in velocity to approximately 8.0 km s^{-1} at a depth of 3–4 km in the black profile is consistent with a transition from partially serpentinized to unaltered peridotite. Figure modified from *Canales et al.* [2000].

that of unaltered peridotite. Thus, the serpentinization process probably has a substantial influence on the tectonic evolution of the oceanic lithosphere [*Escartín et al.*, 1997b]. Because serpentinites have slower compressional and shear-wave velocities than unaltered peridotite [*Miller and Christensen*, 1997], seismic techniques can be used to locate serpentinized portions of the oceanic lithosphere.

Along the Mid-Atlantic Ridge (MAR), south of the Kane Transform (MARK area), serpentinized peridotites crop out in a belt approximately 2 km wide and 20 km long along the western median valley wall [*Karson et al.*, 1987]. These serpentinites likely were exposed during an episode of amagmatic plate separation in which little gabbroic crust was formed and detachment faulting occurred [*Karson and Lawrence*, 1997]. Serpentinites collected along the seafloor in this region using Alvin and Nautile include harzburgites, peridotites containing primarily olivine and orthopyroxene, that are altered to form massive or schistose serpentinites [*Karson et al.*, 1987]. During ODP Leg 153, serpentinized harzburgites were sampled to depths of 200 m in this area [*Cannat et al.*, 1995]. These field observations, coupled with models for similar tectonic regions [e.g., *Tucholke et al.*, 1998] imply that serpentinized peridotites occur down to the unaltered mantle.

Seismic velocity structures south of the MARK area, shown in Figure 15, indicate that a transition to unaltered peridotite occurs at a depth of 3–4 km [*Canales et al.*, 2000]. The grey patterned zone in Figure 15 shows a velocity profiles typical of 0–7 Ma oceanic crust. The black profile (*x* = -6.3 km) is the profile where serpentinized peridotite crops out, while the profile labeled *x* = 7.0 km shows the seismic section for the oceanic lithosphere on the conjugate side of the ridge where no serpentinites are observed. The increase in velocity to roughly 8.0 km s^{-1} at a depth of 3–4 km in the velocity profiles for areas where serpentinized peridotite crops out is consistent with the prediction of a thermally cracked zone in the upper 4 to 6 km of the seafloor. These observations corroborate our conclusion that thermally induced microcracks along grain boundaries contribute to the diffuse alteration of the oceanic lithosphere.

6. CONCLUSIONS

Stresses resulting from anisotropic thermal contraction of olivine during cooling and depressurization are sufficient to create an interconnected permeability network in laboratory samples. Using micromechanical models for grain boundary cracking in single-phase materials, we estimate the critical stress intensity factor necessary for interconnected microfracturing in olivine to be 0.6 ± 0.3 MPa m$^{1/2}$. If natural peridotites have the same fracture toughness as the laboratory samples, we estimate that thermal cracking may occur in the upper 4 to 6 km of the seafloor near transform faults. The likely presence of multiple mineral phases, hydrothermal cooling, increased grain size, and subcritical crack growth would increase the likelihood of cracking and deepen the zone of thermal cracking in the oceanic lithosphere. Conversely, mantle grain sizes less than 1 mm will reduce the likelihood of cracking and yield a thinner zone of thermally cracked rock. Seismic refraction surveys in regions where serpentinites crop out suggest that the transition from altered to unaltered mantle rock occurs at 3–4 km, a value consistent with that estimated here, given the uncertainties in scaling. These results indicate that thermally induced microcracking, and subsequent alteration, plays an essential role in the thermal, chemical, and rheological evolution of the oceanic lithosphere.

Acknowledgments. Xiaohui Xiao provided extensive assistance with the Paterson Rig at MIT. Uli Mok and Yves Bernebé helped with permeability analysis and in the operation of the wide-range permeameter. Mark Behn got us up and running with the thermal and flow model for oceanic spreading centers. Javier Escartín provided images of serpentinized peridotite. We are grateful for insightful reviews by Joanne Fredrich, Teng-Fong Wong, and Joe Cann and discussions with Wenlu Zhu. This research was supported by the NSF grants OCE-0095936 (MIT) and OCE-9907244 (WHOI).

REFERENCES

Burnham, C. W., Importance of volatile constituents, *in Evolution of Igneous Rocks,* edited by H. D. Yoder, pp. 439–482, Princeton University Press, Princeton, N. J., 1979.

Cowan, J. M.., Elasticity of Mantle Phases at High Temperature, *J. of Geophys. Res.,* 95, 439–482, 1999.

Atkinson, B.K., Subcritical crack growth in geological materials, *J. Geophys. Res.,* 89, 4077–4114, 1984.

Atkinson, B.K., and P.G. Meredith, Experimental fracture mechanics data for rocks and minerals, in *Fracture Mechanics of Rock*, edited by B.K. Atkinson, pp. 477–525, Academic Press, London, 1987.

Bauer, S.J., and B. Johnson, Effects of slow uniform heating on the physical properties of Westerly and Charcoal granites, in *Proc. 20th U.S. Symp. Rock Mech,* pp. 7–18, 1979.

Behn, M.D., J. Lin, and M.T. Zuber, Mechanics of normal fault development at mid-ocean ridges, *J. Geophys. Res.,* 107, doi: 10.1029/2001JB000503, 2002.

Bernabé, Y., A wide range permeameter for use in rock physics, *Int. J. Rock Mech. Min. Sci. & Geomech. Abstr.,* 24, 309–315, 1987.

Boas, W., and R.W.K. Honeycombe, The anisotropy of thermal expansion as a cause of deformation in metals and alloys, *Proc. R. Soc. Lond. A,* 188, 427–439, 1947.

Bonatti, E., Serpentinite protrusions in the oceanic crust, *Earth Planet. Sci. Lett.,* 32, 107–113, 1976.

Bouhifd, M.A., D. Andrault, G. Fiquet, and P. Richet, Thermal expansion of forsterite up to the melting point, *Geophys. Res. Lett.,* 23, 1143–1146, 1996.

Brace, W.F., E. Silver, K. Hadley, and C. Goetze, Cracks and pores: A closer look, *Science,* 178, 162–164, 1972.

Brace, W.F., J.B. Walsh, and W.T. Frangos, Permeability of granite under high pressure, *J. Geophys. Res.,* 73, 2225–2236, 1968.

Canales, J.P., J.A. Collins, J. Escartín, and R.S. Detrick, Seismic structure across the rift valley of the Mid-Atlantic Ridge at 23° 20' (MARK area): Implications for crustal accretion processes at slow spreading ridges, *J. Geophys. Res.,* 105, 28,411–28,425, 2000.

Cannat, M., et al., *Proceedings of the Ocean Drilling Program, Initial Reports*, Ocean Drilling Program, College Station, TX, 1995.

Carlson, S.R., M. Wu, and H.F. Wang, Micromechanical Modeling of Thermal Cracking in Granite, in *The Brittle-Ductile Transition in Rocks, The Heard Volume*, edited by A.G. Duba, W.B. Durham, J.W. Handin, and H.F. Wang, pp. 37–48, American Geophysical Union, Washington, D.C., 1990.

Clarke, D.R., Microfracture in brittle solids resulting from anisotropic shape changes, *Acta Metall.,* 28, 913–924, 1980.

Dick, H.J.B., Abyssal peridotites, very slow spreading ridges and ocean ridge magmatism, in *Magmatism in the Ocean Basins*, edited by A.D. Saunders, and M.J. Norry, pp. 71–105, Geol. Soc. Spec. Publ., London, 1989.

Escartín, J., G. Hirth, and B. Evans, Effects of serpentinizaion on the lithospheric strength and the style of normal faulting at slow-spreading ridges, *Earth Planet. Sci. Lett.,* 151, 181–189, 1997a.

Escartín, J., G. Hirth, and B. Evans, Nondilatant brittle deformation of serpentinites: Implications for Mohr-Coulomb theory and the strength of faluts, *J. Geophys. Res.,* 102, 2897–2913, 1997b.

Evans, A.G., Microfracture from thermal expansion anisotropy-I. Single phase systems, *Acta Metall.,* 26, 1845–1853, 1978.

Evans, A.G., and D.R. Clarke, Residual stresses and microcracking induced by thermal contraction inhomogeneity, in *Thermal Stresses in Severe Environments*, edited by D.P.H. Hasselman, and R.A. Heller, pp. 629–648, 1980.

Evans, B., J. Renner, and G. Hirth, A few remarks on the kinetics of static grain growth in rocks, *Int. J. Earth Sciences,* 90, 88–103, 2001.

Fischer, G.J., The determination of permeability and storage capacity: Pore pressure oscillation method, in *Fault Mechanics and Transport Properties of Rocks*, edited by B. Evans, and T.-f. Wong, pp. 187–211, Academic Press, London, 1992.

Fischer, G.J., and M.S. Paterson, Measurement of permeability and storage capacity in rocks during deformation at high temperature and pressure, in *Fault Mechanics and Transport Properties of Rocks*, edited by B. Evans, and T.-f. Wong, pp. 213–252, Academic Press, London, 1992.

Francis, T.J.G., Serpentinization faults and their role in the tectonics of slow spreading ridges, *J. Geophys. Res.,* 86, 11,616–11,622, 1981.

Fredrich, J.T., and T.-f. Wong, Micromechanics of thermally induced cracking in three crustal rocks, *J. Geophys. Res.,* 91, 12,743–12,764, 1986.

Ghahremani, F., and J.W. Hutchinson, Three-dimensional effects in microcrack nucleation in brittle polycrystals, *J. Am. Ceram. Soc.*, 73, 1548–1554, 1990.

Hirth, G., and D.L. Kohlstedt, Experimental constrains on the dynamics of the partially molten upper mantle: Deformation in the diffusion creep regime, *J. Geophys. Res.*, 100, 1981–2000, 1995.

Hooft, E.E.E., R.S. Detrick, D.R. Toomey, J.A. Collins, and J. Lin, Crustal thickness and structure along three contrasting spreading segments of the Mid-Atlantic Ridge, 33.5°N–35°N, *J. Geophys. Res.*, 105, 8205–8226, 2000.

Karato, S.-i., Grain growth kinetics in olivine aggregates, *Tectonophysics*, 168, 255–273, 1989.

Karato, S.-i., M. Toriumi, and T. Fujii, Dynamic recrystallization of olivine single crystals during high-temperature creep, *Geophys. Res. Lett.*, 7, 649–652, 1980.

Karson, J.A., and R.M. Lawrence, Seismic velocities of lower crustal and upper mantle rocks from the slow spreading Mid-Atlantic Ridge, south of the Kane transform zone (MARK), *Proc. Ocean Drill. Program, Sci. Results*, 153, 5–21, 1997.

Karson, J.A., G. Thompson, S.E. Humphris, S.E. Edmond, J.M. Edmond, W.B. Bryan, J.R. Brown, A.T. Winters, R.A. Pockalny, J.F. Casey, A.C. Campbell, G. Klinkhammer, M.R. Palmer, R.J. Kinzler, and M.M. Sulanowska, Along-axis variations in seafloor spreading in the MARK area, *Nature*, 328, 681–685, 1987.

Kuszyk, J.A., and R.C. Bradt, Influence of grain size on the effects of thermal expansion anisotropy in MgTi2O5, *J. Am. Ceram. Soc.*, 56, 420–423, 1973.

Lawn, B., *Fracture of Brittle Solids*, 378 pp., Cambridge University Press, Cambridge, 1993.

Miller, D.J., and N.I. Christensen, Seismic velocities of lower crustal and upper mantle rocks from the slow spreading Mid-Atlantic Ridge, south of the Kane transform zone (MARK), *Proc. Ocean Drill. Program, Sci. Results*, 153, 437–454, 1997.

Miyashiro, A., F. Shido, and M. Ewing, Composition and origin of serpentinites from the Mid-Atlantic Ridge, 24°N and 30°N latitude, *Contrib. Mineral. Petrol.*, 23, 117–127, 1969.

Moore, D.E., D.A. Lockner, Ma Shengli, R. Summers, and J.D. Byerlee, Strengths of serpentinite gouges at elevated temperatures, *J. Geophys. Res.*, 102, 14,786–14,801, 1997.

Murton, B.J., L.J. Redbourn, C.R. German, and E.T. Baker, Sources and fluxes of hydrothermal heat, chemicals and biology within a segment of the Mid-Atlantic Ridge, *Earth Planet. Sci. Lett.*, 171, 201–317, 1999.

O'Hanley, D.S., *Serpentinites, Records of Tectonic and Petrological History*, 277 pp., Oxford University Press, Oxford, 1996.

Olgaard, D.L., and B. Evans, Grain growth in synthetic marbles with added mica and water, *Contrib. Mineral. Petrol.*, 100, 246–260, 1988.

Paterson, M.S., A high temperature high pressure apparatus for rock deformation, *Int. J. Rock Mech. Min. Sci.*, 7, 517–526, 1970.

Phipps Morgan, J., and D. Forsyth, Three-dimensional flow and temperature perturbations due to a transform offset: Effects on oceanic curst and upper mantle structure, *J. Geophys. Res.*, 93, 2955–2966, 1988.

Sclater, J.G., C. Jaupart, and D. Galson, The heat flow through oceanic and continental crust and the heat loss of the Earth, *Rev. Geophys. Space Phys.*, 18, 269–311, 1980.

Shaw, W.J., and J. Lin, Models of ocean ridge lithospheric deformation: Dependence on crustal thickness, spreading rate, and segmentation, *J. Geophys. Res.*, 101, 17,977–17,933, 1996.

Siddiqi, G., *Transport Properties and Mechanical Behavior of Synthetic Calcite-Quartz Aggregates*, Ph.D. thesis, Massachusetts Institute of Technology, Cambridge, MA, 1997.

Simmons, G., and D. Richter, Microcracks in rocks, in *The Physics and Chemistry of Minerals and Rocks*, edited by R.J.G. Sterns, pp. 105–137, Wiley-Interscience, New York, 1976.

Sprunt, E.S., and W.F. Brace, Direct observation of microcavaties in crystalline rocks, *Int. J. Rock Mech. Min. Sci. & Geomech. Abstr.*, 11, 139–150, 1974.

Swain, M.V., and B.K. Atkinson, Fracture surface energy of olivine, *Pure Appl. Geophys.*, 116, 866–872, 1978.

Tucholke, B.E., J. Lin, and M.C. Kleinrock, Megamullions and mullion structure defining oceanic metamorphic core complexes on the Mid-Atlantic Ridge, *J. Geophys. Res.*, 103, 9857–9866, 1998.

Underwood, E.E., *Quantitative Stereology*, 274 pp., Addison Wesley, Reading, MA, 1970.

Van der Wal, D., P.N. Chopra, M. Drury, and J.D. FitzGerald, Relationships between dynamically recrystallized grain size and deformation conditions in experimentally deformed olivine rocks, *Geophys. Res. Lett.*, 20, 1497–1482, 1993.

Wang, H.F., B.P. Bonner, S.R. Carlson, B.J. Kowallis, and H.C. Heard, Thermal stress cracking in granite, *J. Geophys. Res.*, 94, 1745–1758, 1989.

Brian deMartin, MIT/WHOI Joint Program in Oceanography, Marine Geology and Geophysics, 390 Woods Hole Rd., Woods Hole, MA 02345.

Brian Evans, Massachusetts Institute of Technology, Department of Earth, Atmospheric, and Planetary Science, 77 Massachusetts Ave., Cambridge, MA 02139.

Greg Hirth, Woods Hole Oceanographic Institution, Department of Geology and Geophysics, 390 Woods Hole Rd., Woods Hole, MA 02345.

Submarine Lava Flow Emplacement at the East Pacific Rise 9° 50′N:
Implications for Uppermost Ocean Crust Stratigraphy and Hydrothermal Fluid Circulation

Daniel Fornari[1], Maurice Tivey[1], Hans Schouten[1], Michael Perfit[2],
Dana Yoerger[1+], Al Bradley[1+], Margo Edwards[3], Rachel Haymon[4],
Daniel Scheirer[5], Karen Von Damm[6], Timothy Shank[1#], Adam Soule[1]

Meter scale seafloor topography and sidescan backscatter imagery of volcanic terrain along the axis of the fast-spreading northern East Pacific Rise (EPR) near 9° 50′N, coupled with visual and photographic observations provide data that constrain spatial relationships between hydrothermal vents and primary volcanic features and processes along the EPR axis. High-temperature ($\geq$350°C) hydrothermal vents are present in several areas within the EPR axial trough where recent eruptions have been focused and where drainback of lava into the primary eruptive fissure occurred. Chaotic collapse crusts and draped sheet lava surfaces along the margin of eruptive fissures typify sites where drainback primarily occurred. These areas are also coincident with ~10–30 m-wide channels that serve to transport lava across the crestal plateau. Regions of diffuse hydrothermal flow at low temperatures (<35°C) and vent animal communities are concentrated along the primary eruptive fissure that fed the 1991 eruption at the 9° 50′N EPR area. The relationship between seafloor eruption processes and hydrothermal vent locations within the EPR axial trough is linked to the formation of high permeability zones created by: focusing of eruptions along discrete portions of fissures; volcanic episodicity during eruptive phases lasting hours to days; and drainback of lava into the primary fissure at these same locations during waning stages of seafloor eruptions.

---

[1]Woods Hole Oceanographic Institution, Geology & Geophysics Department, Woods Hole, Massachusetts
[1+]Woods Hole Oceanographic Institution, Applied Ocean Physics and Engineering Department, Woods Hole, Massachusetts
[1#]Woods Hole Oceanographic Institution, Biology Department, Woods Hole, Massachusetts

[2]University of Florida, Department of Geological Sciences, Gainesville, Florida
[3]University of Hawaii, Hawaii Institute of Geophysics and Planetology, School of Ocean and Earth Science and Technology, Honolulu, Hawaii
[4]University of California–Santa Barbara, Department of Geological Sciences and Marine Sciences Institute, Santa Barbara, California
[5]Brown University, Department of Geological Sciences, Providence, Rhode Island, now at: U.S. Geological Survey, Menlo Park, California
[6]University of New Hampshire, Complex Systems Research Center, EOS, Durham, New Hampshire

Mid-Ocean Ridges: Hydrothermal Interactions Between the Lithosphere and Oceans
Geophysical Monograph Series 148
Copyright 2004 by the American Geophysical Union
10.1029/148GM08

1. INTRODUCTION

The injection of magma, eruption of lava and tectonic stresses associated with seafloor spreading at the global MOR crest profoundly impact the structure and permeability of the upper ocean crust, and hence the location and style of hydrothermal venting in different spreading environments [e.g., *Von Damm*, 1990, 1995, 2004; *Haymon et al.,* 1991; *Fornari and Embley*, 1995; *Humphris*, 1995; *Alt*, 1995; *Hannington et al.*, 1995; *Wright et al.,* 1995b, 1998, 2002; *Haymon*, 1996; *Wilcock*, 1998; *Perfit and Chadwick*, 1998; *Carbotte and Scheirer*, 2003]. To a first order, heat from magma bodies and intrusions that reach various levels in the ocean crust drive deep and shallow hydrothermal circulation systems that distribute fluids through the crust, to the seafloor, and into the hydrosphere. Vertical to near-vertical fractures, millimeters to <~1 m wide, created by tectonic or magmatic processes serve as pathways within crustal rocks [*Wright et al.,* 1995b, 1998]. Interflow and intraflow permeability within volcanic sequences of the upper oceanic crust further augments and complicates the geometry of fluid flow [e.g. *Fornari et al.,* 1998a]. Volcanic and hydrothermal processes at the mid-ocean ridge (MOR) crest are considered closely linked, although the temporal and spatial relationships between them have been difficult to determine.

The complex nature of submarine lava emplacement processes, the difficulty in directly observing or measuring deep seafloor eruptions, and the causal relationships between volcanic and tectonic processes at the MOR crest have been vexing problems—all of which require innovative technical and conceptual solutions [e.g., *Perfit and Chadwick*, 1998; *Fox et al.,* 2001]. Until recently, the resolution of seafloor imaging sonars has been too coarse to resolve fine-scale (~1–2 m) features required to make accurate spatial correlations between volcanic and hydrothermal features. Only within the past 15 years has sufficiently high-resolution sidescan sonar imaging of the seafloor revealed the plan-view arrangement of individual MOR lava flows and their relationship to tectonic and other features along the ridge crest [e.g., *Haymon et al.,* 1991; *Embley and Chadwick*, 1994; *Embley et al.*, 1995; *Fornari et al.*, 1998a; *Delaney et al.,* 1998; *Bohnenstiehl and Kleinrock*, 1999, 2000; *Sinton et al.,* 2002; *White et al.,* 2002]. The advent of high frequency scanning altimetric sonars and near-bottom multibeam sonars, and their application to producing meter to sub-meter scale bathymetry of small areas of seafloor, has revolutionized the study of seafloor morphology. These data can also provide clues about the structural and morphological responses of the ridge axis to magmatic and tectonic processes occurring over short (e.g., decadal) time periods [e.g., *Yoerger et al.,* 1996, 2002; *Chadwick et al.*, 2001; *Carbotte et al.*, 2003; *Schouten et al.,* 2001, 2002; *Embley et al.*, 2002, *Johnson et al.,* 2002; *Shank et al.*, 2002; *Fornari et al.*, 2003a; *Cormier et al.,* 2003].

A key to understanding complex relationships between volcanic and hydrothermal features at the MOR crest in both space and time is accurate, quantitative mapping of seafloor topography and the ability to repeat the surveys with well-constrained navigation. Such studies should include identifying spatial relationships between hydrothermal vents, and the axial trough or graben walls, primary eruptive fissures and zones of lava drain back, and various lava flow morphologies. Studies of subaerial lava flow morphology, timing of eruptions, flow transitions, channelized flows and textural characterization of flows imaged by land-based radar systems have provided significant insight into eruption history, dynamics and related processes on Iceland and Hawaii [e.g., *Heliker et al.* 1998; *Cashman et al.,* 1999; *Wylie et al.,* 1999; *Crown and Balogna*, 1999; *Polacci et al.,* 1999; *Byrnes and Crown*, 2001, 2003; *Parfitt and Wilson, 1994; Parfitt et al.,* 2002; *Soule et al.,* 2004]. Similar types of studies are critical to understanding submarine volcanic eruptions at the MOR and their relationships to hydrothermal processes [e.g., *Gregg et al.,* 1996, 2000, *Fornari et al.,* 1998a; *Embley et al.*, 1999; *Chadwick et al.*, 1999; 2001; *Sakimoto and Gregg,* 2001; *Umino et al.,* 2002; *Perfit et al.,* 2003; *Chadwick,* 2003; *Embley and Baker*, 2003]. The extreme difficulties associated with observing active submarine eruption processes also require development of conceptual, physical volcanological models based on submarine volcanic morphologies (e.g., lava pillars, flow structures and transitions) and the geochemistry and petrology of submarine lavas [e.g., *Perfit and Chadwick*, 1998; *Sinton et al.,* 2002; *Gregg and Sakimoto,* 2001; *Perfit et al.,* 2003; *Chadwick et al.,* 2003].

In this study, we explore the relationships between primary volcanic processes at a fast-spreading MOR axis and the features developed during seafloor eruptions, and their impact on hydrothermal fluid flow in the upper ocean crust. We use recently collected 120 kHz sidescan sonar imagery [e.g., *Scheirer et al.*, 2000], and micro-bathymetry data derived from a mechanically scanned altimeter on the autonomous vehicle *ABE* [*Yoerger et al.,* 1996; 2002] to map the EPR axis and summit plateau near 9° 50′N (Plate 1). Near-bottom magnetic data are also analyzed to infer local regions of focused hydrothermal fluid flow, similar to observations at the Main Endeavour vent field on the Juan de Fuca Ridge [*Tivey and Johnson*, 2002]. These data sets are interpreted in conjunction with photographic and visual documentation of numerous hydrothermal vents within the axial summit trough (AST) of the EPR [*Haymon et al.,* 1991; *Fornari et al.,*

1998a] to deduce the subsurface volcanic stratigraphy and to conceptually model the complex geometry of hydrothermal fluid flow in the shallow ocean crust [*German and Von Damm*, 2004]. A selection of time-series temperature records from the main high-temperature hydrothermal vents in the 9° 50′N area further substantiates the complexity in shallow subsurface flow between adjacent vents, biological communities living in diffuse flow surrounding the black-smokers, and other vents present in the axial trough between 9° 49′–51′N. These data, together with the data presented in *Von Damm* [2004, this volume], suggest a shallow heat source. The quantitative bathymetric and sidescan sonar maps, along with observational data, provide key evidence for linking the spatial distribution of high- and low-temperature hydrothermal vents within the axial trough in the 9° 50′N area to primary volcanic features, especially the eruptive fissure that fed the recent lava flows.

2. THE EAST PACIFIC RISE 8°–11°N: BACKGROUND

The EPR between 8°N and 11°N is spreading at a full rate of 11 cm/yr [*Carbotte and Macdonald*, 1992]. The area includes: a complete 1st order ridge segment and its two bounding transform faults (right-stepping Clipperton and left-stepping Siqueiros) (Plate 1); two 2nd order segments separated by an overlapping spreading center (OSC) at 9° 03′N; and, multiple finer-scale segments, including 3rd order volcanic segments with boundaries detectable in multibeam bathymetry data, and 4th order segments bounded by smaller, more transient ridge axis discontinuities that are below the resolution of current multibeam sonar systems [e.g., *Haymon et al.*, 1991; *Macdonald et al.*, 1984; *Scheirer and Macdonald*, 1993; *White et al.*, 2002].

Existing regional magnetic, gravity and multibeam/Sea MARC II sonar surveys have completely imaged the ridge flanks beyond 100 km from the axis on both sides [*Macdonald et al.*, 1992; *Carbotte and Macdonald*, 1992; *Cochran*, 1999; *Pockalny et al.*, 1997; *Schouten et al.*, 1999] (Plate 1). Analysis of these data sets has revealed the structure, duration, migration, and temporal behavior of segment boundaries, changes in spreading directions and corresponding changes in ridge segment configuration over the past two million years, the development of faults and abyssal hills, and the distribution of off-axis volcanoes. Near-bottom acoustic and photographic surveys of the ridge crest at 9°–10° N have shown the more recent and fine-scale volcanic segmentation of the ridge crest and its relationship to the distribution of ridge crest hydrothermal and volcanic features [e.g., *Haymon et al.*, 1991; *Fornari et al.*, 1998a; *Kurras et al.*, 2000; *Von Damm*, 2000, 2004; *Von Damm et al.*, 1995; *Engels et al.*, 2003; *White et al.*, 2002] (Plate 2). Fine-scale bathy-

metric mapping available for this area from the late 1980s to 2001 consisted solely of single, near-bottom altimeter profiles acquired during Alvin dives or Argo II lowerings. Improvements in sonar and vehicle technology over the last ~5 years have allowed very detailed (~1–5 m resolution) bathymetric surveys to be carried out along the axis of the MOR [*Kurras et al.*, 2000; *Chadwick et al.*, 2001; *Johnson et al.*, 2002; *Jakuba et al.*, 2002; *Fornari et al.*, 2003a; *White et al.*, 2000] (Plate 3).

The only well-established magmatic events known to have affected this segment of the EPR during the past decade occurred in a region centered near 9° 50′N in ~ March, 1991 and again late in 1991 or early 1992 [*Haymon et al.* 1993; *Rubin et al.*, 1994]. Various studies have described the hydrothermal, geochemical and biological ramifications of the 1991–92 events [*Haymon et al.*, 1993; *Lutz et al.*, 1994; *Perfit et al.*, 1994; *Fornari, et al.*, 1998b; *Shank et al.*, 1998; *Von Damm*, 1995; *Von Damm*, 2000, 2004; *Von Damm and Lilley*, 2004]. *Smith et al.* [2001] document the volcanological and associated hydrothermal effects of the eruption on the small overlapping spreading center (OSC) at 9° 37′N. A previous EPR axial eruption is thought to have taken place during the 1987–1989 time frame in the 9° 17′N region based on freshness of lava flows mapped using Argo I, the high temperature and vapor phase characteristics of F vent (>390°C) in 1991 and its evolution to a hydrothermal brine vent by 1994 [*Haymon et al.*, 1991; *Von Damm et al.*, 1997; *Fornari et al.*, 1998a]. In addition, lack of an AST in the ~9° 14′N to ~9° 20′N region, and the presence of extensive lava lakes and large collapse features along the axis suggest that a relatively large volume lava flow had recently resurfaced this portion of the EPR axis, probably not more than a few years prior to the November 1989 Argo II survey [*Haymon et al.*, 1991; *Fornari et al.*, 1998a].

Multi-channel seismic and expanding spread profile data have allowed for robust determinations of seismic structure [*Vera and Diebold, 1994*; *Toomey et al.* 1994], magma chamber depth [*Detrick et al.*, 1987; *Herron et al.*, 1980; *Kent et al.*, 1993a,b], melt lens width, thickness, and crystallinity [*Kent et al.*, 1990; *Singh et al.*, 1998], and layer 2A thickness [*Christeson et al.*, 1996; *Harding et al.*, 1993] along major portions of the EPR in the study area. Active source seismic refraction experiments to on-bottom receivers have led to two- and three-dimensional seismic velocity models at a variety of scales for many portions of the ridge segment [e.g., *Bazin et al.*, 1998; *Begnaud et al.*, 1997; *Dunn and Toomey*, 1997; *Tian et al.*, 2000; *Toomey et al.*, 1990; *Toomey et al.*, 1994; *van Avendonk et al.*, 1998, 2001]. Seafloor compliance measurements have also provided constraints on shear wave velocity structure and the distribution of melt in the lower crust [*Crawford and Webb*, 2002; *Crawford et al.*, 1999], including sub-

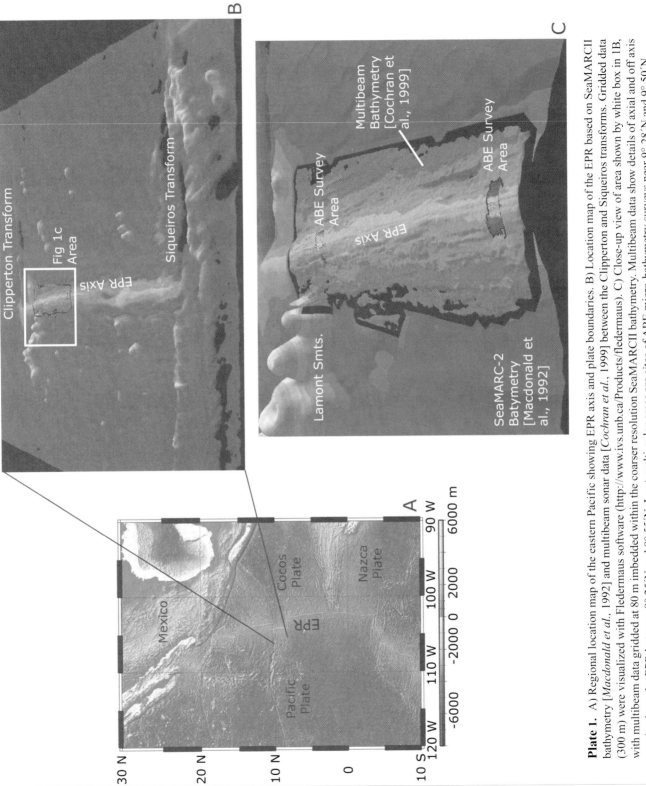

Plate 1. A) Regional location map of the eastern Pacific showing EPR axis and plate boundaries. B) Location map of the EPR based on SeaMARCII bathymetry [*Macdonald et al.*, 1992] and multibeam sonar data [*Cochran et al.*, 1999] between the Clipperton and Siqueiros transforms. Gridded data (300 m) were visualized with Fledermaus software (http://www.ivs.unb.ca/Products/Fledermaus). C) Close-up view of area shown by white box in 1B, with multibeam data gridded at 80 m imbedded within the coarser resolution SeaMARCII bathymetry. Multibeam data show details of axial and off axis terrain along the EPR between 9° 25′N and 9° 55′N. Inset multi-color areas are sites of ABE micro-bathymetry surveys near 9° 28′N and 9° 50′N.

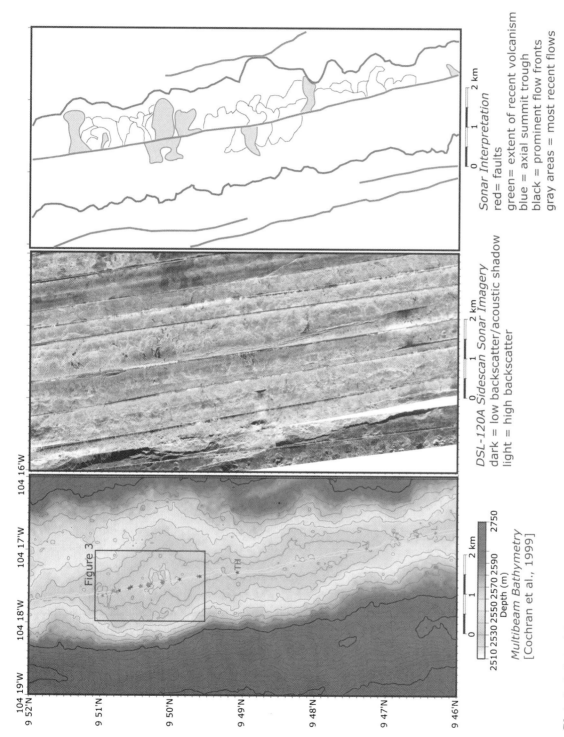

Plate 2. Left- Multibeam bathymetry showing locations of 14 high temperature hydrothermal vents (red stars) and low temperature biological communities (blue stars) along the EPR axis from 9° 46´N to 9° 51´N. Middle- DSL-120A sidescan sonar imagery for the area shown in left panel. Yellow lines show vehicle nadir. White and light gray is high backscatter, dark gray or black is low backscatter or acoustic shadow. Right- interpretative map showing faults (red lines), area encompassing most recent flows (green line bounded area), axial summit trough (AST) trace (blue line), scalloped flow fronts and edges (fine black lines) and most recent flows (purple lines bounding gray areas). The characteristic, shingled backscatter pattern of surficial lava flows on the EPR crest indicate a sustained eruption source within the AST (blue line) for an extensive period of time (thousands of years). The convex outward lobate flow fronts create a shingled pattern of flow lobes that trend away from the AST.

stantiation of *Garmany's* [1989] observation of off-axis lower crustal melt reservoirs.

3. HAWAIIAN FISSURE ERUPTIONS AND MID-OCEAN RIDGE SUBMARINE VOLCANIC FEATURES

In order to provide a volcanological context for the seafloor observations, we present brief background information on Hawaiian eruptive fissures and lava flows and compare and contrast them to MOR submarine lava eruptions and flow characteristics. Many features of submarine volcanic landforms and lava flows have analogs in their subaerial counterparts. Here we examine some features common to both environments and discuss how those features may be related. Fissure eruptions on basaltic shield volcanoes like Kilauea in Hawaii commonly initiate with eruptive activity at a point source that migrates along the length of a fissure, until it all is erupting at once (Plate 4). Eruptive activity focuses quickly, sometimes within hours, to a small number of loci that may remain active for the duration of the eruption. Over most of the fissure length, lava will flow only a short distance perpendicular to the fissure (usually < 1 km). At persistent eruptive vents narrow flows are formed (length/width >> 10), the length of which depends on the volume of lava erupted and length of time that the vent is active. The Maunu Ulu 1974 and Pu'u 'O'o (1983 to present) eruptions are classic examples of persistent vents forming long tube and surface fed flows that extended tens of kilometers from their source vents [*Tilling et al.,* 1987; *Wolfe et al.,* 1987, 1988]. By comparison many submarine lava flows on the EPR in the 9° 50′N area (Plate 3) appear to be ~1–2 km long and with relatively smaller l/w ratios (~1–2) as compared to Kilauea flows.

As subaerial flows extend from their eruptive source they often undergo transitions in surface morphology that reflect changes in the physical properties of the lava and eruption dynamics. In subaerial lava flows, transitions in surface and flow morphology are dependent on a host of interdependent eruption and flow parameters including eruption temperature, volumetric flow rates, cooling rates, crystallization rates, and pre-existing topography. Temporal and spatial changes in these parameters result in a characteristic evolution in surface morphology from pahoehoe to 'a'a over length scales from 0.25–4 km for medium to high effusion rate eruptions [*Soule et al., 2004*]. It is observed that the pahoehoe-to-'a'a transition is initiated within regions of focused flow (i.e. channels) at distances from the vent where lavas have cooled and crystallized sufficiently to fundamentally change the rheology of the lava and consequently the style of lava deformation [e.g. *Cashman, et al.,* 1999; *Soule, et al.,* 2004].

An analogous down-flow transition in surface morphology is observed in submarine flows from proximal sheet flows, to lobates, to pillows. This evolution, while greatly simplified, is observed along some down-flow and cross-flow transects in our existing digital seafloor imagery. In the submarine environment, the parameters controlling surface morphology are not dependent on lava temperature and crystal content, both of which remain relatively constant due to the rapid formation of thick, insulating crusts (e.g., *Gregg and Fornari,* 1998). Analog wax models of submarine lava flow emplacement identify two independent parameters of flow emplacement: the time scales of cooling and advection, and the ratio of those parameters that determines the surface morphology of the flow [*Fink and Griffiths 1992, Griffiths and Fink,* 1992, *Gregg and Fink,* 1998; *Sakimoto and Gregg,* 2001]. Due to the large thermal gradient between molten lava and seawater, cooling rates on submarine lava flows should be similar regardless of variation in the temperature of either the erupting lava or overlying seawater. Thus, the dominant parameter determining submarine flow morphology is the timescale of lava advection. Volumetric effusion rate and pre-existing topography provide the primary controls on this advection rate [e.g. *Gregg and Fink,* 1998; *Sakimoto and Gregg,* 2001].

One of the most distinctive morphological characteristics of the sidescan sonar data in our study area (Plate 2) are the overlapping, convex outward (from the ridge axis) reflectors that comprise the fronts of lava flows erupted at the axis. The bi-lateral symmetry of the reflectors on either side of the axis indicates that the locus of eruption has been quite constant along the axial trough for thousands of years. The scalloped flow-front morphology along volcanically and hydrothermally active portions of the EPR shown in Plates 2 and 3 is analogous to areas adjacent to the rift zones of Mauna Loa and Kilauea volcanoes (Plates 4 and 5). In Hawaii, volcanic activity has been concentrated over hundreds of thousands of years [*Holcomb, 1980, 1987*] and has produced high aspect ratio (flow length/width) flows that inter-finger and overlap. A potential distinction between the submarine and subaerial environments is that the EPR seafloor is actively spreading, thus flow fronts that may have reached 1 km from the vent at the time of eruption are observed at 1.5 km from the vent after 10 kyr. However, the lack of faulting within ~2 km of the EPR axis (equivalent to ~40 kyr spreading distance) relative to areas farther off-axis suggests that the area around 9° 50′N is repaved over a much shorter time interval; therefore, exposed flow fronts can be assumed to have formed nearly in place. The resurfacing rate for Kilauea is estimated to be of the order of ~1 kyr. *Holcomb* [1987] estimated that ~ 70% of Kilauea's surface is younger than ~ 500 yrs; ~90% is younger than ~1.1 kyr. Resurfacing is achieved primarily through tube-fed pahoehoe flows that comprise ~67% of the flows, while 14% consist of surface fed pahoehoe and 16% is 'a'a lava. The sonar imagery shown in Plates 2 and 5 suggest that much of the

EPR axis near 9° 50′N is resurfaced rapidly, probably at rates ≪1ky, because of the abundant scalloped flow fronts and lack of recent faulting of the volcanic carapace.

A more appropriate analogy for individual EPR lava flows may be the overlapping of flow fronts within a single eruptive episode. For example, early episodes of the Pu′u ′O′o eruption produced multiple flow fronts over the timescale of months that look very similar to the overlapping and scalloped-edged flows observed at the EPR in the 9° 50′N area within 0–2 km from the AST (Plates 2, 4b/e, 5a). Pu′u ′O′o flows advanced primarily as 'a'a sheets; while the rheology of the flow may differ at the MOR crest, the mechanism of advance (continuous sheets fed by lava channels) is likely to be similar. In general, Pu′u ′O′o flow fronts show a scalloped margin in plan view and can contain many distinct lobes. Later flows tend to generate flow fronts that in places mirror the existing flow margin, but exploit cusps between the scallops where it is possible to advance (Plate 4b/e).

Discerning the particular lava flow related to each subaerial flow front is difficult, if not impossible, for flows that cannot be confidently traced to their origin or are not directly observed—a luxury that one does not have in the deep ocean. Subaerial eruptive episodes will often produce multiple flow lobes that are emplaced adjacent to each other [e.g., *Kilburn*, 1996]. This results in multiple, lobate flow fronts that represent a single eruptive event. Such emplacement is more common in cooling limited flows, where further advance of the flow-front is inhibited by the cooled and crystallized lava, spurring break-outs from secondary vents ("bocas") along the existing flow path. Due to the rapid rates of cooling in the submarine environment, this process should be enhanced, and distinct flow fronts within the scalloped region may reflect approximately half as many individual eruptive episodes.

Lava drainback is a common feature near the vents of both subaerial and submarine lava flows [e.g. *Kurras et al.,* 2001; *Engels, et al.,* 2003] (Plates 4, and 6–7). Observed drainback occurrences in subaerial lavas include the cyclic filling and draining of Kilauea Iki lava lake [*Eaton et al.,* 1987], Mauna Ulu [*Swanson et al. 1973*; *Tilling et al.,* 1987], and Pu′u ′O′o vents on Kilauea's east rift zone [*Wolfe et al.,* 1987, 1988; *Greenland et al.,* 1988; *Barker et al.,* 2003]. The process of large-scale, subaerial lava lake drainback is believed to occur after vesiculating magma in the conduit expands, causing the lava lake to fill, and then catastrophically degasses providing space for the denser, degassed lava to drain [*Barker et al.,* 2003]. Smaller scale lava drainback is also observed at fissure vent eruptions where no lava lake is present. In those settings, drainback can be accomplished by the same rapid degassing mechanism, or by simply refilling the voided conduit after some volume of lava has been erupted on the surface. The timescale of drainback has been constrained for the Pu′u

′O′o lava lake by examination of geodetic and seismic records during cyclic periods of filling and draining [*Wolfe et al.,* 1987, 1988]. In all cases drainback occurred rapidly, on the order of tens of minutes. No such records exist for smaller volume eruptions, but timescales should be similar for both conduit degassing and void space drainback mechanisms. The short timescale of drainback produces some characteristic surface features that are preserved on the flow surface. On subaerial fissure eruptions, these features include draping of plastically deformed lava crusts into the evacuated vent and along fissure margins (Plate 4c/d). In some cases, this process occurs beneath a 5–25 cm thick pahoehoe crust that does not plastically deform, but breaks in to small (2–10 m^2) plates which founder as the supporting lava is removed. The difference between these two drainback forms results from the duration of the eruption and dynamics of lava at the vent, with the absence of a solidified upper crust signifying shorter eruptive timescales and the lack of a stagnated lava pond at the vent. Very similar lava morphologies indicating drainback are observed at the EPR within lobate, ponded and sheet flows [e.g. *Fornari et al., 1998a*; *Kurras et al.,* 2001; *Engels et al.,* 2003] (Plate 6).

In subaerial lava lakes and ponds, the onset of drainback coincides with violent bubbling that ejects spatter and chews through the existing lava crust. Surficial evidence of foundering crust is likely lost as drainback is generally cyclic and the floor of the lake or pond is resurfaced many times during the eruptive event. At Pu′u ′O′o, a cycle of lake-filling and -drainback occurred eleven times during a period of 50 days and then ceased. The cessation of this pattern is believed to reflect the development of a stable permeability within the conduit achieved either by the draining of blocks of crust into the upper conduit or by the formation of new sets of cracks and fluid pathways as the conduit inflated and deflated during pressurization and gas release events. With stable permeable pathways in place, volatiles can easily escape to the surface precluding the pressurization events necessary to bring lava to the surface.

Collapse and drainage features may also form where lava has not drained back into the vent, but away from the existing volcanic landform, either down-flow or laterally. Here rates of lava removal are controlled by the rheology of the slightly cooler, more crystalline lava, the geometry of fluid pathways within the molten flow interior, and the size of the aperture through which draining occurs. This process is common in subaerial flows where continued advance of the flow front is hindered by the development of a thick solidified crust (e.g. *Mattox et al.* 1993; *Hon et al.* 1994). In most cases, although timescales of drainaway may be similar, volumetric flux rates of drainage are considerably slower in the subaerial environment due to higher viscosities. As a result, features recording these events that are preserved on the flow surface indicate a

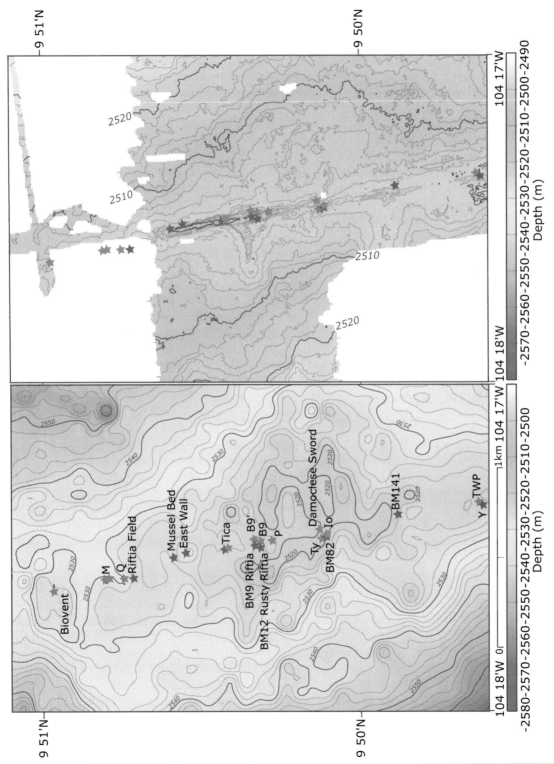

Plate 3. Left- multibeam sonar data (80 m grid, [*Cochran et al.*, 1999]) showing location of all high- (red labels) and low-temperature (blue labels) hydrothermal vents along the EPR axis between 9° 46′–51′N. Right- Detailed 2 m contours based on *ABE* 675 kHz scanning altimetric sonar surveys. Complete coverage of *ABE* micro-bathymetry data over the study area was gridded at 5 m horizontal and 1 m vertical resolution. Vent locations same as on left map. The ~20 m depth difference between the two data sets is attributed to the coarser vertical resolution of multibeam (~10–15 m) and slight differences in assumed velocities and ray paths for the multibeam data, as well as the *ABE* data being much higher frequency altimetric data added to vehicle pressure depth. The *ABE* data are considered much superior and more representative of the correct absolute depth.

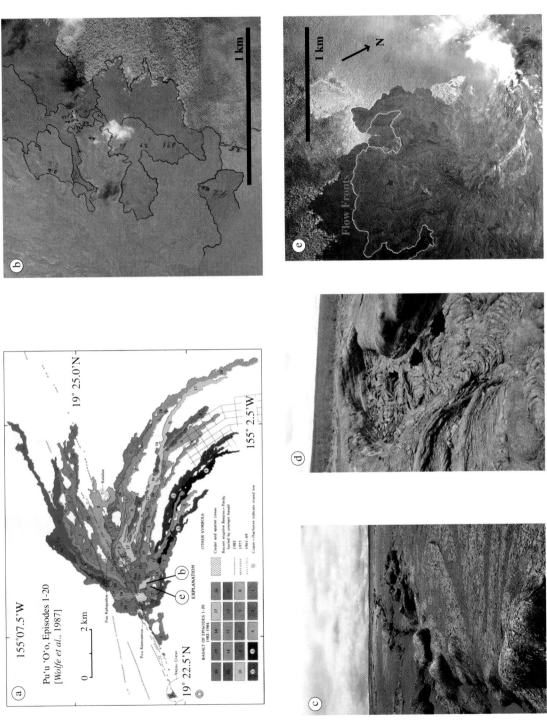

Plate 4. a) Eruptive history of Kilauea volcano, Hawaii [from *Wolfe et al.*, 1987]. b) Aerial photograph showing overlapping, scalloped flow fronts from Puu OO vent Episode 39 of the eruption (source: USGS HVO). c) Eruptive fissure along the east rift zone, showing drainback along the fissure margin. d) Lava channel on Maunu Ulu, note series of collapse pits in the floor of the channel. e) Aerial photograph showing overlapping, scalloped flow fronts from Puu OO vent Episode 40 of the eruption. The active vent is at lower right of the photograph. (source: USGS HVO).

slow sagging of the surface crust rather than foundering and rapid crustal breakup (Plate 4c/d).

At the EPR, eruptions are likely to have similar styles of eruption focusing and short-term (days to <<month?) episodicity within an eruptive phase with waxing and waning of activity. Evidence for episodicity includes multi-storied collapse features (Plate 7-b, and figures in *Engels et al.,* [2003] and *Kurras et al.,* [2002]) that appear to have formed by inflation, lava withdrawal and subsequent additional inflation of the flow, with a final phase of collapse utilizing the original drainback area, thereby exposing two prominent lava crusts within the same collapse feature. Short durations (hours to days) have also been observed for submarine eruptions monitored using hydrophone arrays in the NE Pacific, where the Axial Volcano eruption had several phases and lasted 12 days and the Coaxial eruption lasted 31 days [*Fox et al.,* 1995, 2001; *Embley et al,* 1995; *Chadwick,* 2003].

4. DATA ACQUISITION

Sidescan sonar mapping and *ABE* surveys were carried out during a cruise on R/V Atlantis (AT7–4) in November 2000 [*Schouten et al.,* 2001; 2002] (Plate 1). *ABE* collected near-bottom magnetic data (using a 3-axis fluxgate magnetometer) and micro-bathymetry (using an Imagenex 675 kHz pencil beam altimeter) along track lines spaced ~40–60 m apart at an altitude of ~40 m above the seafloor [*Yoerger et al.,* 1996; *Tivey et al.,* 1997; 1998; *Schouten et al.,* 2001]. *ABE* surveys were navigated using long-baseline (LBL) bottom-moored transponders resulting in navigational accuracies generally better than ~5m (see *Fornari et al.* [1998a] and *Lerner et al.* [1999] for a discussion of LBL navigation techniques); RMS solutions for transponder positions for our surveys were all ≤1.0 m. We used the DSL-120A sonar vehicle, a 120 kHz sidescan sonar system towed at 100 m above the seafloor, capable of producing 1–2 m pixel resolution images of seafloor backscatter amplitude [e.g., *Stewart et al.,* 1994; *Scheirer et al.,* 2000]. Sidescan sonar survey lines were navigated using both LBL transponders and a layback calculation that employed sonar depth and wire out; accuracy is ~5 m.

Plate 3 shows bathymetric maps of the same area around 9° 50′N made using surface ship multibeam sonar and ABE near-bottom altimetry. The ~20 m depth difference between the two is attributed to the coarser vertical resolution of multibeam (~10–15m) and slight differences in assumed velocities and ray paths for the multibeam data, in addition to the *ABE* data being much higher frequency altimetric data added to vehicle pressure depth. The *ABE* data are considered much superior and more representative of the correct absolute depth. Also, we note that the trace of the AST in the sidescan data,

over the 55 km long area mapped in 2001 [*Schouten et al.,* 2002] (Plate 3), is well-correlated with detailed features mapped in 1989 using Argo II [*Haymon et al.,* 1991; *Fornari et al.,* 1998a], and in 2000 using the previous DSL-120 sonar vehicle [AHA-Nemo2 cruise, *Fornari et al.,* 2001; *White et al.,* 2002] (Plates 2 and 3). In some cases, the trough position from 1989 is offset slightly while the overall shape is consistent with sonar reflectors mapped using 2001 data, suggesting that a small (~50 to100 m) navigational shift would correct the mismatch (D. Fornari and J. Escartin, unpublished data). Plates 3 and 5 show the trace of the AST based on analysis of the 2001 sidescan sonar data.

In 2001, we also used a prototype, digital deep sea camera system [I. Macdonald, pers. commun., 2000; *Fornari,* 2003] to image the seafloor along traverses across selected backscatter facies and volcanic contacts displayed in the sidescan imagery. Camera tows were navigated using LBL transponders with positional accuracy of ~5 m. Historical observation and photographic data from over a decade of Alvin diving in this area provided ground truth for interpretation of sonar maps and relationships between hydrothermal vent sites (Plate 2) and volcanic features within and adjacent to the AST [e.g., *Haymon et al.,* 1993; *Von Damm et al.,* 1995; 1996, 2001; *Fornari et al.,* 1998a,b; *Shank et al.,* 1998; *Engels et al.,* 2003; http://www.whoi.edu/marops/vehicles/alvin/epr_photos.html; Soule and Fornari unpublished GIS EPR photo database]. Time-series temperature data were collected using self-recording probes described in detail by *Fornari et al.* [1994, 1996, 1998a,b, 2003b]. Detailed discussion and interpretation of near-bottom magnetic data and the complete sidescan sonar data set, which cover the EPR axial region over an area centered on the AST and extending out to ~4 km on either side, between 9° 26′N to 10°N, will be presented elsewhere (*Schouten et al.,* in prep.; *Tivey et al.,* in prep.; *Fornari et al.,* in prep.).

5. EPR 9° 26′–10°N VOLCANIC HISTORY AND SEAFLOOR MORPHOLOGY

The synoptic view provided by the sidescan imagery shows the EPR axis to be dominated by the scalloped (convex outward on either side of the AST) acoustic reflectors that we interpret to be lava flow surfaces and flow fronts (Plates 2 and 5); a finding that has been corroborated by towed camera surveys and observations from Alvin. This observation is common in the near axis terrain from 9° 55′N to 9° 27′N, extending outward from the axial trough to variable distances— usually 1–2 km—on either side of the trough. Many areas of the EPR crest in the mapped region between 9° 26′N–10°N show extensive re-paving by successive, small volume flows;

these are associated with short (~100–500 m), scalloped flow-front margins consistent with eruptions originating from the axis and flowing onto the upper rise flank out to ~2 km [*Schouten et al.*, 2001, 2002] (Plates 2 and 5). There are few faults within ~2 km of the AST (Plate 2), but they are more common south of the 9° 37′N OSC [*Smith et al.*, 2001]. Most faults have low relief (<~10 m), based on sonar shadow geometry and *ABE* micro-bathymetry data. Most faults parallel the 352° trend of the EPR axis in this area, but some deviate from that azimuth by as much as ~10°, and most faults show some sinuosity along their strike.

Numerous dendritic lava flow channels emanate from the AST in various sections of the ridge axis principally in the 9° 50′N, 9° 48′N, 9° 43′–45′N, 9° 37′N, and 9° 26′–29′N areas (e.g., Plates 2 and 5). These channels are similar to those identified by Cormier et al. [2003] at the southern EPR near 17.5°S. We confirmed with towed camera imagery in 2001 that the low-backscatter channels are floored by smooth-surfaced sheet lava. Sidescan sonar data show that many of the channels originate at the AST rim, however some do not and they are believed to be somewhat older channels that have had their source-proximal ends paved over by more recent flows (Plate 5). Along the EPR axis between 9° 26′N and 9° 29′N, the channelized flows cover areas as great as ~5 km along strike, and coalesce away from the trough at distances of ~500 m to <1000 m [*Schouten et al.*, 2002]. Spacing between sheet flow channels can be from a few hundred meters to several kilometers.

Fields of lava mounds or ridges ~10–30 m high, some comprising coalesced volcanic constructs, are several hundred meters wide and up to ~1 km long. These features usually occur at distances of >1.5 km from the AST (*White et al.*, 2000). In some areas the mounds are concentrated at ~2–2.5 km distance from the AST (e.g., in the region between ~9° 53′–57′N, 9° 43′–47′N and 9° 31′–34′N on the east side of the axis, and at 9° 43′N, 9° 35′–36′N west of the axis). Some pillow mounds are cut by faults and fissures, while in some areas closer to the axis they are not, suggesting their construction is more recent than tectonic features they overprint [*Schouten et al.*, 2001, 2002; *Kurokawa et al.*, 2002].

Sidescan data suggest that the EPR crest has experienced four types of volcanic emplacement processes: (1) axial summit eruptions within a ~<0.5 km wide zone centered along the present trace of the AST (Plates 2 and 5); (2) off-axis transport of lava erupted within or near the AST through channelized surface flows and tubes extending to ~2 km from the axis; (3) local constructional volcanism at distances of <2 km from the axis (Plate 5); and (4) off-axis eruptions at >2 km from the AST. The predominance of lobate flows (59% of the surveyed area) throughout much of the crestal region, and common scalloped flow front morphology along much of the

ridge crest suggest that individual eruptive volumes have been small, similar to that estimated for 1991 EPR flow (~1 x 10^6 m^3, *Gregg et al.*, [1996]).

Our interpretation of lava morphology and related data sets from this ridge segment suggests that the portion of the EPR between ~9° 29′N to 9° 51′N has not experienced large-scale volcanic resurfacing in the past ~30 ka from volumetrically large eruptions that have covered broad areas of the ridge axis. Instead, the resurfacing is accomplished by frequent, small volume (<1–2 x 10^6 m^3) [*Perfit and Chadwick*, 1998] eruptions that do not usually extend much further than ~1–2 km from fissures within or along the AST—the predominant eruptive lineament along the axial zone. Based on sonar reflectivity and overlapping flow fronts and edges, we estimate that much of the EPR crest, between 0–~2 km from the axis, in the 9° 25–55′N area is surfaced by flows that are <1 kyr old. This compares to a mean spreading-based age of ~ 40 kyr for ocean crust at a distance of 2 km from the AST.

6. THE AXIAL SUMMIT TROUGH AND HYDROTHERMAL VENTS AT THE EPR 9° 46′–51′N

The primary fissure system of the 1991 eruptive event is characterized by a relatively continuous, 1–3 m-wide fissure usually located along the center of the AST (Plates 3, 5, and 7a–f) [*Haymon et al.*, 1993; *Gregg et al.*, 1996; *Fornari et al.*, 1998a]. The fissure has an en echelon form and occasionally steps laterally a few meters, such as the right lateral offset of ~2–3 m near the Bio9/9′ vents [*Shank et al.*, 1998]. The depth of the fissure has been observed to be from ~1 m to >3 m, beyond the visual field of view from Alvin's viewports. However, scanning altimeter data show the fissure can extend to a depth of >~7 m, the limit of the sonar's beam. The morphology of the primary eruptive fissure system exhibits many features that reflect its formation and evolution. Often, there is extensive platy talus along the fissure; the talus plates generally are draped along the margin of the fissure, suggesting formation during drainback (Plate 6).

In places, the fissure is filled with sheet flow talus forming a chaotic assemblage of tabular lava fragments that is the substrate for animal communities often found along the axis and sides of the fissure system (e.g., BioMarkers 82, 119 and 141 biological communities [*Shank et al.*, 1998]) (Plate 3). Because of the chaotic nature of the lava flows, primary fissure, and extensive lava pillars along the AST wall throughout the study area, it has been difficult to get a broad, overview of the morphology of the terrain and to relate the vent sites and their edifice morphology to the surrounding volcanic features. The sidescan sonar and micro-bathymetry permit us to correlate the hydrothermal and volcanic features unequivocally for the first time.

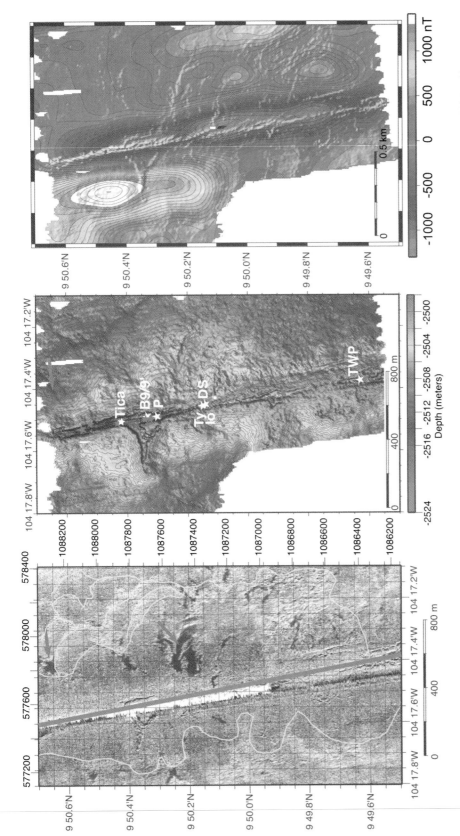

Plate 5. Left map shows a blow-up of the sidescan sonar imagery for the area around 9° 50´N, with prominent channelized sheet flows highlighted by red arrows. Middle map shows detailed micro-bathymetry [*Schouten et al.*, 2001, 2002]. High temperature vents labeled in white, low-temperature diffuse flow vents and animal communities are shown by blue stars. Red arrows point to channelized sheet flows mapped in the DSl-120A sonar data which can also be resolved by the detailed bathymetry. Right map shows reduced to pole magnetic map of the EPR AST based on ABE near-bottom magnetic data collected in 2001–02 on AT7–4 and AT7–12 cruises at same scale as other maps in this figure [*Schouten et al.*, 2001, 2002; *Tivey et al.*, 2003]. Prominent low is coincident with the axis of the axial trough suggesting very thin crust (lava) and presence of shallow dikes. Elongate closed-contour lows around northern vents (Bio9/9´ and P) at 9° 50.3´N and Ty and Io vents near 9° 50.1´N suggest possible 'burn holes', that are smaller but similar to the near-bottom magnetic expression of vents imaged at JDF Main Endeavour Field [*Tivey and Johnson*, 2002]. Because of our 60 m line spacing, more closely spaced lines are required to properly image the smaller EPR vent systems. The prominent magnetic high just west of the AST near 9° 50.4´N is likely to be a primary depo-center for the 1991–92 eruption and is coincident with the large breakout from the AST believed to be related to extensive drain back at this location; proximal to the locus of high-temperature venting at Bio9/9´ and P vents.

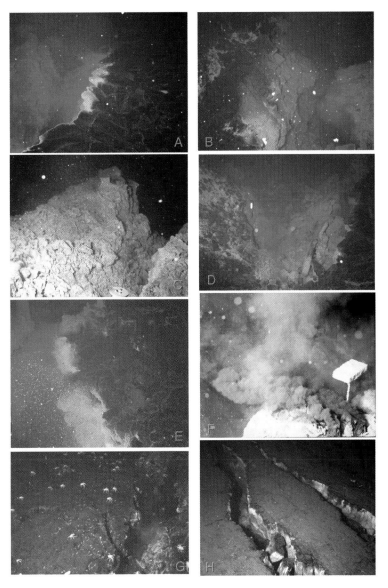

Plate 6. Morphological characteristics of the primary volcanic eruptive fissure system in the 9°50′N region in April 1991 (A-F) as imaged from *Alvin*. Scale in meters across the foreground of each photograph is given in brackets at the end of each description. (A) View along the west edge of the primary eruptive fissure in the 9° 50.1′N region. Note platy sheet flow talus at the margin of the fissure and bacterial coating within the fissure and in cracks in the sheet flow surface [Haymon et al., 1993] [3]. (B) View inside the primary fissure in the 9° 50.0′N area showing bacterial covered fresh lava surfaces and inward dipping plates of sheet lava along left edge of photo [2]. (C) Bacterial mat covered tumulus in the area just north of the Bio9/9′ vent area cleaved by the primary fissure [2]. (D) Talus filled axis of the primary fissure just south of area of photo in (B) where the fissure widens to ~3m [2]. (E) Edge of the primary fissure in the 9° 49.8′N area showing bacteria along the wall and jumbled sheet flow talus on the floor of the AST [3]. (F) Syntactic foam marker (square at right edge of photo) covered with bacterial matter placed a few days earlier in the 'Hole to Hell' area where the Bio9/9′ and P vents would later localize. Note black smoker venting issuing from bare fresh basalt at lower right corner [2]. (G) Tensional cracks in a sheet flow on the floor of the AST near 9° 30′N imaged in 1991 showing extensive diffuse flow and galatheid crabs and mussels (yellow/brown shells at lower right in milky water) [2]. (H) Extension of the cracks seen in photo (G) ~ 30 m further north where no focused low-temperature vent activity was present in 1991. The AST in the 9° 30′N region is ~250 m wide and shows extensive evidence of tensional cracking above recently intruded dikes, however, there is no well developed primary eruptive fissure in this area [*Fornari et al.,* 1998a].

The fundamental physical relationship between the eruptive volcanic fissure, the underlying zone of diking and seafloor exposures of hydrothermal fluid egress from the crust is well defined for the 9° 46′–51′N area. Plates 2, 3 and 5 show the spatial relationships between the high- and low-temperature vents within the AST in this area. Plate 5 shows the locations of three of the most vigorous high-temperature hydrothermal vents near 9° 50′N (Bio9, Bio9′ and P vents), and their close association with two of the most prominent channels in the AST wall adjacent to these vents. These channels interrupt the continuity of the AST wall and are presumed to represent 'breakouts'; where erupting lava has breached the wall and flowed down slope.

Nearly all the hydrothermal vents shown on Plates 2, 3 and 5 are located along the floor of the AST and most occur within or proximal to primary eruptive fissures that fed the 1991 eruption [*Haymon et al.*, 1993; *Rubin et al.*, 1994; *Gregg et al.*, 1996; *Shank et al.*, 1998; *Fornari et al.*, 1998a] (Plates 8–9). This area has been intensively studied since the 1991 volcanic eruption. We have a good understanding of the general location of high- and low-temperature vents and their associated biological communities and of their relationships to ridge segmentation, morphology, and basalt geochemistry [e.g., *Haymon et al.*, 1991, 1993; *Perfit et al.*, 1994; *Lutz et al.*, 1994; *Fornari et al.*, 1998a; *Shank et al.*, 1998; *Mullineaux et al.*, 1998; *Perfit and Chadwick*, 1998; *White et al.*, 2002; *Haymon and White*, in press]. In addition, fluids from these vents have been regularly sampled over the past ~10 years [e.g., *Von Damm*, 2004 (this volume and references therein)]. Continuous time-series temperature measurements have also been acquired providing a historical baseline for the fluid characteristics of individual vents and their variability with time. [e.g., *Fornari et al.*, 1998b; *Fornari and Shank*, 1999, *Scheirer et al.*, submitted, *Shank et al.*, submitted],

There are two distinct clusters of high-temperature vent fields in our study area, one at ~9° 50′N and one at ~9° 47′N (Plates 2–3). The distance between these two areas is ~7 km and simple geometric considerations as well as vent fluid chemistry suggest that vents in these two areas are primarily fed by fluid convection cells, each with its own water-rock reaction zone. Exit fluid compositions are also highly variable between individual vents within each field [e.g., *Von Damm et al.*, 1995, 1996 1997; *Oosting and Von Damm*, 1996; *Von Damm*, 2000, 2004 (this volume); *Ravizza et al.*, 2001; *German and Von Damm*, 2004], as well as between the two adjacent fields, demonstrating that subsurface flow conditions can vary despite close proximity between exit orifices. The high temperature hydrothermal vents located between 9° 49′–51′N fall into two categories based on their spatial relationship with the AST (Plates 3 and 5). Biovent, M and Q

vents are situated along or just outside the AST walls and we presume that fluid pathways are largely controlled by fractures associated with the margins of dikes intruded along the trace of the AST. Tica, Bio9, Bio9′, P, Ty, Io and Tubeworm Pillar (TWP) vents are all located within the floor of the AST, along the primary fissure system. A brief summary of the salient characteristics of each, based on observations made from Alvin between ~1991–2002 follows.

Biovent, the northernmost high-T vent in our study area, is located near 9° 51′N on the western rim of the axial trough (Plates 2, 3 and 8a). Based on towed camera and Alvin dive observations, it is the only high-T vent on the EPR crest between ~9° 51′N and 9° 55′N. In 1995, self-recording temperature probe data [*Fornari et al.*, 1998b] (N.B. probes have resolution of ± 1°C) indicate Biovent fluids were 352°C. In 1997, fluid temperatures at this vent were 343°–345°C and at the end of the recording period, in September 1998, Biovent fluid temperatures had decreased to 341°C [*Fornari et al.*, 1998b; *Von Damm*, 2000]. In January 2002, Biovent fluids measured 345°C.

M vent, the next vent south of Biovent, is located on the eastern wall of the axial trough at 9° 50.8′N (Plates 3, and 9b–c). In 1995, M vent fluid temperature based on self-recording probe data was 358°–360°C (Alvin high-T probe measured 365°C). The time-series fluid temperature record for this vent is extraordinary in that we see no variations greater than a few degrees, or within the 1°C resolution of the HOBO, self-recoding probe [*Fornari et al.*, 1998b]. Alvin high-T probe temperature for M vent in May 1999 was 365°C. For 4 years, M vent fluid temperature recorded at 30 min intervals by the probes was constant at ~358°–360°C. The stability during this period is in marked contrast to the fluid temperature records for high-temperature vents south of M in the Bio-Transect. In January 2002, M vent fluids ranged from 369°–374°C. Q vent, just south of M vent on the eastern AST wall, has not been instrumented successfully although it has been repeatedly sampled for fluids [*Von Damm*, 2000, 2004].

Tica vent lies within the center of the AST floor along the primary eruptive fissure, between the M/Q vent area to the north and the Bio9/9′/P vent complex to the south (Plates 3, 5, and 8d). This hydrothermal site consists of vigorous, low temperature diffuse flow which has supported ~13 discrete patches of tubeworms since 1997. The vent field measures ~30 x 60 m and occurs on pressure ridges formed in lobate and ponded flows in the center of the AST floor. High temperature venting at this site appears to have begun only recently, within the last 2 years. In December 2003, the recorded temperature of fluids from the ~2 m tall black smoker chimney at Tica was 342°C. The ~2 m-wide flat-topped sulfide structure is located towards the southern end of the field with extensive colonies of *Alvinella pompejana* and *Alvinella caudata* polychates thriving from almost the base to the top. *Cyanograea* crabs and *Hesiolyra* polychaetes were also present. In December 2003, the *Riftia pachyptila* populations surrounded Tica vent, extending from the basalt around the base to over the height of the vent. No long-term temperature studies have been carried out at Tica.

Three high-T vents within the AST are located near the northern limit of the BioTransect area ('Hole-to-Hell' locale of *Haymon et al.,* [1993]): Bio9, Bio9′, and P vents [*Shank et al.,* 1998; *Fornari et al.* 1998b; *Von Damm and Lilley,* 2003, *Von Damm,* 2000, 2004 (this volume)] (Plates 2–3, 5, 8b, and 9d–f). Bio9′ and Bio9 vents are located within a ~5 m of each other along the primary eruptive fissure of the 1991 eruption [*Haymon et al.,* 1993; *Fornari et al.,* 1998a,b]. P vent is located ~40 m south of Bio9 vent and consists of a sulfide mound ~1–3 m tall that has grown along the primary eruptive fissure and upon which there are typically three principal chimney structures. Traditionally, fluids and temperature measurements from this small complex are sampled at the southernmost chimney. While Bio9/9′ have had extensive tubeworm communities developed in the areas immediately surrounding the vents and on the adjacent eastern wall of the AST, P vent is largely devoid of Riftia tubeworm communities [*Shank et al.,* 1998]. Both vent areas have had extensive areas of bacterial mats and Alvinellid communities on the surfaces of the high temperature chimneys [*Shank et al.,* 1998; *Von Damm and Lilley,* 2003]. The most extensive time-series exit fluid temperature data exist for the Bio9 vent; and to a lesser extent, Bio9′ and P vents. Selected exit fluid temperature data for the Bio9 vent and other vents within the BioTransect are shown in Plates 10 and 11.

There are three other high-T vents in the BioTransect [*Shank et al.,* 1998]. Ty and Io vents became active in 1997 and are located ~250 m south of P vent, near the Marker 82 diffuse flow biological community [*Shank et al.,* 1998], in an area of jumbled sheet lava and drainback features along the primary eruptive fissure (Plate 5). Ty vent (Plate 9a) is located ~5 m north of Io vent, and in November 1997 it was a ~2 m tall chimney venting fluid at 352°C. The HOBO probe at Ty vent recorded data from November 1997 to September 1998 after which recording stopped due to low battery power [*Scheirer et al.,* submitted]. During the recording period, Ty showed a significant decrease in temperature (~10°C) that also coincided with the onset of the Bio9′ and Bio9 temperature event in late November 1997 (Plate 11). Fluid temperatures at Ty vent continued to decrease until January 1, 1998 and then began a slow, steady rise back to pre-event levels. Alvin high-temperature probe measurements in May 1999 at Ty show the fluid temperature was 352°C; in January 2002, it measured 350°C [*Scheirer et al.,* submitted]. In 1997, when Io first became active, it was a 1–2m tall, ~1 m diameter

stump of diffuse flow and hotter (~>100°C) venting fluids. When visited in May 1999, Io was a ~9 m tall, ~2 m diameter edifice vigorously venting hot, diffuse fluid over its entire surface area. At that time, 348°C fluids were venting from orifices at the top of the structure, including some orifices that had classic "beehive" sulfide caps. Io fluid temperatures had reached 356°C by Jan. 2002 [*Scheirer et al.,* submitted].

Ty, Io and Damoclese Sword vents, located ~300 m south of Bio9/9′ and P vents, are also flanked to the east by a small breakout channel along the east rim of the AST (Plates 3 and 5). There are hints in the sonar data that a similar channel exists to the west of the Ty/Io area but the vehicle nadir obscures the character of the west margin of the AST at this site. The Ty/Io vents are near the middle-eastern side of the AST which is only ~ 60 wide in this location; Damoclese Sword vent is just inside the east wall of the AST. The primary eruptive fissure widens in this area and low-temperature diffuse flow, and vent animal colonization has occurred along the fissure several tens of meters north and south of the black smoker vents since the 1991 eruption; one of the primary biological sites is BioMarker 82 (Plate 5) [*Shank et al.,* 1998].

The southernmost vent in the study area is located at the top of Tube Worm Pillar (TWP) (Plates 5 and 8e–f). TWP is a large lava draped edifice (~10 m tall, ~2–3 m diameter) at the southern end of the BioTransect [*Shank et al.,* 1998] and hosts a large vestimentiferan tubeworm community on its sides and top. TWP was active in 1989 and 1991 (before the 1991 eruption). In 1992, TWP was venting 160°C fluids, and it rapidly became a black smoker by1993 when it was sampled and fluid temperatures were measured at 351°C. Since then, fluid temperatures at this vent have dropped steadily and the number of tubeworms on its sides has decreased dramatically. When visited in January 2002 the exit fluids were only 186°C [*Scheirer et al.,* submitted] and in April 2004 there was no active black-smoker on TWP.

7. DISCUSSION

We hypothesize that breakout channels (Plates 3 and 5) along the AST walls can serve as long-term (10s to 1000s of years) controls on lava distribution at the EPR [*Schouten et al.,* 2001, 2002]. Studies of channelized submarine sheet flows and of the quantitative distribution patterns of various lava morphologies within the AST and around hydrothermal vents provide important first-order constraints on: 1) MOR volcanic effusion rates [e.g., *Gregg et al.,* 1996; *Gregg and Sakimoto,* 2001; *Parfitt et al.,* 2002], 2) the style of volcanic emplacement [e.g., *Embley and Chadwick,* 1994; *Fornari et al.,* 1998a; *Chadwick et al.,* 2001; *Chadwick,* 2003; *Engels et al.,* 2003], and 3) their role in creating shallow crustal fluid flow pathways that utilize volcanically induced permeability

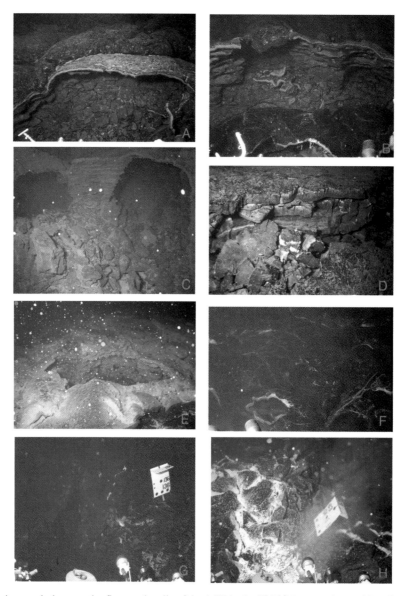

Plate 7. Volcanic morphology on the floor and walls of the AST in the 9° 50′N area as imaged by *Alvin*. Scale in meters across the foreground of each photograph is given in brackets at the end of each description. (A) Collapse blister in lobate flow with 1991 lava mantling the top of the blister [2]. This example underscores the potential for intra-flow permeability in the fast-spreading MOR environment where lava flows can effectively seal off large water-filled cavities in the underlying flows. (B) West rim of the AST showing top surface of fresh, glassy 1991 lava (lower edge of photo) just below the rim [3]. Note that the 1991 flow is also collapsed (bottom center of photo). (C) Lava tube/archways in the wall of the AST [3]. (D) Rim of AST showing thin flow units and small gaps between flows. This is an example of intra-flow permeability. Bottom right corner of photo shows a collapsed portion of the wall, tilted into the AST [2]. (E) Collapse pit just outside the rim of the AST (lower right corner) with extensive coating of bacterial matter. This is an example of inter-flow permeability and underscores the high potential for fluid flow through large drainback and collapse-generated void space within lobate and sheet flows at the EPR in this area [3]. (F) Glassy 1991 lobate lava showing small windows in lobes indicating the flow is partially water-filled; another example of intra-flow permeability [1.5]. (G) Well-defined actively venting primary eruptive fissure adjacent to Biomarker 136 continues to vent since deployment of the marker in 1992 [3]. Note platy talus beneath marker that dips in towards the axis of the fissure. (H) North of Biomarker 63 the fissure system was less well defined adjacent to Biomarker 63, but was also actively venting in 1992 as seen here, and supported a small tube worm community [2]. Twenty-one months later, this fissure was inactive.

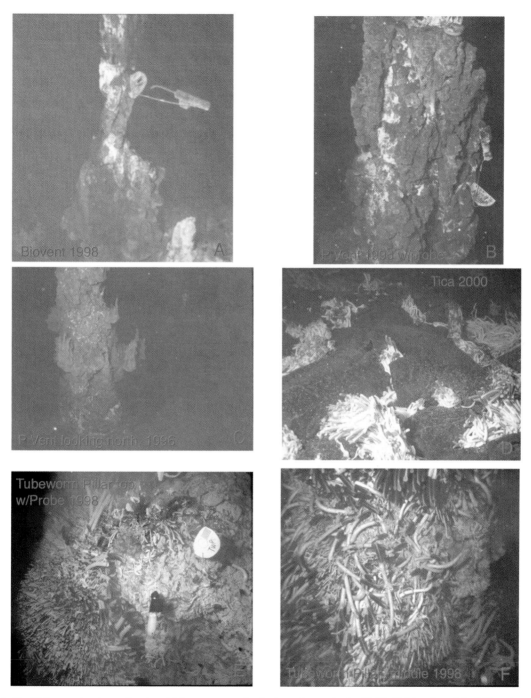

Plate 8. Chimney structures of various high-temperature hydrothermal vents in the 9° 49′–51′N region of the EPR as imaged from *Alvin* (see Plates 2, 3 and 5 for locations). Scale in meters across the foreground of each photograph is given in brackets at the end of each description. (A) Biovent in 1998 [1.5]. (B) P vent with HOBO temperature probe (lower right) in 1996 [2]. (C) P vent mound comprising 3 primary chimneys on top of the sulfide rubble mound constructed along the primary eruptive fissure as viewed from the northwest in 1996. P vent is normally sampled at the southernmost chimney [4]. (D) TICA diffuse flow biological community located at 9° 50.4′N in lobate lavas on the floor of the AST [2]. (E) Top of Tube Worm Pillar in 1998 showing HOBO temperature probe in the black smoker vent at the summit of the ~ 10 m-tall edifice [2]. (F) Middle portion of Tube Worm Pillar in 1998 showing assemblage of Riftia, Tevnia and other biological constituents that are bathed by warm hydrothermal fluids that seep out of the sides of the pillar [2].

structure within and between lava flows. The volcanic stratigraphy within the AST and within the shallow crust plays an
important role in determining the fluid pathways that serve to
mix high-temperature hydrothermal effluent with ambient
seawater—resulting in the diffuse flow fluids that bathe animal communities at sites surrounding the high-T vents. We
suggest that areas where channelized volcanic flows have
breached the rim of the AST are also the primary eruption
centers for the 1991 lava and the sites of extensive post-eruption drain back of lava into the primary fissure. These localized sites where the eruptions focused and drainback occurred
along the fissure in the floor of the AST provide important controls on shallow crustal permeability and vent fluid pathways.
These environments are likely to localize low- and high-temperature hydrothermal venting and may influence reaction
zones for low-temperature fluids, thereby impacting the colonization and development of biological communities.

Many authors have reported on the close association of
fresh volcanic terrain, fissures and faults in an intermediate to
fast-spreading MOR axial zone with hydrothermal venting
[e.g., *Ballard et al.*, 1979; *Francheteau and Ballard*, 1983;
Delaney et al., 1992, 1997, 1998; *Haymon et al.*, 1991, 1993;
Wright et al., 1995a,b; *Wright*, 1998; *Curewitz and Karson*,
1998; *Fornari et al.*, 1998a; *Embley et al.*, 2000; *Chadwick et
al.*, 2001; *Chadwick*, 2003]. What is pertinent to the present
discussion is the extent to which locations of specific
hydrothermal vent sites can be related to primary volcanic
features (e.g., eruptive fissures and vents, collapse features,
channels, lava tubes, etc.) (e.g., Plate 6) or tectonic features,
and if these spatial associations are linked to spreading rate and
ridge crest morphology/structure.

For example, *Embley et al.* [2000] and *Chadwick et al.*
[2001] provided detailed submersible and remotely operated
vehicle (ROV) observational data to link locations of hydrothermal vents formed during the 1993 CoAxial eruption (e.g.,
Source and Floc sites) to fissures and grabens developed
above the intruded dike. In other examples along the Juan de
Fuca, field relationships show that the locations of hydrothermal vents (e.g., Monolith vent) can be related primarily to
narrow zones along fissures and small (~20 m wide) graben
developed as a result of the stress field imparted by the intruded
dikes [*Chadwick et al.*, 2001]. At intermediate spreading
MORs, the deeper depth to the magma lens (~2–3 km) [e.g.,
Morton et al., 1987; *West et al.*, 2001], compared to the shallower magma lens depth (~1.5 km) beneath the EPR [e.g.,
Detrick et al., 1987; *Sinton and Detrick*, 1992; *Kent et al.*,
1993; *Harding et al.*, 1993], and the structural regime in the
~2–4 km wide rift valley—with its prevalence of tensional
fractures and faults—provide a more distributed system of
vertical pathways for hydrothermal fluids to exit the crust.
The recent near-bottom magnetic surveys of the Main Endeav-

our Hydrothermal Field (MEF) [*Tivey and Johnson*, 2002]
point to the presence of well-defined magnetic lows suggesting vertical 'pipes' spaced ~200 m apart where primary
hydrothermal upflow is concentrated in the Endeavour rift
valley. High-resolution near-bottom multibeam data for the
MEF [*Johnson et al.*, 2002; *Tivey et al.*, 2003], when combined
with the *in situ* and magnetic data, clearly show the primary
fault influence on distribution of the vents in this intermediate-spreading MOR setting.

Based on our interpretation of the micro-bathymetry and
sidescan sonar data, and visual and photographic observations of this terrain we concur with *Haymon et al.* [1993] and
Wright et al. [1995a, b] that the first-order control on
hydrothermal vent localization and vertical fluid flow pathways
within the axial trough in the study area are provided by the
1991 eruptive volcanic fissures within the EPR AST between
9° 46´–51´N, and the underlying feeder dikes. Our working
hypothesis is that the seafloor locations of active hydrothermal chimneys in this area are located where the eruption
focused along the fissures and where drainback of magma
occurred. The permeability structure of the uppermost crust
at these sites is likely to be enhanced because of the repeated
cycles of waxing and waning of eruptions within the magmatic conduits and interactions with seawater vapor during
eruptions [*Perfit et al.*, 2003]. Only Biovent, M and Q vents
are located along the bounding wall or at the rim of the AST.
All the other high-temperature vents within the AST are located
along the trace of 1991 eruptive volcanic fissures that lie near
the center of the trough floor (Plates 3 and 5). For the vents near
9° 46´–47´N (Plates 2–3) about half are located within the
AST and the rest are located along the west wall of the trough
[*Von Damm et al.*, 1995].

We suggest that the locations of seafloor vents that comprise
hydrothermal systems at 9° 50´N are being guided by two
primary geologic features. The first are fractures associated
with the vertical to near-vertical dikes that fed the 1991 eruption. The second involves permeable zones created during the
focusing of the eruptions and subsequent drainback of
magma along the widest portions of the eruptive fissure and
the ensuing chaotic mix of wall rock and platey talus fragments of new lava from the eruption that are rafted down the
fissure. The area surrounding Bio9 was described by *Haymon
et al.* [1993] as the 'Hole to Hell' because of the intense and
turbulent hydrothermal flow exiting bare fresh basalt and
extensive evidence for surficial hot rock-water interaction
(Plate 6f), such as would be expected in a zone of active
drainback during the waning stages of a seafloor eruption.
Dike models and observational data suggest that where eruptive fissures are widest the greatest volume of magma erupts,
although many factors can play a role in localization of magmatic flow in an eruptive fissure [e.g., *Delaney and Pollard*,

1982; *Bruce and Huppert*, 1989; *Parfitt and Wilson*, 1994; *Wylie et al.*, 1999]. Although evidence for drainback is often cited in descriptions of submersible-based observations of seafloor volcanic terrain at the EPR and Juan de Fuca it has not been directly related to hydrothermal vent localization [e.g. *Fornari et al., 1998a; Chadwick*, 2003]. *McClain et al.* [1993] and *Rohr* [1994] suggest that volcanic drainback can be important in producing high permeability and anomalously low seismic velocities in shallow crust on the Juan de Fuca. The high-resolution sonar data and visual observations of the EPR AST terrain in the 9° 50′N area, and relationships discussed above provide correlative spatial information that can be used to relate eruption and volcanic processes with hydrothermal vent locations.

We propose that a combination of near surface dike position and narrow zones of permeability created by chaotic post-eruption drain back are providing a primary control on vent location and vertical fluid circulation above the water/rock reaction zone. The depth to the water/rock reaction zone in the study area based on chlorinity and silica geobarometry data [*Von Damm*, 2000, 2004] has varied over time since 1991. The calculated depth of the water/rock reaction zone was as shallow as ~150 m in the 9° 50′N area during the 1991 eruption and can be deep as ~1 km. Von Damm [2004 (this volume)] provides further evidence for the temporally variable nature of the reaction zone beneath the 9° 50′N "Hole-to-Hell" vents and their shallow depth based on the high temperatures of the venting fluids (~>350°C) and the corroborating Cl and Si data; which point to a reaction zone that is within the upper few hundred meters. This depth includes the calculated transition zone between upper extrusive sequences and sheeted dikes determined from geophysical data to be at ~ <200 m depth [*Kent et al.*, 1993; *Harding et al.*, 1993]. As Von Damm [2000, 2004 (this volume)] points out, clearly the heat source has been varying over the decade since the 1991 eruption, and there is ample suggestion based on recent vent chemistry and observations at these vents that magma may have been injected below the AST to shallow crustal levels above the magma lens. Magma injection and fluid flow through primary upper crustal fractures, which are likely to be reactivated frequently by the many micro-earthquakes in this fast-spreading environment, provide the primary control for transferring heat and chemical signatures to venting fluids.

We suggest that secondary control on the shallow fluid flow paths in the upper crust—which can be influential in modulating fluid temperatures and chemistry depending on geometry of flow paths results from the AST morphology, variable nature of seafloor volcanic morphology within the trough and the complex shallow stratigraphy of lava sequences that result in inter-flow and intra-flow porosity/permeabil-

ity. Studies that seek to develop hydrologic and thermal models of the shallow crustal permeability and tectonic/magmatic processes at the MOR [e.g., *Wilcock*, 1998, in preparation; *Yang et al.*, 1998; *Lowell and Germanovich*, 1994; *Germanovich et al.*, 2001] need to factor in the complex geometry imparted by the combination of volcanic eruption and lava distribution processes and the morphology of the axial zone as these have strong influences on the development of the geometry of fluid flow paths in the shallow crust at fast spreading MORs.

Near-bottom magnetic data may be one way to further constrain the nature of the permeability structure in the shallow crust at MORs. The magnetic maps in Plate 5 show a prominent low that is coincident with the axis of the axial trough suggesting very thin crust (lava) and the presence of shallow dikes. Elongate closed-contour lows around the northern vents (Bio9/9′ and P) at 9° 50.3′N, and the Ty and Io vents near 9° 50.1′N suggest possible 'burn holes', that are smaller in diameter, but similar to, the near-bottom magnetic expression of vents imaged at JDF Main Endeavour Field [*Tivey and Johnson*, 2002]. Data shown in Plate 5c were collected along lines spaced 60 m apart; it is evident that more closely spaced lines are required to properly image the smaller EPR vent systems. The prominent magnetic high just west of the AST near 9° 50.4′N is likely to be a primary depo-center for the 1991 eruption and is coincident with the large breakout from the AST believed to be related to extensive drain back at this location, proximal to the locus of high-temperature venting at Bio9/9′ and P vents.

The time-series temperature data available for the EPR 9° 50′N vents provide further insight into the issue of shallow fluid circulation and how closely spaced vents could have fluids with varying chemistry over time. For instance, in March 1995, the HOBO record for Bio9 vent (Plate 10) showed an abrupt temperature increase that has been correlated to a micro-seismic swarm [*Sohn et al.*, 1998, 1999; *Fornari et al.*, 1998b] that presumably resulted from fracturing in the lower crust induced by the evolving hydrothermal fluid circulation pattern and the resulting thermal stresses in lower crustal rocks. Time-series exit fluid temperature data indicate it took ~4 days for the effects of the cracking event to manifest themselves at the surface as an abrupt temperature increase [*Fornari et al.*, 1998]. Wilcock [2004] has modeled these data and shown that the magnitude of the increase and the time delay between the start of the seismic event and the observed increase in fluid temperature at the seafloor can be modeled using a pressure perturbation model. Wilcock's modeling predicts a temperature increase of 50°C in the reaction zone (~1 km depth [*Sohn et al.*, 1998, 1999]) and suggests that the physical dimensions of the crack transporting the fluids from depth is 0.6 mm wide (using laminar flow

Plate 9. Chimney structures of various high-temperature hydrothermal vents in the 9° 49´–51´N region of the EPR as imaged from *Alvin* (see Plates 2-3 and 5 for locations). Scale in meters across the foreground of each photograph is given in brackets at the end of each description. (A) The middle portion of Ty Vent in May 1999 with HOBO temperature probe installed. This photo was taken before the probe was extracted from the chimney wall [3]. (B) M vent in 1994 with HOBO temperature probe installed. The marker at left center is sitting on lobate lava at the rim of the AST, vent edifice is growing from a fracture in the upper portion of the AST eastern wall [2]. (C) Top of M vent in 2002. The tubing from a HOBO probe can be seen just below the right set of black smokers [1]. (D) Bio9 vent complex in 2002. HOBO probe can be seen inserted in chimney orifice in top center of photo [3]. (E) Bio9 vent in 1994 showing HOBO probe installed and developing Riftia tube worm community at left, along the margin of the primary eruptive fissure of the 1991-1992 eruption. (F) Bio9´ vent in 2002 during sampling using a titanium 'majors' bottle [1].

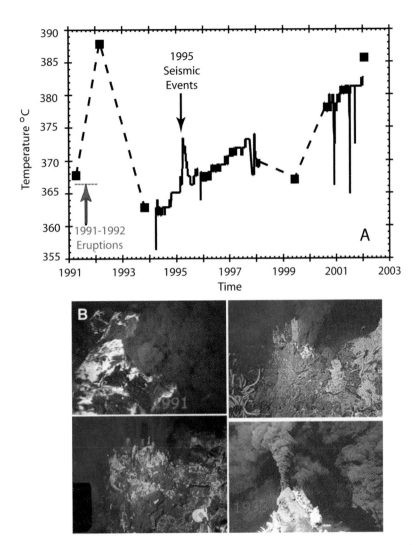

Plate 10. A) Time-series temperature data for the Bio9 high-temperature vent at the EPR near 9° 50.1′N (see Plates 3-5 for location). Data were collected using the DSPL SeaLogger probes (continuous line) and Alvin high-T probe data (black squares), for intervals when the vent was being sampled for fluids and sulfides [*Fornari et al.,* 1998; *Von Damm,* 2004 (this volume)]. Intervals with no data are shown by dashed lines. DSPL HOBO SeaLogger probes have recorded the exit fluid temperature of this and other vents in the area every 30 min over the decade since the 1991–92 eruptions [*Haymon et al.,* 1993; *Rubin et al.,* 1994]. Note that there has been an overall rise in temperature at this vent since the 1991–92 eruptions, with present day temperatures being roughly equal to those measured in April, 1992 immediately after the 1991–92 eruptions. Based on these data and fluid and gas chemistry *Von Damm* [2000], *Lilley et al.,* [2003] and *Von Damm and Lilley,* [2003] suggest magmatic activity at this site is resurgent (K. Von Damm, pers. commun., 2003). B) Alvin bottom photographs showing the physical evolution of the Bio9 vent since 1991. Upper-left photo shows black hydrothermal fluid exiting bare basalt and bacteria covered seafloor during or just following the 1991 eruption [*Haymon et al.,* 199]. Upper-right, and lower-left photos show the small sulfide mound with multiple orifices that typified Bio9 vent during the 1992 to 1993 time interval. Tube worm community that developed around the base of Bio9 is visible in the left corner of upper-right photo. Bottom-right photo shows the smoke from Bio9 vent (right side of photo) and immediately adjacent Bio9′ vent in 1999. By this time both vents had grown to be ~5–7m tall, ~1 m diameter chimneys vigorously venting black smoke. Scales across bottom of upper left and lower right photos are ~1.5 m. Scales across bottom of other two photos is ~2.5 m.

with a velocity of 1.2 m/s). *Von Damm* [2004] points out that the chemical data do not unequivocally constrain whether the cracking resulted from additional emplacement of magma or from hydrothermally induced cooling cracking into deeper, hotter rocks. Long term variations in exit fluid temperatures for the 'Hole-to-Hell' vents show the following trends. Since early 1995, fluid temperatures at the Bio9´ and Bio9 vents have increased from ~365°C to 372°C in November 1997. P vent fluid temperatures increased ~12°C, between November 1997 and May 1999. A general increase in Bio9´ and Bio9 (Plate 8) fluid temperatures of ~7°C was recorded between November 1997 and May 1999 [*Fornari and Shank*, 1999; *Scheirer et al.*, submitted].

Data presented in *Scheirer et al.* (submitted) provide longer time series information than the data shown in Plate 11, however, the temperature event recorded at Bio9 and Bio9´ vents and the adjacent Bio9 Rifta diffuse flow community in the time period November 28 to December 17, 1997 area useful in illustrating how the shallow fluid circulation system feeding adjacent vents is complex and how the setting of the vent (i.e., within the floor of the AST or on the AST wall) impacts primary hydrologic feeders (Plate 11). The November 28 event caused a sharp rise in the exit fluid temperature at Bio9, Bio9´ and in the low temperature biological community adjacent to the Bio9 chimney (Bio9 Riftia). M vent, which is located ~1 km north of Bio9 (Plate 3) on the east wall of the AST saw no change to its fluid temperature which was steady at 357°C. Biovent, also ~1.2 km north of Bio9 saw no change in it's fluid temperature which was steady at 343°C. P vent, only ~40 south of Bio9 also shows no change in its temperature of 372°C despite the close proximity to the Bio9 plumbing system. The probe in Ty vent, ~300 m south of Bio9, does show an effect from the event but it is recorded as a steady drop in the fluid temperatures which had been at 331°C prior to the event and eventually dropped to ~321°C after about a month, whereupon the temperature recovered to pre-event levels (Plate 11). All told, the data suggest a complexity in fluid circulation that is difficult to reconcile with a simplistic fluid circulation model that has strong spatial links between limbs of the plumbing system. While the lack of a thermal effect on the M and Biovent fluids from the November 28 event is easy to understand in light of both the spatial separation between them and Bio9 and their different setting along the AST, the lack of effect on P vent fluids is confounding. The artist's drawing presented in Plate 12 attempts to conceptually show the differences in fluid circulation in the shallow crust beneath the 9° 50´–51´N vents that could cause the types of spatial/temporal variations observed in the time-series temperature data.

The full data set presented in [*Scheirer et al.*, submitted] show that four probes recorded several discrete periods (last-ing a few days at most) where the Bio9´ vent fluid temperature dropped 3°–7°C, and took several weeks to a few months to recover to the value prior to the temperature drop. These types of variations were not observed in the fluid temperature record for the adjacent Bio9 vent during the same periods. Collapse of the Bio9 chimney, was recorded by a probe on February 6, 1998. This resulted in the expulsion of the probe from the vent orifice and its burial in the sulfide rubble that also impacted the Bio9 *Riftia* community just west of the vent (also instrumented with a Vemco, low-temperature recording array; part of that record is shown in Plate 11). In January 2002 Bio9, Bio9´and P vent fluids measured 386°C [*Scheirer et al.*, submitted]. Taken together the diverse fluid temperature responses of these closely spaced vents suggest that they are not well connected hydraulically in the shallow crust.

As mentioned earlier, the very narrow width of the AST and the presence of all but one (Biovent) of the high-temperature hydrothermal vents in the floor or on the wall of the AST in the study area suggests that the up flow limb of hot fluid circulation is tightly constrained to the narrow zone of intrusion beneath the AST (Plate 12). At 21°S EPR, the Rapa Nui vent field consists of a complex of black smoker vents that dominates the western wall of the ~600 m wide axial graben. At Brandon vent, the well-established vertical pathways afforded by the tensional fractures within the upper crust, as a by-product of dike intrusion, are localizing the primary hydrothermal fluid exit from the seafloor. The intensity of venting, the high exit fluid temperatures (>400°C), and the observation of phase-separation at these vents [*Von Damm et al.*, 1998] suggest that the pathways from the rock-water reaction zone are near-vertical; none of the energy is being lost by lateral transport in the hydrothermal plumbing system in the lower or upper crust. Although on a much smaller scale, the 'Brandon' scenario is what we envision for Biovent, M, and Q vents. That is, focused hydrothermal flow along vertical pathways constrained by the tensional fractures above the narrow dike zone, with little complication of the flow geometry imparted by intra-flow and inter-flow permeability, such as is prevalent beneath the AST floor. The high concentrations of magmatic volatiles (e.g., CO_2, He, H) sampled in hydrothermal fluids at Biovent, M and Q vents within ~2 years of the eruptions supports the notion that they were greatly affected by the 1991 volcanic event and suggests direct hydrologic connections between the upper mantle, magma lens and water/rock reaction zone beneath this portion of the axis [*Lilley et al.*, 2003; *Lupton et al.*, 1993; *Von Damm*, 2000].

We do not have micro-bathymetry data from the area around Q, M and Biovent so we cannot speculate on the small-scale morphological associations between their locations and the

AST volcanic structures. However, analysis of the sidescan data for the area between 9° 50′N and 9° 51′N along the AST (Plate 2) suggests that Q, M and Biovent are not directly associated with extensive breakout channels from the AST or drainback, such as has been noted for Bio9/9′, P, Ty, Io and Tube Worm Pillar vents further south (Plate 5).

8. SUMMARY

Focusing of volcanic eruptions, waxing and waning during eruptive phases and subsequent drainback of lava in a deep-sea setting has the potential to create fragmental and jumbled lava with high permeability within a steep sided, laterally restricted conduit that could then be utilized by upwelling, hot hydrothermal fluids (Plate 12). The geometry of shallow crustal permeability within the EPR axial trough in the 9° 50′N region is thought to be mainly controlled by the primary eruptive fissure network within the axial trough and the intersection of it by pipe-like structures where the eruptions have been focused and where drainback of lava localized. Areas of focused drainback are also likely to be centers of magma injection into the shallow crust. Jumbled sheet flows, breakout channels, collapse pits, and lava tubes act to transport and mix hydrothermal fluids in axis-parallel and axis-orthogonal directions within interfingering zones of high permeability. This complex subsurface structure may facilitate the development of hydrologic gradients that are utilized by hydrothermal fluids having various densities and temperatures which lead to hybrid fluids that bathe low-temperature animal communities.

High-resolution bathymetric and sidescan sonar maps of the AST in the EPR 9° 50′N area show spatial correlations between sites where the 1991 eruption is likely to have focused, creating channelized flows that transported the lava away from the AST, and active hydrothermal vents. We suggest that vent locations within the EPR axial trough are linked to formation of discrete, high permeability zones created by focusing of eruptions, cyclicity during eruptions, and subsequent drainback of lava into the primary fissure within the axial trough during waning stages of eruptions. Our new data confirms previous interpretations that current sites of hydrothermal venting are principally related to the trace of the fissure that fed the 1991 eruption. Drainback of lava in a deep-sea setting and magma/seawater interactions [e.g., *Perfit et al.,* 2003] have the potential to create fragmented and jumbled lava with high permeability within a steep sided, laterally restricted conduit that can be utilized by upwelling, hot hydrothermal fluids. The lateral extent of high-permeability regions is complex and may be limited by subsequent infilling of void space by younger lava. Frequent micro-earthquakes at fast-spreading MOR sites act to disrupt hydrothermal mineralization in small

veins within the crustal section which helps maintain permeability. Black-smoker hydrothermal vent locations and differences in time-series temperature history and chemistry between closely spaced vents reflect primary volcanic control on the geometry of fluid circulation in the shallow crust at the EPR axis in the 9° 50′N area.

Acknowledgments. Observations made at the EPR 9° 50′N have been made by many scientists over the last decade, and we thank our colleagues on various Alvin cruises for sharing their ideas and seafloor observations with us. We thank the shipboard and shore based crews of R/V Atlantis and Alvin for their dedication and expertise that resulted in collection of some of the data reported here. The ABE at-sea group—R. Catanach, A. Duester, and A. Billings were instrumental to the success of our 2001 and 2002 data collection efforts, and we thank them for their significant contribution to the acquisition of these important data and the key role they have played in developing the ABE vehicle. Paul Johnson, Akel Sterling and Jenny Engels helped process the DSL-120A sonar data. Dave Dubois helped produce some of the sonar maps. We thank C. Devey, J. Escartin, C. German, D. Naar and W. Wilcock for very constructive reviews that improved the paper. Aerial photographs used in Plate 4 were provided by the U.S. Geological Survey—Hawaii Volcano Observatory. Frank Trusdell of USGS-HVO kindly provided insight and photographs of drainback features on Kilauea volcano. Paul Oberlander helped prepare some of the plates, most notably Plate 12, and his exceptional insight in drawing seafloor perspectives has been very valuable to our research over the years. Field and shore based analyses have been supported by the National Science Foundation under grants OCE-9819261 (H.S, M.A.T and D.J.F.) and OCE-0096468 (D.J.F and T.M.S.), and the Woods Hole Oceanographic Institution's Vetlesen Fund WHOI Contribution #11172.

REFERENCES

Alt, J. C., Subseafloor processes in Mid-Ocean Ridge hydrothermal systems, In: *Physical, Chemical, Biological, and Geological Interactions within Seafloor Hydrothermal Systems* (Humphris, S.E., R.A. Zierenberg, L. Mullineaux and R. Thomson). *American Geophysical Union Monograph 91,* 85–114, 1995.

Ballard, R. D., Holcomb, R. T., and van Andel, T. H., The Galapagos Rift at 86°W: 3. Sheet flows, collapse pits, and lava lakes of the rift valley, *J. Geophys. Res., 84,* 5407–5422, 1979.

Barker, D. H. N., G. L. Christenson, J. A. Austin, and I. W. D. Daiziel, Backarc basin evolution and cordilleran orogenesis: Insights from new ocean-bottom seismograph refraction profiling in Bransfield Strait, Antarctica, *Geology, 31* (2), 107–110, 2003.

Bazin, S., H. van Avendonk, A. J. Harding, J. A. Orcutt, J. P. Canales, and R.S. Detrick, Crustal structure of the flanks of the East Pacific Rise; implications for overlapping spreading centers, *Geophys. Res. Lett., 25,* 2213–2216, 1998.

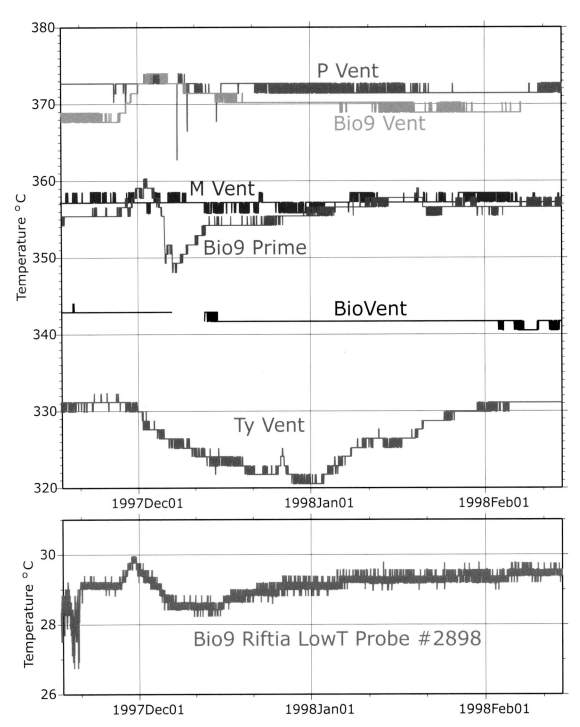

Plate 11. (Top) Detail of high-temperature HOBO probe data [*Fornari, et al.*, 1994, 1996] for the interval between November 1997 and February 1998 showing temperature fluctuations at Bio9, Bio9′. Resolution of HOBO probes is 1°C, data were recorded every 30 minutes [*Fornari et al.,* 1998b]. (Bottom) Vemco low-temperature logger data [*Fornari, et al.,* 1996] for the Bio9 *Riftia* community for the same time period showing temperature increase in late November 1997 that is coincident with the fluctuations in high-temperature fluids from the adjacent vents (see Plate 5 for vent locations). Red bar shows time period corresponding to the high-temperature records shown above. Resolution of Vemco probes is 0.1°C and data were recorded every 15 minutes.

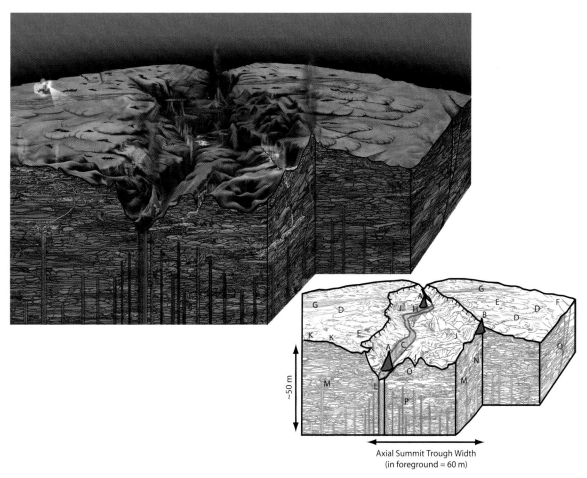

Plate 12. Artist's perspective of the East Pacific Rise crest in the 9° 50´N area based on Alvin observations, ABE and Alvin near-bottom sonar mapping, and DSL-120A sidescan sonar imagery. All red, yellow and blue colors shown in the colored cross-section represent fluid temperatures with red being high-temperature (>350° C) and blue low temperature (<~20°C). The interior structure is inferred based on observations in juvenile subaerial volcanic terrain in Hawaii and Iceland and from seafloor observations made using Alvin and a towed deep-sea camera system. Lettered inset at lower right provides key to various features drawn. A) High temperature vents along primary fissure in the axial summit trough (AST) fed by hot fluids that follow most recent eruptive dikes. B) High temperature vent along the bounding wall of the AST (e.g., M vent, Q vent) that taps fluids flowing through vertical fractures. C) Primary eruptive fissure within the floor of the AST. D) Scalloped flow fronts representing recent eruptions from the AST that flowed down the upper slopes of the EPR crest. E) Collapse pits of various sizes [see *Engels et al.*, 2003], with a focused zone of collapse within ~ 300 m to either side of the AST. F) Small throw (<~5 m high) faults located usually ~1 km from the AST. G) Lava channels on either side of the AST that transported lava from an eruption source within the AST out to 1-2 km from the axis (Alvin, not to scale, shown at left). H) Drainback feature within the floor of the AST located along the primary fissure. This is believed to also be a site where the eruption focused. I) Lava pillar. J) Sites of diffuse hydrothermal flow at low temperature (~ <20° C) along the primary eruptive fissure. K) Sites of diffuse hydrothermal flow at low temperature (~ <20° C) outside the AST. L) Subsurface trace of most recent eruptive dike. Hot (red) hydrothermal fluids are mining the heat from the dike and the magma lens below. M) Older dike that is sustaining hydrothermal flow that eventually feeds sites of off axis diffuse venting. N) Recent dike below the margin of the AST that feeds high-temperature black smoker vent situated along the AST wall (e.g., M or Q vents). O) Area of mixing of seawater and high-temperature fluids within fragmental and porous shallow volcanic layer to create hybrid hydrothermal fluids. P) Intruded dikes below AST of various age. Q) Site of off axis recharge of seawater into the shallow crust. (Drawing by E. Paul Oberlander – WHOI).

Begnaud, M. L., J. S. McClain, G. A. Barth, J. A. Orcutt, and A. J. Harding, Velocity structure from forward modeling of the eastern ridge-transform intersection area of the Clipperton fracture zone, East Pacific Rise, *J. Gepohys. Res.*, *102*, 7803–7820, 1997.

Bohnenstiehl, D. R and M. C. Kleinrock, Fissuring near the TAG active hydrothermal mound 26°N on the Mid-Atlantic Ridge, *J. Vol. Geo. Res.*, *98*, 33–48, 2000.

Bohnenstiehl, D. R and M. C. Kleinrock, Faulting and fault scaling on the median valley floor of the TAG segment, 26°N, Mid-Atlantic Ridge, *J. Geophys. Res.*, *104*, 29,351–29,364, 1999.

Bruce, P. M., and H. E. Huppert, Thermal control of basaltic fissure eruptions, *Nature, 342,* 665–667, 1989.

Byrnes, J. M., and D. A. Crown, Relationships between pahoehoe surface units, topography, and lava tubes at Mauna Ulu, Kilauea Volcano, Hawaii, *J. Geophys. Res. 106,* 2139–2151, 2001.

Carbotte, S., and K. C. Macdonald, East Pacific Rise 8 degrees–10 degrees 30′N; evolution of ridge segments and discontinuities from SeaMarc II and three-dimensional magnetic studies, *J. Geophys. Res.*, *97*, 6959–6982, 1992.

Carbotte, S. M., W. B. F. Ryan, M. Cormier, W. Jin, E. Bergmanis, J. Sinton, and S. White, Magmatic subsidence of the EPR 18°14′S revealed through fault restoration of ridge crest bathymetry, *Geochemistry, Geophysics, Geosystems, 4,* 10.1029/2002GC000337, 2003.

Carbotte, S. M. and D. Scheirer, Variability of ocean crustal structure created along the global midocean ridge, in: Hydrogeology of Oceanic Lithosphere, eds.: E. Davis and H. Elderfield, Cambridge University Press, 2003. Cashman, K. V., C. Thornber, and J. P. Kauahikaua, Cooling and crystallization of lava in open channels, and the transition of Pahoehoe Lava to 'A'a, *Bull. Volcanol., 61,* 306–323, 1999.

Chadwick, W. W., Jr., T. K. P. Gregg, and R. W. Embley, Submarine lineated sheet flows: a unique lava morphology formed on subsiding lava ponds, *Bull. Volcanol., 61,* 194–206, 1999.

Chadwick, Jr., W. W., Quantitative constraints on the growth of submarine lava pillars from a monitoring instrument that was caught in a lava flow, *J. Geophys. Res., 108* (B11), 2534, 2003.

Chadwick, W. W., Jr., D. S. Scheirer, R. W. Embley, and H. P. Johnson, High-resolution bathymetric surveys using scanning sonars: Lava flow morphology, hydrothermal vent and geologic structure at recent eruption sites on the Juan de Fuca Ridge, *J. Geophys. Res., 106,* 16075–16100, 2001.

Christeson, G. L., G. M. Kent, G. M. Purdy and R. S. Detrick, Extrusive thickness variability at the East Pacific Rise, 9°–10°N: Constraints from seismic techniques, *J. Geophys. Res. 101,* 2859–2873, 1996.

Cochran, J. R., D. J. Fornari, B. J. Coakley, R. Herr, and M. A. Tivey, Continuous near-bottom gravity measurements made with a BGM-3 gravimeter in DSV Alvin on the East Pacific Rise crest near 9°30′N and 9°50′N, *J. Geophys. Res., 104,* 10841–10861, 1999.

Cormier, M.-H., Ryan, W. B. F., Shah, A. K., Jin, W., Bradley, A. M, and D. R. Yoerger, Waxing and waning volcanism along the East Pacific Rise on the millennium time scale, *Geology, 31,* 633–636, 2003.

Crawford, W. C., and S. C. Webb, Variations in the distribution of magma in the lower crust and at the Moho beneath the East Pacific Rise at 9°–10°N, *Earth and Planet. Sci. Let.*, *203*, 117–130, 2002.

Crawford, W. C., S. C. Webb, and J. A. Hildebrand, Constraints on melt in the lower crustal and Moho at the East Pacific Rise, 9 degrees 48′N, using seafloor compliance measurements, *Journal of Geophysical Research, B, Solid Earth and Planets, 104* (2), 2923–2939, 1999.

Crown, D. A., and S. M. Baloga, Pahoehoe toe dimensions, morphology, and branching relationships at Mauna Ulu, Kilauea Volcano, Hawaii, *Bull. Volcanol., 61,* 288–305, 1999.

Curwitz, D. and J. A. Karson, Geological consequences of dike intrusion at Mid-Ocean Ridge spreading centers, in: Faulting and magmatism at mid-ocean ridges, W. R. Buck, et al. (eds.), *Am. Geophys. U. Geophys. Monograph 106*, 117–136, 1998.

Delaney, J. R., D. S. Kelley, M. D. Lilley, D. A. Butterfield, J. A. Baross, W. S. D. Wilcock, R. W. Embley and M. Summit, The quantum event of oceanic crustal accretion: Impacts of diking at mid-ocean ridges. *Science 281:* 222–230, 1998.

Delaney, J. R., V. Robigou, R. McDuff, and M. Tivey, Geology of Vigorous Hydrothermal System on the Endeavour Segment, Juan de Fuca Ridge, *J. Geophys. Res.*, *97*, 19,663–19,682, 1992.

Delaney, P. T., and D. D. Pollard, Solidification of basaltic magma during flow in a dike, *Am. J. Sci., 282,* 856–885, 1982.

Detrick, R. S., P. Buhl, E. E. Vera, J. C. Mutter, J. A. Orcutt, J. A. Madsen, and T. M. Brocher, Multi-channel seismic imaging of a crustal magma chamber along the East Pacific Rise, *Nature, 326* (6108), 35–41, 1987.

Dunn, R. A., and D. R. Toomey, Seismological evidence for three-dimensional melt migration beneath the East Pacific Rise, *Nature, 388,* 259–262, 1997.

Eaton, J. P., D. H. Richter, and H. L. Krivoy, *Cycling of Magma Between the Summit Reservoir and Kilauea Iki Lava Lake During the 1959 Eruption of Kilauea Volcano,* 1307–1335 pp., USGS Prof. Paper, 1987.

Embley, R. W., and W. W. Chadwick, Jr., Volcanic and hydrothermal processes associated with a recent phase of seafloor spreading at the northern Cleft segment: Juan de Fuca Ridge, *J. Geophys. Res., 99,* 4741–4760, 1994.

Embley, R. W., Chadwick, W. W., Jonasson, I. R., Butterfield, D. A., and E. T. Baker, Initial results of the rapid response to the 1993 CoAxial event: Relationships between hydrothermal and volcanic processes, *Geophys. Res. Letts., 22,* 143–146, 1995.

Embley, R. W., Chadwick, W.W., Clague, D. and D. Stakes, 1998 eruption of Axial volcano: Multibeam anomalies and seafloor observations, *Geophys. Res. Lett., 26,* 3425–3428, 1999.

Embley, R. W., Chadwick, W. W., Perfit, M. R., Smith, M. C. and J.R. Delaney, Recent eruptions on the CoAxial segment of the Juan de Fuca Ridge: Implications for mid-ocean ridge accretion processes, *J. Geophys. Res., 105,* 16,501–16,525, 2000.

Embley, R. W. et al., Rediscovery and exploration of Magic Mountain, Explorer Ridge, NE Pacific, *Eos Trans. AGU, 83*(47), Fall Meet. Suppl., Abstract T11C–1264, 2002.

Embley, R. W. and E. Baker, 2003.

Engels, J. L., M. H. Edwards, D. J. Fornari, M. R. Perfit, G. J. Kurras, D. R. Bohnenstiehl, A new model for submarine volcanic collapse formation, G^3, 2003

Fink, J. H., and R. W. Griffiths, A laboratory analog study of the surface morphology of lava flows extruded from point and line sources, *Journal of Volcanology and Geothermal Research, 54*, 19–32, 1992.

Fornari, Escartin et al. AST paper in prep

Fornari, D. J., Van Dover, C. L., Shank, T., Lutz, R. and M. Olsson, A versatile, low-cost temperature sensing device for time-series measurements at deep sea hydrothermal vents, *BRIDGE Newsletter, 6*, 37–40, 1994.

Fornari, D. J., and Embley, R. W, Tectonic and volcanic controls on hydrothermal processes at the Mid-Ocean Ridge: An overview based on near-bottom and submersible studies. In: *Physical, Chemical, Biological, and Geological Interactions within Seafloor Hydrothermal Systems* (Humphris, S. E., R. A. Zierenberg, L. Mullineaux and R. Thomson). *American Geophysical Union Monograph 91, 1*–46, 1995.

Fornari, D. J., F. Voegeli, and M. Olsson, Improved low-cost, time-lapse temperature loggers for deep ocean and sea floor observatory monitoring, *RIDGE Events, 7*, 13–16, 1996.

Fornari, D. J., Haymon, R. M., Perfit, M. R., Gregg, T. K. P., Edwards, M. H., Geological Characteristics and Evolution of the Axial Zone on Fast Spreading Mid-Ocean Ridges: Formation of an Axial Summit Trough along the East Pacific Rise, 9°–10°N, *J. Geophys. Res., 103*, 9827–9855, 1998a.

Fornari, D. J., T. Shank, K. L. Von Damm, T. K. P. Gregg, M. Lilley, G. Levai, A. Bray, R. M. Haymon, M. R. Perfit and R. Lutz, Time-Series Temperature Measurements at High-Temperature Hydrothermal Vents, East Pacific Rise 9° 49′–51′N: Monitoring a Crustal Cracking Event, *Earth Planet. Sci. Lett., 160, 419–431, 1998*b.

Fornari, D. J. and T. M. Shank, Summary of High- and Low-T Tiime Series Vent Fluid Temperature Experiments East Pacific Rise 9° 49′–51′N, EXTREME-1 Cruise - May 1999, R/V Atlantis 03–34, Cruise Report, July 1, 1999.

Fornari, D. J., M. Perfit, M. Tolstoy, D. Scheirer and Science Party, *AHA-Nemo2 Cruise Report* cruise data web site: http://science.whoi.edu/ahanemo2, 2001.

Fornari, D. J., A new deep-sea towed digital camera and multi-rock coring system, *EOS, 84*, 69&73, 2003.

Fornari, D. J, M. Tivey, H. Schouten, D. Yoerger, A. Bradley, M. R. Perfit, D. Scheirer, P. Johnson, M. H. Edwards, R. Haymon, S. Humphris, DSL-120A High-Resolution, Near-Bottom Sidescan Sonar Imaging of Mid-Ocean Ridge Crests: East Pacific Rise, Galapagos Rift and Mid-Atlantic Ridge, *EGS/AGU/EUG* Joint Meeting, Nice, France, abstract EAE03-A-02862, 2003a.

Fornari, D. J. Gray, R., Olsson, M., Shank, T., and K. Von Damm, High-T Hydrothermal Vent Monitoring at the East Pacific Rise 9°50′N and a New, Improved 64k MemoryHigh-T Vent Fluid Temperature Logger, *Ridge2000 Events*, 21–25, 2003b.

Fox, C. G., Chadwick, W. W. and Embley, R. W., Direct observation of a submarine volcanic eruption from a sea-floor instrument caught in a lava flow. *Nature, 412*: 727–729, 2001.

Fox, C. G., W. E. Radford, R. P. Dziak, T. K. Lau, H. Matsumoto, and A. E. Schreiner, Acoustic detection of a seafloor spreading episode

on the Juan de Fuca Ridge using military hydrophone arrays, *Geophys. Res. Letts., 22*, 131–134, 1995.

Francheteau, J. and R. D. Ballard, The East Pacific Rise near 21°N, 13°N and 20°S: Inferences for along-strike variability of axial processes of the mid-ocean ridge, *Earth Planet. Sci. Lett., 64*, 93–116, 1983.

Garmany, J., Accumulations of melt at the base of young oceanic crust, *Nature, 340*, 628–632, 1989.

German, C. R. & K. L. Von Damm, Chapter 6.10 Hydrothermal Processes, in The Treatise of Geochemistry, eds. K. K. Turekian & H. D. Holland, *Elsevier*, pp. 181–222, 2004

Germanovich, L. N., R. P. Lowell, and D. K. Astakhov, Temperature-dependent permeability and bifurcations in hydrothermal flow, *J. Gephys. Res., 106*, 473–495, 2001.

Greenland, L. P., Okamura, A. T. and J. B. Stokes, 5. Constraints on the mechanics of the eruption, in: The Puu OO eruption of Kilauea volcano, Hawaii: Episodes 1 Through 20, January 3, 1983, Through June 8, 1984, ed. E. Wolfe, *US Geol. Survey Prof. Paper 1463*, US Government Printing Office, Washington, DC, 1988.

Gregg, T. K. P., D. J. Fornari, M. R. Perfit, W. I. Ridley and M. D. Kurz, Using submarine lava pillars to record mid-ocean ridge eruption dynamics, *Earth Planet. Sci. Lett., 178*, 195–214, 2000.

Gregg, T. K., and D. J. Fornari, Long submarine lava flows: observations and results from numerical modeling, *Journal of Geophysical Research, 103* (B11), 27517–27531, 1998.

Gregg, T. K. P., Fink, Jonathan H., Griffiths, Ross W., Formulation of multiple fold generations on lava flow surfaces: influence of strain rate, cooling rate, and lava composition, *Journal of Volcanology and Geothermal Research, 80*, 281–292, 1998.

Gregg, T. K. P., Fornari, D. J., Perfit, M. R., Haymon, R.M., and Fink, J. H., Rapid Emplacement of a Mid-Ocean Ridge Lava Flow: The East Pacific Rise at 9° 46′–51′N, *Earth Planet. Sci. Express Lett., 144*:, E1–E7, 1996.

Griffiths, R. W. a. F., Jonathan H., The morphology of lava flows in planetary environments: predictions from analog experiments, *Journal of Geophysical Research, 97* (B13), 19,739–19,748, 1992.

Hannington, M. D., Jonasson, I. R., Herzig, P. M., and S. Petersen, Physical and chemical processes of seafloor mineralization at Mid-Ocean Ridges, In: Physical, Chemical, Biological, and Geological Interactions within Seafloor Hydrothermal Systems (Humphris, S. E., R. A. Zierenberg, L. Mullineaux and R. Thomson). *American Geophysical Union Monograph 91*, 115–157, 1995.

Harding, A. J., G. M. Kent, and J. A. Orcutt, A multichannel seismic investigation of upper crustal structure at 9 degrees N on the East Pacific Rise; implications for crustal accretion, *J. Geophys. Res., 98*, 13,925–13,944, 1993.

Haymon, R. M., and S. M. White, Fine-scale segmentation of volcanic/hydrothermal systems along fast-spreading ridge crests, *Earth Planet. Sci. Lett.,* in press.

Haymon, R. M., The response of ridge-crest hydrothermal systems to segmented, episodic magma supply, in: Tectonic, Magmatic, Hydrothermal and Biological Segmentation of Mid-Ocean Ridges, eds: C. J. MacLeod, P. A. Tyler and C. L. Walker, *Geol. Soc. Spec. Publ., 118*, 157–168. 1996.

Haymon, R. M., D. J. Fornari, K. L. Von Damm, M. D. Lilley, M. R. Perfit, and J. M. Edmond, Volcanic eruption of the mid-ocean ridge along the East Pacific Rise crest at 9 degree 45–52′N: Direct submersible observations of seafloor phenomena associated with an eruption event in April, 1991. *Earth Planet. Sci. Lett., 119,* 85–101, 1993.

Haymon, R. M., D. J. Fornari, M. H. Edwards, S. Carbotte, D. Wright and K. C. Macdonald, Hydrothermal vent distribution along the East Pacific Rise crest (9°09′–54′N) and its relationship to magmatic and tectonic processes on fast-spreading mid-ocean ridges, *Earth Planet. Sci. Lett,. 104,* 513–534, 1991.

Heliker, C. C., M. T. Mangan, T. N. Mattox, J. P. Kauahikaua, and R. T. Helz, The character of long-term eruptions: inferences from episodes 50–53 of the Pu′u ′O′o-Kupaianaha eruption of Kilauea Volcano, *Bull. Volcanol. 59,* 381–393, 1998.

Herron, T. J., P. L. Stoffa, and P. Buhl, Magma chamber and mantle reflections; East Pacific Rise, *Geophysical Research Letters, 7* (11), 989–992, 1980.

Holcomb, R. T., Eruptive history and long-term behavior of Kilauea Volcano, in *U.S. Geological Survey Professional Paper 1350,* edited by R. W. Decker, T. L. Wright, and P. H. Stauffer, pp. 261–350, 1987.

Holcomb, R. T., Preliminary geologic map of Kilauea Volcano, Hawaii, *U.S. Geological Survey,* 80–796, 1980.

Hon, K., J. Kauahikaua, R. Denlinger, and K. Mackay, Emplacement and inflation of pahoehoe sheet flows: observations and measurements of active lava flows on Kilauea Volcano, Hawaii, *Geological Society of America Bulletin, v. 106,* p. 351–370, 1994.

Humphris, S. E., Hydrothermal processes at mid-ocean ridges, *U.S. Natl. Rep. Int. Union Geod. Geophys. 1991–1994, Rev. Geophys.,* 33, 71–80, 1995.

Jakuba, M., D. Yoerger, W. Chadwick, A. Bradley, R. Embley, Multibeam sonar mapping of the Explorer Ridge with an Autonomous Underwater Vehicle *Eos Trans. AGU, 83*(47), Fall Meet. Suppl., Abstract T11C-1266, 2002.

Johnson, H. P., S. L. Hautala, M. A. Tivey, C. D. Jones, J. Voight, M. Pruis, I. Garcia-Berdeal, L. A. Gilbert, T. Bjorklund, W. Fredericks, J. Howland, M. Tsurumi, T. Kurokawa, K. Nakamura, K. O' Connell, L. Thomas, S. Bolton, and J. Turner, Survey studies hydrothermal circulation on the northern Juan de Fuca Ridge, *EOS, 83,* 73–79, 2002.

Kent, G. M., A. J. Harding, and J. A. Orcutt, Evidence for a smaller magma chamber beneath the East Pacific Rise at 9°30′N, *Nature, 412,* 145–149, 1990.

Kent, G. M., A. J. Harding, and J. A. Orcutt, Distribution of magma beneath the East Pacific Rise between the Clipperton Transform and the 9 degrees 17′N Deval from forward modeling of common depth point data, *J. Geophy. Res., 98,* 13,945–13,969, 1993a.

Kent, G. M., A. J. Harding, and J. A. Orcutt, Distribution of magma beneath the East Pacific Rise near the 9 degrees 03′N overlapping spreading center from forward modeling of common depth point data, *J. Geophy. Res., 98,* 13,971–13,995, 1993b.

Kilburn, C. R. J., Patterns and predictability in the emplacement of subaerial lava flows and flow fields, in *Monitoring and mitigation of volcano hazards,* edited by R. Scarpa, and R. I. Tilling, pp. 491–537, Springer-Verlag, Berlin, 1996.

Kurokawa, T., M. H. Edwards, P. Johnson, D. J. Fornari, M. Perfit, H. Schouten, and M. A. Tivey, Possible Recent Volcanic Activity on the East Pacific Rise at 9° 32′N, *Eos Trans. AGU, 83(47),* Fall Meet. Suppl., Abstract V52A-1283, 2002.

Kurras, G., D. J. Fornari and M. H. Edwards, Volcanic morphology of the East Pacific Rise crest 9°49′–52′N: Implications for extrusion at fast spreading mid-ocean ridges, *Mar. Geophys. Res.,* 21, 23–41, 2000.

Lerner, S., D. Yoerger and T. Crook, Navigation for the Derbyshier Phase2 Survey, *Woods Hole Oceanographic Institution Technical Report,* 24 pp., May, 1999.

Lilley, M. D., Butterfield, D. A., Lupton, J. E., and E. J. Olson, Magmatic events can produce rapid changes in hydrothermal vent chemistry, *Nature, 422,* 878–881, 2003

Lowell, R. P. and L. N. Germanovich, On the temporal evolution of high-temperature hydrothermal systems at ocean ridge crests, *J. Geophys, Res., 99,* 565–575, 1994.

Lupton, J. E., E. T. Baker, M. J. Mottl, F. J. Sansone, C. G. Wheat, J. A. Resing, G. J. Massoth, C. I. Measures, and R. A. Feely, Chemcial and physical diversity of hydrothermal plumes along the East Pacific Rise, 8°45′N to 11°50′N, *Geophysical Research Letters,* 20, 2913–2916, 1993.

Lutz, R. A., T. M. Shank, D. J. Fornari, R. M. Haymon, M. D. Lilley, K. L. Von Damm & D. Desbruyeres, Rapid growth at deep-sea vents, *Nature, 371,* 663–664, 1994.

Macdonald, K., J.-C. Sempere, and P. J. Fox, East Pacific Rise from Siqueiros to Orozco fracture zones; along-strike continuity of axial neovolcanic zone and structure and evolution of overlapping spreading centers, *J. Geophy. Res. 89,* 6049–6069, 1984.

Macdonald, K. C., P. J. Fox, S. Miller, S. Carbotte, M. H. Edwards, M. Eisen, D. J. Fornari, L. Perram, R. Pockalny, D. Scheirer, S. Tighe, C. Weiland and D. Wilson, The East Pacific Rise and its flanks 8–18°N: History of segmentation, propagation and spreading direction based on SeaMARC II and Sea Beam studies, *Mar. Geophys. Res., 14,* 299–344, 1992.

Mattox, T., C. Heliker, J. Kauahikaua, and K. Hon, Development of the 1990 Kalapana flow field, Kilauea Volcano, Hawaii, *Bulletin of Volcanology,* 55 (407–413), 1993.

McClain, J. S., Begnaud, M. L., Wright, M. A., Fondrk, J. and K. Von Damm, Seismicity and tremore in a submarine hydrothermal field: the northern juan de Fuca Ridge, *Geophys. Res Lett., 20,* 1883–1886, 1993.

Morton, J. L., M. L. Holmes, and R. A. Koski, Volcanism and massive sulfide formation at a sedimented spreading center, Escanaba Trough, Gorda Ridge, northeast Pacific Ocean, *Geophys. Res. Letts., 14,* 769–772, 1987.

Mullineaux, L. S., S. W. Mills and E. Goldman, Recruitment variation during a pilot colonization study of hydrothermal vents (9°50′N, East Pacific Rise). *Deep-Sea Res. II* 45: 441–464, 1998.

Naar, D. F. and R. N. Hey, Recent Pacific-Easter-Nazca plate motions, in: *Evolution of Mid-Ocean Ridges, Geophys. Monogr. Ser.,* 57, ed. J. Sinton, 9–30, AGU, Washington, DC., 1989.

Oosting, S. E. and K. L. von Damm, Bromide/chloride fractionation in seafloor hydrothermal fluids from 9–10°N East Pacific Rise, *Earth Planet. Sci. Lett.,* 144, 133–145, 1996.

Parfitt, E. A., T. K. P. Gregg, and D. K. Smith, A comparison between subaerial and submarine eruptions at Kilauea Volcano, Hawaii: implications for the thermal viability of lateral feeder dikes, *J. Volcanol. Geotherm. Res., 113,* 213–242, 2002.

Parfitt, E. A, and L. Wilson, The 1983–86 Pu'u 'O'o eruption of Kilauea Volcano, Hawaii; a study of dike geometry and eruption mechanisms for a long-lived eruption, *J. Volcanol. Geotherm. Res., 59,* 179–205, 1994.

Perfit, M. R., J. R. Cann, D.J. Fornari et al., Interaction of seawater and lava during submarine eruptions at mid-ocean ridges, 2003 nature paper

Perfit, M. R. and W. C. Chadwick, Magmatism at mid-ocean ridges: Constraints from volcanological and geochemical investigations, in: Faulting and magmatism at mid-ocean ridges, W.R. Buck, et al. (eds.), *Am. Geophys. U. Geophys. Monograph* 106, 59–116, 1998.

Perfit, M. R., Fornari, D. J., Smith, M., Bender, J., Langmuir, C. H., Haymon, R. M., Across-axis spatial and temporal diversity of MORB on the crest of the East Pacific Rise between 9°30′–9°32′N, *Geology, 22,* 375–379, 1994.

Pockalny, R. A., P. J. Fox, D. J. Fornari, K. C. Macdonald, and M. R. Perfit, Tectonic reconstruction of the Clipperton and Siqueiros Fracture zones: Evidence and consequences of plate motion change for the last 3 Myr, *J. Geophys. Res., 102,* 3167–3182, 1997.

Polacci, M., K. V. Cashman, and J. P. Kauahikaua, Textural characterization of the pahoehoe – 'a'a transition in Hawaiin basalt, *Bull. Volcan., 60,* 595–609, 1999.

Ravizza, G. , J. Blusztain, K. L. Von Damm, A. M. Bray, W. Bach, and S. R. Hart, Sr isotope variations in vent fluids from 9° 46′–9°54′N East Pacific Rise: evidence of a non-zero Mg fluid component, *Geochem. Cosmochem. Acta, 65,* 729–739, 2001.

Rohr, K. M., Increase of seismic velocities in upper oceanic crust and hydrothermal circulation in the Juan de Fuca plate, *Geophys. Res. Lett., 21,* 2163–2166, 1994.

Rubin, K. H., J. D. MacDougall, and M. R. Perfit, $^{210}Po/^{210}Pb$ dating of recent volcanic eruptions on the sea floor, *Nature, 468,* 841–844, 1994.

Sakimoto, S. E. H. and Gregg, T. K. P., Channeled flow: Analytic solutions, laboratory experiments, and applications to lava flows. *J. Geophys. Res., 106(5):* 8629–8644, 2001.

Scheirer, D. S., D. J. Fornari, S. E. Humphris, and S. Lerner, High-Resolution Seafloor Mapping Using the DSL-120 Sonar System: Quantitative Assessment of Sidescan and Phase-Bathymetry Data from the Lucky Strike Segment of the Mid-Atlantic Ridge, *Mar. Geophys. Res. 21,* 121–142, 2000.

Scheirer, D. S., and K. C. Macdonald, Variation in cross-sectional area of the axial ridge along the East Pacific Rise; evidence for the magmatic budget of a fast spreading center, *J. Geophy. Res., 98,* 7871–7885, 1993.

Scheirer, D. S., T. M. Shank, and D. J. Fornari, Temperature variations at diffuse-flow hydrothermal vent sites along the East Pacific Rise, *Earth Planet. Sci. Lett.,* (submitted).

Schouten, H., M. A. Tivey, D. J. Fornari, and J. R. Cochran, Central Anomaly Magnetization High: Constraints on the volcanic construction and architecture of seismic layer 2A at a fast-spreading Mid-Ocean Ridge, the EPR at 9?30′–50′N, *Earth Planet. Sci. Lett., 169,* 37–50, 1999.

Schouten, H., Tivey, M. A., and D. J. Fornari, <u>AT7-4 Cruise Report</u>, at: http://imina.soest.hawaii.edu/HMRG/EPR/index.htm under AT7-4 Cruise Report, 2001.

Schouten, H., M. Tivey, D. Fornari, D. Yoerger, A. Bradley, P. Johnson, M. Edwards, and T. Kurokawa, Lava Transport and Accumulation Processes on EPR 9 27′N to 10N: Interpretations Based on Recent Near-Bottom Sonar Imaging and Seafloor Observations Using ABE, Alvin and a new Digital Deep Sea Camera, *Eos Trans. AGU, 83(47),* Fall Meet. Suppl., Abstract T11C-1262, 2002.

Shank, T. S., D. J. Fornari, and R. Lutz, Periodicities and variability of high and low-temperature hydrothermal venting along the Biotransect (9° 50′N) on the East Pacific Rise: three years of continuous synchronous temperature monitoring, *Eos, Trans. Amer. Geophys. Union, 78, F739, 1997.*

Shank, T. M., D. J. Fornari, K. L. Von Damm, M. D. Lilley, R. M. Haymon and R. A. Lutz, Temporal and spatial patterns of biological community development at nascent deep-sea hydrothermal vents along the East Pacific Rise, *Deep Sea Res. II* 45, 465–515, 1998.

Shank, T. M., Hammond, S., D. J. Fornari, et al., Time-Series Exploration and Biological, Geological, and Geochemical Characterization of the Rosebud and Calyfield Hydrothermal Vent Fields at 86°W and 89.5°W on the Galapagos Rift, *Eos,* T11C-1257, 2002.

Shank, T. M., D. S., Scheirer, and D. J.Fornari, Influences of temperature and chemistry on habitat selection at hydrothermal vents, *Limnology and Oceanography,* (submitted).

Singh, S. C., G. M. Kent, J. S. Collier, A. J. Harding, and J. A. Orcutt, Melt to mush variations in crustal magma properties along the ridge crest at the southern East Pacific Rise, *Nature (London), 394* (6696), 874–878, 1998.

Sinton, J., Bergmanis, E., Rubin, K. et al., volcanic eruptions on mid-ocean ridges: New evidence from the superfast sprading East Pacific Rise, 17°–19°S, *J. Geophys. Res., 107,* 10.1029/2000JB000090, 2002.

Sinton, J. M., and R. S. Detrick, Mid-ocean ridge magma chambers, *J. Geophys. Res., 97,* 197–216, 1992.

Smith, M., M. R. Perfit, D. J. Fornari, W. Ridley, I., M. H. Edwards, G. Kurras, and K. L. Von Damm, Segmentation and Magmatic Processes at a Fast Spreading Mid-Ocean Ridge: Detailed Geochemistry and Mapping of the East Pacific Rise Crest at the 9° 37′N OSC, G^3, 2, paper #2000GC000134, 2001.

Sohn R., D. J. Fornari, K. L. Von Damm, S. Webb, and J. Hildebrand, Seismic and hydrothermal evidence for a propagating cracking event on the East Pacific Rise crest at 9° 50′N, *Nature, 396,* 159–161, 1998.

Sohn, R. A., J. A. Hildebrand, and S. C. Webb, A microearthquake survey of the high-temperature vent fields on the volcanically active East Pacific Rise, *J. Geophys. Res., 104* (11), 25,367–25,378, 1999.

Soule, S. A., K. V. Cashman, and J. P. Kauahikaua, Examining flow emplacement through the surface morphology of three rapidly emplaced, solidified lava flows, Kilauea Volcano, Hawai'I, *Bull. Volcanology,* 66, 1–14, 2004.

Stewart, W. K., Chu, D., Malik, S., Lerner, S., and H. Singh, Quantitative seafloor characterization using a bathymetric sidescan sonar, *J. Ocean. Engin. 19*, 599–610, 1994.

Swanson, D. A., Pahoehoe flows from the 1969–1971 Mauna Ulu eruption, Kilauea Volcano, Hawaii, *Geological Society of America Bulletin, v. 84*, p. 615–626, 1973.

Tian, T., W. S. D. Wilcock, D. R. Toomey, and R. S. Detrick, Seismic heterogeneity in the upper crust near the 1991 eruption site on the East Pacific Rise, 9 degrees 50′N, *Geophysical Research Letters, 27* (16), 2369–2372, 2000.

Tilling, R. I., Christiansen, R. L., Duffield, W. A. et al., The 1972–1974 Maunu Ulu eruption, Kilauea volcano: an example of quasi-steady state magma transfer, in: Decker, R. W, Wight, T. L. and P. H. Stauffer, eds., *USGS Prof. Paper 1350*, 405–469.

Tivey, M., D. Fornari, H. Schouten, D. Yoerger, A. Bradley, P. Johnson, R. Embley, High-Resolution Magnetic Field and Bathymetric Imaging of Hydrothermal Vent Areas using Autonomous Underwater Vehicles, Remotely Operated Vehicles and Submersibles, EGS/AGU/EUG Joint Meeting, Nice, France, abstract EAE03-A-02858, 2003.

Tivey, M. A. and H. P. Johnson, Crustal magnetization reveals subsurface structure of Juan de Fuca Ridge hydrothermal fields, *Geology*, 30, 979–982, 2002.

Tivey, M. A., H. P. Johnson, A. Bradley, and D. Yoerger, Thickness measurements of submarine lava flows determined from near-bottom magnetic field mapping by autonomous underwater vehicle, *Geophys. Res. Lett.*, 25, 805–808, 1998.

Tivey, M. A., A. Bradley, D. Yoerger, R. Catanach, A. Duester, S. Liberatore and H. Singh, Autonomous underwater vehicle maps seafloor, *EOS*, 78, 229–230, 1997(a).

Tivey, M. A., A. Bradley, D. Yoerger, R. Catanach, A. Duester, S. Liberatore and H. Singh, Autonomous underwater vehicle maps seafloor, *Earth in Space*, 10, 10–14, 1997(b).

Toomey, D. R., G. M. Purdy, S. C. Solomon, and W. S. D. Wilcock, The three-dimensional seismic velocity structure of the East Pacific Rise near latitude 9 degrees 30′ N, *Nature, 347* (6294), 639–645, 1990.

Toomey, D. R., S. C. Solomon, and G. M. Purdy, Tomographic imaging of the shallow crustal structure of the East Pacific Rise at 9°30′N, *J. Geophy. Res*, 99, 24,135–24,157, 1994.

Umino, S., Obata, S., Lipman, P., et al., Emplacement and inflation structures of submarine and subaerial pahoehoe lavas from Hawaii. Hawaiian Volcanoes: Deep Underwater perspectives, American Geophys. Mono. 128, 85–102, 2002.

van Avendonk, H. J. A., A. J. Harding, and J. A. Orcutt, Contrast in crustal structure across the Clipperton transform fault from travel time tomography, *J. Geophy. Res, B, , 106*, 10,961–10,981, 2001.

van Avendonk, H. J. A., A. J. Harding, J. A. Orcutt, and J. S. McClain, A two-dimensional tomographic study of the Clipperton transform fault, *J. Geophy. Res., 103*, 17,885–17,899, 1998.

van Dover, C. L., S. E. Humphris, D. Fornari, C. M. Cavanaugh, R. Collier, S. K. Goffredi, J. Hashimoto, M. Lilley, A.-L. Reysenbach, T. Shank, K. L. Von Damm, A. Banta, R. M. Gallant, D. Gotz, D. Green, J. Hall, T. L. Harmer, L. A. Hurtado, P. Johnson, Z. P. McKiness, C. Meredith, E. Olson, I. L. Pan, M. Turnipseed, Y. Won, C. R. Young III, and R. C. Vrijenhoek, Biogeography and

Ecological Setting of Indian Ocean Hydrothermal Vents, *Science, 294*, 818–823, 2001.

Vera, E. E., and J. B. Diebold, Seismic imaging of oceanic layer 2A between 9°30′N and 10°N on the East Pacific Rise from two-ship wide-aperture profiles, *Journal of Geophysical Research, 99*, 3031–3041, 1994.

Vera, E. E., J. C. Mutter, P. Buhl, J. A. Orcutt, A. J. Harding, M. E. Kappus, R. S. Detrick, and T. M. Brocher, The structure of 0- to 0.2-m.y.-old oceanic crust at 9 degrees N on the East Pacific Rise from expanded spread profiles, *J. Geophy. Res.*, 95, 15,529–15,556, 1990.

Vera, E. E., and J. B. Diebold, Seismic imaging of oceanic layer 2A between 9°30′N and 10°N on the East Pacific Rise from two-ship wide-aperture profiles, *Journal of Geophysical Research, 99*, 3031–3041, 1994.

Von Damm, K. L., Evolution of the Hydrothermal System at East Pacific Rise 9°50′N: Geochemical Evidence for Changes in the Upper Oceanic Crust, in: German, C. R., Lin, J., Parson, L., eds., *AGU Monograph.*

Von Damm, K. L., Lilley, M. D., Shanks, W. C.III, et al., Extraordinary phase separation and segregation in vent fluids from the Southern East Pacific Rise, *Earth Planet. Sci. Lett.*, 206, 365–378, 2003a.

Von Damm, K. L., and M. D. Lilley, Diffuse flow hydrothermal fluids from 9°50′N East Pacific Rise: origin, evolution, and biogeochemical controls, in: *The Subsurface Biosphere at Mid-Ocean Ridges,* AGU monograph RIDGE Theoretical Institute), W. Wilcock e al., eds., in press, 2004.

Von Damm, K. L., Chemistry of hydrothermal vent fluids from 9–10°N, East Pacific Rise: "Time zero" the immediate post-eruptive period, *J. Geophys. Res.* 105, 11203–11222, 2000.

Von Damm, K. L., Bray, A. M., Buttermore, L. G. Oosting, S.E.The geochemical controls on vent fluids from the Lucky Strike vent field, Mid-Atlantic Ridge, *Earth and Planet. Sci. Let.,* 160, 521–536, 1998

Von Damm, K. L, Buttermore L. G., Oosting S. E, Bray A. M., Fornari DJ, Lilley M.D., *et al.*, Direct observation of the evolution of a seafloor 'black smoker' from vapor to brine. *Earth Planet. Sci. Lett.*, 149, 101–111, 1997.

Von Damm, K. L., A. M. Bray, L. G. Buttermore, M. D. Lilley, E. J. Olson and E. McLaughlin, Chemical changes between high temperature and diffuse flow fluids at 9° 50′N East Pacific Rise, *Eos Trans. AGU*, 77 (46), F403–404, 1996.

Von Damm, K. L., Temporal and compositional diversity in seafloor hydrothermal fluids, *Rev. of Geophys.*, U.S. National Report to International Union of Geodesy and Geophysics, 1297–1305, 1995.

Von Damm, K. L., S. E. Oosting, R. Kozlowski, L. G. Buttermore, D. C. Colodner, H. N. Edmonds, J. M. Edmond, and J. M. Grebmeier, Evolution of East Pacific Rise hydrothermal vent fluids following a volcanic eruption, *Nature, 375,* 47–50, 1995.

Von Damm, K. L., Seafloor Hydrothermal Activity: Black Smoker Chemistry and Chimneys, *Annu. Rev. Earth Planet. Sci., 18*, 173–204, 1990

Wallace, P. J. and A. T. Anderson, Jr., Effects of eruption and lava drainback on the H2) contents of basaltic magmas at Kilauea Volcano, *Bull. Volcanol.*, 59, 327–344, 1998.

West, M., Menke, W., Tolstoy, M., Webb, S., and R. Sohn, Magma storage beneath Axial volcano on the Juan de Fuca mid-ocean ridge, *Nature, 413,* 833–836, 2003.

West, M., W. Menke, M. Tolstoy, S. C. Webb, and R. A. Sohn, Magma storage beneath Axial volcano on the Juan de Fuca mid-ocean ridge, *Nature, 413,* 833–836, 2001.

White S. M., R. M. Haymon, D. J. Fornari, K. C. Macdonald, and M.R. Perfit, Volcanic Structures and Lava Morphology of the East Pacific Rise, 9°–10°N: Constraints on Volcanic Segmentation and Eruptive Processes at Fast-Spreading Ridges, *J. Geophy. Res, 107,* 10,1029, 2001JB000571, 2002.

Wilcock, W. S. D., The physical response of mid-ocean ridge hydrothermal systems to local earthquakes, *Geochemistry, Geophysics, Geosystems, In press,* 2004.

Wilcock, W. S. D., Cellular convection models of mid-ocean ridge hydrothermal circulation and the temperatures of black smoker fluids, *J. Geophys. Res., 103,* 2585–2596, 1998.

Wolfe, E. W., Garcia, M. O., Jackson, D. B., et al., The Puu OO eruption of Kilauea volcano, episodes 1–20, January 3, 1983 to June 8, 1984, in: Decker, R. W, Wight, T. L. and P. H. Stauffer, eds., *USGS Prof. Paper 1350,* 471–508,1987.

Wolfe, E. W., Neal, C. A., Banks, N. G. and Duggan, T. J., Geologic observations and chronology of eruptive events. In: E. W. Wolfe (Eds.) *USGS Prof. Paper 1463.* The Puu Oo eruption of Kilauea Volcano, Hawaii: Episodes 1 through 20, January 3, 1983, through June 8, 1984, pp. 1–98, 1988.

Wright, D. J., Haymon, R. M., and D. J. Fornari, Crustal fissuring and its relationship to magmatic and hydrothermal processes on the East Pacific Rise crest (9°12′–54′N), *J. Geophys. Res., 100,* 6097–6120, 1995a.

Wright, D. J., Haymon, R. M., and K. C. Macdonald, Breaking new ground: Estimates of crack depth along the axial zone of the East Pacific Rise (9° 12–54′N), *Earth Planet. Sci. Lett., 134,* 441–457, 1995b.

Wright, D. J., Formation and development of fissures at the East Pacific Rise: Implications for faulting and magmatism at Mid-Ocean ridges, in: Faulting and magmatism at mid-ocean ridges, W.R. Buck, et al. (eds.), *Am. Geophys. U. Geophys. Monograph 106,* 137–151, 1998.

Wright, D. J., R. M. Haymon, S. M. White, and K. C. Macdonald, Crustal fissuring on the crest of the southern East Pacific Rise at 17° 15′–40′S, *J. Geophys. Res., 107,* 10,1029/2001JB000544, 2002.

Wylie, J. J., K. R. Helfrich, B. Dade, J. R. Lister, and J. F. Salzig, Flow localization in fissure eruptions, *Bull. Volcanol., 60,* 432–440, 1999.

Yang, J. K. Latychev, and R. N. Edwards, Numerical computation of hydrothermal fluid circulation in fractured Earth structures, *Geophys. J. Int., 135,* 627–649, 1998.

Yoerger, D., A. Bradley, R. Bachmayer, R. Catanach, A. Duester, S. Liberatore, H. Singh, B. Walden and M. A. Tivey, Near-bottom magnetic surveys of the Coaxial Ridge segment using the Autonomous Benthic Explorer survey vehicle, *RIDGE Events, 7,* 5–9, 1996.

Yoerger, D.R., R. Collier, and A.M. Bradley, Hydrothermal Vent Plume Discovery and Survey with an Autonomous Underwater Vehicle, *Eos Trans. AGU, 83(47),* Fall Meet. Suppl., Abstract T11C-1261, 2002.

Daniel J. Fornari, Woods Hole Oceanographic Institution, Geology and Geophysics Department, Clark South 172, MS 24, 286 Woods Hole Road, Woods Hole, MA 02543. (dfornari@whoi.edu)

Hydrothermal Processes at Mid-Ocean Ridges: Results From Scale Analysis and Single-Pass Models

Robert P. Lowell

School of Earth and Atmospheric Sciences, Georgia Institute of Technology, Atlanta, Georgia

Leonid N. Germanovich

School of Civil and Environmental Engineering, Georgia Institute of Technology, Atlanta, Georgia

Hydrothermal processes at ocean ridge crests involve a complex interplay between the dynamics of heat supply and the evolution of crustal permeability that results in a broad spectrum of issues to challenge hydrothermal modelers. Among these are: (1) the detailed relationships between heat supply and hydrothermal heat transfer, (2) the mechanisms controlling vent stability and long-term evolution of hydrothermal vents, including the relationship between focused and diffuse flow, (3) temporal variability related to magmatic and tectonic events, (4) boiling, phase separation and the evolution of vent chlorinity, and (5) the linkages among heat transfer, fluid flow, geochemical and biological processes in controlling character and evolution of vent fields. Here we address a number of these issues in the context of scale analysis and single-pass models. Scale analysis shows that plume heat transfer in a homogeneous system with an imposed heat flux from below does not correspond to observed black smoker flow. Single-pass models show that black smoker venting requires a conductive boundary layer no more than a few tens of meter thick above a magma chamber and a discharge zone with permeability between 10^{-13} and 10^{-11} m^2. Long-term vent stability requires that the thin conductive boundary be maintained, probably by some combination of magma replenishment and crack propagation, and a broad recharge zone to prevent clogging by anhydrite. Event plumes may be driven by dike emplacement provided the dike locally increases the permeability by two orders of magnitude and provides a significant fraction of event plume heat.

1. INTRODUCTION

Seafloor hydrothermal systems are an integral component of Earth's dynamic heat engine by which energy brought towards the seafloor by mantle convection is transferred to the ocean. These systems are of global importance and their discovery more than two decades ago has revolutionized the understanding of thermal, geochemical and biological processes on Earth. Seafloor hydrothermal systems transport nearly 25% of Earth's total heat flux, and approximately 33% of the heat flux through the ocean floor [*Williams and Von Herzen*, 1974; *Sclater et al.*, 1980; *Stein and Stein*, 1994]. As a result of chemical reactions between the oceanic crust and hydrothermal seawater, some elements are removed from seawater while hydrothermal discharge is significantly enriched in other elements [*Wolery and Sleep*, 1976; *Edmond et al.*, 1979; *Thompson*, 1983]. Hydrothermal fluids also serve as

Mid-Ocean Ridges: Hydrothermal Interactions Between the Lithosphere and Oceans
Geophysical Monograph Series 148

an energy resource for complex chemosynthetic biological ecosystems [*Jannasch and Wirsen*, 1979; *Jannasch*, 1983, 1995; *Shank et al.*, 1998]. The discovery of chemosynthetic ecosystems at seafloor hydrothermal vents has led to a new awareness of life in extreme environments and has stimulated the discussion of the origin of life on Earth [e.g., *Corliss et al.*, 1981; *Baross and Hoffman*, 1985; *Holm*, 1992; *Nisbet and Fowler*, 1996] and other planetary bodies in the solar system [e.g., *McCollum*, 1999]. *Lowell et al.* [1995] and *Kelley et al.* [2002] provide useful reviews of seafloor hydrothermal processes. *Karson* [2002] reviews the general crustal structure of fast and intermediate spreading ridges.

1.1. Styles of Seafloor Hydrothermal Activity

Research on seafloor hydrothermal processes over the past 30 years has progressed from examining the role of hydrothermal transport on depressing conductive heat flux values from that expected for lithospheric cooling [*Langseth and Von Herzen*, 1970; *Talwani et al.*, 1971; *Bodvarsson and Lowell*, 1972; *Lowell*, 1975; *Anderson et al.*, 1977; *Sclater et al.*, 1980; *Stein and Stein*, 1994] to the recognition of a variety of flow regimes and the high level of complexity that may occur within these regimes. The most spectacular of these regimes is the occurrence of high-temperature axial hydrothermal systems in which seawater percolating into the crust is heated by hot rock and magma, undergoes chemical reactions with the crustal rock, and emerges through chimney-like structures as mineral-laden fluids at temperatures ~ 350 to 400°C. These high-T systems, termed "black smokers" were first discovered on the East Pacific Rise (EPR) [*Spiess et al.*, 1981] and have now been found at oceanic spreading centers and back-arc basins of the world ocean comprising a full range of spreading rates (see Figure 1 in *Baker and German* [2004]). High-temperature venting is often accompanied by nearby low temperature diffuse flow venting [*Delaney et al.*, 1992; *Schultz et al.*, 1992; *Rona and Trivett*, 1992; *Von Damm and Lilley*, 2004], and the relationship between these flows is unclear [*Lowell et al.*, 2003]. Moreover, high-temperature flows may exhibit large transient episodes of heat output termed "event plumes" [*Baker et al.*, 1987, 1989, 1995; *Baker*, 1998]. The schematic in Figure 1 depicts crust-mantle relationships beneath ridge crests and possible hydrothermal circulation pathways for high-temperature ridge crest systems.

As one moves away from the ridge axis, styles of hydrothermal circulation change. In the off-axis regime, circulation and fluid discharge appears to occur at much lower temperature (T ~ a few tens of degrees or less) [*Davis et al.*, 1992, 1996; *Shipboard Scientific Party*, 1997; *Elderfield and Schultz*, 1996]; however, these systems may be responsible for a large

fraction of the hydrothermal heat loss and chemical recycling [*Stein et al.* 1995; *Elderfield and Schultz*, 1996].

A near-axis circulation system that is essentially unexplored occurs between the high-temperature ridge-axis system and the low-temperature off axis system. Recent seismic experiments [*Dunn et al.*, 2000] indicate that the entire crust cools significantly within a few kilometers of the ridge axis. Such cooling would require that hydrothermal circulation deepen to Moho levels within the near-axis region. At present, high-temperature venting in the near-axis region has not been observed. It is not known whether deep near-axis circulation contributes to the observed axial heat output, exists as a separate, yet unobserved high-temperature system, or discharges fluids at relatively low temperature. *Mottl* [2003] and *Nicolas et al.* [2003] begin to address near-axis circulation issues.

Finally, the recent discovery of the Lost City vent field [*Kelley et al.*, 2001] located on 1.5 m.y old crust, 15 km west of the eastern inner corner high of the Atlantis transform fault at 30°N on the Mid-Atlantic Ridge (MAR) [*Cann et al.*, 1997; *Blackman et al.*, 1998] points to a new class of hydrothermal system of considerable interest. The Lost City field appears to be driven by the heat released during serpentinization of mantle peridotite [*Kelley et al.*, 2001]. Conditions conducive to serpentinization, i.e., low magma budget and high fracture/fault permeability that facilitates access of seawater to large volumes of upper mantle peridotites are common at slow spreading ridges. Moreover, hydrothermal fluids exhibiting chemical signatures of serpentinization reactions are commonly observed

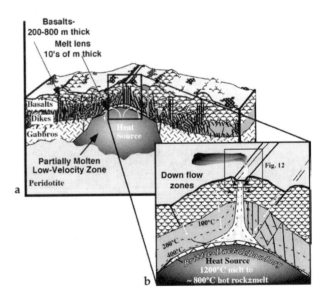

Figure 1. Schematic drawing of crust and mantle beneath mid-ocean ridges showing the main layering of oceanic crust, a subsurface magma body, and high-temperature hydrothermal circulation system tapping heat from the magma. [from *Kelley et al.*, 2002].

along the MAR [*Charlou et al.*, 1991; *Rona et al.*, 1992; *Charlou and Donval*, 1993; *Bougault et al.*, 1998; *Gracia et al.*, 2000], and possibly along slow spreading ridges in general. It is therefore likely that hydrothermal systems driven by exothermal serpentinization reactions are a quantitatively significant component of hydrothermal heat and chemical fluxes of slow-spreading ocean ridges, which comprise more than half the global ≈ 60,000 km length of the ocean ridge system in the Atlantic, Indian and Arctic oceans.

1.2. Some Major Issues in Seafloor Hydrothermal Research Related to Modeling

The complexity of seafloor hydrothermal circulation raises a number of important issues that need to be addressed not only by continued laboratory and field studies, but also must be complemented by mathematical modeling. These issues include (1) the detailed relationships between heat supply and hydrothermal heat transfer, (2) the mechanisms controlling vent stability and long-term evolution of hydrothermal vents, including the relationship between focused and diffuse flow, (3) temporal variability related to magmatic and tectonic events, (4) boiling, phase separation and the evolution of vent chlorinity, (5) the linkages among heat transfer, fluid flow, geochemical and biological processes in controlling character and evolution of vent fields, and (6) Archean hydrothermal activity and evolution of life.

To place these issues a conceptual framework we view high-temperature ridge crest hydrothermal systems in terms of episodic cycles consisting of an initial or "birth" phase, followed by a quasi-steady "living" phase, and culminating in a declining "dying" phase (Figure 2). The "birth" phase coincides with a pulse of magmatic input (e.g., a dike) and rapid growth of thermal output to some high level consistent with temperature and heat flux from mature black-smoker type systems. Magmatic inputs may generate new permeability that will affect the heat transfer characteristics of the system and short-lived "event plumes" may occur (see below). Depending upon the seafloor pressure, either subcritical or supercritical boiling and phase separation may occur during the "birth phase" of the system as a result of the close contact between seawater and a magmatic heat source. As a result of the density differences between the more-saline liquid and relatively low salinity vapor phases the phases travel at different velocities and may become separated in space, with dense saline liquids sinking toward the base of the system. During the quasi-steady "living" phase, heat transport from subsurface magma maintains relatively steady vent temperatures, heat output, and fluid chemistry, though some fluctuations may result from magmatic and/or tectonic activity. Depending upon the volume and replenishment rate of magma, the living

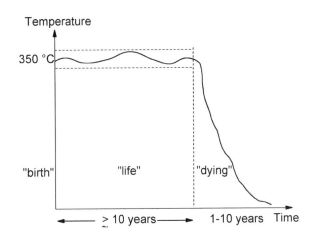

Figure 2. Conceptual model of temperature-time behavior of high temperature hydrothermal discharge at an ocean ridge crest. Hydrothermal episode begins with rapid onset or "birth" following fresh magmatic input, a quasi-steady "life" phase that may range from years to several decades, and a "dying" phase as the heat source is exhausted. A analogous figure could be drawn for heat output versus time. [from *Lowell and Germanovich*, 1994].

phase can last at least decades (as at East Pacific Rise (EPR) 21°N or the Main Endeavor Field (MEF) on the Juan de Fuca Ridge (JDF)), but may only last a few years. Boiling and phase separation may occur at times during the "living" phase, particularly if small magmatic inputs episodically occur. The system may also vent mixtures of liquid and vapor or brine phases that were generated during earlier episodes of boiling and phase separation. Eventually, however, as the heat source becomes exhausted, the system enters the "dying" phase. We have relatively little data on the dying phase of mature systems, but model calculations (see section *3.2*) suggest a time scale of years. When the main heat input comes from dike emplacement, the entire system cycle may last only a few years [e.g., *Baker et al.*, 1998; *Baker et al.*, 2004].

The efficiency of heat transfer, whether from magma or serpentinization reactions, to circulating hydrothermal fluid is not well understood. For high-temperature magma-driven systems, the rate of local, short-time magma supply is unknown and the relative importance of magma supply and dike emplacement to faulting and other tectonic processes that may generate permeable cracks for fluid circulation is uncertain. For systems driven by serpentinization reactions, the rate of serpentinization must be coupled to mechanisms controlling the rate of fluid access to fresh rock and these rates are essentially unknown.

Neither the long-term stability nor the temporal evolution of seafloor hydrothermal systems is well understood. As will be shown in section *3.2*, systems driven by conductive heat transfer across an impermeable layer separating liquid magma from the hydrothermal circulation system should undergo noticeable decay in a matter of years. Simple heat balance models indicate that the heat input to ocean ridges as a result of lithospheric creation and seafloor spreading is less than the rate of hydrothermal heat loss, even at fast spreading ridges [e.g., *Macdonald et al.*, 1980]. Hydrothermal systems must ultimately be transient. Little is known concerning the details of vent field decay, however.

Hydrothermal precipitates from the TAG hydrothermal field on the slow-spreading Mid-Atlantic Ridge (MAR) indicate a long history ($> 10^5$ years) of high-temperature activity; and that at the currently active TAG mound, episodic high-temperature venting has occurred every few thousand years, with current activity beginning $\approx$ 80 years ago [*Lalou et al.*, 1998]. Temperature and chemical data at MARK and TAG fields on the MAR [*Edmond et al.*, 1995], GR 14 on the Gorda ridge [*McClain et al.*, 2002], the vents at 21°N EPR [*Campbell et al.*, 1988] have exhibited stability on a timescale of years. Recent sampling of the 21°N vents in 2002 shows some changes in vent fluid chemistry and temperature [*Von Damm et al.*, 2002]; however, the changes are not as dramatic as at sites of active magmatic and tectonic activity.

On the other hand, systems are known to change dramatically following a magmatic event or an earthquake. At the fast spreading EPR between 9 and 10°N, high temperature venting, vent fluid chemistry, and biological ecosystems have undergone dramatic changes following magmatic eruptions in 1991 and 1992 [*Haymon et al.*, 1993; *Von Damm, et al.*, 1995, 1997, *Von Damm*, 2000; *Shank et al.*, 1998; *Von Damm and Lilley*, 2003; *Von Damm*, 2004]. Large, rapid pulses of hydrothermal activity termed "event plumes" have been discovered on the intermediate spreading Juan de Fuca (JDF) and Gorda Ridges in the northeast Pacific [e.g., *Baker et al.*, 1987, 1995, *Baker*, 1998]. Detailed time series of temperatures Bio9 vent on the EPR 9°50.2′N between 1991 and 1997 recorded a number of long term temperature fluctuations, including a rapid increase and slow relaxation to pre-event values following a "cracking event" [*Sohn et al.*, 1998; *Fornari et al.*, 1998]. Similar temperature fluctuations also occurred at the MEF on the JDF following an earthquake swarm [*Johnson et al.*, 2000], and changes in vent chemistry there suggest the event was volcanic in origin [*Lilley et al.*, 2003]. Limited data are available to describe the decay of hydrothermal output following a magmatic event (e.g., at "A" at 9°46.5′N on the EPR [*Von Damm*, 2002, personal communication]) and event plumes (see below) [*Butterfield et al.*, 1997; *Baker et al.*, 1998; 2004].

Another aspect of the stability and evolution of seafloor hydrothermal systems concerns the relationship between focused and diffuse flow. The processes by which ascending high-temperature hydrothermal fluids become focused into discrete vents at the seafloor are not well understood. At some sites, such as the Galapagos, the ascending high-temperature fluids are not focused at all and emerge as low temperature diffuse flow as a result of conductive cooling and mixing with cooler seawater [*Corliss et al.*, 1979; *Edmond et al.*, 1979]. At others, such at the "V" vent on the East Pacific Rise at 9° 47′ N, discharge evolved from 78°C in 1991 to black smoker flow by 1997 [*Von Damm*, 2000]. Finally, sites of vigorous high-temperature focused flow are often accompanied by nearby regions of low temperature diffuse flow. The TAG hydrothermal field on the Mid-Atlantic Ridge [e.g., *Edmond et al.*, 1995; *James and Elderfield*, 1996; *Humphris and Tivey*, 2000], the Endeavor segment of the Juan de Fuca Ridge [*Delaney et al.*, 1992 *Schultz et al.*, 1992; *Kelley et al.*, 2002], and the 9°–10°N area of EPR [*Von Damm and Lilley*, 2003] serve as examples.

Event plumes (Figure 3), which release 10^{14}–10^{17} J of heat to the water column in a matter of days [*Baker et al.*, 1987, 1989, 1995; *Baker*, 1998b], are another form of temporal variability and place additional constraints on seafloor hydrothermal processes. Such heat output over a matter of days corresponds to a thermal power output ranging between 10^9–10^{11} W, which is as much as 100 times normal black smoker power output. Dike injection [*Lowell and Germanovich*, 1995; *Germanovich et al.*, 2000], heat transfer from lavas [*Butterfield et al.*, 1997; *Palmer and Ernst*, 1998], sudden evacuation of a hypothetical high-temperature fluid reservoir [*Baker et al.*, 1987; *Wilcock*, 1997], and cracking/sealing mechanisms associated with black smoker venting [*Cann and Strens*, 1989; *Cathles*, 1993] have all been proposed as mechanisms for the generation of event plumes.

Seafloor hydrothermal systems not only transfer heat but also chemicals. Upon heating to temperatures of 150–200°C during recharge, anhydrite precipitates from seawater [*Bischoff and Seyfried*, 1978] and may clog fractures and pore space [*Mottl*, 1983; *Sleep*, 1991; *Lowell and Yao*, 2002]. Seawater Mg is exchanged for Ca from the rock, which may enhance the process [*Bischoff and Dickson*, 1975]. As seawater undergoes

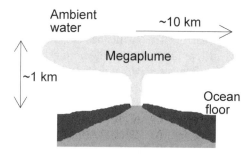

Figure 3. Schematic of event plume above a mid-ocean ridge [from *Lowell and Germanovich*, 1995].

further reactions with crustal rocks, the resulting hydrothermal fluid becomes enriched in Ca, Si, metals and other trace elements [e.g., *Von Damm*, 1995]. As the ascending hydrothermal fluids approaches the seafloor cooling and mixing with cooler relatively un-reacted seawater may result in subsurface precipitation of quartz, anhydrite, metallic sulfides, etc. [e.g., *Embley et al.*, 1988; *Richards et al.*, 1989; *Janecky and Shanks*, 1988] along with the formation of chimneys and sulfide ore deposits on the seafloor [*e.g., Haymon*, 1983; *Hannington et al.*, 1995].

To date, mathematical models of seafloor hydrothermal activity do not simultaneously treat heat transfer, water-rock reactions, and the evolution vent fluid chemistry. Moreover, the occurrence of "snowblower" vents following magmatic and tectonic events [*Haymon et al.*, 1993; *Juniper et al.*, 1995; *Holden et al.* 1998; *Delaney et al.*, 1998; *Summit and Baross*, 1998; *Johnson et al.*, 2000], points to the existence of an extensive subsurface biosphere. The biota use minerals as an energy resource and precipitate minerals in the subsurface, thus affecting fluid flow patterns [e.g., *Juniper and Sarrizin*, 1995; *Taylor and Wirsen*, 1997]. The coupling between biological and fluid transport processes has not yet been quantified.

Chlorinity of vent fluids is either greater than or less than ambient seawater in almost all cases with extreme values ranging from about 5% to about 200 % of seawater [*Von Damm, 1995*]. It is generally accepted that most the vent chlorinity variations result from boiling and either sub- or super-critical phase separation [e.g., *Berndt and Seyfried*, 1990; *Von Damm et al.*, 1997, 2002; *Von Damm*, 2004], although some variations in vent fluid chlorinity may result from the precipitation of Cl-bearing mineral [*Seyfried et al.*, 1986]. Phase separation is clearly recognized in vent fluids following magmatic events such as occurred at EPR 9°–10° N in 1991 [*Von Damm et al.* 1995, 1997], or at places where seafloor pressures and vent temperatures place the vent fluid on or near the two-phase boundary (e.g., Axial [*Butterfield et al.*, 1990], Lucky Strike and Menez Gwen on the MAR [*Fouquet et al.*, 1994; *Langmuir et al.*, 1997]). At sites currently venting at liquid-phase temperatures and pressures (e.g., Monolith on the JDF [*Butterfield et al.*, 1994] and EPR 21° N [*Von Damm et al.*, 2002]), deviations of vent chlorinity from seawater may represent the effects mixing between hydrothermal seawater and phase-separated fluid (either brine or vapor) that formed during an earlier episode of phase separation. Despite its importance, however, little mathematical modeling has been done on the dynamics of phase separation in a NaCl-H$_2$O fluid. Estimates have been made of the rate of brine formation [*Lowell and Germanovich*, 1997], and recently NaCl-H$_2$O heat pipes [*Bai et al.*, 2003] and phase separation near a dike [*Lewis and Lowell*, 2004] have begun to be investigated.

Finally, considerable attention has been paid to seafloor hydrothermal systems as sites for the origin of life on Earth [e.g. *Baross and Hoffman*, 1985; *Holm*, 1992; *Nisbet and Fowler*, 1996]. Hence there is a need to explore the characteristics of seafloor hydrothermal activity during the Archean. Evidence of early hydrothermal systems comes from the Barberton complex, South Africa [*deRonde et al.*, 1994], Pilbarra craton, Australia [*Rasmussen*, 2000], and other Precambrian cratons. It is generally well understood that heat loss from oceanic lithosphere was greater during the Archean as a result of greater spreading rates, greater ridge length, or some combination of the two [e.g., *Bickle*, 1978, *Abbott and Hoffman*, 1984] and that hydrothermal heat loss would have been greater as well [*Isley*, 1995; *Sleep and Zahnle*, 2001; *Lowell and Keller*, 2003]. Details concerning seafloor hydrothermal heat loss during the Archean are uncertain, however. For example, thicker oceanic crust [*Sleep and Windley*, 1982], shallower ridge depths [*Galer*, 1991; *Isley*, 1995], younger subduction ages [*Bickle,* 1978; *Abbott and Hoffman*, 1984], and a different ocean environment coupled with extrusion of hotter komatiite lavas [*Nisbet and Fowler*, 1996] could have all affected seafloor hydrothermal flow.

In the next section of this paper, we will discuss briefly the main observational constraints and parameters that are involved in modeling seafloor hydrothermal systems. Because the observational data is incomplete and important parameters, such as permeability are highly uncertain, we advocate the development of relatively simple mathematical models. In section 3 we discuss traditional approaches (cellular and single-pass) to modeling seafloor hydrothermal systems and derive results based on scale analysis to elucidate the basic behavior of high-temperature ridge crest hydrothermal systems. In section 4 we will introduce additional features to the basic single-pass model to better understand system behavior. Finally, we will suggest some possible future directions for mathematical modeling of seafloor hydrothermal processes.

2. OBSERVATIONAL CONSTRAINTS AND MODEL PARAMETERS

Meaningful mathematical models of seafloor hydrothermal circulation must be constrained by observational data. The available data varies widely from vent field to vent field, and in general the ability to formulate detailed mathematical models of any particular vent field is limited by the lack of data. Of fundamental importance for models of high-temperature circulation are vent temperatures and the heat flux (either of individual vents or, preferably, integrated over the entire field); but also important are the discharge and recharge areas of the vent field. Unfortunately, recharge zones have yet to be identified; but they are generally thought to be more extensive than discharge areas [*Lowell and Yao*, 2002]. Table 1 provides the temperature, heat flux data, and discharge area for several vent fields.

High-temperature systems generally require magmatic heat sources [*Strens and Cann*, 1982; *Lowell and Rona*, 1985; but see also *Lister*, 1974, 1983; *Wilcock and Delaney*, 1996]; consequently, information on the temperature, area extent, depth, and thickness of subsurface magma bodies, along with data on vent field spacing provide useful modeling constraints. The liquidus temperature T_M for basaltic magma is $\approx 1200°C$ [*Sinton and Detrick*, 1992]. For many vent fields, data on magmatic extent is unavailable; as a rule of thumb, however, at fast and intermediate spreading ridges, subsurface magma lends to be located at a depth of 1–2 km, with a cross-axis extent of the same order [*Detrick et al.*, 1987; *Sinton and Detrick*, 1992; *Detrick et al.* 2002]. The thickness of the liquid magma lens is typically several tens of meters [*Kent et al.*, 1990; *MacLeod*

and Yaouancq, 2000]. Although at fast spreading ridges the magma lens may extend along axis for tens of kilometers [*Detrick et al.*, 1987; *Sinton and Detrick*, 1992], vent fields are typically spaced approximately 1–3 km apart [*Gente et al.*, 1991; *Delaney et al.*, 1992; *Wilcock and Fisher*, 2003; *Kelley et al.*, 2002]. The area A_b available for heat extraction by a single vent field is thus ~ 1 km^2 [*Lowell and Germanovich*, 1994]. Finally, although vent chemistry [e.g. *Von Damm*, 1995] and trace element data [e.g., *Seyfried and Ding*, 1995] and time series data indicating the thermal and chemical response of the system to magmatic and/or tectonic events [*Schultz et al.*, 1992; *Sohn et al.* 1998; *Fornari et al.*, 1998; *Baker et al.*, 1998; *Johnson et al.*, 2000; *Lilley et al.*, 2003; *Seyfried et al.*, 2003; *Von Damm*, 2004] provide important constraints on mathematical models, these data have not yet been directly incorporated into models.

Mathematical models, described in terms of the governing equations expressing conservation of mass, momentum and

Table 1. Thermal Data From Seafloor Hydrothermal Systems

Location[1]	T_{max} (°C)	Vent Heat Flux (MW)	Integrated Heat Flux (GW)	Vent Field Area (m^2)
Axial (JDF)[a]	326	2.4-6.4	0.8	10^4
Endeavor (JDF)[b]	400	70-364	1-10	10^5
South Cleft (JDF)[c]	285	15-55	0.6	10^4
North Cleft (JDF)[d]	327	----	0.8-2.5[2]	----
Co-Axial (JDF)[e]	294[3]	----	0.3-30[3]	----
11°N (EPR)[f]	347	2.4-25	----	3 x 10^3
21°N (EPR)[g]	355	114-311	----	----
TAG (MAR)[h]	366	225	0.5-1	3 x 10^4
Broken Spur (MAR)[i]	365	27	0.3	1.5 x 10^4
Rainbow (MAR)[j]	364	---	1-5	3 x 10^4
Kairei (CIR)[k]	360	----	0.075-0.125	3 x 10^3

[1]JDF, Juan De Fuca; EPR, East Pacific Rise; MAR, Mid-Atlantic Ridge; CIR, Central Indian Ridge; MT, Mariana Trough. [2] The high estimate is from 1999 [*Gendron et al.* 1994] whereas the lower estimate is a multi-year average, indicating declining heat flux over time [*Baker*, 1994]. [3] The Co-Axial site is an event plume site consists of three separate vent areas. High temperature venting occurs only at the "source site" [*Butterfield et al.*, 1997]; the range of heat flux estimates shows the decline in heat output following the event plume [*Baker et al.*, 1998]. Sources are: [a]*Rona and Trivett* [1992], *Baker* et al. [1990]; [b]*Delaney* et al. [1992], *Rosenberg et al.* [1988], *Bemis* et al. [1993], *Ginster et al.* [1994], *Thompson* et al. [1992]; [c]*Baker and Massoth* [1986], *Baker and Massoth* [1987], *Normark* [1987]; [d]*Baker* [1994], *Baker and Cannon* [1993], *Baker et al.* [1993], *Gendron et al.* [1994]; [e]*Butterfield et al.* [1997], *Baker et al.* [1998]; [f]*McConachy* et al. [1986], *Little* et al. [1987]; [g]*Macdonald et al.* [1980], *Converse* et al. [1984]; [h]*Rona* et al. [1986], *Rona* et al. [1993b], *Edmond et al.* [1990], *Rudnicki and Elderfield* [1992]; [i]*Murton et al.* [1995], *Murton et al.* [1999]; [j]*German and Lin* [2004], *Thurnherr and Richards* [2001], *Thurnherr et al.* [2002]; [k]*Gamo et al.* [2001], *Hashimoto et al.* [2001], *Rudnicki and German* [2002].

energy (see section 3) contain a number of rock and fluid physical parameters that influence the solutions. These include the fluid properties of density ρ_f, specific heat capacity c_f, thermal expansion coefficient α, and kinematic viscosity ν as well as rock properties of thermal conductivity λ, porosity ϕ, and permeability k. Table 2 lists the values and ranges of the main fluid and rock parameters, along with data derived from observational constraints, that are used in the models below.

Of the physical parameters that affect hydrothermal circulation, the most important one, and the least well determined is the rock permeability k. At any given vent field this parameter may vary over several orders of magnitude. It is likely to be both heterogeneous and anisotropic. Measurements made in DSDP and ODP boreholes give permeability values ranging from 10^{-18} m^2 in sheeted dikes to as high as 10^{-10} m^2 in pillows (see review by *Fisher* [1998]). Crustal permeability has also been estimated from field studies in ophiolites [*Nehlig and Juteau*, 1988; *van Everingdon*, 1995, *Germanovich et al.*, 2000], and by mathematical modeling studies [*Lowell and Germanovich*, 1994; *Pascoe and Cann*, 1995; *Wilcock and McNabb*, 1996; *Cherkaoui et al.*, 1997; *Germanovich et al.*, 2000], yielding values ranging between 10^{-13} and 10^{-8} m^2. (See also sections 3 and 4 below). The high values of permeability and the broad range of values estimated from field and modeling studies, indicate that permeability in oceanic crust is fracture-controlled [*Brace*, 1980] with crack distribution occurring on many scales [*Turcotte*, 1992]. Because k may vary over several orders of magnitude and is a highly uncertain parameter, it dominates the uncertainties in all other rock, fluid and thermodynamic parameters in the governing equations. As a result, we develop simple models using k as a parameter and generally neglect the uncertainties in the other parameters.

3. MATHEMATICAL MODELS OF RIDGE CREST HYDROTHERMAL CONVECTION-BASIC STATE

Mathematical models of seafloor hydrothermal circulation may be employed to explain observational data as well as to elucidate and constrain physical processes. These models fall into two broad classes: (1) cellular convection or porous medium models, which consider hydrothermal convection in a bulk fluid-saturated permeable layer and (2) single-pass or "pipe" models, which consider restricted flow paths consisting of recharge element, a discharge element joined by a heating zone. As a variant of the single-pass model, *Lister* [1974, 1983] proposed a model whereby heat is extracted during downward migration of fractures into hot rock. Cellular and single-pass modeling concepts have been reviewed in *Lowell* [1991] and *Lowell et al.* [1995]. The essential features of these

modeling constructs, together with an assessment of the strengths, weaknesses, and overall utility of each approach are highlighted below. We outline more recent results based on the single-pass approach in section *4*.

3.1. Cellular Convection Models

Cellular convection models require numerical solutions to coupled, non-linear partial differential equations expressing fundamental conservation laws, together with an equation of state. These equations are usually solved in a more or less homogeneous permeable layer or simple geometric 2- or 3-D constructs, subject to boundary conditions on the temperature and fluid flow. Typically, the upper boundary is assumed to be either impermeable or at constant pressure, and isothermal; the lateral boundaries are assumed to be impermeable and insulated, and the lower boundary is assumed to be impermeable with either a fixed heat flux or fixed temperature condition. Often, only steady state solutions are determined.

In cellular convection models, convective circulation patterns depend upon the Rayleigh number Ra, which must exceed a critical value Ra_c for convection to occur [*Lapwood*, 1948; *Nield*, 1968]. Heat transfer in the supercritical convective regime is expressed in terms of a dimensionless Nusselt number Nu, which expresses the ratio of heat transfer in a convecting system to that by heat conduction alone. A discussion of techniques for modeling cellular convection is beyond the scope of this paper; The basics are reviewed in *Lowell* [1991], and more extensive discussions of theoretical results and applications to geothermal systems are found in *Combarnous and Bories* [1975], *Cheng* [1978], *Elder* [1981], and *Norton* [1984].

3.1.1. Applications. The earliest application of the cellular convection model was to simulate surface heat flow distributions [e.g., *Williams et al.*, 1974; *Ribando et al.*, 1976; *Fehn and Cathles*, 1979; *Green et al.*, 1981; *Becker and Von Herzen*, 1983; *Fehn et al.*, 1983]. *Patterson and Lowell* [1982] and *Brikowski and Norton* [1989] investigated hydrothermal circulation near a ridge-axis magma body. These two themes have been at the root of many later modeling efforts, including a number of studies of off-axis convection [*Lowell*, 1980; *Fisher et al.*, 1990, 1994; *Fisher and Becker*, 1995; *Davis et al.*, 1996, 1997; *Yang*, 2002; *Wang et al.*, 1997] and heat transfer from a dike [*Cherkaoui et al.*, 1997; *Lowell and Xu*, 2000].

Cellular convection models have also been applied to ridge-axis circulation. *Wilcock* [1998] showed that the venting temperature of a plume rising from the base of a porous layer with constant permeability would range between 0.5 an 0.65 of the basal temperature and suggested that sealing of highly permeable pillows by mineral precipitation was necessary for black smoker venting. *Rabinowicz et al.* [1999] consider con-

Table 2. Symbols, Definitions and Values (Parameters and Vent Field Observables) Used to Constrain Mathematical Models

Symbol	Definition	Value
	English Symbols	
a	Thermal diffusivity	10^{-6} m^2
A_b	Heat uptake area at base of hydrothermal system	$\sim 10^6$ m^2
A_d	Vent field discharge area	$\sim 10^3 - 10^4$ m^2
A_w	Area of dike wall for event plume formation	$\sim 10^6 - 10^7$ m^2
b	Crack width	
c_{pf}	High temperature specific heat of fluid	6×10^3 J(kg°C)$^{-1}$
c_{pm}, c_p	Rock-fluid mixture/magma specific heat	10^3 J(kg°C)$^{-1}$
C_e	Solubility of mineral in fluid	
d	Dike width	~ 1 m
E_D	Energy content of dike	
$F(T)$	Temperature dependent function in thermoelastic problem	
g	Acceleration due to gravity	10 ms^{-2}
h	Height of convecting layer	10^3 m
H	Plume heat flux in cellular convection model	
H_{ep}	Event plume heat flux	$\sim 10^{10} - 10^{11}$ W
H_0	Basal heat flux density	H_t/A_b
H_t	Vent field heat output	$\sim 10^8 - 10^9$ W
$\hat{H}$	Total driving head	
k	Rock permeability	
k_{res}	Residual rock permeability	
l	Total path length of single pass flow	
L	Horizontal scale of convection cell	
$\hat{L}$	Latent heat of fusion of magma	10^5 Jkg^{-1}
M	Mass of sulfide ore deposit	
Nu	Nusselt number	
P	Pressure	
q	Mass flow rate per unit area	
Q	Total mass flux	
R	Integrated flow resistance	
Ra	Rayleigh number	
Ra_c	Critcal Rayleigh number	
t	Time	
T	Temperature	
T_M	Magma temperature	1200 °C
ΔT	Temperature difference across convecting layer	
$\vec{u}$	Specific discharge (Darcian velocity)	
V	Dike volume	
z	Vertical Cartesian coordinate	
	Greek Symbols	
α	Thermal expansion coefficient of fluid	$\sim 10^{-3}$ °C^{-1}
χ	Fe concentration in vent fluid	$\sim 10^{-4}$ kgkg^{-1}
δ	Conductive thermal boundary layer in convecting system	
δ_M	Conductive boundary layer between magma and base of hydrothermal system	
ϕ	Porosity	$\sim 0.01-0.1$
γ	Thermoelastic closure parameter	

Table 2. (cont.)

Symbol	Definition	Value
η	Crack geometry and distribution parameter	
λ_m	Thermal conductivity of rock fluid mixture	2.0 W (m-°C)$^{-1}$
ν	Kinematic viscosity	~10^{-7} m^{2}s^{-1}
ρ	Magma density	3 x 10^3 kgm^{-3}
ρ_f	Fluid density	10^3 kgm^{-3}
ρ_m	Density of rock-fluid mixture	
τ	Time scale for various processes	
ξ	Mass fraction of iron in ore deposit	1/2
	Subscripts	
0	Reference value	
d	Discharge zone	
r	Recharge zone	

vection in a vertical slot and suggest that a sloping bottom may stabilize high-temperature hydrothermal output. *Schoofs and Hanson* [2000] developed a numerical model to describe the depletion of brine layer at the base of a vigorously convecting system. *Jupp and Schultz* [2000] use numerical models to show that the thermodynamic properties of water may control venting at temperatures of ~ 400°C. *Fontaine et al.* [2001] investigated the role of anhydrite precipitation on the evolution of permeability, temperature, and fluid flow patterns in a convecting slot.

A drawback to the cellular convection models is their scale of resolution and the application of an isothermal upper boundary condition. Cellular convection models tend to address features of convective flow on a scale of hundreds of meters. They have difficulty resolving small-scale features (~ 1–10 m) such as faults and fractures that have high permeability and thin thermal boundary layers. The isothermal upper boundary also precludes the direct modeling of high-temperature venting.

3.1.2. Fundamental equations and scaling laws. In the context of this paper, it is useful to discuss cellular convection in terms of simple scaling laws because it allows a useful comparison between cellular and the single-pass model described below and in section 4. To achieve this, we first write the basic conservation equations. In Cartesian coordinates, conservation of mass, momentum, and energy for an incompressible, single-phase fluid in a homogeneous isotropic Darcian fluid-saturated porous medium are given by

$$\frac{\partial\left(\phi\rho_f\right)}{\partial t} + \nabla\bullet\left(\rho_f\vec{u}\right) = 0 \tag{1}$$

$$\vec{u} = -\frac{k}{\rho_f\nu}\left[\nabla P - \rho_f\vec{g}\right] \tag{2}$$

$$\rho_m c_{pm}\frac{\partial T}{\partial t} + \rho_f c_{pf}\vec{u}\bullet\nabla T = \nabla\bullet\lambda_m\nabla T \tag{3}$$

respectively. In equations (1) through (3) $\vec{u}$ is the specific discharge (Darcian velocity), P is the pressure, $\vec{g}$ is the acceleration due to gravity, and t is the time, respectively. The product $\rho_f\vec{u} = \vec{q}$ is the mass flux per unit area. The subscripts m and f refer to properties of the rock-fluid mixture and the fluid, respectively. Thermodynamic and transport properties of the fluid are treated as functions of temperature and pressure. These equations are supplemented by an equation of state relating the density of seawater to its temperature. In its simplest form,

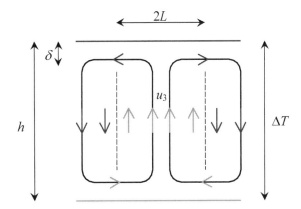

Figure 4. Schematic of vigorous cellular convection in a porous layer heated from below showing pertinent length scale L of circulation cell and δ of the conductive boundary layer.

$$\rho_f = \rho_{f0}\left[1 - \alpha\left(T - T_0\right)\right] \tag{4}$$

where the subscript 0 refers to a reference value of the parameter.

For a vigorously convecting porous layer heated from below $Ra >> Ra_c$. Then circulation can be described in terms of cells of horizontal scale L consisting of pairs of ascending and descending thermally buoyant plumes (also with horizontal scale L), with heat transfer occurring by conduction across thin boundary layers (of scale δ) near the top and base of the system (Figure 4). The pertinent length scales δ and L, the vertical velocity u_z, and the Nusselt number can be determined using scale analysis of equations (1) through (4). We consider two separate cases. First we assume that there is a fixed temperature difference ΔT applied across the layer; then we consider the situation in which a constant heat flux H_0 is maintained at the base of the layer.

Constant ΔT scaling: Following *Bejan* [1995] we write the following scaling relations:

$$\rho_{f0}c_{pf}u_z\Delta T/h \sim \lambda_m \Delta T/L^2 \tag{5}$$

$$u_z \sim k\alpha g \Delta T / v \tag{6}$$

These relations stem from the conservation of energy and momentum equations (2) and (3), respectively. We further write

$$\rho_{f0}c_{pf}u_z\Delta TL \sim \lambda_m \Delta TL/\delta \tag{7}$$

which states that the heat transported upward by the plume must equal the heat conducted across the top thermal boundary layer. Equations (5) through (7) yield the scaling relationships:

$$u_z \sim \left(a/h\right)Ra \tag{8}$$

$$L \sim hRa^{-1/2} \tag{9}$$

$$\delta \sim hRa^{-1} \tag{10}$$

where $a = \lambda_m/(\rho_{f0}c_{pf})$ is the effective thermal diffusivity, and the Rayleigh number $Ra = \alpha gk\Delta Th/av$. In addition the Nusselt number is given by

$$Nu = \frac{\lambda_m \Delta T/\delta}{\lambda_m \Delta T/h} \sim Ra \tag{11}$$

The scalings given by equations (8) through (11) are valid provided $Ra >> Ra_c$, where $Ra_c = 4\pi^2$ [*Lapwood*, 1948]. The most important results of this simple scaling are that (a) the higher the Rayleigh number the greater the heat transport by an individual thermal plume, (b) vent spacing ($\sim L$) can be estimated, and (c) the bulk permeability of the high-temperature system can be estimated from knowing the vent temperature and heat output. Upon substituting equations (8) and (9) into the left hand side of equation (7) we obtain H, the heat output per unit length of a single 2-D plume

$$H \sim \lambda_m \Delta TRa^{1/2} \tag{12}$$

Equation (12) shows that the heat output of a single plume increases with increasing Rayleigh number. This simple result comes from the condition of constant temperature at the base of the system. This condition corresponds to an assumption of an infinite reservoir of heat at the base the hydrothermal system. For a 100 MW hydrothermal system, acting as a 2-D plume with a length of 10^3m, $H = 10^5$ W/m. If the system operates at $\Delta T = 400°C$, $\lambda_m = 2.0$ W $(m°C)^{-1}$, equation (12) gives $Ra \sim 10^4$. With $h = 10^3$ m, inserting this value of Ra into equation (9) yields $L \sim 10$ m. Finally, using the parameter values in Table 2 with this value of Ra gives the permeability $k \sim 3 \times 10^{-13}$ m^2. Similar permeability estimates have also been provided by other analyses [*Lowell and Germanovich*, 1994; *Wilcock and McNabb*, 1996; *Pascoe and Cann*, 1995]

Constant H_0 scaling: If we envision that the hydrothermal system is driven by heat conducted from a magma body across an impermeable boundary layer, then a better representation of the convective system might be the assumption of constant heat flux H_0 at the base of the layer. In this case the equations (5) and (6) remain formally identical, but equation (7) becomes

$$\rho_f c_{pf}u_z\Delta TL \sim \lambda_m \Delta TL/\delta \sim H_0L \tag{13}$$

Equation (13) highlights the fact that the heat transport is limited by the basal input. For this system the Rayleigh number becomes $Ra = k\alpha gH_0h^2/(\lambda_m av)$ [*Nield*, 1968]], where $H_0h/\lambda_m = \Delta T_0$ is the steady state temperature difference that would exist across the layer in the absence of convection. The effect of thermal convection in this case is to lower the mean temperature difference across the porous layer while transporting the same heat as in the absence of convection (i.e., the Nusselt number is unity). These results are seen in the scaling relationships. That is,

$$\Delta T/\Delta T_0 \sim Ra^{-1/2} \tag{14}$$

$$u_z \sim \left(a/h\right)Ra^{1/2} \tag{15}$$

$$\delta \sim hRa^{-1/2} \qquad (16)$$

$$L \sim hRa^{-1/4} \qquad (17)$$

These results contrast in a significant way from the results for constant ΔT, where increasing Rayleigh number meant increasing plume heat transport because of the increased fluid velocity. Here, equations (13) and (17) together show that the condition of constant basal heat flux means that increasing the Rayleigh number results in decreasing the heat transport in a single plume. This occurs because increasing Ra results in more plumes, but the heat flux from below is fixed.

The critical Rayleigh number for a layer with fixed basal heat flux is 17.7 [*Nield*, 1968]. As shown in the following subsection, the basal heat flux H_0 needed to drive a'100 MW black smoker system is $\sim$ 100 W/m^2. Using this value, the parameter values considered in the previous analysis, and assuming $\Delta T = 400°C$ in equation (14), we obtain $Ra \sim 10^4$. From this value we find $k \sim 2 \times 10^{-15}$ m^2. From equation (17) with this value of Ra, we find $L \sim 100$ m. Then upon substituting equation (17) into equation (13), we find the plume heat transport $H \sim 1.5 \times 10^4$ W/m. Thus for a convecting porous layer driven by constant heat flux, plume heat transport is nearly an order of magnitude less than observed in typical black smoker systems. To observe a 100 MW hydrothermal system from this model, one would require the coalescence of ≈ 10 individual plumes spaced approximately 100 m apart.

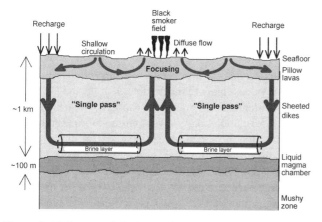

Figure 5. (a) Cartoon of single-pass hydrothermal circulation cell at a mid-ocean ridge (simplified from Figure 1). Single-pass primarily refers to the deep circulation system in which fluid circulates downward into to ocean crust, flows more-or-less horizontally near the top of the magma chamber at the base of the sheeted dikes and ascends back to the surface. Focused high-temperature flow is thought to occur in the main single-pass limb; diffuse flow may occur as a result of mixing of the deep circulation with shallower circulation in the pillow basalts. [from *Germanovich et al.*, 2000]

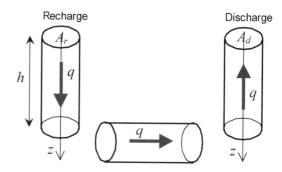

Figure 5. (b) An even simpler representation of single-pass circulation as flow through three pipes.

The permeability derived from the constant heat flux scaling is also much lower than for the constant ΔT scaling. The lower permeability is needed to maintain the high observed vent temperature from the constant heat flux model. In the following section, the results from the cellular convection model with constant heat flux are contrasted with the results form the single-pass model.

The simple scaling analyses performed here suggest that homogeneous cellular convection models may not necessarily provide a useful means for modeling high-temperature ridge axis hydrothermal systems. Though the model with an isothermal lower boundary condition gives reasonable results for permeability and heat transport, such a boundary condition is inappropriate for treating heat flux from a subjacent crystallizing magma body. On the other hand, although the constant heat flux boundary condition is physically more realistic, scale analysis suggests that plume heat transport is too low.

3.2. Single-Pass Modeling

Another relatively simple approach is to consider the single-pass (Figure 5a) or pipe model approach (Figure 5b). Single-pass, or pipe models provide a mechanism for examining the general behavior of the hydrothermal system without considering details of the temperature and velocity distribution. Rather, hydrothermal circulation is modeled in terms of simple recharge, discharge and heating elements. Cold fluid enters the recharge zone and descends to depth where it is heated; the heated fluid ascends to the surface where it discharges. Because the fluid only travels through the system once, the models are referred to as single-pass. The single-pass model can be viewed as a special case of the cellular convection model, in which a heterogeneous distribution of permeability restricts the flow paths to pipe-like zones embedded in an essentially impermeable matrix. Although these models greatly oversimplify many real world complexities, they are very useful for elucidating the basic physics of ridge crest hydrothermal systems. Note that the single-pass model, by restricting the flow geometry, allows the heat flux from a significant area of the sub-seafloor magma body to be collected into a single discharge zone [e.g., *Lowell and Burnell*, 1991 (and see below)]. As a result, one of the difficulties presented by cellular convection models with a heat flux boundary condition can be circumvented.

Although single-pass models are used as a matter of convenience, conceptual models of hydrothermal circulation at ridge crests commonly invoke recharge through some set of faults or fractures, water-rock chemical reactions and heat uptake at depth, and ascent through another set of localized faults and fractures (Figure 1). Comparison of vent chemistry with seawater/basalt reaction experiments [e.g., *Seyfried and Mottl*, 1982] and with determination of hydrothermal fluid residence times using short-lived isotopes [*Kadko et al.*, 1986] is consistent with single-pass flow.

Details of how these pipe elements fit in the real world is highly speculative. It is not known whether circulation is along-axis, cross-axis or both. If it is cross-axis, one must presume that faults striking along axis are listric, but it is not obvious how the recharge and discharge faults are linked. If flow is along-axis, dikes contacts may furnish the main permeability; but it is not clear why fluid travels along a thin, quasi-horizontal path along the base of the system where it reacts and takes up heat, nor what controls the discharge sites. The transition zone between base of the hydrothermal system and liquid magma must be extremely complicated geologically. Generally this zone has been identified with the base of the sheeted dikes and the upper level gabbros [e.g., *Richardson et al.*, 1987; *Nehlig and Juteau*, 1988; *Nehlig*, 1994; *Gillis and Roberts*, 1999; *Manning et al.*, 2000].

Pipe models were initially applied to seafloor hydrothermal circulation studies to explain conductive heat flow anomalies [*Bodvarsson and Lowell*, 1972; *Lowell*, 1975]. Later models of this type were used to investigate the formation of sulfide ore deposits at seafloor spreading centers, which are thought to be analogous to ore deposits found in ophiolites [*Strens and Cann*, 1982; 1986; *Cann et al.*, 1985; *Lowell and Rona*, 1985]. More recent studies have focused on the basic heat transfer requirements for black smoker venting and its temporal evolution [*Lowell and Burnell*, 1991; *Lowell and Germanovich*, 1994], the formation of event plumes [*Lowell and Germanovich*, 1995; *Germanovich et al.*, 2000], the rate of brine formation and depletion [*Lowell and Germanovich*, 1997], and mineral precipitation and thermoelastic processes and the evolution of permeability [*Lowell et al.*, 1993; *Martin and Lowell*, 2000; *Germanovich et al.*, 2001; *Lowell and Yao*, 2002; *Lowell et al.*, 2003]. In this paper will illustrate

the single pass approach to some of these problems in an ultra-simple manner.

The pertinent equations for pipe models are written in terms of integrated properties of the system such as the total resistance to flow, total mass flux and heat supply, etc. The formalism has a direct analogy with electrical circuits where the total mass flow rate, Q, is identified with electric current, the pressure head driving the flow, , is identified with the applied potential difference, and the flow resistance, R, is identified with the electrical resistance [e.g., *Elder*, 1981].

3.2.1. Fundamental equations and basic results. The integrated version of Darcy's Law, which is the equivalent of Ohm's Law in circuit theory, for buoyancy driven circulation in a porous medium is given by [Germanovich et al., 2000, 2001]

$$Q = \frac{\widehat{H}}{R} = \frac{\rho_{f0} g \int_0^h \alpha(z)\left(T_d(z) - T_r(z)\right) dz}{\int_0^l \frac{v}{kA} ds} \quad (18)$$

where T_d and T_r refer to the temperature distribution in the discharge and recharge limbs of the model, A is the cross-sectional area perpendicular to the flow path, and l is the total length of the flow path, respectively. An ultra-simplistic approach assumes that $T_r = 0$, $T_d(z) = T_d$ (a constant), and that the flow resistance is dominated by the discharge limb. Then equation (18) reduces to a simple scaling relationship

$$Q \sim \frac{\rho_{f0} \alpha_d g k_d T_d A_d}{v_d} \quad (19)$$

where the parameter values in the discharge limb are assigned constant values. Note that equation (19), for the single pass model, is identical in form to the scaling relationship in equation (6), for the cellular convection model, integrated over the area of the discharge zone.

To express conservation of energy with a basal heat flux boundary condition, we first assume that heat is transferred from a subsurface magma body to the base of the hydrothermal system by conduction across an impermeable thermal boundary layer of thickness δ_M and horizontal basal area A_b. We then assume that the heat conducted from the magma is carried, without conductive heat loss, by hydrothermal circulation to the seafloor, where it discharges through the vent field area A_d. Thus

$$H_0 A_b = \frac{\lambda_m \left(T_M - T_d\right) A_b}{\delta_M} = c_f Q T_d = H_t \quad (20)$$

where T_M is the magma temperature, and H_t is the observed heat output of the hydrothermal system, which is typically $\sim 10^2 - 10^3$ MW as indicated in Table 1. Upon recognizing that $Q = \rho_{f0} u_z A_d$, one observes that the relationship (20) is formally identical to the scaling law in equation (13) for the cellular convection model. Moreover, with $A_b \sim 10^6$ m^2 (see Table 2), we find $H_0 \sim 10^2 - 10^3$ Wm^{-2} as used in the previous section.

The simple algebraic equations (19) and (20), together with observational and parameter constraints from Tables 1 and 2, provide a number of useful constraints on Q, k, and δ_M in magma-driven high-temperature hydrothermal systems. We find:

$$2 \leq \delta_M \leq 20 \text{ meters}$$
$$40 \leq Q \leq 400 \text{ kgs}^{-1} \qquad (21)$$
$$10^{-13} \leq k_d \leq 10^{-11} \text{m}^2$$

Equations (21) indicate that the magma-hydrothermal boundary layer must be thin and the bulk permeability of the discharge zone must be high. The calculated bulk permeability is several orders of magnitude greater than measured in ODP boreholes in sheeted dikes (see section 2). The mass fluxes given by equation (21) are of the order observed in high-temperature systems. These results, though based on an ultra-simple mathematical model, do not change appreciably upon invoking somewhat more complex single-pass models [e.g., *Lowell and Burnell,* 1991; *Lowell and Germanovich,* 1994].

3.2.2. Stability and decay of a hydrothermal system. Another simple result that follows from the single pass model reflects the fact that magma at ≈ 1200°C is at its liquidus [see *Sinton and Detrick,* 1992]. Thus heat conducted from the magma to the hydrothermal system results in freezing of the magma. If the frozen magma is plated upon the upper boundary of the magma lens, the conductive thermal boundary layer thickens as $t^{1/2}$ [*Carslaw and Jaeger,* 1959; *Lowell,* 1991; *Lowell and Germanovich,* 1994]. Substituting $\delta_M \sim (at)^{1/2}$ into equation (20) yields $H_t \sim t^{-1/2}$ and combining equations (19) and (20) shows that both Q and $T_d \sim t^{-1/4}$ (see also *Lowell and Germanovich,* 1994). Such rapid, predictable decay of a seafloor hydrothermal system is not observed. Although the TAG system on the Mid-Atlantic Ridge is known to be episodic, it current active phase is thought to be several decades long [*Lalou et al.,* 1993, 1998]; and repeated temperature measurements at the 21°N sites on EPR have indicated considerable stability for more than two decades [*Campbell et al.,* 1988; *Von Damm et al.,* 2002]. The steady state results given by equation (21) thus indicate that to maintain constant heat output and vent temperature, the thin conductive boundary layer must be maintained during the steady state phase of the system.

Two primary mechanisms for maintaining a thin conductive boundary layer have been proposed: (a) downward crack propagation and (b) magma replenishment. The concept of downward crack propagation, initially quantified by *Lister* [1974, 1983] has received continued support [*Seyfried and Ding,* 1995; *Wilcock and Delaney*; 1996; *Gillis and Roberts,* 1999; *Alt and Teagle,* 2000; *Kelley et al.,* 2002]. Because Lister's approach neglects rock compressive stresses as cracks propagate into the crust, further work on this basic idea is needed. Some investigators have suggested that thermal stresses associated with dike emplacement and cooling may be preferable to Lister's original mechanism [*Bodvarsson,* 1982; *Lowell and Germanovich,* 1994]. The quasi-steady state character of some seafloor hydrothermal systems has also been linked to magma replenishment [*Lowell and Germanovich,* 1994; *Humphris and Cann,* 2000]. These authors show that magma replenishment at rates similar to those observed at basaltic shield volcanoes could sustain hydrothermal output on decade-long time scales. The recent observation of a magma body located roughly 2.5 km beneath the MEF on the Juan de Fuca Ridge [*Detrick,* 2002] and the response of the MEF to a recent magmatic event [*Lilley et al.* 2003] further supports the idea that magma replenishment is a likely mechanism for maintaining quasi-steady hydrothermal output. The recent observation of magma beneath MEF is especially intriguing because seismic refraction experiments did not find evidence of magma [*White and Clowes,* 1990] beneath the weak reflector observed by *Rohr et al.* [1988]. The MEF had been considered the archetype for crack propagation [*Wilcock and Delaney,* 1996].

4. EXTENSIONS OF THE SINGLE-PASS MODEL

One advantage of the single-pass model is that it can be used to address other problems of interest without adding undue complexity. Such problems include estimates of the rates of ore deposition [*Lowell and Rona,* 1985], the roles of thermal stresses and mineral precipitation on the evolution of permeability [*Germanovich and Lowell,* 1992; *Lowell et al.,* 1993; *Germanovich et al.,* 2001; *Martin and Lowell,* 2000; *Lowell and Yao,* 2002], the formation of brines [*Lowell and Germanovich,* 1997], and issues related to heat transfer from dikes and the generation of event plumes [*Lowell and Germanovich,* 1995; *Germanovich et al.,* 2000]. In accordance with the approach taken in this paper, we will address some of these issues in an ultra-simple fashion, similar to that above.

4.1. Sulfide Deposits

One of the primary features of seafloor hydrothermal venting is the formation of chimney structures as well as larger ore bodies such as the Endeavor structure or the TAG hydrothermal mound. The processes of hydrothermal ore deposit formation are quite complex in detail [e.g., *Haymon*, 1983; *Goldfarb et al.*, 1983; *Tivey*, 1995; *Hannington*, 1995]. If we assume, however, that the hydrothermal ore deposit consists mainly of pyrite (FeS_2), that a fraction ($\xi \approx 1/2$) of the mass of the deposit is Fe, and that all the hydrothermal fluid goes into forming the deposit, the single pass model provides some insight into the time of formation of both chimneys and larger ore deposits structures.

The concentration of Fe in hydrothermal vent fluids is quite variable-ranging from less than 0.4 mmol/kg to greater than 10 mmol/kg [*Von Damm*, 1995]. Assuming an average value of $\approx$ 2 mmol/kg, the concentration χ by mass becomes $\chi \approx 10^{-4}$ kg(Fe)/kg(H_2O). Then for a deposit of mass M derived from a hydrothermal system with flow rate Q, the time of formation τ is given simply by the relation

$$\tau = \frac{\xi M}{\chi Q} \tag{22}$$

Thus a chimney with a mass $M = 3 \times 10^3$ kg discharging at a rate $Q = 10$ kgs^{-1} would form in about 15 days. This is an underestimate, of course, because the calculation has assumed that all the discharge has gone into forming the chimney. Nevertheless, the calculation points to rapid chimney growth, which in fact has been observed [e.g., *Goldfarb et al.*, 1983; *Hekinian et al.*, 1983].

On the other hand a body such as the TAG mound, which contains approximately 4.5×10^9 kg and is discharging $\approx$ 250 kgs^{-1} would form in ~ 3000 years. This time is again an underestimate. Though the time to form such a large structure as TAG is short, geologically, it is still long compared to the suspected lifetime of a hydrothermal episode. The simple calculation thus suggests that the TAG mound is a product of several hydrothermal episodes, which is consistent with recent-geochronologic studies of mound samples [*Lalou et al.*, 1993, 1998] interpretations of the mound structure [*Humphris and Tivey*, 2000].

4.2. Mineral Precipitation and the Evolution of Permeability

As seawater enters the crust at recharge zones it begins to undergo a number of transformations. The fluid reacts with the crustal basalts at low-temperatures fix alkalis in celadonite and nontronite [*e.g. see Alt*, 1995], and as the temperature rises to approximately 150°C anhydrite ($CaSO_4$) begins to precipitate from seawater [*Bischoff and Seyfried*, 1978]. With continued heating from 150°C to the reaction zone temperatures of 350–400°C, Mg and SO_4 are quantitatively removed from seawater and, as a result of water-rock reactions, the fluid become enriched Ca, K, S, and metals such as Fe, Cu, Zn

[see reviews by *Alt*, 1995; *Von Damm* 1995]. During these high-temperature water-rock reactions, the hydrothermal fluid reaches equilibrium with quartz at high temperature and pressure [e.g., *Mottl*, 1983; *Von Damm*, 1995]. As the ascending hydrothermal fluid enters the upper crust mixing with low-temperature relatively unaltered seawater, together with conductive cooling, , may lead to precipitation of minerals such as quartz, anhydrite, and metal sulfides resulting in the clogging of permeable pathways [*Alt et al.*, 1986; *Nehlig and Juteau*, 1988; *Gillis and Robinson*, 1990; *Fouquet et al.*, 1998]. Finally, as the fluid vents at the seafloor rapid mineral precipitation leads to the formation of sulfide chimneys and massive sulfide ore deposits [e.g., *Haymon*, 1983; *Goldfarb et al.*, 1983; *Tivey*, 1995; *Hannington et al.*, 1995].

Basalt-seawater reactions have been studied in the laboratory [e.g., *Bischoff and Dickson*, 1975; *Hajash*, 1975; *Mottl and Seyfried*, 1980; *Seyfried and Ding*, 1993, 1995] and the evolution of hydrothermal fluids has been modeled using equilibrium reaction path models [e.g., *Bowers et al.*, 1985, 1988; *Janecky and Shanks*, 1988]. Some modeling has also been done to describe the formation of sulfide chimneys [*Turner and Campbell*, 1987; *Tivey and McDuff*, 1990; *Tivey*, 1995]. To date, however, no full reaction transport models of seafloor hydrothermal systems have been developed that simultaneously provide heat, mass, and chemical fluxes of hydrothermal vents. Such an effort is well beyond the scope of this paper, and may not be possible in full detail at present. Here we use the single pass model to describe the effects of mineral precipitation on the evolution of permeability in a high-temperature hydrothermal system. This is important because permeability changes can strongly affect heat output and temperature of hydrothermal vents. As examples, we consider simple models of anhydrite precipitation in the recharge zone and quartz precipitation in the discharge zone.

4.2.1. Anhydrite precipitation in the recharge zone. To determine the rate of precipitation of anhydrite from seawater as a result of heating in the recharge zone, we consider the rate of change in porosity resulting from the precipitation of anhydrite. Assuming that precipitation is controlled by the solubility of $CaSO_4$ in seawater as a function of temperature (i.e. the gradient reaction of *Phillips*, 1990), then

$$\frac{d\phi}{dt} = \frac{q}{\rho_m} \frac{\partial C_e}{\partial T} \frac{dT}{dz} \tag{23}$$

where ϕ is the porosity, ρ_m is the density of the precipitating mineral, $q = \rho_f u_z$ is the mass flow rate per unit area (Q/A) and C_e is the temperature dependent solubility of the mineral in solution under the assumption of thermodynamic equilibrium, respectively. We have neglected the dependence of C_e on pressure, because this term is usually much smaller than the temperature effect [*Kennedy*, 1950; *Fournier and Potter*, 1982; *Fournier et al.*, 1982].

To determine dT/dz in equation (23), we consider a simple steady state, one-dimensional temperature profile in the recharge zone. Using equation (3) we write

$$\rho_{f0} c_f u_z \frac{dT}{dz} = \lambda_m \frac{d^2T}{dz^2} \qquad (24)$$

The solution to equation (24) is subject to boundary conditions

$$\begin{aligned} T(0) &= 0 \\ T(h) &= T_0 \end{aligned} \qquad (25)$$

The solution to equation (24) with boundary conditions (25) is

$$T(z) = T_0 \frac{1 - \exp\left(c_f q z / \lambda_m\right)}{1 - \exp\left(c_f q h / \lambda_m\right)} \qquad (26)$$

Upon calculating the derivative dT/dz from equation (26) and substituting it into equation (23) we obtain

$$\frac{d\phi}{dt} = \frac{c_f q^2 T_0}{\lambda_m \rho_m} \frac{\partial C_e}{\partial T} \left\{ \frac{\exp\left(c_f q z / \lambda_m\right)}{1 - \exp\left(c_f q h / \lambda_m\right)} \right\} \qquad (27)$$

Because the temperature gradient term is constant in time, equation (27) can be simply integrated to get the rate of change of porosity. As the porosity decreases, the flow rate will

decrease also, so strictly speaking equation (27) only gives the initial rate of porosity decrease. Nevertheless, equation (27) yields two significant results. First, for large values of q, the term in brackets reaches its largest negative value (-1) at $z = h$, thus indicating that the larger the flow rate, the greater the depth of anhydrite precipitation [*Lowell and Yao*, 2002]. Secondly, the rate of porosity decrease depends upon the square of the mass flow rate per unit area, q. Integrating equation (27) for term in brackets equal to -1 and assuming q and $\partial C_e/\partial T$ are constant gives

$$\phi = \phi_0 - \left\{ \frac{c_f T_0 q^2}{\lambda_m \rho_m} \frac{\partial C_e}{\partial T} \right\} t \qquad (28)$$

where ϕ_0 is the initial porosity.

Upon substituting parameter values from Table 2, the time τ to completely close the porosity can be estimated. Figure 6 shows that the time to completely close the porosity as a result of anhydrite precipitation in the recharge zone is extremely rapid. Because we have neglected the response of the flow rate to porosity decrease, Figure 6 gives an underestimate of the sealing time; however, many vent systems appear to be relatively stable for at least several years [*Campbell et al.*, 1988; *Edmond et al.*, 1995; *Von Damm et al.*, 2002], and the rapid sealing indicated by the calculations here does not appear to occur. A reasonable solution to this paradox situation comes from recognizing that the flow rates used in Figure 6 correspond to flow rates determined from $q = Q/A_d$, where Q has been taken from equation (21). Since total mass flux Q must be constant throughout the system, small values of q in the recharge zone can be obtained by requiring that the area of recharge A_r is much larger than A_d. The more detailed cal-

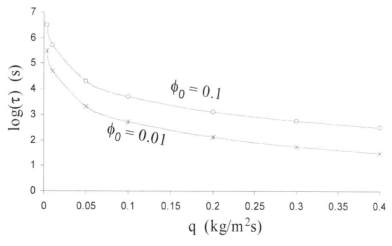

Figure 6. Plot of permeability sealing time τ resulting from mineral precipitation as a function of flow rate per unit area q, for different values of initial porosity ϕ_0.

culations in *Lowell and Yao* [2002], where the effect of the decrease in flow rate to a decrease in porosity in considered, suggests that area of the recharge zones are 100 times greater than that of the discharge zones.

4.2.2. Quartz Precipitation in the Discharge Zone. After undergoing high-temperature water-rock reactions, hydrothermal fluid is typically saturated in SiO_2 with respect to quartz. Thus if the fluid cools during ascent, quartz should precipitate from solution and reduce the porosity and permeability of the discharge zone. This process has been modeled in some detail by Martin and Lowell [2000], however, one can obtain some insight using the results obtained here. Equation (27) for the initial rate of porosity decrease is still formally valid, but in the discharge zone q is negative (Figure 5b) and the term in brackets reaches its maximum value of +1 at the seafloor ($z = 0$). Figure 6 thus also holds for quartz precipitation near the seafloor, suggesting that hydrothermal venting should be rapidly sealed as a result of quartz precipitation. Again, this is not observed, however the solution to the paradox is different from that for anhydrite. The mass flow rate in the discharge zone is reasonably well constrained, so the difficulty must lie with the other key assumption in the model—namely, thermodynamic equilibrium. Quartz precipitation is typically controlled by reaction kinetics. The results obtained here using simple single-pass modeling also shows importance of kinetics on quartz precipitation.

4.3. Event Plumes

Event plumes (Figure 3) are characterized by the rapid release of heat from the crust by a buoyant plume that ascends

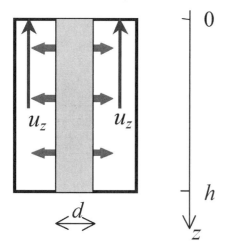

Figure 7. Schematic of vertical Darcian flow u_z near a dike of width d during an event plume. Arrows denote heat transfer from dike to hydrothermal fluid. [modified from *Germanovich et al.*, 2000].

into the water column. The heat content of event plumes is estimated to range from 10^{15} to 10^{17} J and the duration of even plume is estimated to range between 10^5 and 10^6 s [*Baker et al.*, 1987, 1989, 1995; *Baker*, 1998]. If we assume that the smaller event plumes are released over a shorter time period and larger event plumes occur on the longer time scale, the rate of heat output H_{ep} in event plumes ranges from $\sim 10^{10} - 10^{11}$ W (i.e., 1 to 2 orders of magnitude greater than black smokers). Without examining any specifics of the mode of formation, and assuming that event plume fluids vent at $\approx 400°C$, the simple single-pass analysis in section *3* gives mass flux $Q \sim 4 \times 10^3 - 4 \times 10^4$ kg/s and a permeability $k \sim 10^{-11} - 10^{-9}$ m^2.

We have suggested that the enhanced permeability needed for event plume formation is created along dike margins as a result of dike emplacement, with the dike itself also acting as a heat source [*Lowell and Germanovich*, 1995; *Germanovich et al.*, 2000]. The discharge area A_d of 10^3 to 10^4 m^2 then corresponds to a region extending ~ 1 m from the dike wall and between 10^3 and 10^4 m along strike.

To estimate the viability of the dike as a heat source we calculate the heat content of the dike E_D (Figure 7)

$$E_D = \left(\rho \hat{L} + \rho c_p \Delta T\right)V \qquad (29)$$

where $\hat{L}$ is the latent heat of fusion, ρ is the rock density, c_p is its specific heat, ΔT is the temperature difference between the magma and the hydrothermal fluid, and $V = dA_w$ is the dike volume, where d is the dike width and A_w is the vertical cross-sectional area of the dike wall. For parameter values listed in Table 2, and V ranging between 10^6 and 10^7 m^3, for a 1 m wide dike, 1 km in height and between 1 and 10 km in length, we obtain

$$E_D \sim 4 \times 10^{15} - 4 \times 10^{16} \text{ J} \qquad (30)$$

Equation (30) indicates that the heat content of a ~ 1 m-wide dike approximates the heat content estimated for an event plume. About 25% of this heat would be used to heat the rock in a ~ 1 m zone near the dike by $\sim 300°$, however. Moreover, the conductive cooling time $\tau_c \sim d^2/a$ for a 1 m-wide dike is ~ 10 days, so the heat supply rate is $\sim 4 \times 10^9 - 4 \times 10^{10}$ W, which is on the low end of the scale observed for event plumes. Thus, the simple analysis presented here suggests that if dike emplacement provides enhanced permeability, the dike could also supply most of the heat content observed in all but the largest event plumes. Some additional heat presumably comes from a pre-heated fluid reservoir of fluid within the crust.

A key issue for modeling event plumes stems from the observation of the low ^{3}He/heat ratios in event plumes relative to those of underlying chronic plumes of waters above

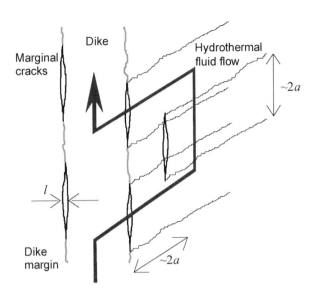

Figure 8. Diagram depicting quasi-vertical network of marginal cracks near a dike margin as envisioned for thermoelastic effects.

lavas [*Lupton et al.*, 1999]. We suggest that because the dike-generated event plumes require some pre-heated reservoir fluid, this would account in part for the lower ^{3}He/heat ratio. Moreover, because fluids initial in contact with the dike fall into the vapor-halite equilibrium regime [*Lewis and Lowell*, 2004], the formation of halite at the dike wall may initially inhibit degassing of helium from the dike.

4.4. Thermal Stresses and Bifurcations

The permeability of the oceanic crust not only responds to mineral dissolution and precipitation, and to magmatic and tectonic processes, but also to thermal stresses. *Bodvarsson and Lowell* [1972] first mentioned the possible importance of thermal stresses on hydrothermal systems. *Lister* [1974, 1983], *Bodvarsson* [1982], *Lowell* [1990], *Germanovich and Lowell* [1992], *Lowell et al.* [1993], and *Martin and Lowell* [1997, 2000], and *Germanovich et al.* [2000, 2001] have further expanded on this idea. *Lowell et al.* [1993] show that thermal stresses may modify permeability on an order of magnitude shorter time scale than mineral precipitation. An interesting feature of thermal stresses is the possibility that they lead to bifurcations in hydrothermal systems (i.e, a system can exist in more than one steady state and be driven from one steady state to another as a result of a perturbation). This feature, highlighted in *Germanovich et al.* [2001], can be elucidated rather simply within the single-pass model framework.

Fracture-based permeability (Figure 8) can be generally written in the form

$$k = \eta b^3 \qquad (31)$$

where η is a factor that may depend upon crack density, crack length, crack network geometry, etc., and b is the crack width [e.g., *Bear*, 1972; *Germanovich et al.*, 2001]. Under the action of thermal stresses, the change in crack width b is proportional to the temperature difference between the fluid in the cracks T and the far-field reference T_0 [e.g., *Lowell*, 1990]. Upon substituting this relationship into equation (31) and recognizing that permeability may exist on different scales [*Turcotte*, 1992], we write [*Germanovich et al.*, 2000, 2001]:

$$k(T) = k_0 \left[1 - \gamma \left(T - T_i \right) \right]^3 + k_{res} \qquad (32)$$

where γ is a factor expressing the thermoelastic effect on main permeability network (given by k_0), and k_{res} is a residual permeability that exists if the main permeability is closed by thermal stresses. Upon substituting equation (32) into equation (19) for k_d and that result into equation (20), we obtain

$$k(T) = H_t / F(T) \qquad (33)$$

where $F(T)$ is a temperature dependent function of the form

$$F(T) = CT^2 / v(T) \qquad (34)$$

In equation (34), C represents a combination of parameters that is independent of temperature, and $v(T)$ represents the temperature dependent kinematic viscosity. Assuming the kinematic viscosity is inversely proportional to the temperature to some power n, $F(T)$ can be written

$$F(T) = CT^{n+2} \qquad (35)$$

Now if pulse of magmatic activity causes a new episode of hydrothermal circulation and the temperature in the discharge zone rises above a critical value (which depends upon γ), then the main permeability in the discharge zone will become closed. Equation (33) will then become

$$k_{res} = H_t / F(T) \qquad (36)$$

On the other hand, if as a result of fresh heat input, the main permeability in equation (33) does not become closed, the system will evolve to a state given by

$$k_1 = H_t / F(T) \qquad (37)$$

where $k_{res} << k_1 \leq k_0$. Because the total heat output H_t is the same in both cases, equation (36) corresponds to a high-temperature solution, and equation (37) corresponds to a low temperature solution. Upon substituting the equation (33) into equations (36) and (37) we can obtain the ratio of the temperatures in the two solutions

$$\frac{T_h}{T_l} = \left(\frac{k_1}{k_{res}} \right)^{\frac{1}{n+2}} \qquad (38)$$

Testing of this formula awaits more complete time series data on hydrothermal systems; however, the most important result of this simple analysis is that the state of a ridge crest hydrothermal system may depend upon small differences in initial condition at the ridge crest. The lack of high-temperature discharge in some ridge crest environments (e.g., Galapagos Spreading Center and Reykjanes Ridge) may be a possible consequence of thermoelastic processes.

4.5. Other Examples

The previous sections of this paper provide a number of examples of problems that can be addressed by relatively simple scaling or single-pass model approaches, but these examples do not exhaust the possible applications. Additional examples that have been addressed include an analysis of the thickness and evolution of brine layers [Lowell and Germanovich, 1997] and how precipitation of anhydrite in the shallow parts of discharge zones upon mixing between hydrothermal seawater and cooler secondary circulations may control the relationship between focused and diffuse discharge [Lowell et al., 2003].

We speculate that other important issues may be addressed by these types of models as well. For example, there is currently some controversy as to whether event plumes are driven by dikes as proposed in section 4.3 or by lavas as proposed by Palmer and Ernst [1998] and Butterfield et al. [1997]. Relatively simple models of heat and chemical exchange between lavas and seawater would be useful for elucidating the viability of the lava flow model. Simple models addressing the response of hydrothermal systems to tectonic and magmatic events would also be useful, as would additional work on secondary circulation near high-temperature focused discharge zones. Near-axis hydrothermal circulation has yet to be addressed.

Finally, although full reactive-transport models are needed to incorporate water-rock reactions and vent chemistry, it may be possible to incorporate some additional chemical constraints into relatively simple models. For example, the high, relatively stable concentrations of incompatible trace elements such as boron and lithium in vent fluids suggest that high-temperature hydrothermal fluids are subject to low water-rock ratios and thus are continually accessing fresh unaltered rock [Seyfried and Ding, 1995]. To incorporate such a chemical constraint into simple scaling or single-pass models would provide considerable insight in to hydrothermal processes.

5. CONCLUSIONS AND SUGGESTIONS FOR FUTURE WORK

Important insights into vigorous high-temperature hydrothermal circulation at mid-ocean ridges can be obtained using scaling relationships and relatively simple single pass models. These simple models show that the conductive impermeable boundary layer separating liquid magma from the hydrothermal circulation system must be thin (~ a few tens of meters or less) and that the bulk permeability is high (~ 10^{12} m^2). In addition the models show that some mechanism to maintain the thin boundary is required to prevent rapid decay of hydrothermal output. Finally, we presented a number of simple extensions of the basic single pass model. These models suggest that: (1) large sulfide deposits such as TAG likely indicate multiple hydrothermal events; (2) anhydrite precipitation in the recharge zone will rapidly clog permeability and reduce hydrothermal output unless the areal extent of recharge is ~100 times that of the discharge zone; (3) quartz precipitation in the discharge zone will rapidly clog permeability unless the kinetics of quartz precipitation is taken into account; (4) event plumes can be driven by dike emplacement provided the dike generates enhanced permeability near its margin and supplies a significant fraction of the event plume heat; and (5) temperature dependent permeability related to thermoelastic stresses can lead to different hydrothermal steady states.

This paper highlights the utility of scale analysis and mathematical modeling using the single-pass approach. Many important problems remain to be addressed that will ultimately require the development of robust numerical models. These include such issues as (a) two-phase flow and supercritical phase separation, (b) coupled reactive transport, (c) biogeochemical processes in hydrothermal systems, (d) the response of hydrothermal systems to magma replenishment, dike emplacement, and tectonic events, (e) heat transfer near magma-hydrothermal interfaces, (f) processes related to mixing between deep-seated hydrothermal fluids and cooler, shallowly circulating seawater. Even with the development of numerical models, however, scale analysis and the single-pass model approach will continue to provide useful insights into seafloor hydrothermal processes.

Acknowledgments. We thank the reviewers Carol Stein, Bill Seyfried, and Mark Rudnicki for their thoughtful reviews of this manuscript. We also thank Karen Von Damm, Ed Baker, Bill Seyfried, and Chris German and others for many thoughtful and enlightening discussions regarding various hydrothermal data and their potential for constraining mathematical models. This work was supported by NSF Grant OCE-0221974.

REFERENCES

Abbott, D. and S. Hoffman, Archaean plate tectonics revisted–Part1: Heat flow, spreading rate, and the age of subducting lithosphere, and their effects on the origin and evolution of continents, *Tectonics, 3*, 429–448, 1984.

Alt, J. C., Subseafloor processes in mid-ocean ridge hydrothermal systems, in *Seafloor Hydrothermal Systems*, ed. by S.E. Humphris, R. A. Zierenberg, L. S. Mullineaux, and R. E. Thomson, *Geophys. Monogr. 91*, p. 85–114, Amer. Geophys. Union, Washington, DC, 1995.

Alt, J. C., J. Honnorez, C. Laverne, and R. Emmermann, Hydrothermal alteration of a 1 km section through the upper oceanic crust, Deep Sea Drilling Project Hole 504B: Mineralogy, chemistry, and evolution of seawater-basalt interactions, *J. Geophys. Res., 91*, 10,309–10,335, 1986.

Alt, J. C. and D. A. H. Teagle, Hydrothermal alteration and fluid fluxes in ophiolites and oceanic crust, in Dilek, Y., Moores, E., Elthon, D. and Nicolas, A. (eds): *Ophiolites and Oceanic Crust: New Insights from Field Studies and the Ocean Drilling Program*: Boulder, CO, Geological Society of America Special Paper 349, pp. 273–282, 2000.

Anderson, R. N., M. G. Langseth, and J. G. Sclater, The mechanisms of heat transfer through the floor of the Indian Ocean, *J. Geophys. Res., 82*, 3391–3409, 1977.

Bai, W., W. Xu, and R. P. Lowell, The dynamics of submarine geothermal heat pipes, *Geophys. Res. Lett., 30(3)*, 1108, doi: 10.1029/2002GL016176, 2003.

Baker, E. T., A six-year time series of hydrothermal plumes over the Cleft segment of the Juan de Fuca Ridge, *J. Geophys. Res., 99*, 4889–4904, 1994.

Baker, E. T., Patterns of event and chronic hydrothermal venting following a magmatic intrusion: new perspectives from the 1996 Gorda Ridge eruption, *Deep Sea Res. II, 45*, 2599–2618, 1998.

Baker, E. T. and G. A. Cannon, Long-term monitoring of hydrothermal heat flux using moored temperature sensors, Cleft segment, Juan de Fuca Ridge, *Geophys. Res. Lett., 20*, 1993.

Baker, E. T. and C. R. German, On the global distribution of hydrothermal vents, in *The thermal structure of oceanic crust and the dynamics of hydrothermal circulation*, ed. by C. R. German, J. Lin, and L. Parsons, 2004 (this volume).

Baker, E. T., J. W. Lavelle, R. A. Feely, G. J. Massoth, and S. L. Walker, Episodic venting of hydrothermal fluids from the Juan de Fuca Ridge, *J. Geophys. Res., 94*, 9237–9250, 1989.

Baker, E. T., R. P. Lowell, J. A. Resing, R. A. Feely, R. W. Embley, G. J. Massoth, and S. L. Walker, Decay of hydrothermal output following the 1998 seafloor eruption at Axial Volcano: Observations and models, *J. Geophys. Res., 109*, B01 205 doi: 10.1029/2003JB002618, 2004.

Baker, E. T. and G. J. Massoth, Hydrothermal plume measurements: A regional perspective, *Science, 234*, 980–982, 1986.

Baker, E. T. and G. J. Massoth, Characteristics of hydrothermal plumes form two vetn fields on the Juan de Fuca Ridge, northeast Pacific Ocean, *Earth Planet. Sci. Lett., 85*, 59–73, 1987.

Baker, E. T., G. J. Massoth, and R. A. Feely, Cataclysmic hydrothermal venting on the Juan de Fuca Ridge, *Nature, 329,* 149–151, 1987.

Baker, E. T., G. J. Massoth, R.A. Feely, G. A. Cannon, and R. E. Thomson, The rise and fall of the CoAxial hydrothermal site, 1993–1996, *J. Geophys. Res., 103*, 9791–9806, 1998.

Baker, E. T., G. J. Massoth, R.A. Feely, R. W. Embley, R. E. Thomson, and B. J. Burd, Hydrothermal event plumes from the CoAxial eruption site, Juan de Fuca Ridge, *Geophys. Res. Lett., 22*, 147–150, 1995.

Baker, E. T., G. J. Massoth, S. L. Walker, and R. W. Embley, A method for quantitatively estimating diffuse and discrete hydrothermal discharge, *Earth Planet. Sci. Lett., 118*, 235–249, 1993.

Baker, E. T., R. E. McDuff, and G. J. Massoth, Hydrothermal venting from the summit of a ridge axis seamount: Axial Volcano, Juan de Fuca Ridge, *J. Geophys. Res., 95*, 12,843–12,854, 1990.

Baross, J. A. and S. E. Hoffman, Submarine hydrothermal vents and associated gradient environments as sites for the origin and evolution of life, *Origins of Life, 15*, 327–345, 1985.

Bear, J. *Dynamics of Fluids in Porous Media*, 764p., Elsevier, New York, 1972.

Becker, K., and R. P. Von Herzen, Heat transfer through the sediments of the mounds hydrothermal area, Galapagos spreading center at 86W, *J. Geophys. Res., 88*, 995–1008, 1983.

Bejan, A., *Convection Heat Transfer*, 2nd ed., John Wiley & Sons, New York, 623p.,1995.

Bemis, K. G., R. P. Von Herzen, and M. J. Mottl, Geothermal heat flux from hydrothermal plumes on the Juan de Fuca Ridge, *J. Geophys. Res., 98*, 6351–6366, 1993.

Berndt, M. E., and W. E. Seyfried Jr., Boron, bromine, and other trace elements as clues to the fate of chlorine in mid-ocean ridge vent fluids, *Geochim. Cosmochim. Acta, 54*, 2235–2245, 1990.

Bickle, M. J., Heat loss from the Earth: a constraint on Archaean tectonics from the relation between geothermal gradients and the rate of plate production, *Earth Planet. Sci. Lett., 40*, 301–315, 1978.

Bischoff, J. L. and F. W. Dickson, seawater-basalt interaction at 200ºC and 500 bars: implications for origin of seafloor heavy metal deposits and regulation of seawater chemistry, *Earth Planet. Sci. Lett., 25*, 385–397, 1975.

Bischoff, J. L. and W. E. Seyfried, Jr., Hydrothermal chemistry of seawater from 25º to 350ºC, *American J. Sci., 278*, 838–860, 1978.

Blackman, D. K., J. R. Cann, B. Janssen, and D. K. Smith, Origin of extensional core complexes: evidence from the Mid-Atlantic Ridge at Atlantis fracture zone, *J. Geophys. Res., 103*, 21,315–21,334, 1998.

Bodvarsson, G., Terrestrial energy currents and transfer in Iceland, in *Continental and Oceanic Rifts Geodyn. Ser.*, vol. 8, edited by G. Palmason, pp. 271–282, AGU, Washington, D.C., 1982.

Bodvarsson, G., and R. P. Lowell, Ocean-floor heat flow and the circulation of interstitial waters, *J. Geophys. Res., 77*, 4472–4475,

1972.

Bougault, H., M. Aballea, J. Radford-Knoery, J. L. Charlou, P. Jean-Baptiste, P. Appriou, H. D. Needham, C. German, and M. Miranda, FAMOUS and AMAR segments on the Mid-Atlantic Ridge: ubiquitous hydrothermal Mn, CH_4, d^3He signals along the rift valley walls and rift offsets, *Earth Planet., Sci Lett*, *161*, 1–17, 1998.

Bowers, T. S., and H. P. Taylor, Jr., An integrated chemical and stable-isotope model of the origin of midocean ridge hot springs, *J. Geophys. Res.*, *90*, 12,583–12,606, 1985.

Bowers, T. S., K. L. Von Damm, and J. M. Edmond, Chemical evolution of mid-ocean ridge hot springs, *Geochim. Cosmochim. Acta*, *49*, 2239–2252,1985.

Bowers, T. S., A. C. Campbell, C. I. Measures, A. Spivack, M. Khadem, and J. M. Edmond, chemical controls on the composition of vent fluids at 13°–11°N and 21°N, East Pacific Rise, *J. Geophys. Res.*, *93*, 4522–4536, 1988.

Brace, W. F., Permeability of crystalline and argillaceous rocks, *Int. J. Rock Mech. Min. Sci. Geomech. Abstr.*, *17*, 241–251, 1980.

Brikowski, T. and D. Norton, Influence of magma chamber geometry on hydrothermal activity at mid-ocean ridges, *Earth Planet. Sci. Lett.*, *93*, 241–255, 1989.

Butterfield, D. S., G. J. Massoth, R. E. McDuff, J. E. Lupton, and M.D. Lilley, The geochemistry of hydrothermal fluids from ASHES vent field, Axial Seamount, Juan de Fuca Ridge: Subseafloor boiling and subsequent fluid-rock interactions, *J. Geophys. Res.*, *95*, 12,895–12,921, 1990.

Butterfield, D. A., R. E. McDuff, M. J. Mottl, M. D. Lilley, J. E. Lupton, and G. J. Massoth, Gradients in the composition of hydrothermal fluids from Endeavour Ridge vent field: Phase separation and brine loss, *J. Geophys. Res.*, *99*, 9561–9583, 1994.

Butterfield, D. A., I. R. Jonasson, G. J. Massoth, R. A. Feely, K. K. Roe, R. E. Embley, J. F. Holden, R. E. McDuff, M. D. Lilly, and J. R. Delaney, Seafloor eruptions and evolution of hydrothermal fluid chemistry, *Philos. Trans. R. Soc. London, Ser. A*, *355*, 369–386, 1997.

Campbell, A. C., T. S. Bowers, C. I. Measures, K. K. Falkner, M. Khadem, and J. M. Edmond, A time series of vent fluid compositions from 21oN, East Pacific Rise (1979, 1981, 1985), and the Guaymas Basin, Gulf of California (1982, 1985), *J. Geophys. Res.*, *93*, 4537–4549, 1988.

Cann, J. R., D. K.Blackman, D.K. Smith, E. McAllister, B. Janssen, S. Mello, E. Averinos, A. R. Pascoe, and J. Escartin, Corrugated slip surfaces formed at ridge-transform intersections on the Mid-Atlantic Ridge, *Nature*, *385*, 329–332, 1997.

Cann, J. R., and M. R. Strens, Modeling periodic megaplume emissions by black smoker systems, *J. Geophys. Res.*, *94*, 12,227–12,238, 1989.

Cann, J. R., M. R. Strens, and A. Rice, A simple magma-driven thermal balance model for the formation of volcanogenic massive sulphides, *Earth Planet. Sci. Lett.*, *76*, 123–134, 1985.

Carslaw, H. S. and J. C. Jaeger, *Conduction of Heat in Solids*, 2nd ed., Clarendon Press, Oxford, UK, 510 pp., 1959.

Cathles, L., A capless 350C flow zone model to explain megaplumes, salinity variations, and high-temperature veins in ridge crest hydrothermal systems, *Econ. Geol.*, *88*, 1977–1988, 1993.

Charlou, J. L., H. Bougault, P. Appriou, T. Nelsen, and P. Rona, Different TDM/CH_4 hydrothemal plume signatures: TAG site at 26°N and serpentinized ultrabasic diapir at 15°05' on the Mid-Atlantic Ridge, *Geochim. Cosmochim. Acta*, *55*, 3209–3223, 1991.

Charlou, J. L., and J. P. Donval, Hydrothermal methane venting between 12°N and 26°N along the Mid-Atlantic Ridge, *J. Geophys. Res.*, *98*, 9625–9642, 1993.

Cheng, P., Heat transfer in geothermal systems, in *Advances in Heat Transfer*, ed. by T.F. Irvine, Jr. and J.P. Hartnett, v. *14*, 1–105, Academic Press, Troy, MA, 1978.

Cherkaoui, A. S. M., W. S. D. Wilcock, and E. T. Baker, Thermal fluxes associated with the 1993 diking event on the CoAxial segment, Juan de Fuca Ridge: A model for convective cooling of a dike, *J. Geophys. Res.*, *102*, 24,,887–24,902, 1997.

Combarnous, M. A. and S. A. Bories, Hydrothermal convection in saturated porous media, *Adv. Hydrosci.*, *10*, 231–307, 1975.

Converse, D. R., H. D. Holland, and J. M. Edmond, Flow rates in the axial hot springs of the East Pacific Rise (21N): Implications for the heat budget and the formation of massive sulfide deposits, *Earth Planet. Sci. Lett.*, *69*, 159–175, 1984.

Corliss, J. B., et al., Submarine thermal springs on the Galapagos Rift, *Science, 203*, 1073–1083, 1979.

Corliss, J. B., J. A. Baross, and S. E. Hoffman, An hypothesis concerning the relationship between submarine hot springs and the origin of life on Earth, *Oceanol. Acta*, *4*, 59–69, 1981.

Davis, E. E., et al., FlankFlux: An experiment to study the nature of hydrothermal circulation in young oceanic crust, *Can. J. Earth Sci.*, *29*, 925–952, 1992.

Davis, E. E., D. S. Chapman, and C. B. Forster, Observations concerning the vigor of hydrothermal circulation in young oceanic crust, *J. Geophys. Res.*, *101*, 2927–2942, 1996.

Davis, E. E., K. Wang, J. He, D. S. Chapman, H. Villinger, and A. Rosenberger, An unequivocal case for high Nuselt number hydrothermal convection in sediment-buried igneous oceanic crust, *Earth Planet. Sci. Lett.*, *146*, 137–150, 1997.

Delaney, J. R., V. Robigou, R. E. McDuff, and M. K. Tivey, Geology of a vigorous hydrothermal system on the Endeavor segment, Juan de Fuca Ridge, *J. Geophys. Res.*, *97*, 19,663–19,682, 1992.

Delaney, J. R., D. S. Kelley, M. D. Lilley, D. A. Butterfield, J. A. Baross, W. S. D. Wilcock,, R. W. Embley, and M. Summit, The quantum event of oceanic crustal accretion: Impacts of diking at mid-ocean ridges, *Science*, *281*, 222–230, 1998.

DeRonde, C. E. J., M. J. de Wit, and E. T. C. Spooner, Early Archean (>3.2Ga) Fe-oxide-rich, hydrothermal discharge vents in the Barberton greenstone belt, South Africa, *Geol. Soc. Am. Bull.*, *106*, 86–104, 1994.

Detrick, R. S., P. Buhl, E. Vera, J. Mutter, J. Orcutt, J. Madsen, and T. Brocher, Multi-channel seismic imaging of a crustal magma chamber along the East Pacific Rise, *Nature*, *326*, 35–41, 1987.

Detrick, R. S., S. Carbotte, E. Van Ark, J. P. Canales, G. Kent, A. Harding, J. Diebold, and M. Nedimovic, New multichannel seismic constraints on the crustal structure of the Endeavour Segment, Juan de Fuca Ridge: Evidence for a crustal magma chamber, *EOS Trans. AGU*, *83(47)*, Fall Meet. Suppl. Abstract T12B–1316, 2002.

Dunn, R. A., D. R. Toomey, and S. C. Solomon, Three-dimensional seismic structure and physical properties of the crust an shallow mantle beneath the East Pacific Rise at 9°30′N, *J. Geophys. Res.,* *105*, 23537–23555, 2000.

Edmond, J. M., C. Measures, R. E. McDuff, L. H. Chan, R. Collier, B. Grant, L. I. Gordon, and J. B. Corliss, Ridge crest hydrothermal activity and the balances of the major and minor elements in the ocean: The Galapagos data, *Earth Planet. Sci. Lett., 46,* 1–18, 1979.

Edmond, J. M., A. C. Campbell, M. R. Palmer, C. R. German, G. P. Klinkhammer, H.N. Edmonds, H. Elderfield, G. Thompson, and P. A. Rona, Time series studies of vent fluids fro the TAG and MARK sites (1986, 1990) Mid-Atlantic Ridge and a mechanism for Cu/Zn zonation in massive sulphide orebodies, in *Hydrothermal Vents and Processes, Special Publication,* Geological Society of London, 1995.

Elder, J. W. *Geothermal Systems,* 508 pp., Academic, Troy, Mass., 1981.

Elderfield, H. and A. Schultz, Mid-ocean ridge hydrothermal fluxes and the chemical composition of the ocean, *Ann. Rev. Earth Planet. Sci., 24,* 191–224, 1996.

Embley, R. W., I. R. Jonasson, M. R. Perfit, J. M. Franklin, M. A. Tivey, A. Malahoff, M. F. Smith, and T. J. G., Francis, Submersible investigations of an extinct hydrothermal system on the Galapagos Ridge: sulfide mounds, stockwork zone and differentiated lavas, *Can. Mineral., 26,* 517–540, 1988.

Embley, R. W., W. W. Chadwick, Jr., I. R. Jonasson, D. A. Butterfield, and E. T. Baker, Initial results of the rapid response to the 1993 CoAxial event: relationships between hydrothermal and volcanic processes, *Geophys. Res. Lett., 22,* 143–146, 1995.

Fehn, U., and L. M. Cathles, Hydrothermal convection at slow-spreading mid-ocean ridges, *Tectonophysics, 55,* 239–260, 1979.

Fehn, U., K. E. Green, R. P. Von Herzen, and L. M. Cathles, Numerical models for the hydrothermal field at the Galapagos spreading center, *J. Geophys. Res., 88,* 1033–1048, 1983.

Fisher, A. T., K. Becker. T. N. Narasimhan, M. G. Langseth, and M. J. Mottl, Passive, off-axis convection through the southern flank of the Costa Rica rift, *J. Geophys. Res., 95,* 9343–9370, 1990.

Fisher, A. T., K. Becker, and T. N. Narasimhan, Off-axis hydrothermal circulation: Parametric tests of a refined model of processes at Deep Sea Drilling Project/Ocean Drilling Program site 504, *J. Geophys. Res., 99,* 3097–3021, 1994.

Fisher, A. T., Permeability within basaltic oceanic crust, *Rev Geophys., 36,* 143–182, 1998.

Fisher, A. T. and K. Becker, Correlation between seafloor and basement relief: Observational constraints and numerical examples and implications for upper crustal permeability, *J. Geophys. Res., 100,* 12641–12657, 1995.

Fontaine, F. Jh,, M. Rabinowicz, and J. Boulègue, Permeability changes due to mineral diagenesis in fractured crust: implications for hydrothermal circulation at mid-ocean ridges, *Earth Planet. Sci. Lett., 184,* 407–425, 2001.

Fornari, D. J., T. Shank, K. L. Von Damm, T. K. P. Gregg, R. M. Haymon, M. Lilley, G. Levai, A. Bray, M. R. Perfit, and R. Lutz, A dike intrusion or crustal fracturing event inferred from time-series temperature measurements at high-temperature hydrothermal vents on the East Pacific Rise 9°49–51′N, *Earth Planet. Sci. Lett., 160,* 419–431, 1998.

Fouquet, Y., J.-L. Charlou, I. Costa, J.-P. Donval, J. Radford-Knoery, H. Pelle, H. Ondreas, N. Lourenco, M. Segonzac, and M. K. Tivey, A detailed study of the Lucky Strike hydrothermal site and the discovery of a new hydrothermal site: Menez Gwen: preliminary results of the DIVA1 cruise (5–29 May, 1994) *Inter-Ridge News, 3,* 14–17, 1994.

Fouquet, Y., K. Henry, R. Knott, and P. Cambon, Geochemical section of the TAG hydrothermal mound in *Proc. Ocean Drill. Program Sci. Results, 158,* 363–387, 1998.

Fournier, R. O., and R. W. and Potter II, An equation correlating the solubility of quartz in water from 25°C to 900°C at pressures up to 10,000 bars, *Geochim. Cosmshim Acta, 46,* 1969–1974, 1982.

Fournier, R. O., R. J. Rosenbauer, and J. L. Bischoff, The solubility of quartz in aqueous sodium chloride solution at 350C and 180 to 500 bars, *Geochim. Cosmochim. Acta, 46,* 1975–1978, 1982.

Galer, S. J. G., Interrelationships between continental freeboard, tectonics and mantle temperature, *Earth Planet. Sci. Lett., 105,* 214–228, 1991.

Gamo, T., *et al.,* Chemical characteristics of newly discovered black-smoker fluids and associated hydrothermal plumes at the Rodriguez Triple Junction, Central Indian Ridge. *Earth Planet. Sci. Lett., 193,* 371–379, 2001.

Gendron, J. F., J. F. Todd, R. A. Feely, E. T. Baker, and D. Kadko, Excess ^{222}Rn over the Cleft segment, Juan de Fuca Ridge, *J. Geophys. Res., 99,* 5007–5015, 1994.

Gente, P., J. M. Auzende, V. Renard, Y. Fouquet, and D. Bideau, detailed geological mapping by submersible of the East Pacific Rise axial graben near 13N, *Earth Planet. Sci. Lett., 78,* 224–236, 1986.

German, C. R. and J. Lin, The thermal structure of the oceanic crust, ridge spreading, and hydrothermal circulation: How well do we understand their interconnection, in *The thermal structure of oceanic crust and the dynamics of hydrothermal circulation,* ed. by C. R. German, J. Lin, and L. Parsons, 2004 (this volume).

Germanovich, L. N. and R. P. Lowell, Percolation theory, thermoelasticity, and discrete hydrothermal venting in the earth's crust, *Science, 255,* 1564–1567, 1992.

Germanovich, L. N., R. P. Lowell, and D. K. Astakhov, Stress dependent permeability and the formation of seafloor event plumes, *J. Geophys. Res., 105,* 8341–8354, 2000.

Germanovich, L. N., R. P. Lowell and D. K. Astakhov, Bifurcations in seafloor hydrothermal flow, *J. Geophys. Res., 106,* 473–496, 2001.

Gillis, K. M. and M. D. Roberts, Cracking at the magma-hydrothermal transition: evidence from the Troodos Ophiolite, Cyprus, *Earth Planet. Sci. Lett., 169,* 1999.

Gillis, K. M., and P. T. Robinson, Patterns and processes of alteration in the lavas and dykes of the Troodos ophiolite, Cyprus, *J. Geophys. Res., 95,* 21,523–21,548, 1990.

Ginster, U., M. J. Mottl, and R. P. Von Herzen, Heat flux from black smokers on the Endeavor and Cleft segments, Juan de Fuca Ridge, *J. Geophys. Res., 99,* 4937–4950, 1994.

Goldfarb, M. S., D. R. Converse, H. D. Holland, and J. M. Edmond, The genesis of hot spring deposits on the East Pacific Rise, 21°N, in: *Econ Geol. Mongraph 5,* ed.by H. Ohmoto and B.Skinner, 184–197, 1983.

Gracia, E., J. L. Charlou, J. Radford-Knoery, and L. M. Parson, Non-transform offsets along the Mid-Atlantic Ridge south of the Azores (38°N–34°N): Ultramafic exposures and hosting of hydrothermal vents, *Earth Planet. Sci. Lett.*, *177*, 89–103, 2000.

Green, K. E., R. P. Von Herzen, and D. L. Williams, The Galapagos Spreading Center at 86W: A detailed geothermal field study, *J. Geophys. Res.*, *86*, 979–986, 1981.

Hajash. A. Hydrothermal processes along mid-ocean ridges: an experimental investigation, *Contrib. Mineral. and Petrol.*, *53*, 205–226, 1975.

Hannington, M. D., I. R. Jonasson, P. M. Herzig, and S. Petersen, Physical and chemical processes of seafloor mineralization at mid-ocean ridges, in *Seafloor Hydrothermal Systems*, ed. by S. E. Humphris, R. A. Zierenberg, L. S. Mullineaux, and R. E. Thomson, *Geophys. Monogr. 91*, p. 115–157, Amer. Geophys. Union,

Hashimoto, J., S. Ohta, T. Gamo, H. Chiba, T. Yamaguchi, S. Tsichida,, T. Okudaira, H. Wanabe, T.Yamanaka, and M. Kitazawa, Hydrothermal vents and associated biological communities in the Indian Ocean, *InterRidge News*, *10*, 21–22, 2001.

Haymon, R. M. Growth history of hydrothermal black smoker chimneys, *Nature*, *301*,695–698, 1983.

Haymon, R. M., et al., Volcanic eruption of the mid-ocean ridge along East Pacific Rise crest at 945'–52'N: Direct submersible observations of seafloor phenomena associated with an eruption event in April 1991, *Earth Planet. Sci. Lett.*, *119*, 85–101, 1993.

Hekinian, R., et al., Intense hydrothermal activity at the rise axis of the East Pacific Rise near 13N: submersible witnesses the growth of a sulfide chimney, *Mar. Geophys. Res.*, *6*, 1–14, 1983.

Holden, J. F., M. Summit, and J. A. Baross, Thermophilic and hyperthermophilic microorganisms in 3–30 deg C hydrothermal fluids following a deep-sea volcanic eruption. *FEMS Microbiology and Ecology*, *25*, 33–41, 1998.

Holm, N. G. ed., Marine hydrothermal systems and the origin of life, *Origins of Life and Evolution of the Biosphere*, *22*, no. 1–4, 5–242, 1992.

Humphris, S. E., and J. R. Cann, Constraints on the energy and chemical balances of the modern TAG and ancient Cyprus seafloor sulfide deposits, *J. Geophys. Res.*, *105*, 28,477–28–488, 2000.

Humphris, S. E. and Tivey, M. K., A synthesis of geological and geochemical investigations of the TAG hydrothermal fluid: Insights into fluid flow and mixing processes in a hydrothermal system, in: Dilek, Y., Moores, E., Elthon, D. and Nicolas, A. (eds): *Ophiolites and Oceanic Crust: New Insights from Field Studies and the Ocean Drilling Program*: Boulder, CO, Geological Society of America Special Paper 349, pp. 213–235, 2000.

Isley, A. E. Hydrothermal plumes and the delivery of iron to banded iron formation, *J. Geol.*, *103*, 169–185, 1995.

James, R. H. and H. Elderfield, Chemistry of ore-forming fluids and mineral formation rates in an active hydrothermal sulfide deposit on the Mid-Atlantic Ridge, *Geology*, *24*, 1147–1150, 1996.

Janecky, D. R., and W. E. Seyfried, Jr., Formation of massive sulfide deposits on oceanic ridge crests: Incremental reaction models for mixing between hydrothermal solutions and seawater, *Geochim. Cosmochim. Acta*, *48*, 2723–2738, 1984.

Janecky, D. R., and W. C. Shanks, III, Computational modeling of chemical and sulfur isotopic reaction processes in seafloor hydrothermal systems: Chimneys, massive sulfides, and subja-

cent alteration zones, *Canadian Mineral.*, *26*, 805–826, 1988.

Jannasch, H. W., Microbial processes at deep-sea hydrothermal vents, in *Hydrothermal Processes at Sea Floor Spreading Centers*, ed. by P. A. Rona, K. Bostrom, L. Laubier, and K. L. Smith, p. 677–709, Plenum Press, New York, 1983.

Jannasch, H. W., Microbial interactions with hydrothermal fluids, in *Seafloor Hydrothermal Systems*, ed. by S. E. Humphris, R. A. Zierenberg, L. S. Mullineaux, and R. E. Thomson, *Geophys. Monogr. 91*, p. 273–296, Amer. Geophys. Union, Washington, DC, 1995.

Jannasch, H. W. and C. O. Wirsen, Chemosynthetic primary production at East Pacific sea floor spreading centers, *Bioscience*, *29*, 592–598, 1979.

Johnson, H. P., M. Hutnak, R. P. Dziak, C. G., Fox, I. Urcuyo, J. P. Cowen, J. Nabelek, and C. Fisher, Earthquake-induced changes in a hydrothermal system on the Juan de Fuca mid-ocean ridge, *Nature*, *407*, 174–177, 2000.

Juniper, S. K., P. Martineau, J. Sarrazin, and Y. Gelinas, Microbial-mineral floc associated with nascent hydrothermal activity on CoAxial segment, Juan de Fuca Ridge, *Geophys. Res. Lett.*, *22*, 179–182, 1995.

Juniper, S. K. and J. Sarrazin, Interaction of vent biota and hydrothermal deposits: present evidence and future experimentation, in *Seafloor Hydrothermal Systems*, ed. by S. E. Humphris, R. A. Zierenberg, L. S. Mullineaux, and R. E. Thomson, *Geophys. Monogr. 91*, p. 178–193, Amer. Geophys. Union, Washington, DC, 1995.

Jupp, T. and A. Schultz, A thermodynamic explanation for black smoker temperatures, *Nature*, *403*, 880–883, 2000.

Kadko, D. R., J. A. Baross, and J. Alt, The magnitude and global implications of hydrothermal flux, in *Seafloor Hydrothermal Systems*, ed. by S. E. Humphris, R. A. Zierenberg, L. S. Mullineaux, and R. E. Thomson, *Geophys. Monogr. 91*, p. 273–296, Amer. Geophys. Union, Washington, DC, 1995.

Kadko, D., R. Koski, M. Tatsumoto, and R. Bouse, An estimate of hydrothermal fluid residence times and vent chimney growth rates based on ^{210}Pb/Pb ratios and mineralogic studies of sulfides dredged from the Juan de Fuca Ridge, *Earth Planet. Sci. Lett.*, *76*, 35–44, 1986.

Karson, J. A., Geologic structure of the uppermost oceanic crust created at fast- to intermediate-rate spreading centers, *Ann. Rev. Earth Planet. Sci.*, *30*, 347–384, 2002.

Karson, J. A., and P. A. Rona, Block-tilting, transfer faults, and structural control of magmatic and hydrothermal processes in the TAG area, Mid-Atlantic Ridge 26N, *Geol. Soc. Am. Bull.*, *102*, 1635–1645, 1990.

Kelley, D. S. et al., An off-axis hydrothermal vent field near the Mid-Atlantic Ridge at 30°N, *Nature*, *412*, 145–149, 2001.

Kelley, D. S., J. A. Baross, and J. R. Delaney, Volconoes, fluids and life at mid-ocean ridge spreading centers, *Annu. Rev. Earth Planet Sci.*, *30*, 385–491, 2002.

Kelley, D. S., J. R. Delaney, and D. A. Yoerger, Geology and venting characteristics of the Mothra Hydrothermal Field, Endeavor Segment, Juan de Fuca Ridge, *Geology*, *29*, 959–962, 2001.

Kennedy, G. C., A portion of the system silica-water, *Econ. Geol.*, *45*,629–653, 1950.

Kent, G. M., A. J. Harding, and J. A. Orcutt, Evidence for a small magma chamber beneath the East Pacific Rise at 930'N, *Nature*, *344*, 650–652, 1990.

Kinoshita, M., O. Matsubayashi, and R. P. Von Herzen, Sub-bottom temperature anomalies detected by long-term temperature monitoring at the TAG, *Geophys. Res. Lett.*, *23*, 3467–3470, 1996.

Lalou, C., J.-L. Reyss, E. Brichet, M. Arnold, G. Thompson, Y. Fouquet, and P. A. Rona, New age data from Mid-Atlantic Ridge hydrothermal sites: TAG and Snake Pit chronology revisited, *J. Geophys. Res., 98*, 9705–9713, 1993.

Lalou, C., J-L. Reyss, E. Brichet, P. A. Rona, and J. Thompson, Hydrothermal activity on a 10^5 year time scale at a slow-spreading ridge, TAG hydrothermal field, Mid-Atlantic Ridge 26N, *J. Geophys. Res.*, *100*, 17,855–17,862, 1995.

Lalou, C., J. L. Reyss, and E. Brichet, Age of sub-bottom sulfide samples at the TAG active mound, in *Proc. Ocean Drilling Prog., Sci. Results*, ed. by P. M. Herzig, S. E. Humphris et al., *158*, 111–117, College Station, TX, 1998.

Langmuir, C. et al., Hydrothermal vents neara a mantle hot spot: the Lucky Strike vent field at 37°N on the Mid-Atlantic Ridge, *Earth Planet Sci. Lett.*, *148*, 69–91, 1997.

Langseth, M. G., Jr., and R. P. Von Herzen, Heat flow through the floor of the world oceans, in *The Sea*, vol. 4, part 1, edited by A. E. Maxwell, pp. 299–352, Wiley-Interscience, New York, 1970.

Lapwood, E. R., Convection of a fluid in a porous medium, *Proc. Camb. Phil Soc.*, *44*, 508–521, 1948

Lewis, K. C., and R. P. Lowell, Mathematical modeling of phase separation of seawater near and igneous dike, *Geofluids, 4*, 197–209, 2004.

Lilley, M. D., D. A. Butterfield, J.E. Lupton, and E. J. Olson, Magmatic events produce rapid changes in hydrothermal vent chemistry, *Nature*, *422*, 878–881, 2003.

Lister, C. R. B., On the penetration of water into hot rock, *Geophys. J. R. Astron. Soc.*, *39*, 465–509, 1974.

Lister, C. R. B., The basic physics of water penetration into hot rocks, in *Hydrothermal Processes at Seafloor Spreading Centers*, edited by P. A. Rona, K. Bostrom, L. Laubier, and K. L. Smith, Jr., pp. 141–168, Plenum, New York, 1983.

Little, S. A., K. D. Stolzenbach, and R. P. Von Herzen, Measurements of plume flow from a hydrothermal vent field, *J. Geophys. Res.*, *92*, 2587–2596, 1987.

Lowell, R. P., Circulation in fractures, hot springs, and convective heat transport on mid-ocean ridge crests, *Geophys. J. R. Astron. Soc., 40*, 351–365, 1975.

Lowell, R. P., Topographically driven sub-critical hydrothermal convection in the oceanic crust. *Earth Planet. Sci. Letters, 49*, 21–28, 1980.

Lowell, R. P., Modeling continental and submarine hydrothermal systems, *Rev. Geophys., 29*, 457–476, 1991.

Lowell, R. P. and D. K. Burnell, A numerical model for magma-hydrothermal boundary layer heat transfer in the oceanic crust, *Earth and Planet. Sci. Lett., 104*, 59–69, 1991.

Lowell, R. P. and L. N. Germanovich, On the temporal evolution of high-temperature hydrothermal systems at ocean ridge crests, *J. Geophys. Res., 99*, 565–575, 1994

Lowell, R. P., and L. N. Germanovich, Dike injection and the formation of megaplumes at ocean ridges, *Science, 267*, 1804–1807, 1995.

Lowell, R. P. and L. N. Germanovich, Evolution of a brine-saturated layer at the base of a ridge crest hydrothermal systems, *J. Geophys. Res., 102*, 10245–10255, 1997.

Lowell. R. P. and S. M. Keller, High-temperature seafloor hydrothermal circulation over geologic time and Archean banded iron formations, *Geophys. Res. Lett., 30(7)*, 1391, doi:10.1029/2002GL016536, 2003.

Lowell, R. P., and P. A. Rona, Hydrothermal models for the generation of massive sulfide ore deposits, *J. Geophys. Res., 90*, 8769–8783, 1985.

Lowell, R. P., P. A. Rona and R. P. Von Herzen, Seafloor hydrothermal systems, *J. Geophys. Res., 100*, 327–352, 1995.

Lowell, R. P., P. Van Cappellen, and L. N. Germanovich, Silica precipitation in fractures and the evolution of permeability in hydrothermal upflow zones, *Science, 260*, 192–194, 1993.

Lowell, R. P. and W. Xu, Sub-critical two-phase seawater convection near a dike, *Earth Planet. Sci. Lett., 174*, 385–396, 2000.

Lowell, R. P. and Y. Yao, Anhydrite precipitation and the extent of hydrothermal recharge zones at ocean ridge crests, *J. Geophys. Res., 107(B9)*, 2183, 10.1029/2001JB001289 2002.

Lowell, R. P., Y. Yao, and L. N. Germanovich, On the relationship between focused and diffuse flow in seafloor hydrothermal systems, *J. Geophys. Res., 108(B9)*, 2424, doi: 10.1029/2002JB002371, 2003.

Lupton, J. E., E. T. Baker, and G. J. Massoth, Helium, heat, and the generation of hydrothermal event plumes at mid-ocean ridges, *Earth Planet. Sci. Lett., 171*, 343–350, 1999.

Macdonald, K. C., K. Becker, F. N. Spiess, and R. D. Ballard, Hydrothermal heat flux of the "black smoker" vents on the East Pacific, *Earth Planet. Sci. Lett., 48,* 1–7, 1980.

MacLeod, C. J. and G. Yaouancq, A fossil melt lens in the Oman ophiolite: Implications for magma chamber processes at fast spreading ridges, *Earth Planet. Sci. Lett., 176*, 357–373, 2000.

Manning, C. E., C. J. McLeod, and P. E. Weston, Lower-crustal cracking front at fast–spreading ridges: Evidence from the East Pacific Rise and the Oman Ophiolite, *Geol. Soc. Amer. Special Paper 349*, 261–273, 2000.

Martin, J. T. and R. P. Lowell, On thermoelasticity and silica precipitation in hydrothermal systems: Numerical modeling of laboratory experiments, *J. Geophys. Res., 102*, 12,095–12,107, 1997.

Martin, J. T. and R. P. Lowell, Precipitation of quartz during high-temperature fracture-controlled hydrothermal upflow at ocean ridges: equilibrium vs. linear kinetics, *J. Geophys. Res., 105*, 869–882, 2000.

Massoth, G. J., E. T. Baker, J. E. Lupton, R. A. Feely, D. A. Butterfield, K. Von Damm, K. K.. Roe, and G. T. Lebon, Temporal and spatial variability of hydrothermal manganese and iron at Cleft segment, Juan de Fuca Ridge, *J. Geophys. Res., 99*, 4905–4923, 1994.

Massoth, G. J., E. T. Baker, R. A. Feely, J. E. Lupton, R. W. Collier, R. W. Gendron, K. K. Roe, S. M. Maenner, and J. A. Resing, Manganese and iron in hydrothermal plumes resulting from the 1996 Gorda Ridge event, *Deep-Sea Res. II, 45*, 2683–2712, 1998.

McClain, J. S. R. A. Zierenberg, D. A. Clague, K. L. Von Damm, J. R. Voight, and E. J. Olson, Stability and localization of a hydrothermal field on a rift vally wall: GR-14 on the northern Gorda Ridge, 2002, EOS trans. AGU, 83(47), Fall Meet. Suppl., Abstract T11B-1243, 2002.

McCollum, T. M., Methanogensis as a potential source of chemical energy for primary biomass production by autotrophic organisms in hydrothermal systems on Europa, J. Geophys. Res., 104, 30,729–30,742, 1999.

McConachy, T. F., R. D. Ballard, M. J. Mottl, and R. P. Von Herzen, Geological form and setting of a hydrothermal vent field at latitude 1056'N, East Pacific Rise: A detailed study using Angus and Alvin, Geology, 14, 295–298, 1986.

Mottl, M. J., Metabasalts, axial hot springs, and the structure of hydrothermal systems at mid-ocean ridges, Geological Society of America Bulletin, 94, 161–180, 1983.

Mottl, M. J., Partitioning of energy and mass fluxes between mid-ocean ridge axes and flanks at high and low temperature, in Energy and Mass Transfer in Marine Hydrothermal Systems, ed. by P. E. Hallbach, V. Tunnicliff, and J. R. Hein, p. 271–286, Dahlem University Press, Berlin, 2003.

Mottl, M. J. and W. E. Seyfried, Jr., Sub-seafloor hydrothermal systems rock-vs. seawater- dominated, in Seafloor Spreading Centers: Hydrothermal Systems, ed. by P. A. Rona and R. P. Lowell, Benchmark Papers in Geology, v. 56, p. 66–82, Dowden, Hutchinson and Ross, Stroudsberg, PA, 1980.

Murton, B. J., L. J. Redourn, C. R. German, and E. T. Baker, Sources and fluxes of hydrothermal heat, chemicals and biology within a segment of the Mid-Atlantic Ridge, Earth Planet. Sci. Lett., 171, 301–317, 1999.

Murton, B. J., C. Van Dover, and E. Southward, Geological setting and ecology of the Broken Spur hydrothermal vent field: 29°10'N on the Mid-Atlantic Ridge, in Hydrothermal Vents and Processes, ed. by L. M. Walker and C. L. Dixon, Geol. Soc. Special Publ. 87, p. 33–42, 1995.

Nehlig, P., Fracture and permeability analysis in magma-hydrothermal transition zones in the Samail ophiolite (Oman), J. Geophys. Res., 99, 589–601, 1994.

Nehlig, P., and T. Juteau, Flow porosities, permeabilities and preliminary data on fluid inclusions and fossil thermal gradients in the crustal sequence of the Samail ophiolite (Oman), Tectonophysics, 151, 199–221, 1988.

Nicolas, A., D. Mainprice, and F. Boudier, High-temperature seawater circulation throughout crust of oceanic ridges: A model derived from the Oman ophiolite, J. Geophys. Res., 108(B8), 2371, doi:10.1029/2002JB002094, 2003.

Nield, D. A., Onset of thermohaline convection in a porous medium, Water Resour. Res., 4, 553–560, 1968.

Nisbet, E. G. and C. M. R. Fowler, The hydrothermal imprint on life: Did heat-shock proteins, metalloproteins and photosynthesis begin around hydrothermal vents? In Tectonic, Magmatic, Hydrothermal and Biological Segmentation of Mid-Ocean Ridges, ed. by C. J. MacLeod, P. A. Tyler, and C. L. Walker, Geol. Soc. Special Publ. 118, p 239–251, 1996.

Normark, W. R., J. L. Morton, and S. L. Ross, Submersible observations along the southern Juan de Fuca Ridge, J. Geophys. Res., 92, 11283–11290, 1987.

Norton, D., Theory of hydrothermal systems, Ann. Rev. Earth Planet. Sci., 12, 155–177, 1984.

Palmer, M. R., and G. G. J. Ernst, Generation of hydrothermal megaplumes by cooling of pillow basalts at mid-ocean ridges, Nature, 393, 643–647, 1998.

Pascoe, A. R., and J. R. Cann, Modeling diffuse hydrothermal flow in black smoker vent fields, in Hydrothermal Vents and Processes, ed. by L. M. Walker and C. L. Dixon, Geol. Soc. Special Publ. 87, p. 159–173, 1995.

Patterson, P. L., and R. P. Lowell, Numerical models oh hydrothermal circulation for the intrusion zone at an ocean ridge axis, in The Dynamic Environment of the Ocean Floor, edited by K. A. Fanning and F. T. Manheim, pp. 471–492, University of Miami, Coral Gables, Fla. 1982.

Phillips, O. M. Flow and Reaction in Permeable Rocks, Cambridge Univ. Press, New York, 285 p., 1991.

Rasmussen, B., Filamentous microfossils in a 3,235-million-year-old volcanogenic massive sulphide deposit, Nature, 405, 676–679, 2000.

Ribando, R. J., K. E. Torrance, and D. L. Turcotte, Numerical models for hydrothermal circulation in the oceanic crust, J. Geophys. Res., 81, 3007–3012, 1976.

Rabinowicz, M., J.-C. Sempere, and P. Genthon, Thermal convection in a vertical permeable slot: Implications for hydrothermal circulation along mid-ocean ridges, J. Geophys. Res., 104, 29275–29292, 1999.

Richards, H. G., J. R. Cann, and J. Jensenius, Mineralogical zonation and metasomatism of the alteration pipes of Cyprus sulfide deposits, Econ. Geol., 84, 91–115, 1989.

Richardson, C. J., J. R. Cann, H. G. Richards and J. G. Cowan, Metal-depleted root zones of the Troodos ore-forming hydrothermal systems, Cyprus, Earth Planet. Sci. Lett., 84, 243–253, 1987.

Rohr, K. M. M., B. Milkereit, and C. J. Yonath, Asymmetric deep crustal structure across the Juan de Fuca Ridge, Geology, 16, 533–537, 1988.

Rona, P. A., H. Bougault, J. L. Charlou, P. Appriou, T. A. Nelson, J. H. Trefry, G. L. Eberhart, A. Barone, and H. D. Needham, Hydrothermal circulation, serpentinization, and degassing at a rift valley-fracture zone intersection: Mid-Atlantic Ridge near 15° N, 45° W, Geology, 20, 783–786, 1992.

Rona, P.A., M. D. Hannington, C. V. Raman, G. Thompson, M. K. Tivey, S. E. Humphris, L. Lalou, and S. Petersen, Major active and relict seafloor hydrothermal mineralization: TAG hydrothermal field, Mid-Atlantic Ridge 26N, 45W, Econ. Geol., 88, 1989–2017, 1993.

Rona, P. A., G. Klinkhammer, T. A. Nelsen, J. H. Trefry, and H. Elderfield, Black smokers, massive sulfides and vent biota at the Mid-Atlantic Ridge, Nature, 321, 33–37 1986.

Rona, P. A., and D. A. Trivett, Discrete and diffuse heat transfer at ASHES vent field, Axial Volcano, Juan de Fuca Ridge, Earth Planet. Sci. Lett., 109, 57–71, 1992.

Rona, P. A., L. Widenfalk, and K. Bostrom, Serpentinized ultra-mafics and hydrothermal activity at the Mid-Atlantic Ridge crest near 15°N, *J. Geophys. Res.*, *92*, 1417–1427, 1987.

Rosenberg, N. D., J. E. Lupton, D. Kadko, R. Collier, M. D. Lilley, and H. Pak, Estimation of heat and chemical fluxes from a seafloor hydrothermal vent field using radon measurements, *Nature, 344*, 604–607, 1988.

Rosenberg, N. D. F. J. Spera, and R. M. Haymon, The relationship between flow and permeability field in seafloor hydrothermal systems, *Earth Planet. Sci. Lett., 116*, 135–153, 1993.

Rudnicki, M. D. and H. Elderfield, Theory applied to the Mid-Atlantic Ridge hydrothermal plumes: the finite difference approach, *J. Volcanol. Geotherm. Res.*, *50*, 161–172, 1992.

Rudnicki, M. D. and C. R. German, Temporal variability of the hydrothermal plume above the Kairei vent field, 25°S, Central Indian Ridge, *Geochem. Geophys. Geosyst.*, *3* (2), 10.1029/2001GC000240, 2002.

Schoofs, S. and U. Hanson, Depletion of a brine layer at the base of ridge-crest hydrothermal systems, *Earth Planet. Sci. Lett.*, *180*, 341–353, 2000.

Schultz, A., J. M. Delaney, and R. E. McDuff, On the partitioning of heat flux between diffuse and point source venting, *J. Geophys. Res., 97*, 12,229–12,314, 1992.

Schultz, A., P. Dicksen, and H. Elderfield, Temporal variations in the diffuse hydrothermal flow at TAG, *Geophys. Res. Lett.*, *23*, 3471–3474, 1996.

Sclater, J. G., C. Jaupart, and D. Galson, The heat flow through oceanic and continental crust and the heat loss of the Earth, *Rev. Geophys.*, *18*, 269–311, 1980.

Seyfried, Jr., W. E., M. E. Berndt, and D. R. Janecky, Chloride depletions and enrichments in hydrothermal fluids: constraints from experimental basalt alteration studies, *Geochim. Cosmochim. Acta*, *50*, 469–475, 1986.

Seyfried, Jr., W. E. and K. Ding, The effect of redox on the relative solubilities of copper and iron in Cl-bearing aqueous fluids at elevated temperature and pressures: An experimental study with application to subseafloor hydrothermal systems, *Geochim. Cosmochim Acta*, 57, 1905–1918, 1993.

Seyfried, Jr., W. E. and K. Ding, Phase equilibria in subseafloor hydrothermal systems: A review of the role of redox, temperature, pH and dissolved Cl on the chemistry of hot spring fluids at mid-ocean ridges, in *Seafloor Hydrothermal Systems*, ed. by S. E. Humphris, R. A. Zierenberg, L. S. Mullineaux, and R. E. Thomson, *Geophys. Monogr. 91*, p. 248–272, Amer. Geophys. Union, Washington, DC, 1995.

Seyfried, Jr., W. E., and M. J. Mottl, Hydrothermal alteration of basalt by seawater under seawater dominated conditions, *Geochim. Cosmochim. Acta, 46*, 985–1002, 1982.

Seyfried, Jr. W. E., J. S. Seewald, M. E. Berndt, K. Ding, and D. I. Foustoukos, Chemistry of hydrothermal vent fluids from the Main Endeavor Vent Field, northern Juan de Fuca Ridge: Geochemical controls in the aftermath of June 1999 seismic events, *J. Geophys. Res.*, *108(B9)* 2429, doi: 10.1029/2002JB001957, 2003.

Shank, T. M., D. J. Fornari, K. L. Von Damm, M. D. Lilley, R. M. Haymon, and R. A. Lutz, Temporal and spatial patterns of biological community development at nascent deep-sea hydrothermal vents along the East Pacific Rise, *Deep Sea Res. II*, *45*, 465–515, 1998.

Shanks, W. C. III, J. L. Bischoff, and R. J. Rosenbauer, Seawater sulfate reduction and sulfur isotope fractionation in basaltic systems: Interaction of seawater with fayalite and magnetite at 200–350°C, *Geochim. Cosmohim. Acta.*, *45*, 1977–1995, 1981.

Shipboard Scientific Party, Introduction and summary: Hydrothermal circulation in the oceanic crust and its consequences on the eastern flank of the Juan de Fuca Ridge, *Proc. Ocean Drill. Program, Initial Rep.*, *168*, 1–15, 1997.

Sinton, J. M. and R. S. Detrick, Mid-ocean ridge magma chambers, *J. Geophys. Res.*, *97*, 197–216, 1992.

Sleep, N. H., Hydrothermal circulation, anhydrite precipitation and thermal structure at ridge axes, *J. Geophys. Res. 96*, 2375–2387, 1991.

Sleep, N. H. and B. F. Windley, Archean plate tectonics: constraints and inferences, *J. Geol.*, *90*, 363–379, 1982.

Sleep, N. H. and K. Zahnle, Carbon dioxide cycling and implications for climate on ancient earth, *J. Geophys. Res.*, *106*, 1373–1399, 2001.

Sohn, R. A., D. J. Fornari, K. L. Von Damm, J. A. Hildebrand, and S. C. Webb, Seismic and hydrothermal evidence for a cracking event on the East Pacific Rise at 9° 50'N, *Nature*, *396*, 159–161, 1998.

Spiess, F. N., et al., East Pacific Rise: Hot springs and geophysical experiments, *Science, 207*, 1421–1433, 1980.

Stein, C. A., and S. Stein, Constraints on hydrothermal heat flux through the oceanic lithosphere from global heat flow, *J. Geophys. Res., 99*, 3081–3095, 1994.

Stein, C. A., S. Stein, and A. Pelayo, Heat flow and hydrothermal circulation, in *Seafloor Hydrothermal Systems*, ed. by S. E. Humphris, R. A. Zierenberg, L. S. Mullineaux, and R. E. Thomson, *Geophys. Monogr. 91*, p. 425–445, Amer. Geophys. Union, Washington, DC, 1995.

Strens, M. R. and J. R. Cann, A model of hydrothermal circulation in fault zones at mid-ocean ridge crests, *Geophys. J. Roy. Astr. Soc.*, *71*, 225–240, 1982.

Summit, M. and J. A. Baross, Thermophilic sub-seafloor microorganisms from the 1996 Nort Gorda Ridge eruption, *Deep Sea Res. II*, *45*, 2751–2766, 1998.

Talwani, M., C. C. Windisch, and M. G. Langseth, Reykjanes Ridge crest: A detailed geophysical study, *J. Geophys. Res.*, *76*, 473–517, 1971.

Taylor, C. D. and C. O. Wirsen, Microbiology and ecology of filamentous sulfur formation, *Science*, *277*, 1483–1485, 1977.

Thompson, G., Basalt-seawater interaction, in Hydrothermal Processes at Seafloor Spreading Centers, edited by P. A. Rona, K. Boström, L. Laubier, and K. L. Smith, *NATO Conf. Ser. IV, Mar. Sci.*, *12*, 225–278, 1983.

Thompson, R. E., J. R. Delaney, R. E. McDuff, D. R. Janecky, and J. S. McClain, Physical characteristics of the Endeavor Ridge hydrothermal plume during July 1988, *Earth Planet. Sci. Lett.*, *111*, 141–154, 1992.

Thurnherr, A. M., and K. J. Richards, Hydrography and high-temperature heat flux of the Rainbow hydrothermal site (36°14'N, Mid-Atlantic Ridge), *J. Geophys. Res.*, *106*, 9411–9426, 2001.

Thurnherr, A. M., K. J. Richards, C. R. German, G. F. Lane-Serff, and K. G. Speer, Flow and mixing in the rift valley of the Mid-

Atlantic Ridge, *J. Phys. Oceanogr.*, *32*, 1763–1778, 2002.

Tivey, M. K., Modeling chimney growth and associated fluid flow at seafloor hydrothermal vent sites, in *Seafloor Hydrothermal Systems*, ed. by S. E. Humphris, R. A. Zierenberg, L. S. Mullineaux, and R. E. Thomson, *Geophys. Monogr. 91*, p. 158–177, Amer. Geophys. Union, Washington, DC, 1995.

Tivey, M. K. and R. E. McDuff, Mineral precipitation in the walls of a black smoker chimney: a quantitative model of transport and chemical reactions, *J. Geophys. Res.*, *95*, 12617–12637, 1990.

Travis, B. J., D. R. Janecky, and N. D. Rosenberg, Three-dimensional simulation of hydrothermal systems at mid-ocean ridges, *Geophys. Res. Lett., 18*, 1441–1444, 1991.

Turcotte, D. L., *Fractals and Chaos in Geology and Geophysics*, 221p., Cambridge University Press, New York, 1992.

Turner, J. S. and I. H. Campbell, A laboratory and theoretical study of the growth of "black smoker chimneys, *Earth. Planet. Sci. Lett., 82*, 36–48, 1987.

van Everdingen, D. A., Fracture characteristics of the Sheeted Dike Complex, Troodos ophiolite, Cyprus: Implications for permeability of oceanic crust, *J. Geophys. Res.100, 19*,957–19,972, 1995.

Von Damm, K. L., Controls on the chemistry and temporal variability of seafloor hydrothermal fluids, in *Seafloor Hydrothermal Systems*, ed. by S. E. Humphris, R. A. Zierenberg, L. S. Mullineaux, and R. E. Thomson, *Geophys. Monogr. 91*, p. 222–247, Amer. Geophys. Union, Washington, DC, 1995.

Von Damm, K. L., Chemistry of hydrothermal vent fluids from 9°–10°N, East Pacific Rise: "Time Zero," the immediate posteruptive period, *J. Geophys. Res., 105*, 11,203–11,222, 2000.

Von Damm, K. L., Evolution of the hydrothermal system at East Pacific Rise 9°50′N: Geochemical evidence for changes in the oceanic crust, in *The thermal structure of oceanic crust and the dynamics of hydrothermal circulation*, ed. by C. R. German, J. Lin, and L. Parsons, 2004 (this volume).

Von Damm, K. L., L. G. Buttermore, S. E. Oosting, A. M. Bray, D. J. Fornari, M. D. Lilley, and W. C. Shanks. III, Direct observation of the evolution of a seafloor 'black smoker' from vapor to brine, *Earth Planet. Sci. Lett., 149*, 101–111, 1997.

Von Damm, K. L. and Lilley, M. D., Diffuse flow hydrothermal fluids from 9°50′N East Pacific Rise: Origin, evolution and biogeochemical controls, in *The Subsurface Biosphere at Mid-Ocean Ridges*, edited by W. S. D. Wilcock, E. F DeLong, D. S. Kelley, J. A. Baross, and S. C. Cary, *AGU Monog.*, 2004 (in press).

Von Damm, K. L., S. E. Oosting, R. Kozlowski, L. G. Buttermore, D. C. Colodner, H. N. Edmonds, J. M. Edmond, and J. M. Greb-meier, Evolution of East Pacific Rise hydrothermal vent fluids following a volcanic eruption, *Nature, 375*, 47–50, 1995.

Von Damm, K. L., C. M. Parker, R. M. Gallant, and J. P. Loveless, Chemical evolution of hydrothermal fluids from EPR 21°N: 23 years later in a phse separating world, *EOS Trans. AGU, 83*(47), Fall Meet. Suppl., Abstract V61B–1365, 2002.

Wang, K., J. He, and E. E. Davis, Influence of basement topography on hydrothermal circulation in sediment-buried igneous oceanic crust, *Earth Planet Sci. Lett., 146*, 151–164, 1997.

White, D. J., and R. M. Clowes, Shallow crustal structure beneath the Juan de Fuca Ridge from 2-D seismic refraction tomography, *Geophys. J. Int., 100*, 349–367, 1990.

Wilcock, W. S. D., A model for the formation of transient event plumes above mid-ocean ridge hydrothermal systems, *J. Geophys. Res., 102*, 12,109–12,121, 1997.

Wilcock, W. S. D., Cellular convection models of mid-ocean ridge hydrothermal circulation and the temperatures of black smoker fields, *J. Geophys. Res., 103*, 2585–2596, 1998.

Wilcock, W. S. D. and J. R. Delaney, Mid-ocean ridge sulfide deposits: Evidence for heat extraction from magma chambers or cracking fronts?, *Earth Planet. Sci. Lett., 145*, 49–64, 1996.

Wilcock, W. S. D. and A. T. Fisher, Geophysical constraints on the sub-seafloor environment near mid-ocean ridges, in *The Subsurface Biosphere at Mid-Ocean Ridges AGU Monog.*, 2003 (in press)

Wilcock, W. S. D., and A. McNabb, Estimates of crustal permeability on the Endeavor segment of the Juan de Fuca mid-ocean ridge, *Earth Planet. Sci. Lett., 138*, 83–91, 1995.

Williams, D. L., R. P. Von Herzen, J. G. Sclater, and R. N. Anderson, The Galapagos Spreading Center: Lithospheric cooling and hydrothermal circulation, *Geophys. J. Roy. Astr. Soc., 38*, 587–608, 1974.

Williams, D. L., and R. P. Von Herzen, Heat loss from the Earth: New estimate, *Geology, 2*, 327, 1974.

Wolery, T. J., and N. H. Sleep, Hydrothermal circulation and geochemical flux at mid-ocean ridges, *J. Geology, 84*, 249–275, 1976.

Yang, J., Influence of normal faults and basement topography on ridge-flank hydrothermal fluid circulation, *Geophys. J. Int., 151*, 83–87, 2002.

Leonid N. Germanovich, School of Civil and Environmental Engineering, Georgia Institute of Technology, Atlanta, Georgia 30332–0355

Robert P. Lowell, School of Earth and Atmospheric Sciences, Georgia Institute of Technology, Atlanta, Georgia 30332–0340

On the Global Distribution of Hydrothermal Vent Fields

Edward T. Baker

NOAA/Pacific Marine Environment Laboratory, Seattle, Washington

Christopher R. German

Southampton Oceanography Centre, Southampton, UK

The "magmatic budget hypothesis" proposes that variability in magma supply is the primary control on the large-scale hydrothermal distribution pattern along oceanic spreading ridges. The concept is simple but several factors make testing the hypothesis complex: scant hydrothermal flux measurements, temporal lags between magmatic and hydrothermal processes, the role of permeability, nonmagmatic heat sources, and the uncertainties of vent-field exploration. Here we examine this hypothesis by summarizing our current state of knowledge of the global distribution of active vent fields, which presently number ~280, roughly a quarter of our predicted population of ~1000. Approximately 20% of the global ridge system has now been surveyed at least cursorily for active sites, but only half that length has been studied in sufficient detail for statistical treatment. Using 11 ridge sections totaling 6140 km we find a robust linear correlation between either site frequency or hydrothermal plume incidence and the magmatic budget estimated from crustal thickness. These trends cover spreading rates of 10–150 mm/yr and strongly support the magma budget hypothesis. A secondary control, permeability, may become increasingly important as spreading rates decrease and deep faults mine supplemental heat from direct cooling of the upper mantle, cooling gabbroic intrusions, and serpentinization of underlying ultramafics. Preliminary observations and theory suggest that hydrothermal activity on hotspot-affected ridges is relatively deficient, although paucity of data precludes generalizing this result. While the fullness of our conclusions depends upon further detailed study of vent field frequency, especially on slow-spreading ridges, they are consistent with global distributions of deep-ocean ^{3}He, an unequivocally magmatic tracer.

1. INTRODUCTION

After 25 years of seafloor exploration, hydrothermal venting is known to discharge along divergent plate boundaries in every ocean, at all spreading rates, and in a diversity of geological settings. Ever since the discovery of submarine hydrothermal venting, and indeed even before that, prescient researchers have speculated on the global distribution of vent sites and the geologic conditions that control it. Eight years prior to the historic Galápagos discoveries [*Corliss et al.,* 1977], *Boström et al.* [1969] noted that iron- and manganese-rich sediments were preferentially found near "active" oceanic ridges. Moreover, both the extent and Fe+Mn content of these

Mid-Ocean Ridges: Hydrothermal Interactions Between the Lithosphere and Oceans
Geophysical Monograph Series 148
Copyright 2004 by the American Geophysical Union
10.1029/148GM10

sediments increased with increasing spreading rate, suggesting that volcanism and "mantle outgassing" followed a similar pattern. This observation was the germ of a simple but challenging hypothesis: Hydrothermal activity increases linearly with the magmatic budget. (See *Haymon* [1996] for a succinct history of the development of this hypothesis.) The "magmatic budget" hypothesis was perhaps first articulated by *Francheteau and Ballard* [1983], who envisioned a simple geometry of increased magma supply and hydrothermal activity at segment centers bounded by transform faults.

Testing of this idea was limited for almost a decade by the lack of detailed segment-scale surveys of hydrothermal activity. *Crane et al.* [1985] attempted the first multisegment plume survey, identifying several broad regions of hydrothermally warmed water along the Juan de Fuca Ridge (JDFR). Quantitative support of the link between magma supply and venting was first supplied by *Haymon et al.* [1991], who visually surveyed most of a 2nd-order segment on the East Pacific Rise (EPR), and by *Baker and Hammond* [1992], who used continuous plume mapping to ascertain the distribution of vent sites over an entire 1st-order segment, the JDFR. Accumulation of plume surveys, mostly on fast-spreading ridges, led to the proposal that, on a relative scale, hydrothermal activity increased linearly with spreading rate [*Baker et al.*, 1995, 1996]. *German and Parson* [1998] explored the extension of this relation to the slow-spreading Mid-Atlantic Ridge (MAR). They concluded that while the MAR results generally supported the model, the interplay of magmatic and tectonic processes could create dramatic departures, both positive and negative, from the over-arching linear trend.

In this paper, which includes a wealth of new hydrothermal data gathered since our previous review a decade ago [*Baker et al.*, 1995], we present a comprehensive summary of the distribution of confirmed (from seafloor observations) or inferred (from water column measurements) active hydrothermal sites. We compare distributions on "fast" (rift valley absent) and "slow" (rift valley present) ridges, including a consideration of ultraslow and hotspot-affected ridges, to test the "magmatic budget hypothesis" described above.

While testing of this hypothesis seems straightforward, in fact several factors complicate the task. The most obvious difficulty is that quantitative knowledge of hydrothermal heat or chemical fluxes is rare and imprecise. Thus, we must use qualitative indices for hydrothermal activity, such as plume incidence p_h (the fraction of ridge crest length overlain by a significant hydrothermal plume [*Baker and Hammond*, 1992]), or vent site frequency F_s (sites/100 km of ridge length). Large-scale distributions of plumes are most easily mapped using optical sensors, which are simple, inexpensive, and sensitive. The distributions described here are from light backscattering measurements (or light attenuation measurements transformed

to light backscattering), given in terms of nephelometric turbidity units (NTUs) [*APHA*, 1985], determined from a laboratory calibration using formazine [*Baker et al.*, 2001a]. ΔNTU is the plume optical anomaly in excess of the NTU value of local ambient deep waters.

Another hindrance is the differing temporal scales of magma cycling and hydrothermal venting. Ideally, monitoring of hydrothermal activity at sites spanning the full range of the magmatic budget (or its proxy, spreading rate) at oceanic ridges would test this hypothesis effectively. Unfortunately, the temporal scale required to obtain statistically valid data, perhaps roughly the time required to accrete the full breadth of the neovolcanic zone at each site, is impossibly long. A reasonable alternative is to substitute spatial sampling for temporal sampling. In principle, a survey of multiple tectonic segments of the same spreading rate should provide a statistically significant measure of the hydrothermal activity characteristic of that spreading rate or magmatic budget. The actual ridge length necessary for a reliable survey remains to be established but almost certainly increases as spreading rate decreases. In practice, we consider ~200 km to be a minimum survey size for fast ridges, with longer surveys required for slow ridges.

A third complication is that hydrothermal circulation requires the confluence of a sufficient heat source and adequate permeability. While the availability of magmatic heat along the global ridge system is roughly proportional to the spreading rate, changes in the crustal permeability of the neovolcanic zone at either the local or regional scale are presently unconstrained even in relative terms [e.g., *Fisher*, 1998]. If bulk permeability is a strong function of spreading rate, for example, its effect on the development of hydrothermal circulation could blur the relationship between magma supply and hydrothermal activity [e.g., *German and Parson*, 1998].

The generation of hydrothermal circulation by heat sources other than magma cooling introduces more uncertainty. On slow- and ultraslow-spreading ridges where the basaltic magma flux is weakest, heat from gabbroic intrusions, from cooling of the lithospheric mantle [*Cannat et al.*, in press; *Bach et al.*, 2002], and from exothermic serpentinization of ultramafic rocks [*Kelley et al.*, 2001; *Schroeder et al.*, 2002; *Lowell and Rona*, 2002] may increase the vent field population. Serpentinization, for example, may "contaminate" entire segments of slow-spreading ridge with plumes of dissolved products such as CH_4 [*Charlou and Donval*, 1993], complicating the identification of magma-driven hydrothermal venting. Vent fields driven solely by serpentinization reactions will likely not be readily detected using our well established, optically based survey methods, however. Such fluid discharge is rich in dissolved gases but metal-poor, so that the precipitation of Fe-Mn oxyhydroxides following release into the water column is minimal [*Kelley et al.*, 2001].

Finally, the hydrothermal data set to be considered must be representative of activity on ridges of all spreading rates. While it may be premature to characterize the available data as truly globally representative, we now have at least preliminary hydrothermal surveys from ridge sections across the entire spectrum of spreading rates.

2. THE GLOBAL VENT FIELD DISTRIBUTION

2.1. Vent Field Locations

Large-scale, systematic searches for undiscovered vent sites have been organized with increasing frequency since the early 1990s [*Baker et al.,* 1995; *German et al.,* 1995] and now span the global range of spreading rates. The practical difficulty of comprehensively imaging vent fields means that the most efficient searches rely on inferring their presence from water column observations, a technique that carries varying levels of uncertainty. The enrichment of hydrothermal fluids in several key chemical tracers (e.g., Mn, Fe, CH_4, H_2, ^3He) relative to deep ocean waters offers an unambiguous method for detecting hydrothermal discharge even kilometers away from seafloor vent sites. As early as 1985 it became clear that optical properties (light transmission or backscattering) could be effective proxies for chemical anomalies in plumes [*Baker et al.*, 1985; *Klinkhammer et al.*, 1986; *Nelsen et al.,* 1986/87]. This conformity validates many recent plume surveys, which are often ancillary projects combined with geophysical or rock-sampling operations that do not include plume water sampling. These projects almost uniformly measure only optical properties of the water column, because these data are sensitive, economical, simple to collect, and almost invariably provide a reliable indicator of underlying hydrothermal activity. Interpretative difficulties can arise, however, especially on slow-spreading ridges, which generally exhibit greater relief and thicker sediment accumulation than fast ridges. Under these conditions, the potential for a false positive (as from sediment resuspension around complex bathymetry) is increased.

A review of the literature, existing vent-field databases, and unpublished sources enumerates ~280 sites of active hydrothermal venting on spreading ridges (including back-arc spreading centers), volcanic arcs, and intraplate volcanoes (Figure 1a). Details of each site including (where available) location, water depth, ridge spreading rate at the site location, and literature reference are available on the Inter-RIDGE web site (http://www.interridge.org). Some 145 of these sites have been confirmed by visual observation or imagery, while another ~130 are inferred solely from water-column observations. Confirmed vent sites range from isolated

patches of low-temperature diffuse flow to enormous sulfide constructs hosting multiple high-temperature chimneys. Precise enumeration of these vent sites is necessarily subjective in the absence of any clear definition of a "vent field." Discharge along fast- and intermediate-rate ridges can be common for kilometers along axis, and investigators sometimes give unique names to vent sites only 10s or 100s of meters apart. In this paper, we have aggregated some of these crowded areas into single sites. Vent sites inferred from plume observations range from plumes unequivocally identified as hydrothermal from a combination of optical, hydrographic, and diagnostic chemical measurements, to minor optical anomalies that may ultimately prove unrelated to hydrothermal discharge. As with seafloor enumerations, a single plume observation reported here may arise from a single isolated source or from multiple but closely spaced vent fields extending for several kilometers along axis.

Despite these caveats, the data distribution along ridges shows several robust trends. Over half (118) of our 222 known or inferred ridge vent sites occur on the heavily surveyed eastern Pacific ridges. Only one-fifth as many (24) have been discovered on the MAR, and all confirmed sites there fall within a narrow latitudinal band between 15° and 38°N, reflecting historic funding/research priorities within the international community. In the southern hemisphere, only two confirmed sites, both near the Rodriguez Triple Junction in the Central Indian Ocean, have been observed along a continuous ~30,000 km chain of ridge crest that includes the southern MAR, all of the SW, Central, and SE Indian Ocean ridges, and the Pacific Antarctic Ridge/East Pacific Rise south of 38°S. Plume surveys, however, have so far identified 25 targets along that same stretch of ridge, almost entirely on Indian Ocean ridges. While the focus of this paper is on vent sites along divergent plate boundaries, we note that intensive surveying efforts on volcanic arcs, especially since 1999, have found 58 confirmed or inferred sites, 20% of the global total [*Ishibashi and Urabe,* 1995; *de Ronde et al.,* 2003; *Embley et al.*, 2004]. In general, these sites remain understudied relative to ridge sites.

The usefulness of this distribution depends largely on its accuracy and completeness. Based on the spatial density of acquired data, we identify three types of surveyed areas along the global ridge system (Figure 1b). The most reliable statistics come from densely surveyed areas, where detailed water column and/or seafloor investigations provide excellent control on where vent fields are, and, equally importantly, are not. Human or photographic imaging of the seafloor provides the most meticulous surveying, but is an exceedingly inefficient mode of data-collection when compared to plume surveys and presently accounts for only a small fraction of mapped ridges. In most densely surveyed areas, hydrothermal plumes

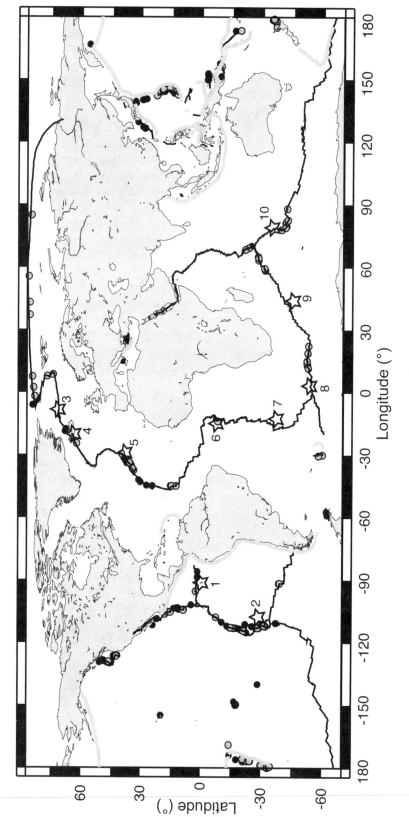

Figure 1a. Distribution of 144 known (black dots) and 133 inferred (gray dots) hydrothermal fields. Some individual fields contain several closely spaced (~100 m) "sites" distinctly named in the literature. Solid black lines are the midocean ridge and transform faults, gray lines are subduction zones. Hotspots (open stars) on or near (<500 km) the midocean ridge include 1, Galápagos; 2, Easter; 3, Jan Mayen; 4, Iceland; 5 Azores; 6, Ascension; 7, Tristan de Cunha; 8, Bouvet; 9 Crozet; and 10, Amsterdam–St. Paul.

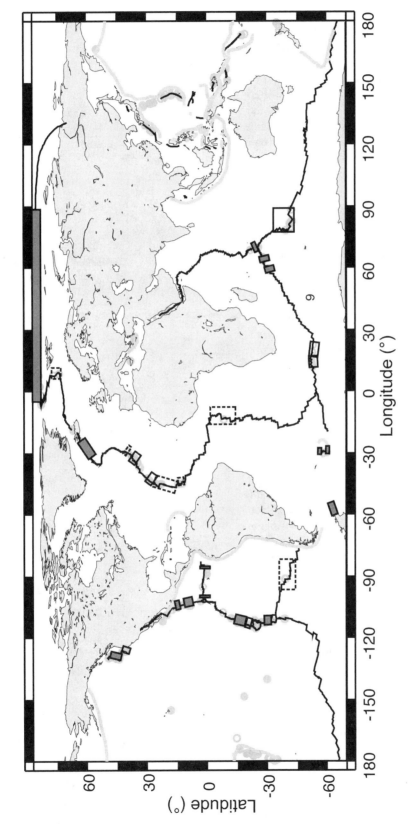

Figure 1b. Vent site distribution overlain with boxes indicating the quality of existing survey data along oceanic spreading ridges: densely surveyed (solid boxes), moderately well surveyed (open boxes), or sparsely surveyed (dotted line boxes).

have been mapped either continuously (e.g., by a towed instrument package) or by closely spaced vertical profiles. These areas cover a total of ~4900 km of ridge axis and are most common along the eastern Pacific ridges, but also include the Reykjanes Ridge south of Iceland, areas around the Rodriguez Triple Junction in the Indian Ocean, the western Gakkel Ridge, and other scattered sections (Figure 1b). Moderately surveyed areas have undergone systematic searches for hydrothermal activity, but the spatial density of the data is lower than in densely surveyed areas and, thus, is not comprehensive. Within this category, therefore, it is quite probable that more active vent sites may exist than those that have actually been discovered to date. These moderately surveyed areas are most common on slow- and intermediate-rate spreading ridges and total ~4900 km (Figure 1b). Sparsely surveyed areas have been sampled in sufficient detail to locate the presence of confirmed and/or likely vent sites, but considerable uncertainty remains about the full extent of hydrothermal activity in those areas. This category totals ~3500 km of ridges and includes the Chile Rise and several large sections in the Atlantic Ocean (Figure 1b). Together, these three categories of surveyed areas still total only ~13,000 km of ridge length, or ~20% of the ~67,000 km [*Bird et al.,* 2003] of global ridges (including ~6500 km in back-arc basins). Other ridge vent and plume sites fall outside these areas, of course, because they were discovered serendipitously, rather than as part of any systematic, large-scale hydrothermal survey.

2.2. Vent Field Statistics

If we restrict our attention to vent fields on ridges (oceanic and back arc), a histogram of the water depths at which confirmed and inferred vent fields occur describes a distribution range from 200 to 4300 m, with highest frequencies at depths between 2200 and 2800 m (Figure 2a). Depths of those sites inferred from plume surveys are strongly skewed toward deeper waters, reflecting a concentration of recent plume surveys along ultraslow ridges [*German et al.*, 1998a; *Bach et al.*, 2002; *Edmonds et al.*, 2003]. Only one site on an ultraslow ridge has actually been observed, at a confirmed depth of 4100 m on the Gakkel Ridge [*Edmonds et al.*, 2003]. The depth reported for each inferred site in Figure 2a is the depth of the shallowest feature on the underlying rift-valley floor close to the location of the plume observation. Plotting frequency of vent sites against spreading rate results in a complex histogram with peaks at slow-, intermediate-, fast-, and superfast-spreading rates (Figure 2b). From this figure it is clear that the ratio of confirmed to inferred sites increases from the slowest to the fastest spreading rates, a testimony to the historical pattern of seafloor hydrothermal exploration. This distribution is also quite different from that for the cumulative length of ridge

crests themselves vs. spreading rate (using data from *DeMets et al.* [1990] and *Bird* [2003]), which shows a general decline in ridge length at increasing spreading rates (Figure 2c). For example, ridges spreading faster than 120 mm/yr comprise only 9% of the total length of the global ridge system, but account for 27% (60) of the presently recognized ridge vent

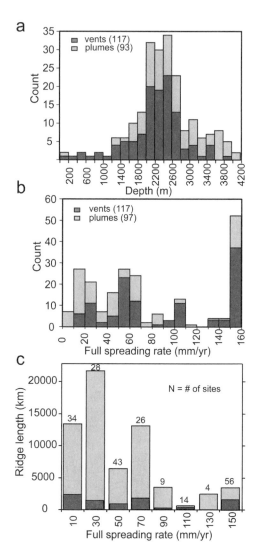

Figure 2. (a) Frequency distribution of the depth of vent sites located along midocean ridge and back-arc basin spreading centers. (b) Frequency distribution of the full spreading rate at the vent site locations. (c) Frequency distribution for the full spreading rate of the global spreading center system in 20 mm/yr increments [*DeMets et al.*, 1990; *Bird*, 2003]. Numbers above each bar indicate the number of vent+plume sites in each spreading rate category. Dark bars at the bottom of each column indicate the approximate length of densely surveyed plus moderately surveyed ridge axis (from Figure 1b) in each spreading rate bin. This total includes more ridge sections than used in Table 1.

field total, a sum consistent with the 26% of total crustal volume generated each year along these ridges. Note, however, that fast-spreading ridges are currently oversampled compared to the rest of the ridge system. About 30% of ridges spreading faster than 100 mm/yr have been surveyed at least "densely" or "moderately," compared to only 12% of the ridges spreading slower than 80 mm/yr (Figure 2c).

3. "FAST" RIDGES

"Fast" ridges are defined here as those having a narrow and shallow (relief on the order of 100 m or less) axial valley. This category typically ranges from intermediate (55–80 mm/yr) to superfast (>140 mm/yr) spreading rates. We use this definition because it describes ridge sections where virtually all detectable hydrothermal discharge is concentrated in or near a narrow neovolcanic zone, and where hydrothermal plumes can be readily dispersed off axis by local currents. These qualities allow efficient and credible surveying for hydrothermal plumes (and their seafloor sources), and make it likely that the plume distribution will be a reliable indicator of the location and extent of the seafloor sources (i.e., plumes are unlikely to be transported long distances along axis).

Among the best places to demonstrate the capability of optical plume transects to denote the spatial density of seafloor vent fields is the EPR between 13.5° and 18.67°S (Figure 3), the focus of several submersible, remotely or autonomously operated vehicle, and plume survey campaigns that have systematically surveyed for vent sites [Renard et al., 1985; Urabe et al., 1995; Auzende et al., 1996; Haymon et al., 1997; Lupton et al., 1997; Embley et al., 1998; Von Damm et al., 1999; Lupton et al., 1999; Wright et al., 2002]. A continuous plume transect here using an optical light backscattering sensor on a conductivity-temperature-depth-optical (CTDO) package in the "tow-yo" mode [Baker et al., 1995] found $p_h = 0.60$ [Baker and Urabe, 1996]. Comparing the ΔNTU distribution with discrete particulate (Fe) and dissolved (CH_4) chemical tracers confirms that the optical signal reliably described the distribution of all significant hydrothermal plumes on this section of the EPR (Figure 3). How well does the plume data reflect the seafloor sources? Water column data identify approximately six discrete plume maxima between 14° and 17°S, plus a continuous and variable plume from 17.33° to 18.67°S (Figure 3a). While submersible and remotely/autonomously operated vehicle expeditions have transited only ~10% of this ridge section, these expeditions have nevertheless located active vent fields under five of the discrete plumes and throughout the 17.33°–18.67°S area. This substantial agreement between the location of the mapped vent fields and hydrothermal plumes implies that no significant vent fields went undetected by the plume surveys, though

future volcanic activity will certainly create new fields even as some old fields grow extinct.

In addition to the EPR section described above, continuous plume data have been acquired along three other multi-segment sections of fast-spreading ridge: the Juan de Fuca/Explorer Ridge [Baker and Hammond, 1992; E.T. Baker, unpublished data], the EPR 8.67°–11.83°N [Baker et al., 1994], and the EPR 27.5°–32.3°S [Baker et al., 2002] (Table 1). To compare the plume distributions along all four ridge sections, we compile transects of vertically integrated ΔNTU ($\Sigma\Delta$NTU). $\Sigma\Delta$NTU is calculated by gridding the two-dimensional ΔNTU transects (e.g., Figure 3a) at cell sizes of $x = 0.03$° of latitude (~3 km), $y = 25$ m, then vertically summing ΔNTU in each x interval from the seafloor to plume top. The $\Sigma\Delta$NTU plots (Figure 4) illustrate that as spreading rate increases not only does p_h increase, but so does the along-axis extent of the largest plume features. On superfast ridge sections, vent sources can be so extensive that plumes are continuous for upwards of 100 km along axis.

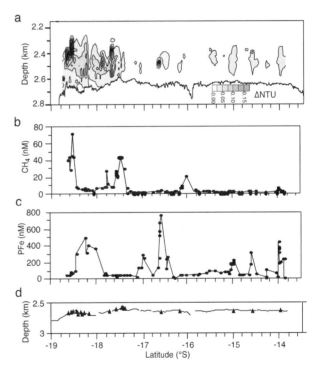

Figure 3. Along-axis transect of (a) light-scattering, (b) dissolved CH_4, and (c) particulate Fe from 13.5° to 18.67°S on the East Pacific Rise. Light-scattering data collected with a series of continuous CTDO tow-yos. Discrete CH_4 and Fe samples (solid circles) from bottle samples and vertical casts. Note that some plumes are volatile rich (e.g., near 18.5°S and 17.67°), while others are metal rich (e.g., near 18°, 16.5°, and 14°S). Location of known vent areas (solid triangles) shown in (d).

Two other fast ridge sections without continuous plume data still meet the criteria of having multisegment extent and a sufficiently high density of individual plume profile data to reliably calculate p_h: the EPR 15.33°–18.5°N [*Baker et al., 2001b*] and the South East Indian Ridge (SEIR) 77°–88°E [*Scheirer et al., 1998*]. For these types of data sets, *Scheirer et al.* [1998] argued that sufficiently closely spaced vertical profiles represent a statistically valid subsampling of the continuous plume distribution, such that the p_h value is equivalent to the per-

centage of vertical profiles that detect a hydrothermal plume. The SEIR section is unusual in that about one third of the ridge length studied bisects the Amsterdam–St. Paul Plateau, a hotspot location with a crustal thickness of ~10 km [*Scheirer et al., 2000*]. Hydrothermal plumes there were mapped using 81 vertical light-scattering/temperature profiles collected at an average spacing of 18 km with Miniature Autonomous Plume Recorders (MAPRs [*Baker and Milburn, 1997*]) attached to dredges and rock cores. *Scheirer et al.* [1998] rec-

Table 1. Determinations of Site Frequency, Plume Incidence, and Magmatic Budget for Selected Ridge Sections

Location	ID #	km	# of active sites[a]	F_s sites/ 100 km	p_h: transect[b]	p_h: % of profiles[c]	Avg. full spreading rate (mm/yr)[d]	Crustal thickness (km)[e]	Magmatic budget (km³/(Myr km))	Refs.[f]
"Fast" Ridges										
EPR 14°–19°S	1	540	21	3.89	0.60		145.0	6.3	914	1
EPR 27°–32°S	2	610	14	2.30	0.54		148.0	6.3	932	2
EPR 9°–13°N	3	300	10	3.33	0.38		101.4	6.3	639	3
EPR 15°–18°N	4	350	6	1.71	0.36		86.0	6.3	542	4
JDFR+Explorer	5	480	22	4.53	0.21		55.0	6.3	347	5
SEIR normal	6	1050	6	0.57		0.17	66.0	6.3	416	6
"Slow" Ridges										
MAR 27°–30°N	7	330	1	0.30			24.0	6.3	151	7
MAR 36°–38°N (min)		230	5	2.17			23.9	6.3	151	8
MAR 36°–38°N (max)		230	10	4.35			23.9	6.3	151	8
MAR 36°–38°N (avg)	8	230	7.5	3.26			23.9	6.3	151	8
SWIR 58°–66°E	9	450	6	1.33	0.12		14.0	4	56	9
SWIR 10°–24°E (min)		950	2	0.21		0.058	11.2	4	45	10
SWIR 10°–24°E (max)		950	8	0.84		0.13	11.2	4	45	10
SWIR 10°–24°E (avg)	10	950	5	0.53		0.087	11.2	4	45	10
Gakkel R. (min)		850	9	1.06	0.75		8.5	4	34	11
Gakkel R. (max)		850	10	1.18	0.75		8.5	4	34	11
Gakkel R. (avg)	11	850	9.5	1.12	0.75		8.5	4	34	11
Hotspot Ridges										
Reykjanes R.	12	750	1	0.13	0.012		19.1	10	191	12
SEIR Hotspot (min)		445	2	0.45		0.034	66.0	10	660	6
SEIR Hotspot (max)		445	4	0.90		0.069	66.0	10	660	6
SEIR Hotspot (avg)	13	445	3	0.67		0.052	66.0	10	660	6

[a]Site listings available at http://www.interridge.org

[b] p_h calculated from contoured plume data.

[c] p_h calculated from percent of profiles detecting a plume.

[d] *DeMets et al.* [1990].

[e] Average crustal thickness for "normal" ridge (6.3 km), ultraslow-spreading ridges (4 km), and hotspot-affected ridges (10 km) [*White et al., 1992, 2001*].

[f] (1) *Baker and Urabe* [1996]; (2) *Baker et al.* [2002]; (3) *Baker et al.* [1994]; (4) *Baker et al.* [2001b]; (5) *Baker and Hammond* [1992], E.T. Baker, unpublished data; (6) *Scheirer et al.* [1998]; (7) *Murton et al.* [1994]; (8) *Langmuir et al.* [1993], *Fouquet et al.* [1995], *German et al.* [1996a], *Barriga et al.* [1998], *Fouquet et al.* [2002]; (9) *German et al.* [1998a]; (10) *Bach et al.* [2002], *Baker et al.*, in press; (11) *Edmonds et al.* [2003], *Baker et al.*, in press; (12) *German et al.* [1994]. Additional references available at http://www.interridge.org.

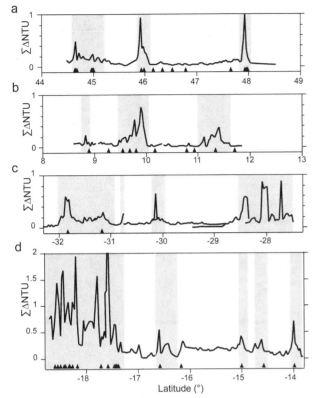

Figure 4. Vertically integrated ΔNTU transects from sections of fast ridges. ΣΔNTU values were determined by gridding continuous ΔNTU data into cells of 0.03° latitude along axis by 25 m thick from the top of the plume to the seafloor, then summing ΔNTU values vertically in each latitude bin. Shaded areas identify regions where light-scattering plumes are continuous along axis. The fraction of the total surveyed areas covered by these regions (p_h) increases with increasing spreading rate: (a) Juan de Fuca Ridge (55 mm/yr), p_h = 0.23; (b) northern East Pacific Rise (103 mm/yr), p_h = 0.38; (c) southern EPR (148 mm/yr), p_h = 0.60; (d) southern EPR (148 mm/yr), p_h = 0.54. Locations of known vent fields (solid triangles) indicated along the bottom of each transect.

ognized plumes on 11 of those 81 profiles (including one in the caldera of an off-axis volcano), yielding a maximum p_h of 0.14 for the entire area. For reasons we discuss below, hydrothermal circulation along ridge sections that overlie mantle hotspots may be markedly affected by their characteristic thickened crust and higher mantle temperatures [e.g., *Chen*, 2003]. We have partitioned our data for the SEIR survey, therefore, into those ridge sections intersecting the hotspot influence and those adjacent to it [*Scheirer et al.*, 2000]. Outside of the hotspot-influenced region, 36 profiles along 1050 km of ridge axis produced 6 plume sites, for a p_h = 0.17.

The data summarized above span spreading rates from intermediate to superfast, and if the long-term magma budget is the primary control on hydrothermal activity we should expect a robust correlation between the magma delivery rate, V_m, and both p_h and F_s. We estimate V_m from

$$V_m = T_c u_s$$

where T_c is the nominal crustal thickness of 6.3 ± 0.9 km for ridges with a full spreading rate (u_s) > 20 mm/yr [*White et al.*, 1992, 2001] (Table 1). Plotting F_s (Table 1) against V_m yields a scattered relationship that reflects not only the observed distribution of hydrothermal activity, but also the current intensity of seafloor exploration effort and the uncertainty in enumerating vent fields on fast-spreading ridges (Figure 5a). The JDFR has been explored for over 20 years, while a submersible has yet to visit the EPR 15.33°–18.5°N or SEIR areas. We can improve the statistics dramatically, however, by binning these six sections according to superfast, fast, and intermediate spreading rates so that each data point represents at least 500 km of ridge length (Figure 5c). The more robust relationship obtained by plotting p_h against V_m demonstrates the integrating value of multisegment plume surveys in determining relative hydrothermal activity (Figure 5b). The evidence is compelling in either case: multisegment hydrothermal activity along "fast" ridges is primarily controlled by the magmatic budget.

4. "SLOW" RIDGES

We define "slow" ridges as those with deep, spacious, and generally enclosed rift valleys, commonly with several hundreds of meters of relief, typically spreading at rates between 20 and 55 mm/yr (we discuss ultraslow-spreading ridges separately, below). These characteristics create a challenging environment for identifying and locating vent sites. The frequent absence of a well-defined neovolcanic zone and the presence of deep and enduring faults that facilitate the discharge of hydrothermal circulation over a broad expanse of seafloor make survey strategies far more complex than on fast ridges. Even when plumes are detected, their source is typically not obvious because the rift valley relief may trap and disperse a plume throughout much or all of an individual tectonic segment [e.g., *German et al.*, 1998b; *Bougault et al.*, 1998].

The first systematic investigations for submarine hydrothermal activity on any slow-spreading ridge were conducted on the northern MAR (11°–26°N) in the 1980s [*Klinkhammer et al.*, 1985; *Bougault et al.*, 1990; *Charlou and Donval*, 1993]. Those preliminary investigations relied upon vertical casts within individual segments of the MAR rift valley to detect midwater chemical anomalies (e.g., total dissolvable (TD)Mn, dissolved CH_4) indicative of hydrothermal discharge from the seafloor. This "point-sampling" approach found midwater TDMn anomalies in nine segments of the MAR between 11°

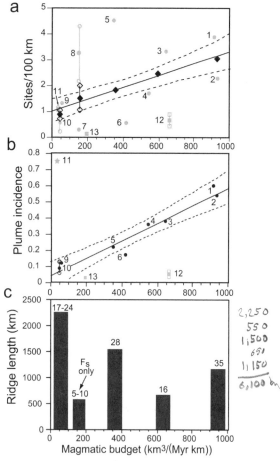

to locate the precise source of venting (Figure 6) [e.g. *Charlou et al.*, 1991; *Aballéa et al.*, 1998; *Chin et al.*, 1998; *German et al.*, 2002].

Unlike on fast ridges, plume surveys on slow ridges have been either too imprecise or too short to yield reliable estimates of p_h. (Note that we are here adopting a more rigorous interpretation than we have used earlier [e.g., *Baker et al.*, 1995, 1996], where MAR p_h values were estimated from vent site frequency.) Lengthy surveys have been conducted from 27° to 30°N [*Murton et al.*, 1994] and 35.7° to 38°N [*German et al.*, 1996a], but in both cases using only a single transmissometer mounted on the Towed Ocean Bottom Instrument (TOBI) sidescan vehicle towed 150–500 m above bottom. Plumes have been mapped in detail along the Broken Spur [*German et al.*, 1999] and Rainbow [*German et al.*, 1996b] segments, but both surveys extended for no more than ~50 km along axis.

Figure 5. (a) Scatter plot of hydrothermal sites/100 km of ridge length (F_s) vs. magmatic budget (V_m). Numbers refer to ridge sections identified in Table 1. Sections 8, 10, 11, and 12 show high and low estimates with open symbols, mean estimate with solid symbol. Squares indicate hotspot-affected ridges. Black diamonds shown binned data (not including sections 12 or 13), with least-squares regression fit of $F_s = 1.01 + 0.0023V_m$ ($r^2 = 0.97$). 95% confidence bands for prediction of F_s from V_m given by dotted lines. (b) Plume incidence (p_h) vs. V_m, with symbols as in (a). Least-squares regression fit of $p_h = 0.043 + 0.00055V_m$ ($r^2 = 0.93$), excluding sections 11–13 (see text). Note also there is no data point for slow-spreading (non-hotspot-affected) ridges. (c) Total ridge length and number of vent sites in each of the five bins in (a).

and 26°N [*Klinkhammer et al.*, 1985], and midwater CH_4 enrichments in 14 segments within the same region [*Charlou and Donval*, 1993]. Subsequent detailed studies within this region discovered major vent fields such as TAG [*Rona et al.*, 1986], Snake Pit [*ODP Leg 106 Scientific Party*, 1986], and Logatchev [*Krasnov et al.*, 1995]. Other "point-sampling" surveys have identified more than 10 additional sites between 40°N and 8°S where chemical and/or optical anomalies indicate the presence of hydrothermal plumes but are insufficient

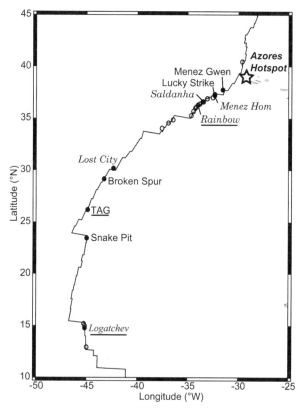

Figure 6. Known (solid circles) and inferred (open circles) vents sites along the northern Mid-Atlantic Ridge. Sites sited on axial neovolcanic highs are in plain type, sites hosted at least partially in ultramafic rocks are in italics, and sites apparently controlled by cross-cutting fault populations are underlined. Unlike the situation on most fast-spreading ridges, many of the inferred vent sites are based on single vertical profiles and have a high degree of uncertainty.

While we have insufficient data to reliably determine p_h, we can calculate minimum estimates of F_s for the two MAR sections mentioned above (we discuss the Reykjanes Ridge below). From 27° to 30°N only the Broken Spur vent field is known [*German et al.*, 1999], yielding a minimum F_s of 0.30. Multiple surveys in the 35.7°–38°N section have located at least three high-temperature hydrothermal sites (Menez Gwen [*Fouquet et al.*, 1995], Lucky Strike [*Langmuir et al.*, 1993], and Rainbow [*German et al.*, 1996b; *Fouquet et al.*, 1997]), and two low-temperature sites (Mt. Saldanha [*Barriga et al.*, 1998] and Menez Hom [*Fouquet et al.* 2002]). Five other sites have been inferred from water column profiles only (in the South Lucky Strike, north FAMOUS, AMAR, and South AMAR segment sections [*German et al.*, 1996a; *Chin et al.*, 1998]), but these inferences are based on very limited data. F_s values thus range from 2.2 to 4.3. Merging the 27°–30°N and 35.7°–38°N sections results in an F_s of 1.5 over 560 km of ridge crest (Figure 5a). *German and Parson* [1998] undertook a similar analysis over a more extended ridge length with consequently weaker data constraints. They found a vent field spacing of ~130 km from 11°–30°N ($F_s = 0.77$), shrinking to every 30 km from 35.7°–38°N ($F_s = 3.3$). Merging these two estimates yields a mean F_s of 1.3, consistent with our analysis of the combined 27°–30° and 35.7°–38°N regions ($F_s = 1.5$).

These MAR surveys have documented that hydrothermal activity on slow ridges can be influenced not only by the magmatic budget but by local tectonic processes as well. Sites closely associated with axial neovolcanic activity include Menez Gwen, Lucky Strike, Broken Spur, and Snake Pit (Figure 6). Other sites occur not at magma-rich segment centers but at the confluence of cross-cutting fault populations on segment walls (TAG, Logatchev), or in segment-end non-transform offsets (NTOs) (Rainbow, Mt. Saldanha, Menez Hom, and other unnamed sites). The apparent increased incidence of venting between 35.7°N and 38°N, compared to the Kane-Atlantis (24°–30°N) section (Figure 6), led to a "tectonic control of venting" hypothesis [*German et al.*, 1996a; *German and Parson*, 1998; *Gràcia et al.*, 2000; *Parson et al.*, 2000]. This hypothesis argues that obliquity of the axial strike in the 35.7°–38°N area results in a greater number of short 2nd-order ridge segments and larger NTOs than found farther south. Crustal permeability is enhanced within the NTOs by deeply penetrating and long-lived faults, allowing seawater to mine heat not only from crustal magma but also from gabbroic intrusions or cooling in the lithospheric mantle [*Cannat et al.*, in press], and from exothermic serpentinization [*Schroeder et al.* 2002; *Lowell and Rona*, 2002]. Recent drilling at 14°–16°N on the MAR, for example, suggests the crust there may be 25% gabbroic [*Kelemen*, 2003], and crystallization of such rocks at depths to 20 km might provide a deep

heat source in basalt-poor ridge segments. The Rainbow vent field perhaps best exemplifies this hypothesis: it is situated at the intersection of the ridge-axis and an NTO, at 36.25°N on the MAR, and vents 362°C "black smoker" fluids with distinctive chemical compositions indicative of "contamination" by the products of serpentinization reactions at depth [*Holm and Charlou*, 2001; *Douville et al.*, 2002; *Charlou et al.*, 2002]. Sites such as the Lost City Field at 30°N [*Kelly et al.*, 2001] and Mt. Saldanha at 36.51° [*Barriga et al.*, 1998] appear to be driven almost purely by serpentinization. Their low temperature (<100°C) and metal-poor (and thus optically invisible) discharge makes systematic searches for additional sites of this type, which may be pervasive along all slow-spreading ridges, a demanding future challenge.

If processes other than variability in the magmatic budget play a significant role in controlling the distribution of hydrothermal activity on slow ridges, we might expect the MAR not to agree with the trend of fast ridges on a plot of F_s vs. V_m. While F_s for the 27°–30°N and 35.7°–38°N MAR sections are quite different, their weighted mean value closely follows the fast ridge trend (Figure 5a). Thus, the available data are insufficient to reject the magmatic budget hypothesis for slow ridges. While significant departures from the linear trend are clearly observed along individual short segments, what is indisputable is that more extensive and systematic surveys along slow-spreading (20–55 mm/yr) ridges are required to fully test the hypothesis. Currently, the database for these ridges is smaller than for any other ridge class, even though they constitute ~40% of the global ridge total (Figure 5c).

5. ULTRASLOW RIDGES

Ultraslow ridges are of particular interest because both geochemical and geophysical inferences indicate that the amount of melt generated in the mantle beneath ridges decreases abruptly as spreading rates drop below ~20 mm/yr [*Reid and Jackson*, 1981; *White et al.*, 2001]. Magmatism becomes discontinuous and mantle peridotite is emplaced directly to the seafloor over broad areas, creating a class of ridge fundamentally different from all faster-spreading ridges [*Dick et al.*, 2003]. If the spatial density of hydrothermal circulation is directly related to magma budget, then the frequency of vent fields on ultraslow ridges should be even less than that predicted by the spreading rate alone.

In the last few years, detailed plume surveys have been conducted along three long sections of ultraslow-spreading ridge: two sections of the eastern South West Indian Ridge (SWIR) between 58° and 66°E (14–16 mm/yr full rate) [*German et al.*, 1998a]; the western SWIR, 10°–23°E (8–14 mm/yr) [*Bach et al.*, 2002; *Baker et al.*, in press]; and the Gakkel Ridge, 7°W–86°E (6–11 mm/yr) [*Edmonds et al.*, 2003]. Detailed analyses of plume and vent field distributions along these ridge sections suggest that hydrothermal activity along each is similar.

The only ultraslow ridge sections where plumes have been mapped using a continuous tow method are two segments of the eastern SWIR between 58° and 66°E [*German et al.*, 1998a]. This study located six possible hydrothermal sites using dual-pass tracks of TOBI, with an array of MAPRs spanning a 300 m-thick layer above and below the package. The TOBI data mapped laterally continuous, above-bottom optical anomalies with distinct geographical limits, strongly indicative of hydrothermal rather than erosional origins (Plate 1a). The total plume extent of 50 km within the 420 aggregate km of ridge length studied results in $p_h = 0.12$, consistent with a low F_s of 1.33 (Figure 5). For the 10°–23°E SWIR, p_h was calculated from the percentage of vertical MAPR profiles that recorded a plume. Of the 86 profiles collected on cruises in 2000/2001 and 2003 (Plate 1b), at least five and a maximum of 11 detected hydrothermal plumes, yielding a p_h of 0.06 to 0.13 and F_s values of 0.21–0.84 (Figure 5), although these maximum values are not tightly constrained. There is an unusually high level of uncertainty here, because Antarctic Bottom Water flowing into this section from the bounding fracture zones appears to induce widespread resuspension of bottom sediments, complicating the identification of hydrothermally derived optical anomalies [*Baker et al.*, in press].

The Gakkel Ridge, the slowest-spreading and perhaps deepest section of the global ridge system, crosses ~1800 km of the Arctic Ocean from Greenland to Siberia. The Arctic Mid-Ocean Ridge Expedition in 2001 collected 145 MAPR profiles over 850 km of the western half of the ridge (average spacing 6.6 km), and 114 displayed light-scattering (and often temperature) anomalies characteristic of hydrothermal plumes [*Edmonds et al.*, 2003]. Calculating the p_h value either from the fraction of MAPR profiles that detected a hydrothermal plume (0.82) or from the axial plume coverage based on contouring the gridded data set (0.75, Plate 1c) yields the highest p_h yet documented on any lengthy ridge section (Figure 5b). Because of the remarkable hydrographic characteristics over the Gakkel Ridge and the unusual bathymetric characteristics of ultraslow ridges in general, however, this exceptional p_h value is not representative of the relative spatial density of hydrothermal sites. The water column below ~3500 m is effectively isopycnal within the Gakkel Ridge, allowing plumes to rise above the axial bathymetry (Plate 1c). The capacious and continuous axial valley traps many of these plumes, permitting some to disperse coherently for up to 200 km [*Edmonds et al.*, 2003; *Baker et al.*, in press]. This situation is a cautionary lesson about the limitations of interpreting p_h in axial valleys of great relief, an even more extreme example than the Rainbow plume on the MAR, which has been

traced at least 50 km downstream from its source [*German et al.*, 1998b; *Thurnherr et al.*, 2002]. Careful analysis indicates that despite an extreme p_h, only 9–10 hydrothermal fields are active on the Gakkel Ridge [*Edmonds et al.*, 2003; *Baker et al.*, in press], so F_s = 1.1–1.2.

Adding the results from ultraslow ridges to Figure 5 shows, except for the anomalous p_h value of the Gakkel Ridge, good agreement with the trend of faster-spreading ridges. Values for p_h on the SWIR are ~0.1 and F_s values on all three sections are ~1. V_m values for these ridges are computed using a nominal crustal thickness of 4 km [*White et al.*, 2002], though seismic observations [*Muller et al.*, 1999; *Jokat et al.*, 2003] have found thicknesses of < 2 km. Moreover, dredging and magnetic surveys of ultraslow ridges indicate that large sections may actually have near-zero crustal thickness, unless serpentinized peridotite is regarded as "crust" [*Dick et al.*, 2003]. The V_m estimates for ultraslow ridges (Table 1) are, therefore, maximum values.

The presence of ΔNTU plumes over both volcanic and avolcanic areas on these ridges (Plate 1) indicates that hydrothermal activity on ultraslow ridges cannot be simply related to the magmatic budget. Whereas at least six of the nine identified sites along the Gakkel Ridge occur on axial volcanic highs [*Edmonds et al.*, 2003; *Baker et al.*, in press], sites on the SWIR appear more controlled by the tectonic environment. Enhanced permeability created by the long-lived and deeply penetrating fault planes that are a hallmark of ultraslow ridges [*Dick et al.*, 2003] can provide access to nonvolcanic heat sources such as direct cooling of the upwelling mantle and exothermic serpentinization. For example, the strongest plume on the western SWIR, at ~13.3°E (Plate 1b), was found along the wall of a large fault block with a 1200 m footwall extending from valley floor to the crest of the rift valley wall [*Bach et al.*, 2002]. Fossil hydrothermal deposits in the western SWIR were primarily found on the rift valley walls, further emphasizing the role of tectonism in controlling hydrothermal activity there [*Bach et al.*, 2002; *Dick et al.*, 2003]. On the eastern SWIR, several bathymetric highs identify locations of focused volcanism [*Münch et al.*, 2001; *Sauter et al.*, 2002; *Cannat et al.*, 2003], but at least half the detected plumes occur in the weakly volcanic zones (as determined from sidescan sonar imagery). None were observed around the summits of the major bathymetric highs [*German et al.*, 1998a]. Analyses of core-top sediments have confirmed the presence of geologically recent vent activity near plume signals at both volcanically and tectonically dominated sites on the eastern SWIR [*German*, 2003], consistent with inferences from the western SWIR that hydrothermal discharge occurs in both volcanic and nonvolcanic settings.

Clearly, we should anticipate a greater complexity in the nature of hydrothermal activity on weakly volcanic crust than

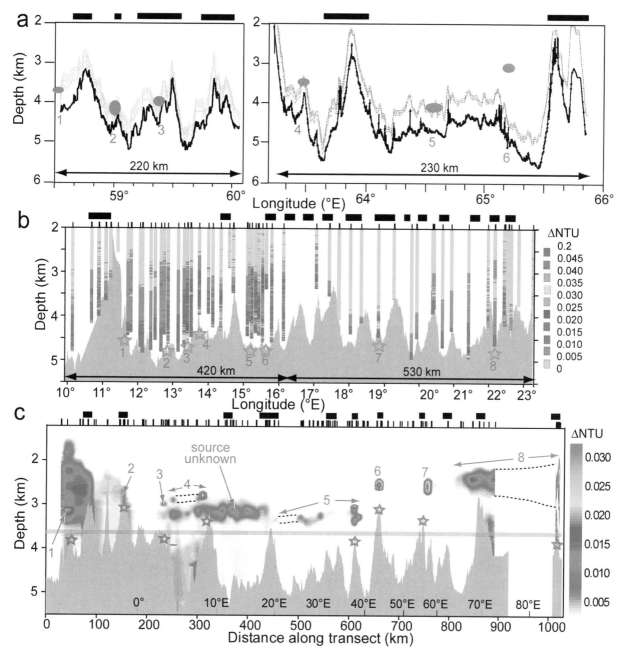

Plate 1. Along-axis transects of ΔNTU for the (a) SWIR 58.5°–66°E, (b) SWIR 10°–23°E, and (c) Gakkel Ridge (4°W–86°E). Black bars above each panel show approximate location of volcanic centers [*Grindlay et al.,* 1998; *Sauter et al.,* 2002; *Cannat et al.,* 2003; *Dick et al.,* 2003; *Michael et al.,* 2003]. In (a) the MAPR paths (light blue) intersected six incidences of ΔNTU (red symbols) while following the bathymetry (heavy black line).This diagram shows only one of two parallel tracks; the plume at 65.12°E was detected on the other track, which followed a slightly different bathymetry. In (b) ΔNTU data are displayed as individual profiles to avoid contouring artifacts. Not all casts were made at the rift axis, so profile depths may be deeper or shallower than the bathymetric profile at the same longitude. Stars mark sites where hydrothermal activity is probable. In (c) MAPR data was sufficiently dense to contour [*Baker et al.,* in press]. Numbers mark individual plumes and the possible seafloor source location of each is shown by the underlying stars; source location of the extensive plume from 250–450 km along section, centered at ~3200 m, is unknown. Most of the ΔNTU anomaly west of ~150 km has no thermal expression and is thus nonhydrothermal [*Baker et al.,* in press]. Supplementary x-axis scale gives longitude along the transect. Pink line indicates approximate top of the bottom isopycnal layer.

would ever have been predicted from studies of fast-spreading ridges alone. Improved understanding of this aspect of heat transfer and hydrothermal circulation will only be disclosed by direct observation of seafloor processes at ultra-slow-spreading ridges, a prime target for InterRIDGE-coordinated research in the future.

6. HOTSPOT-AFFECTED RIDGES

Ridge sections apparently underlain by mantle melt anomalies, or hotspots, make up an intriguing subset of the global ridge system. Three hotspot-associated ridge sections have been systematically surveyed for plume distributions: the Reykjanes Ridge from 57.75° to 63.15°N, almost a full radial transect of the Iceland hotspot; the SEIR from 35.6° to 40.2°N, a 445 km section passing over the top of the Amsterdam–St. Paul hotspot; and 35.7°–38°N on the MAR, a short section on the southern fringe of the Azores hotspot. The first two sections appear unusually deficient in hydrothermal activity compared with other sections of similar magma budget (with the exception of the 27°–30°N MAR section, of which only 50 km has been densely surveyed). *German et al.* [1994] collected 174 optical/chemical profiles using a CTDO-rosette along 750 km of the Reykjanes Ridge, at intervals of 4–18 km. Despite this intense sampling, only the Steinahóll vent field at 63.1°N was located, an F_s of only 0.013 (Figure 5a). Based on the ~10 km lateral extent to which this Steinahóll plume could be traced, these results translate to a p_h = 0.012 (Figure 5b). On the SEIR section, two certain and two possible plumes were detected on 58 MAPR profiles, corresponding to an F_s of 0.45–0.90 and a p_h value of 0.03–0.07 (Figure 5).

Mantle upwelling beneath both these ridge sections has abnormally thickened the oceanic crust to at least ~10 km [*Scheirer et al.*, 2000; *Smallwood and White*, 1998]. An early hypothesis for the Reykjanes Ridge hydrothermal results [*German et al.*, 1994, 1996c], consistent with the first sidescan sonar images of the Reykjanes Ridge [*Parson et al.*, 1993], proposed that thermal gradients beneath the present-day ridge may be sufficiently high, compared to elsewhere along the MAR, to markedly reduce the depth of brittle fracturing of the ocean crust and consequent penetration of seawater. This hypothesis is consistent with a crustal thermal model [*Phipps Morgan and Chen*, 1993; *Chen*, 2003] that explains the combination of a shallow magma chamber at 57.74°N on the Reykjanes Ridge [*Sinha et al.*, 1997] and the lack of hydrothermal plumes [*German et al.*, 1994] as the result of a positive mantle-temperature anomaly (ΔT ~ 40°–70°C) and inefficient hydrothermal cooling. The model calculations for a shallow, steady state magma chamber on a slow-spreading ridge require that convective cooling by hydrothermal discharge be only ~25% of that expected on a slow-spreading ridge not

affected by a hotspot. Similar model calculations have not been run for the Amsterdam–St. Paul SEIR section, but its unusually low p_h value suggests that thickened crust there may similarly depress hydrothermal cooling.

The 35.7°–38°N section of the MAR seems to contradict this hypothesis, since F_s here is 2–3 times higher than expected for its magmatic budget (Figure 5a). A gravity analysis along this section of the MAR, however, found crustal thicknesses south of 38°N to be increased by < 2 km relative to normal (~6 km) crust, shrinking to no increase south of 37°S [*Escartín et al.*, 2001]. While this area was underlain by a melt anomaly at 5–10 Ma, at present there is no crustal thickness or mantle temperature anomaly signature here [*Cannat et al.*, 1999; *Escartín et al.*, 2001]. Instead, moderate crustal thickness variations suggest magma focusing at the segment centers, especially at Lucky Strike and Menez Gwen [*Escartín et al.*, 2001]. Magma focusing, combined with fracturing of the ridge axis and establishment of short 2nd order segments linked by broad, ultramafic-exposing NTOs that accommodate the oblique orientation of the plate boundary [*Parson et al.*, 2000], may explain the relatively high F_s found here.

Determining if these results signify a systematic hotspot effect or simply a present-day sampling artifact will require a broader consideration of other hotspot-affected ridges. Of the 47 hotspots listed by *Richards et al.* [1988], only a few lie within 500 km of a ridge axis; beyond that distance hotspot influence upon ridge morphology appears negligible [*Ito and Lin*, 1995]. According to this criterion, possibly influential hotspots, besides Iceland and the St. Paul–Amsterdam system, include Galápagos, Easter, Jan Mayen, Azores, Ascension, Tristan de Cunha, Bouvet, and Crozet (Figure 1). Bouvet and Crozet are near the ultraslow spreading SWIR where the incidence of venting is already low and so the hypothesized "hotspot effect" on venting may be difficult to discern at these locations. Tristan de Cunha is distant from the MAR axis, while the effect of the Easter Island mantle plume may be further complicated by tectonic interactions with the adjacent microplate. We conclude, therefore, that the most attractive candidates deserving of future, detailed plume/vent surveys are the Galápagos, Jan Mayen, and Ascension hotspots, together with more detailed observations around the Iceland, Azores, and Amsterdam–St. Paul hotspots.

7. DISCUSSION

7.1. Global Trends

While we have hydrothermally surveyed, to some degree, about 20% of the global ridge system (Figure 1b), only 13 ridge sections totaling half that distance (Figure 5) are suitable for examining the magmatic budget hypothesis described in the

Introduction. These sections span the entire global spectrum of the ridge magmatic budget (as calculated from spreading rates), however, and describe a consistent and robust relationship. The first-order trend in plots of V_m vs. both p_h and F_s is a decline in hydrothermal activity with decreasing magma budget, though inconsistencies occur in each plot as well. In the p_h plot (Figure 5b), most of the data points (ignoring the heavily biased Gakkel Ridge data and the hotspot-affected ridges) define a robust trend with a least-squares regression of

$$p_h = 0.043 + 0.00055 V_m$$

with $r^2 = 0.93$. This result agrees with earlier predictions [*Baker and Hammond*, 1992; *Baker et al.*, 1996].

Considerably more scatter exists in the F_s plot (Figure 5a). We interpret this scatter as a function of three overlapping uncertainties: increasing difficulty in defining the boundaries of discrete vent fields as spreading rate increases, wide variability in the effort expended in finding discrete vent sites on different ridge sections, and pronounced differences in the length of surveyed ridge sections (Table 1). After binning these data to mitigate these uncertainties as far as presently possible, the resultant least-squares regression is

$$F_s = 1.0 + 0.0023 V_m$$

with $r^2 = 0.97$.

These robust linear relations offer sturdy support for the hypothesis that crustal magmatic budget is the primary influence on the large-scale (multisegment) distribution of hydrothermal activity. The fact that the activity on ultraslow ridges is appreciably greater than zero, especially considering that we may have conservatively overestimated the true melt supply at the sampled ridges, suggests that ultraslow ridges may in fact be more efficient producers of vent fields than other ridges. We can test this idea by normalizing F_s to the time-averaged delivery of magma [*Baker et al.*, in press]. To normalize F_s, we calculate F_m, sites/(1000 km³ Myr), for five spreading rate bins (equivalent to the magmatic budget bins in Figure 5a) as

$$F_m = 10^3 N/(L u_s T_c)$$

where for each bin N is the number of vent fields observed, L is the total ridge length (km), u_s is the weighted average full spreading rate (mm/yr), and T_c is the nominal crustal thickness of 4 km for ultraslow ridges and 6.3 km for all other ridges [*White et al.*, 1992, 2001] (Table 1). F_m for ultraslow ridges are minimum values considering that sections of these ridges (especially where the spreading rate <12 mm/yr) may actually have near-zero crustal thickness [*Dick et al.*, 2003]. F_m steadily

increases from superfast to slow ridges then increases sharply for ultraslow ridges (Figure 7). For a given magmatic budget, ultraslow ridges appear 2–4 times as efficient as fast-spreading ridges in creating vent fields. The trend of increasing F_m with decreasing u_s is consistent with the expected addition of heat from sources other than crustal cooling of basaltic magma at slow and ultraslow ridges. We caution again, though, that data from slow ridges is uncomfortably scarce.

Additional work on slow and ultraslow ridges will be required to establish whether the apparent trend between F_m and u_s can be verified. In particular, the data for slow (20–55 mm/yr) ridges are so limited that our calculated F_m is speculative. If this trend is confirmed by future surveys, it will indicate that as spreading rates decrease the incidence of hydrothermal activity can be increased by factors other than the long-term magmatic budget. These may include an increased bulk permeability effected by the penetration of deep and enduring faults, strongly three-dimensional magma delivery, and additional heat sources such as direct cooling of the upper mantle, cooling gabbroic intrusions, and serpentinization.

Despite the robust trends in Figure 5, it is important to remember that these results do not demand that the fluxes of hydrothermal heat or chemicals are also similarly related to the magmatic budget. Many additional quantitative flux measurements, at a variety of spreading rates, will be needed to confirm that inference.

The global trend established in Figure 5a can also be used to estimate the total number of vent fields presently active on

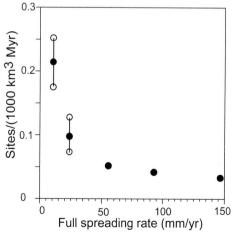

Figure 7. Site frequency F_s normalized to the magma delivery rate vs. spreading rate for the binned data in Figure 5a. Solid circles indicate mean values, open circles the uncertainty range (see Figure 5a). Approximately same trend would result using the p_h data in Figure 5b (excluding the biased Gakkel Ridge result and the hotspot-affected ridge sections).

the ridge system. If we recast the binned data in Figure 5a as a plot of F_s vs. u_s, the least-squares regression is

$$F_s = 0.88 + 0.015u_s.$$

Applying this relation to the global distribution of ridge spreading rates binned at 20 mm/yr intervals yields a histogram of predicted vent sites with a total population of 1060 (with 95% confidence limits of 992–1153) (Figure 8). In contrast to the dominance of slow-spreading ridges in the global distribution of spreading rates (Figure 2c), vent field populations differ by only about a factor of 2 across spreading rate categories, except in the 100–140 mm/yr bins where total ridge length is <2% of the global sum.

7.2. A Millennial-Scale View of Global Hydrothermal Venting

A recurring concern in attempting to relate hydrothermal activity to geological variability is that hydrothermal data are unavoidably time aliased. Mean hydrothermal activity over a much longer period might be distributed quite differently than at present, especially at slow spreading rates. This problem can be partially addressed by comparing the vent field distribution with the oceanic distribution of ^{3}He, an unequivocal and conservative tracer of magmatic degassing along ridge crests [*Craig and Lupton*, 1981]. Overlaying the global ridge system with the deep-water (2000–3000 m) $\delta(^3$He%) pattern [*Geosecs Atlantic, Pacific, and Indian Ocean Expeditions*, 1987; *Jamous et al.*, 1992; *Lupton*, 1995, 1998; *Rüth et al.*, 2000] reveals a strong, but not perfect, correlation between u_s and $\delta(^3$He%) (Figure 9). The most intense $\delta(^3$He%) plume

in the ocean spreads westward from the fastest spreading ridge segments in the ocean, and other intense plumes originate on the fast-spreading northern EPR and the intermediate-rate JDFR. The Atlantic has lower $\delta(^3$He%) than either the Pacific or Indian Oceans [*Geosecs Atlantic, Pacific, and Indian Ocean Expeditions*, 1987; *Rüth et al.*, 2000]. This general pattern suggests that we have not grossly misrepresented the differences in hydrothermal activity between fast- and slow-spreading ridges, and demonstrates that this relation holds over at least the last ~10^3 yr. This timescale is equivalent to both the turnover time of the oceans' thermohaline circulation and, coincidentally, the time for a water volume equal to the global ocean to be cycled through hydrothermal plumes [*Kadko et al.*, 1995; German et al., preprint, 2004].

Past this basic distinction between fast and slow ridges in Figure 9, the distribution of $\delta(^3$He%) becomes more complex. Values of $\delta(^3$He%) along fast-spreading ridges in the eastern Pacific show discrete highs rather than a uniform distribution corresponding to the spreading rate trend. Unexpectedly low $\delta(^3$He%) values occur all along the ridge from the Indian Ocean triple junction to the EPR–Chile Rise triple junction. These features are not solely the result of long-wavelength fluctuations in the distribution of vent fields, but instead may be generated by patterns of oceanographic advection and ventilation. Deep currents in the eastern Pacific and vigorous vertical and circumpolar mixing around Antarctica are two primary controls on the $\delta(^3$He%) distribution [*Farley et al.*, 1995; *Lupton*, 1998]. These complications, in fact, emphasize the value of more complete knowledge about the global distribution of vent fields. Without accurate knowledge of ^{3}He sources along the global ridge system, the utility of $\delta(^3$He%) distributions as a tracer of ocean advection and mixing will remain limited.

8. CONCLUSIONS

In the last quarter-century we have progressed from the discovery of vent fields, to conceptually straightforward hydrothermal surveys along many fast-spreading ridges, to the challenge of cataloging hydrothermal activity throughout Earth's oceans. This progress allows us to test, on a global scale, the proposition that hydrothermal activity increases linearly with the magmatic budget. About 20% of the ~67,000 km global ridge system has now been explored, to some degree, for hydrothermal venting, along with ~3000 km of submarine volcanic arcs and a few intraplate volcanoes. We can precisely locate ~145 confirmed vent sites and have indications of another ~130 based on water column observations alone. Fast ridges (no rift valley), spreading at full rates >55 mm/yr, presently account for 125 sites, slow ridges (rift valley, 20–55 mm/yr) for 55 sites, and ultraslow ridges (<20 mm/yr) for 34

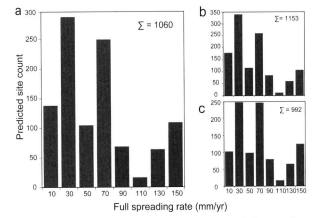

Figure 8. (a) Histogram of predicted vent site population as a function of spreading rate in 20 mm/yr bins (e.g., 0–20, 20–40, etc.). Total population on ridges is estimated as 1060. (b) and (c) show 95% confidence limits for high and low estimates, respectively.

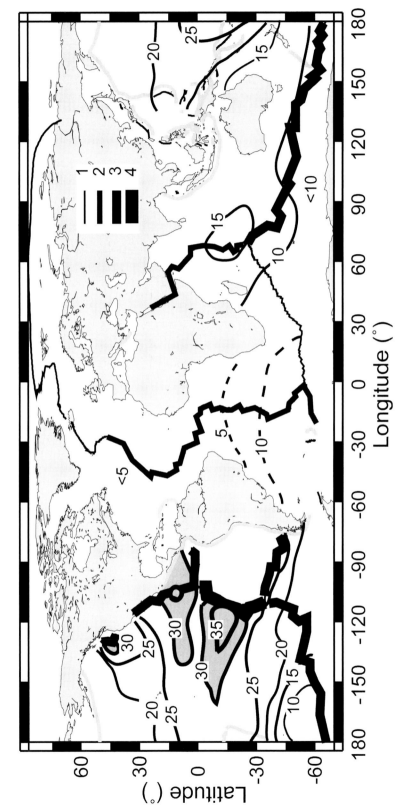

Figure 9. Deep-water (~2000–3000 m) distribution of $\delta(^3\mathrm{He})\%$ values [*Geosecs Atlantic, Pacific, and Indian Ocean Expeditions,* 1987; *Lupton,* 1995, 1998; *Jamous et al.,* 1992; *Rüth et al.,* 2000] compared to ridge full spreading rates ($\delta(^3\mathrm{He})\% = 100(R/R_{\mathrm{air}} - 1)$; $R = {}^3\mathrm{He}/{}^4\mathrm{He}$). Areas with $\delta(^3\mathrm{He})\%$ values >30 are shaded. The MOR is divided into four rate categories corresponding to increasing line thickness: 1, <20 mm/yr; 2, 20–<50 mm/yr; 3, 50–100 mm/yr; 4, >100 mm/yr.

sites. Vent site exploration has long been focused in the eastern Pacific and northern Atlantic ridges. Along a 30,000 km span of ridge from the equatorial Mid-Atlantic Ridge, through the Indian Ocean, to 38°S on the East Pacific Rise we know of only two confirmed sites, separated by <150 km, although more than 20 further likely targets have been identified from plume studies. If our present distribution of sites is representative across all spreading rates, then the expected total global population of active vent fields on the ridge system is ~1000.

Of the explored ridge, on only 13 sections totaling ~7400 km do we have reasonable confidence in our estimates of the relative frequency of hydrothermal activity. Using either plume incidence, p_h, or site frequency, F_s, and regressing the estimates from each section (except two hotspot-affected ridge sections) against the crustal magmatic budget, V_m, yields statistically robust linear trends spanning spreading rates from 10–150 mm/yr. These direct correlations are consistent with the hypothesis that V_m, rather than any spreading-rate-dependent variability in the bulk crustal permeability, for example, is the principal control on the distribution of hydrothermal activity. This conclusion is supported by the oceanic pattern of deep ?(^{3}He%), an unequivocal hydrothermal indicator, which agrees with a ridge source distribution that is a function of spreading rate.

In addition to this first-order trend, the data also suggest that ultraslow and slow ridges support 2–4 times as many vent sites, for a given V_m, than do faster ridges. We interpret this result as an indication that ultraslow ridges, as well as slow ridges to a lesser degree, use deep and enduring faults, strongly three-dimensional magma delivery, and additional heat sources such as direct cooling of the upper mantle, cooling gabbroic intrusions, and serpentinization to supplement the heat supplied by the crustal cooling of basaltic melt.

It is important to remember that these conclusions assume that the average vent field heat flux does not vary systematically with spreading rate. While flux measurements are rare and arduous to obtain, the steady accumulation of information on the tectonic setting, size, and nature of vent fields will eventually permit a reliable test of this assumption.

What are some of the critical problems that call for our attention in the next decade? Several key questions evolve from this review:

1. Are the geologic processes that control venting on slow ridges more similar to those on fast ridges or ultraslow ridges? Slow ridges have fewer systematic hydrothermal surveys than any other ridge type (~10% of their total length), and none suitable for calculating p_h. The question of magmatic vs. tectonic control is still an open one. Current "volcanic-tectonic-hydrothermal cycle" models are based on studies at fast ridges and require testing in

other environments. Our inventory of surveys along slow-spreading ridges needs considerable expansion.

2. Are hotspot-affected ridge sections systematically deficient in convective hydrothermal cooling compared to other sections of similar spreading rate? Additional work at the Reykjanes Ridge and more comprehensive surveys over the Galápagos and Azores hotspots are needed. Surveys of discovery should be conducted over hotspots such as Galápagos, Jan Mayan, and Ascension.

3. Is heat from sources other than crustal cooling of basalt melt an important contributor to hydrothermal circulation at ultraslow ridges? If so, does it materially influence the fluid chemistry and ecology of these sites?

4. Will surveys along the contiguous half of the global ridge that is virtually unexplored support the conclusions presented here? Exploration, spurred by such fundamental questions as vent biogeography, should remain a key objective of InterRIDGE investigators. Back-arc basins such as the Lau Basin and the Mariana Trough, where spreading rates often vary substantially along axis over comparatively short distances, also remain under-surveyed to-date.

With continued attention to both spatial and temporal variability, perhaps in another decade we can derive a confident and quantitative relationship between hydrothermal activity and the magmatic budget at ridges of all types.

Acknowledgments. This review was supported by the NOAA VENTS Program (ETB) and by NERC (CRG). We expressly acknowledge the countless colleagues who have contributed time and energy into the collection of these data. The impetus for this paper was the 2001 InterRIDGE Theoretical Institute at the University of Pavia in Italy. We thank the conveners and participants for channeling our energies into this effort. Thanks also to helpful reviews from R. Haymon, L. Parson, and J. Lin. PMEL contribution number 2544.

REFERENCES

Aballéa, M., J. Radford-Knoery, P. Appriou, H. Bogault, J. L. Charlou, J. P. Donval, J. Etoubleau, Y. Fouquet, C. R. German, and M. Miranda, Manganese distribution in the water column near the Azores Triple Junction along the Mid-Atlantic Ridge and in the Azores domain, *Deep-Sea Res. I, 45,* 1319–1338, 1998.

A.P.H.A. (American Public Health Assoc.), *Standard Methods for the Examination of Water and Wastewater*, 16th ed. A.P.H.A., A.W.W.A, and W.P.C.F. joint publication, Washington D.C., 1268 pp., 1985.

Auzende, J.-M., V. Ballu, R. Batiza, D. Bideau, J.-L. Charlou, M. H. Cormier, Y. Fouquet, P. Geistdoerfer, Y. Lagabrielle, J. Sinton, and P. Spadea, Recent tectonic, magmatic, and hydrothermal activity on the East Pacific Rise between 17°S and 19°S: Submersible observations, *J. Geophys. Res., 101,* 17,995–18,010, 1996.

Bach, W., N. R. Banerjee, H. J. B. Dick, and E. T. Baker, Discovery of ancient and active hydrothermal systems along the ultra-slow spreading Southwest Indian Ridge, 10°–16°E, *Geochem. Geophys. Geosyst., 3,* 2001GC000279, 2002.

Baker, E. T., and S. R. Hammond, Hydrothermal venting and the apparent magmatic budget of the Juan de Fuca Ridge, *J. Geophys. Res., 97,* 3443–3456, 1992.

Baker, E. T., and G. J. Massoth, Characteristics of hydrothermal plumes from two vent fields on the Juan de Fuca Ridge, northeast Pacific Ocean, *Earth Planet. Sci. Lett., 85,* 59–73, 1987.

Baker, E. T., and H. B. Milburn, MAPR: A new instrument for hydrothermal plume mapping, *RIDGE Events, 8,* 23–25, 1997.

Baker, E. T., and T. Urabe, Extensive distribution of hydrothermal plumes along the superfast-spreading East Pacific Rise, 13°50'–18°40'S, *J. Geophys. Res., 101,* 8685–8695, 1996.

Baker, E. T., J. W. Lavelle, and G. J. Massoth, Hydrothermal particle plumes over the southern Juan de Fuca Ridge, *Nature, 316,* 342–344,1985.

Baker, E. T., R. A. Feely, M. J. Mottl, F. J. Sansone, C. G. Wheat, J. A. Resing, and J. E. Lupton, Hydrothermal plumes along the East Pacific Rise, 8°40' to 11°50'N: Plume distribution and relationship to the apparent magmatic budget, *Earth Planet. Sci. Lett., 128,* 1–17, 1994.

Baker, E. T., C. R. German, and H. Elderfield, Hydrothermal plumes over spreading-center axes: Global distributions and geological inferences, in *Seafloor Hydrothermal Systems: Physical, Chemical, Biological, and Geological Interactions, Geophys. Monogr. Ser., 91,* edited by S. Humphris, R. Zierenberg, L. S. Mullineaux, and R. Thomson, pp. 47–71, AGU, Washington D.C., 1995.

Baker, E. T., Y. J. Chen, and J. Phipps Morgan, The relationship between near-axis hydrothermal cooling and the spreading rate of midocean ridges, *Earth Planet. Sci. Lett., 142,* 137–145, 1996.

Baker, E. T., M.-H. Cormier, C. H. Langmuir, and K. Zavala, Hydrothermal plumes along segments of contrasting magmatic influence, 15°20'–18°30' N, East Pacific Rise: Influence of axial faulting, *Geochem. Geophys. Geosyst., 2,* doi:2000GC000165, 2001b.

Baker, E. T., D. A. Tennant, R. A. Feely, G. T. Lebon, and S. L. Walker, Field and laboratory studies on the effect of particle size and composition on optical backscattering measurements in hydrothermal plumes, *Deep–Sea Res. I, 48,* 593–604, 2001a.

Baker, E. T., R. N. Hey, J. E. Lupton, J. A. Resing, R. A. Feely, J. J. Gharib, G. J. Massoth, F. J. Sansone, M. Kleinrock, F. Martinez, D. F. Naar, C. Rodrigo, D. Bohnenstiehl, and D. Pardee, Hydrothermal venting along Earth's fastest spreading center: East Pacific Rise, 27.5°–32.3°S, *J. Geophys. Res., 107,* doi:10.1029/2001JB000651, 2002.

Baker, E. T., H. N. Edmonds, P. J. Michael, W. Bach. H. J. B. Dick, J. E. Snow, S. L. Walker, N. R. Banerjee, and C. H. Langmuir, Hydrothermal venting in magma deserts: The ultraslow-spreading Gakkel and South West Indian Ridges, *Geochem., Geophys., Geosyst.,* in press.

Barriga, F. J. A. S., Y. Fouquet, A. Almeida, M. Biscoito, J.-L. Charlou, R. L. P. Costa, A. Dias, A. M. S. F. Marques, J. M. A. Miranda,

K. Olu, F. Porteiro, M. G. P. S. Queiroz, Discovery of the Saldanha Hydrothermal Field on the FAMOUS Segment of the MAR (36°30'N), *Eos Trans. AGU, 79*(45), Fall Meet. Suppl., F67, 1998.

Bird, P., An updated digital model of plate boundaries, *Geochem. Geophys. Geosyst., 4,* 1027, doi:10.1029/2001GC000252, 2003.

Boström, K., M. N. A. Peterson, O. Joensuu, and D. E. Fisher, Aluminum-poor ferromangoan sediments on active ocean ridges, *J. Geophys. Res., 74,* 3261–3270, 1969.

Bougault, H., J. L. Charlou, Y. Fouquet, H. D. Needham, Activité hydrothermale et structure axiale des dorsales Est-Pacifique et médio-Atlantique, *Oceanol. Acta, vol. special 10,* 199–207, 1990.

Bougault, H., M. Aballéa, J. Radford-Knoery, J. L. Charlou, P. Jean Baptiste, P. Appriou, H. D. Needham, C. German, and M. Miranda, FAMOUS and AMAR segments on the Mid-Atlantic Ridge: ubiquitous hydrothermal Mn, CH_4, δ^3He signals along the rift valley walls and rift offsets, *Earth Planet. Sci. Lett., 161,* 1–17, 1998.

Cannat, M., A. Briais, C. Deplus, J. Escartín, J. Georgen, J. Lin, S. Mercouriev, C. Meyzen, M. Muller, G. Pouliquen, A. Rabain, and P. da Silva, Mid-Atlantic Ridge-Azores hotspot interactions: along-axis migration of a hotspot-derived event of enhanced magmatism 10 to 3 Ma ago, *Earth Planet. Sci. Lett., 173,* 257–269, 1999.

Cannat, M., C. Rommevaux-Jestin, and H. Fujimoto, Melt supply variations to a magma-poor ultra-slow spreading ridge (Southwest Indian Ridge 61° to 69°E), *Geochem. Geophys. Geosyst., 4,* 9104, doi:10.1029/2002GC000480, 2003.

Cannat, M., J. Cann, and J. Maclennan, Some hard rock constraints on the supply of heat to mid-ocean ridges, in *The Thermal Structure of the Ocean Crust and Dynamics of Hydrothermal Circulation,* edited by C. German, J. Lin, and L. Parson, this volume.

Charlou, J. L., and J.-P. Donval, Hydrothermal methane venting between 12°N and 26°N along the Mid-Atlantic Ridge, *J. Geophys. Res., 98,* 9625–9642, 1993.

Charlou, J. L., H. Bougault, P. Appriou, T. Nelsen, and P. Rona, Different TDM/CH_4 hydrothermal plume signatures: TAG site at 26°N and serpentinized ultrabasic daipir at 15°05'N on the Mid-Atlantic Ridge, *Geochim. Cosmochim. Acta, 55,* 3209–3222, 1991.

Charlou, J. L., J. P. Donval, Y. Fouquet, P. Jean-Baptiste, and N. Holm, Geochemistry of high H_2 and CH_4 vent fluids issuing from ultramafic rocks at the Rainbow hydrothermal field (36°14'N, MAR), *Chem. Geol., 191,* 345–359, 2002.

Chen Y. J., Influence of the Iceland mantle plume on crustal accretion at the inflated Reykjanes Ridge: Magma lens and low hydrothermal activity?, *J. Geophys. Res., 108* (B11), 2524, doi:10.1029/2001JB000816, 2003.

Chin, C. S., G. P. Klinkhammer, and C. Wilson, Detection of hydrothermal plumes on the Northern Mid-Atlantic Ridge: results from optical measurements, *Earth Planet. Sci. Lett., 162,* 1–13, 1998.

Corliss, J. B., J. Dymond, L. I. Gordon, J. M. Edmond, R. P. von Herzen, R. D. Ballard, K. Green, D. Williams, A. Bainbridge, K. Crane, and T. H. van Andel, Submarine thermal springs on the Galápagos Rift, *Science, 203,* 1073–1083, 1979.

Craig, H., and J. E. Lupton, Helium-3 and mantle volatiles in the ocean and the oceanic crust, in *The Sea,* edited by C. Emiliani, New York, Wiley, *7,* 391–428, 1981.

Crane, K., F.A. Aikman III, R. Embley, S. Hammond, A. Malahoff, and J. Lupton, The distribution of geothermal fields on the Juan de Fuca Ridge, *J. Geophys. Res., 90*, 727–744, 1985.

DeMets, C., R. G. Gordon, D. F. Argus, and S. Stein, Current plate motions, *Geophys. J. Int., 101*, 425–478, 1990.

de Ronde, C. E. J., G. J. Massoth, E. T. Baker, and J. E. Lupton, Submarine hydrothermal venting related to volcanic arcs, Giggenbach Memorial Volume, in *Volcanic, geothermal and ore-forming fluids: Rulers and witnesses of processes within the Earth*, edited by S. F. Simmons and I. Graham, *Soc. of Econ. Geolog., 10*, 91–110, 2003.

Dick, H. J. B., J. Lin, and H. Schouten, An ultraslow-spreading class of ocean ridge, *Nature, 426*, 405–412, 2003.

Douville, E., J. L. Charlou, E. H. Oelkers, P. Bienvenu, C. F. J. Colon, J. P. Donval, Y. Fouquet, D. Prieur, and P. Appriou, The rainbow vent fluids (36°14'N, MAR): the influence of ultramafic rocks and phase separation on trace metal content in Mid-Atlantic Ridge hydrothermal fluids, *Chem. Geol., 184*, 37–48, 2002.

Edmonds, H. N., P. J. Michael, E. T. Baker, D. P. Connelly, J. E. Snow, C. H. Langmuir, H. J. B. Dick, R. Mühe, C. R. German, and D. W. Graham, Discovery of abundant hydrothermal venting on the ultraslow-spreading Gakkel Ridge, Arctic Ocean, *Nature, 421*, 252–256, 2003.

Embley, R. W., J. E. Lupton, G. Massoth, T. Urabe, V. Tunnicliffe, D. A. Butterfield, T. Shibata, O. Okano, M. Kinoshita, and K. Fujioka, Geological, chemical, and biological evidence for recent volcanism at 17.5°S: East Pacific Rise, *Earth Planet. Sci. Lett., 163*, 131–147, 1998.

Embley, R. W., E. T. Baker, W. W. Chadwick, Jr., J. E. Lupton, J. A. Resing, G. J. Massoth, and K. Nakamura, Explorations of Mariana Arc volcanoes reveal new hydrothermal systems, *Eos Trans. AGU, 85*(4), 37, 40, 2004.

Escartín, J., M. Cannat, G. Pouliquen, and A. Rabain, Crustal thickness of V-shaped ridges south of the Azores: Interaction of the Mid-Atlantic Ridge (36°–39°N) and the Azores hot spot, *J. Geophys. Res., 106*, 21,719–21,735, 2001.

Farley, K. A., E. Maier-Reimer, P. Schlosser, and W. S. Broecker, Constraints on mantle ^{3}He fluxes and deep-sea circulation from an oceanic general circulation model, *J. Geophys. Res., 100*, 3829–3839, 1995.

Fisher, A. T., Permeability within basaltic oceanic crust, *Rev. Geophys., 36*, 143–182, 1998.

Fouquet, Y., H. Ondréas, J.-L. Charlou, J.-P. Donval, J. Radford-Knoery, I. Costa, N. Lourenço, and M. K. Tivey, Atlantic lava lakes and hot vents, *Nature, 377*, 201, 1995.

Fouquet, Y., J. L. Charlou, H. Ondréas, J. Radford-Knoery, J. P. Donval, E. Douville, R. Apprioual, P. Cambon, H. Pell, J. Y. Landur, and A. Normand, Discovery and first submersible investigations on the Rainbow hydrothermal field on the MAR (36°14'N), *Eos Trans. AGU, 78*(46), Fall Meet. Suppl., F832, 1997.

Fouquet, Y., J. L. Charlou, and F. Barriga, Modern seafloor hydrothermal deposits hosted in ultramafic rocks, *Geol. Soc. Am. Abstracts with Programs, 34*(6), A 194–7, 2002.

Francheteau, J., and R. Ballard, The East Pacific Rise near 21°N, 13°N and 20°S: inferences for along-strike variability of axial

processes of the Mid-Ocean Ridge, *Earth Plant. Sci. Lett., 64*, 93–116, 1983.

Geosecs Atlantic, Pacific, and Indian Ocean Expeditions, 7, Shore-based Data and Graphics, National Science Foundation, U.S. Govt. Print. Office, Washington, D.C., 1987.

German, C. R., Hydrothermal activity on the eastern SWIR (50°–70°E): Evidence from core-top geochemistry, 1887 and 1998, *Geochem. Geophys. Geosyst., 4*(7), 9102 doi:10.1029/2003GC000522, 2003.

German, C. R., and L. M. Parson, Distributions of hydrothermal activity along the Mid-Atlantic Ridge: Interplay of magmatic and tectonic controls, *Earth. Planet. Sci. Lett., 160*, 327–341, 1998.

German, C. R., J. Briem, C. Chin, M. Danielsen, S. Holland, R. James, A. Jónsdóttir, E. Ludford, C. Moser, J. Olafsson, M. R. Palmer, and M. D. Rudnicki, Hydrothermal activity on the Reykjanes Ridge: The Steinahóll Vent-field at 63°06'N, *Earth Planet. Sci. Lett., 121*, 647–654, 1994.

German, C. R., E. T. Baker, and G. Klinkhammer, The regional setting of hydrothermal activity, in *Hydrothermal Vents and Processes*, edited by L. M. Parson, C. L. Walker, and D. R. Dixon, *Geol. Soc. Spec. Pub. No. 87*, 3–15, 1995.

German, C. R., L. M. Parson, and HEAT Scientific Team, Hydrothermal exploration at the Azores Triple-Junction: Tectonic control of venting at slow-spreading ridges?, *Earth Planet. Sci. Lett., 138*, 93–104, 1996a.

German C. R., G. P. Klinkhammer, and M. D. Rudnicki, The Rainbow hydrothermal plume, 36°15'N, MAR, *Geophys. Res. Lett., 23*, 2979–2982, 1996b.

German, C. R., L. M. Parson, B. J. Murton, and H. D. Needham, Hydrothermal activity and ridge segmentation on the Mid-Atlantic Ridge: A tale of two hot-spots? in *Ridge Segmentation*, edited by C. McLeod, C. L. Walker and P. Tyler, *Geol. Soc. Spec. Pub. 118*, 169–184, 1996c.

German, C. R., E. T. Baker, C. Mevel, K. Tamaki, and the FUJI Scientific Team, Hydrothermal activity along the southwest Indian Ridge, *Nature, 395*, 490–493, 1998a.

German, C. R., K. J. Richards, M. D. Rudnicki, M. M. Lam, J. L. Charlou, and FLAME Scientific Party, Topographic control of a dispersing hydrothermal plume, *Earth Planet. Sci. Lett., 156*, 267–273, 1998b.

German, C. R., M. D. Rudnicki, and G. P. Klinkhammer, A segment-scale survey of the Broken Spur hydrothermal plume, *Deep-Sea Res. I, 46*, 701–714, 1999.

German, C. R., D. P. Connelly, A. J. Evans, L. M. Parson, Hydrothermal activity on the southern Mid-Atlantic Ridge, *Eos Trans. AGU, 83*(47), Fall Meet. Suppl., Abstract V61B–1361, 2002.

Gràcia, E., J. L. Charlou, J. Radford-Knoery, and L. M. Parson, Non-transform offsets along the Mid-Atlantic Ridge south of the Azores (38°N–34°N): ultramafic exposures and hosting of hydrothermal vents, *Earth Planet. Sci. Lett., 177*, 89–103, 2000.

Grindlay, N. R., J. A. Madsen, C. Rommevaux-Jestin, and J. Sclater, A different pattern of ridge segmentation and mantle Bouguer gravity anomalies along the ultra-slow spreading Southwest Indian Ridge (15°30'E to 25°E), *Earth Planet. Sci. Lett., 161*, 243–253, 1998.

Haymon, R.M., The response of ridge-crest hydrothermal systems to segmented, episodic magma supply, in *Tectonic, Magmatic, Hydrothermal, and Biological Segmentation of Mid-Ocean Ridges*, edited by C. J. MacLeod, P. A. Tyler, and C. L. Walker, *Geol. Soc. Spec. Pub. 118*, 157–168, 1996.

Haymon, R. M., D. J. Fornari, M. H. Edwards, S. Carbotte, D. Wright, and K. C. Macdonald, Hydrothermal vent distribution along the East Pacific Rise crest (9°09'–54'N) and its relationship to magmatic and tectonic processes on fast-spreading mid-ocean ridges, *Earth Planet. Sci. Lett., 104*, 513–534, 1991.

Haymon, R. M. et al., Distribution of fine-scale hydrothermal, volcanic, and tectonic features along the EPR crest, 17°15'–18°30'S: Results of near-bottom acoustic and optical surveys, *Eos Trans. AGU, 78*(46), Fall Meet. Suppl., Abstract F705, 1997.

Holm, N. G., and J.-L. Charlou, Initial indications of abiotic formation of hydrocarbons in the Rainbow ultramafic hydrothermal system, Mid-Atlantic, *Earth Plant. Sci. Lett., 191*, 1–8, 2001.

Ishibashi, J., and T. Urabe, Hydrothermal activity related to arc-backarc magmatism in the western Pacific, in *Backarc Basins: Tectonics and Magmatism*, edited by B. Taylor, pp. 451–495, Plenum Press, New York, 1995.

Ito, G., and J. Lin, Oceanic spreading center-hotspot interactions: Constraints from along-isochron bathymetric and gravity anomalies, *Geology, 23*, 657–660, 1995.

Jamous, D., L. Mémery, C. Andrié, P. Jean-Baptiste, and L. Merlivat, The distribution of helium 3 in the deep western and southern Indian Ocean, *J. Geophys. Res., 97*, 2243–2250, 1992.

Jokat, W., O. Ritzmann, M. C. Schmidt-Aursch, S. Drachev, S. Gauger, and J. Snow, Geophysical evidence for reduced melt production on the Arctic ultraslow Gakkel mid-ocean ridge, *Nature, 423*, 962–965, 2003.

Kadko, D., J. Baross, and J. Alt, The magnitude and global implications of hydrothermal flux, in *Seafloor Hydrothermal Systems: Physical, Chemical, Biological, and Geological Interactions, Geophys. Monogr. Ser., 91*, edited by S. Humphris, R. Zierenberg, L. S. Mullineaux, and R. Thomson, pp. 446–466, AGU, Washington D.C., 1995.

Kelemen, P., Igneous crystallization beginning at 20 km beneath the Mid-Atlantic Ridge, 14°–16°N, *Eos Trans. AGU, 84*(46), Fall Meet. Suppl., Abstract V22H–03, 2003.

Kelley, D. S., J. A. Karson, D. K. Blackman, G. Fruh-Green, J. Gee, D. A. Butterfield, M. D. Lilley, E. J. Olson, M. O. Schrenk, K. K. Roe, G. Lebon, P. Rizzigno, J. Cann, B. John, D. K. Ross, D. Hurst, and G. Sasagawa, An off-axis hydrothermal vent field near the Mid-Atlantic Ridge at 30°N, *Nature, 412*, 145–149, 2001.

Klinkhammer, G., P. Rona, M. Greaves, and H. Elderfield, Hydrothermal manganese plumes in the mid-Atlantic Ridge rift valley, *Nature, 314*, 727–731, 1985.

Klinkhammer, G., H. Elderfield, M. Greaves, P. Rona, and T. Nelsen, Manganese geochemistry near high-temperature vents in the Mid-Atlantic rift valley, *Earth Planet. Sci. Lett., 80*, 230–240, 1986.

Krasnov, S. G., G. A. Cherkashev, T. V. Stepanova, B. N. Batuyev, A. G. Krotov, B. V. Malin, M. N. Maslov, V. F. Markov, I. M. Poroshina, M. S. Samovarov, A. M. Ashadze, and I. K. Eromlayev, Detailed geographical studies of hydrothermal fields in the North Atlantic, in *Hydrothermal Vents and Processes*, edited by L. M. Parson,

C. L. Walker, and D. R. Dixon, *Geol. Soc. Spec. Pub. 87*, 43–64, 1995.

Langmuir, C. H., D. Fornari, D. Colodner, J.-L. Charlou, I. Costa, D. Desbruyeres, D. Desonie, T. Emerson, A. Fiala-Medioni, Y. Fouquet, S. Humphris, L. Saldanha, R. Sours-Page, M. Thatcher, M. Tivey, C. Van Dover, K. Von Damm, K. Weiss, and C. Wilson, Geological setting and characteristics of the Lucky Strike Vent Field at 37°17'N on the Mid-Atlantic Ridge, *Eos, Trans. AGU, 74* (Fall Supplement), 99, 1993.

Lowell, R. P., and P. A. Rona, Seafloor hydrothermal systems driven by the serpentinization of peridotite, *Geophys. Res. Lett., 29*, doi:10.1029/2001GL014411, 2002.

Lupton, J. E., Hydrothermal plumes: Near and far field, in *Seafloor Hydrothermal Systems: Physical, Chemical, Biological, and Geological Interactions, Geophys. Monogr. Ser., 91*, edited by S. Humphris, R. Zierenberg, L. S. Mullineaux, and R. Thomson, pp. 317–346, AGU, Washington D.C., 1995.

Lupton, J. E., Hydrothermal helium plumes in the Pacific Ocean, *J. Geophys. Res., 103*, 15,853–15,868, 1998.

Lupton, J. E., J.-I. Ishibashi, and D. A. Butterfield, Gas chemistry of hydrothermal fluids along the southern East Pacific Rise, 13.5°–18.5°S, *Eos Trans. AGU, 78*(46), Fall Meet. Suppl., F706, 1997.

Lupton, J. E., D. Butterfield, M. Lilley, J-I. Ishibashi, D. Hey, and L. Evans, Gas chemistry of hydrothermal fluids along the East Pacific Rise, 5°S to 32°S, *Eos Trans. AGU, 80*(46), Fall Meet. Suppl., F1099, 1999.

Michael, P. J., C. H. Langmuir, H. J. B. Dick, J. E. Snow, S. L. Goldstein, D. W. Graham, K. Lehnert, G. Kurras, W. Jokat, R. Mühe, and H. N. Edmonds, Magmatic and amagmatic seafloor generation at the ultraslow-spreading Gakkel Ridge, Arctic Ocean, *Nature, 423*, 956–961, 2003.

Muller, M. R., T. A. Minshull, and R. S. White, Segmentation and melt supply at the Southwest Indian Ridge, *Geology, 27*, 867–870, 1999.

Münch, U., C. Lalou, P. Halbach, and H. Fujimoto, Relict hydrothermal events along the super-slow Southwest Indian spreading ridge near 63°56'E—mineralogy, chemistry and chronology of sulfide samples, *Chem. Geol., 177*, 341–349, 2001.

Murton, B. J., G. Klinkhammer, K. Becker, A. Briais, D. Edge, N. Hayward, N. Millard, I. Mitchell, I. Rouse, M. Rudnicki, K. Sayanagi, H. Sloan, and L. Parson, Direct evidence for the distribution and occurrence of hydrothermal activity between 27°N–30°N on the Mid-Atlantic Ridge, *Earth Planet. Sci. Lett., 125*, 119–128, 1994.

Nelsen, T. A., G. P. Klinkhammer, J. H. Trefrey, and R. P. Trocine, Real-time observations of dispersed hydrothermal plumes using nephelometry: examples from the Mid-Atlantic Ridge, *Earth. Planet. Sci. Lett., 81*, 245–252, 1986/87.

ODP (Ocean Drilling Program) Leg 106 Scientific Party, Drilling the Snake Pit hydrothermal sulfide deposit on the Mid-Atlantic Ridge, Lat. 23°N, *Geology, 14*, 1004–1007, 1986.

Parson, L. M., B. J. Murton, R. C. Searle, D. Booth, J. Evans, P. Field, J. Keeton, A. Laughton, E. McAllister, N. Millard, L. Redbourne, I. Rouse, A. Shor, D. Smith, S. Spencer, C. Summerhayes, and C. Walker, En-echelon axial volcanic ridges at the Reykjanes

Ridge—A life-cycle of volcanism and tectonics, *Earth Plant. Sci. Lett., 117*, 73–87, 1993.

Parson L., E. Gràcia, D. Coller, C. German, and D. Needham, Second-order segmentation; the relationship between volcanism and tectonism at the MAR, 38°N–35°40'N, *Earth Planet. Sci. Lett., 178,* 231–251, 2000.

Phipps Morgan, J., and Y. J. Chen, The genesis of oceanic crust: Magma injection, hydrothermal circulation, and crustal flow, *J. Geophys. Res., 98*, 6283–6297, 1993.

Reid, I., and H. R. Jackson, Oceanic spreading rate and crustal thickness, *Mar. Geophys. Res., 5,* 165–172, 1981.

Renard, V., R. Hekinian, J. Francheteau, R. D. Ballard, and H. Backer, Submersible observations at the axis of the ultra-fast-spreading East Pacific Rise (17°30' to 21°30'S), *Earth Planet. Sci. Lett., 75*, 339–353, 1985.

Richards, M. A., B. H. Hager, and N. H. Sleep, Dynamically supported geoid highs over hotspots: Observation and theory, *J. Geophys. Res., 93*, 7690–7708, 1988.

Rona, P. A., G. Klinkhammer, T. A. Nelsen, J. H. Trefry, and H. Elderfield, Black smokers, massive sulfides and vent biota at the mid-Atlantic Ridge, *Nature, 321*, 33–37, 1986.

Rüth, C., R. Well, and W. Roether, Primordial ^{3}He in South Atlantic deep waters from sources on the Mid-Atlantic Ridge, *Deep-Sea Res. I, 47*, 1059–1075, 2000.

Sauter D., L. Parson, V. Mendel, C. Rommevaux-Jestin, O. Gomez, A. Briais, C. Mevel, and K. Tamaki, TOBI sidescan sonar imagery of the very slow-spreading Southwest Indian Ridge: evidence for along-axis magma distribution, *Earth Plant. Sci. Lett., 202*, 511–512, 2002.

Scheirer, D. S., E. T. Baker, and K. T. M. Johnson, Detection of hydrothermal plumes along the Southeast Indian Ridge near the Amsterdam–St. Paul hotspot, *Geophys. Res. Lett., 25*, 97–100, 1998.

Scheirer, D. S., D. W. Forsyth, J. A. Conder, M. A. Eberle, S.-H. Hung, K. T. M. Johnson, and D. W. Graham, Anomalous seafloor spreading of the Southeast Indian Ridge near the Amsterdam–St. Paul Plateau, *J. Geophys. Res., 105*, 8243–8262, 2000.

Schroeder, T., B. John, and B. R. Frost, Geologic implications of seawater circulation through peridotite exposed at slow-spreading ridges, *Geology, 30*, 367–370, 2002.

Sinha, M. C., D. A. Navin, L. M. MacGregor, S. Constable, C. Peirce, A. White, G. Heinson, and M. A. Inglis, Evidence for accumulated melt beneath the slow-spreading Mid-Atlantic ridge, *Phil. Trans. R. Soc. Lond. A, 355*, 233–253, 1997.

Smallwood, J. R., and R. S. White, Crustal accretion at the Reykjanes Ridge, 61°–62°N, *J. Geophys. Res., 103*, 5185–5201, 1998.

Thurnherr, A. M., K. J. Richards, C. R. German, G. F., Lane-Serff, and K. G. Speer, Flow and mixing in the rift valley of the Mid-Atlantic Ridge, *J. Phys. Oceanog., 32*, 1763–1778, 2002.

Urabe, T., E. T. Baker, J. Ishibashi, R. A. Feely, K. Marumo, G. J. Massoth, A. Maruyama, K. Shitashima, K. Okamura, J. E. Lupton, A. Sonada, T. Yamazaki, M. Aoki, J. Gendron, R. Green, Y. Kaiho, K. Kisimoto, G. Lebon, T. Matsumoto, K. Nakamura, A. Nishizawa, O. Okano, G. Paradis, K. Roe, T. Shibata, D. Tennant, T. Vance, S. L. Walker, T. Yabuki, and N. Ytow, The effect of magmatic activity on hydrothermal venting along the superfast-spreading East Pacific Rise, *Science, 269*, 1092–1095, 1995.

Von Damm, K. L., M. K. Brockington, A. M. Bray, K. M. O'Grady, and SouEPR Science Party, SouEOR 98: Extraordinary phase separation and segregation in vent fluids from the Southern East Pacific Rise, *Eos Trans. AGU, 80*(46), Fall Meet. Suppl., F1098–F1099, 1999.

White, R. S., D. McKenzie, and R. K. O'Nions, Oceanic crustal thickness from seismic measurements and rare earth element inversions, *J. Geophys. Res., 97*, 19,683–19,715, 1992.

White, R. S., T. A. Minshull, M. J. Bickle, and C. J. Robinson, Melt generation at very slow-spreading oceanic ridges: Constraints from geochemical and geophysical data, *J. Petrology, 42*, 1171–1196, 2001.

Wright, D. J., R. M. Haymon, S. M. White, and K. Macdonald, Crustal fissuring on the crest of the southern East Pacific Rise at 17°15'–40'S, *J. Geophys. Res., 107*, doi:10.1029/2001JB000544, 2002.

Edward T. Baker, NOAA, Pacific Marine Environmental Laboratory, 7600 Sand Point Way NE, Seattle, WA 98115-6349, U.S.A. (e-mail: edward.baker@noaa.gov).

Christopher R. German, Southampton Oceanographic Centre, Empress Dock, Southampton, SO14 3ZH, UK.

Ultramafic-Hosted Hydrothermal Systems at Mid-Ocean Ridges: Chemical and Physical Controls on pH, Redox and Carbon Reduction Reactions

W. E. Seyfried, Jr., D. I. Foustoukos and D. E. Allen

Department of Geology and Geophysics, University of Minnesota, Minneapolis, Minnesota

Experimental, theoretical and field investigations of hydrothermal alteration processes in ultramafic systems at mid-ocean ridges, indicate that these systems have the capacity to buffer pH at surprisingly low values ($pH_{T,P}$ = 4.9–5.2), which profoundly affects fluid chemistry. Sluggish reaction kinetics of olivine at elevated temperatures and pressures, (e. g., 400°C, 500 bars), together with SiO_2 and Ca dissolution from coexisting pyroxene minerals, enhance the stability of tremolite and talc accounting for the observed acidity. Moreover, oxidation of ferrous silicate components in unstable minerals, especially pyroxenes, generates high $H_{2(aq)}$ concentrations, which together with the relatively low pH, increase Fe solubility, consistent with the Fe-rich nature of vents fluids issuing from ultramafic-hosted hydrothermal systems at Rainbow and Logatchev at 36°N and 14°N, respectively, on the Mid-Atlantic Ridge. The high dissolved Cu and Ni concentrations, and low $H_2S_{(aq)}$ of these vent fluids, indicate redox buffering by magnetite-bornite-chalcocite-heazelwoodite (Ni_2S_3)-fluid equilibria, as indicated by experimental and theoretical data. Data show that dissolved Cu is particularly sensitive to temperature change, while $H_2S_{(aq)}$ and Fe are affected less by this, although Fe is highly sensitive to pH and dissolved chloride. Dissolved chloride concentrations observed for both the Rainbow and Logatchev hydrothermal systems depart significantly from seawater and suggest supercritical phase separation in subseafloor reaction zones. The relatively high temperatures required for this, together with the high rates of fluid flow at Rainbow, indicate a magmatic heat source. The most unusual feature of fluids issuing from the Rainbow and Logatchev hydrothermal systems, however, involves high dissolved concentrations of methane and other hydrocarbon species, and detectable carbon monoxide. Experimental data indicate that reducing conditions and mineral catalytic effects may account for this, although the reported $CO_{(aq)}$ at Rainbow is well below predicted levels, suggesting re-equilibration at lower temperatures.

Mid-Ocean Ridges: Hydrothermal Interactions Between the Lithosphere and Oceans
Geophysical Monograph Series 148
Copyright 2004 by the American Geophysical Union
10.1029/148GM11

1. INTRODUCTION

It has long been recognized that serpentinized peridotites are prominent features in rift valleys of slow spreading ridges, especially where ridge-fracture zone intersection occurs [*Bougault et al.*, 1993; *Cannat*, 1993; *Cannat et al.*,

1992; *Cannat and Seyler*, 1995]. Geologic settings such as these likely enhance crustal permeability permitting seawater access to lower crust and upper mantle lithologies [*Bougault et al.*, 1998; *Rona et al.*, 1992]. Expressions of hydrothermal activity involving serpentinized peridotite have been long apparent from the anomalous concentrations of methane, and δ^3He in hydrothermal plumes in the water column-overlying ridge segments of the MAR [*Bougault et al.*, 1998; *Charlou et al.*, 1988; *Charlou et al.*, 1997; *Charlou et al.*, 1998]. With the discovery of high temperature vent fluids at the Rainbow and Logatchev systems at 36°N and 14°N on the MAR, respectively, however, the compositional evolution of seawater during reaction with ultramafic rocks has become more apparent [*Charlou et al.*, 1997; *Douville et al.*, 2002].

The Rainbow vent fluids reveal temperatures as high as 364°C and contrast sharply with the composition of vent fluids issuing from basaltic systems in a number of important ways [*Douville et al.*, 2002]. Firstly, these fluids are characterized by unusually high concentrations of methane, in keeping with plume studies, as noted above, and reveal dissolved H_2 concentrations 1–2 orders of magnitude greater than typical for basaltic systems (Figure 1). The high methane concentrations may result from Fischer Tropsch-Type (FTT) synthesis in which magmatic and/or seawater-derived $CO_{2(aq)}$ is reduced to methane and other reduced carbon phases. Indeed, *Holm and Charlou* [2001] have reported dissolved hydrocarbons in Rainbow vent fluids where the carbon chain lengths range from 16 to 29 atoms. Similarly, fluids venting from the ultramafic-hosted Logatchev system (14°45'N), where temperatures as high as 352°C have been reported [*Charlou et al.*, 2002], also reveal high $H_{2(aq)}$ and methane concentrations.

In addition to the high dissolved hydrocarbon and hydrogen concentrations of Rainbow and Logatchev vent fluids, these fluids also contain high dissolved transition metal concentrations. In fact, the Rainbow vent fluids contain the highest dissolved Fe concentrations of any vent system yet discovered (Figure 2). The relatively high dissolved $H_{2(aq)}$ concentrations undoubtedly contribute to the high Fe, although other factors are likely important [*Ding and Seyfried*, 1992]. The high dissolved Fe concentrations also contrast sharply with results of theoretical models assuming full equilibrium involving an ultramafic system of minerals at temperatures from 350–400°C, 500 bars [*Wetzel and Shock*, 2000]. Indeed, model results predict virtually no dissolved Fe in the aqueous fluid coexisting with peridotite and its alteration products. Either hydrothermal alteration processes occur at conditions distinct from those chosen for the computer simulation, or processes other than full equilibrium need to be considered to model accurately the composition of vent fluids at Rain-

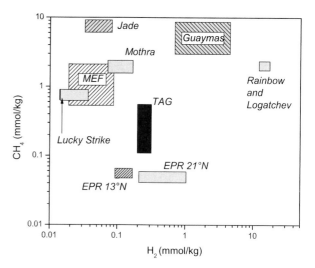

Figure 1. Distribution of dissolved $H_{2(aq)}$ and $CH_{4(aq)}$ in selected hydrothermal vent fluids at mid-ocean ridges. Vent fluids at the Main Endeavour Field and Guaymas Basin, Gulf of California, are clearly influenced by thermogenic decomposition of sedimentary organic matter, which tends to enrich the fluids in $CH_{4(aq)}$ and $H_{2(aq)}$, although dissolved $H_{2(aq)}$ is still controlled by rock-fluid interaction effects and maintained at relatively low values. The high concentrations of both and $H_{2(aq)}$ and $CH_{4(aq)}$ in fluids issuing from the Rainbow and Logatchev systems are unusual for non-sedimented hydrothermal systems and are likely the result of oxidation of ferrous silicates and carbon reduction reactions, respectively (see text) [*Charlou et al.*, 2002; *Charlou et al.*, 1993; *Donval et al.*, 1997; *Evans et al.*, 1988; *Holm and Charlou*, 2001; *Kelley et al.*, 1997; *Lilley et al.*, 1993; *Lilley et al.*, 1982; *Lilley et al.*, 1992; *Merlivat et al.*, 1987; *Von Damm*, 1991; *Von Damm and Bischoff*, 1987; *Welhan and Craig*, 1979; *Welhan and Craig*, 1983; *Welhan and Lupton*, 1987; *Welhan and Schoell*, 1988].

bow [*Allen and Seyfried*, 2003a]. Unlike Rainbow vent fluids, dissolved Fe concentrations in vent fluids at Logatchev are not unusually elevated [*Douville et al.*, 2002], although still significant (Figure 2).

One of the more interesting aspects of the chemistry of vent fluids at Rainbow and Logatchev, however, is the high dissolved Cu concentrations and distinctive mFe/mCu (molar) ratios [*Douville et al.*, 2002]. *Seyfried and Ding* [1993], for example, conducted a series of redox and pH buffered experiments involving phases in the $FeO-Fe_2O_3-CuO-HCl-H_2S-H_2O$ system where for pyrite-bearing assemblages and dissolved chloride equivalent to that of seawater, it was emphasized that the mFe/mCu ratio of the coexisting aqueous fluid increased with increasing reductive capacity. The observed increase in mFe/mCu ratio was largely the result of decreasing concentrations of dissolved Cu. Clearly, phase relations in ultramafic systems are reducing, but other aspects of fluid-mineral equilibria must off-

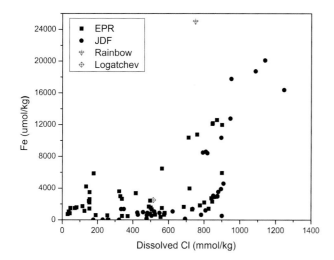

Figure 2. Dissolved Fe versus dissolved chloride for selected vent fluids at mid-ocean ridges. The observed increase in Fe with chloride is largely due to the tendency of Fe to from strong aqueous complexes with chloride at elevated temperatures. Redox and especially pH, however, also enhance Fe solubility (see text). Data shown are from the following sources: [*Bowers et al.*, 1988; *Butterfield and Massoth*, 1994; *Butterfield et al.*, 1994; *Campbell et al.*, 1988a; *Campbell et al.*, 1988b; *Charlou et al.*, 2002; *Charlou et al.*, 1996; *Douville et al.*, 2002; *Evans et al.*, 1988; *Michard*, 1990; *Oosting and Von Damm*, 1996; *Palmer*, 1992; *Trefry et al.*, 1994; *Von Damm*, 1995; *Von Damm*, 2000; *Von Damm et al.*, 1995].

set this to account for the observed high Cu. This takes on added significance, however, in light of recent observations of the temporal evolution of the composition of metaliferrous sediment in the immediate vicinity of the Rainbow system [*Cave et al.*, 2002]. These data show that the flux of Cu and Fe has remained elevated and relatively constant for approximately the past 8,000 years, which suggests constancy, as well, in the composition of these species in the vent fluid source.

Here we investigate pH and redox-controlling reactions that may play a role in accounting for the dissolved concentrations of Cu and Fe in fluids issuing from vents associated with ultramafic-hosted hydrothermal systems. These reactions may not only constrain Fe and Cu mobility during hydrothermal alteration processes, but may also influence mineral catalysis, which may play a role in the generation of reduced carbon species so prevalent in these vent fluids. In addition to redox, however, processes that contribute to chloride variability are also investigated, since, like redox, chloride variability affects metal mobility and provides clues to subsurface conditions so important to the overall chemical evolution of heat and mass transfer in hydrothermal systems at mid-ocean ridges.

2. REVIEW OF pH CONTROLLING REACTIONS DURING HYDROTHERMAL ALTERATION IN ULTRAMAFIC SYSTEMS

Allen and Seyfried [2003a] conducted experiments in the MgO-CaO-FeO-Fe$_2$O$_3$-Si O$_2$-Na$_2$O-H$_2$O-HCl system at 400°C and 500 bars to understand better compositional controls during hydrothermal alteration of ultramafic rocks. In particular, these investigators emphasized the lack of olivine reactivity at the conditions at which the experiments were performed, which is consistent with the well-known increase in olivine stability with increasing temperature (Figure 3). Olivine-serpentine conversion rate data [*Martin and Fyfe*, 1970] also reveal strong temperature dependence, with rate maxima at approximately 250°C (700–1000 bars). Thus, at temperatures lower and greater than this mass transfer is inhibited, which diminishes the role of olivine as a constraint on fluid chemistry. The same is not the case for orthopyroxene (i.e., enstatite), however, which, based on kinetic data [*Martin and Fyfe*, 1970], reveals a continuous increase in conversion rate with temperature for the range of temperatures investigated. Thus, for experiments involving both minerals at elevated temper-

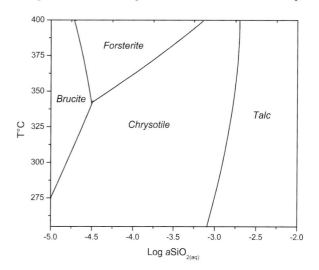

Figure 3. Phase equilibria in the MgO-SiO$_2$-H$_2$O system at elevated temperatures and 500 bars. The expansion of the forsterite field with increasing temperature is well known, and tends to inhibit forsterite dissolution and replacement by hydrous alteration phases. Although reaction kinetics and mineral solubility effects are distinctly different concepts, it is well established that mineral reaction rates slow for near equilibrium conditions. The broadening of the forsterite stability field expands the range over which the reaction rate can be expected to slow. These data provide a conceptual backdrop to the forsterite-serpentine conversion rate data reported by *Martin and Fyfe* [1970]. The figure was constructed with thermodynamic data for minerals and aqueous species using the SUPCRT92 database [*Johnson et al.*, 1992].

atures (>350°C), over expression of enstatite dissolution is to be expected. This is precisely what was observed in experiments performed by *Allen and Seyfried* [2003a], even when the relative abundance of olivine greatly exceeded enstatite as is typical in abyssal peridotite. An ion activity diagram that depicts phase relations in the $MgO-CaO-SiO_2-H_2O-HCl$ system at 400°C, 500 bars, illustrates this best (Figure 4). Here, numbered symbols "3" and "4" indicate experiments composed of enstatite and diopside (50:50 mass ratio) and olivine, enstatite and diopside (76:17:7), respectively. Fluids from both experiments indicate the existence of excess $SiO_{2(aq)}$, which is provided by pyroxene dissolution. This is particularly significant in the case of the olivine bearing experiment ("4") in that it indicates a clear inability of olivine to react with dissolved SiO_2 and generate chrysotile, in keeping with equilibrium phase relations. Thus, relatively SiO_2-rich fluids allow formation of SiO_2-rich alteration phases; talc and tremolite. This, in turn, provides a source of acidity, especially if the aqueous fluid contains relatively high Mg^{+2} and/or Ca^{+2} con-

centrations, as in the case for seawater derived hydrothermal fluid. It has long been recognized, for example, that seawater derived Mg^{+2} provides an important source of acidity for fluid coexisting with Mg-bearing hydrous alteration phases. In association with Ca-bearing minerals (see Table 1), however, as is typically the case for both basaltic and ultramafic systems, the acid generating potential of Mg^{+2} is transferred to Ca^{+2} provided there exists sufficient dissolved SiO_2 to render stable hydrous Ca-silicates, such as tremolite. Reaction (1) depicts one mechanism by which Mg^{+2} for Ca^{+2} exchange occurs, while reaction (2) illustrates the pH buffering provided by tremolite coexisting in SiO_2-bearing aqueous fluids:

$$2 \text{ Diopside} + 2 \text{ Enstatite} + Mg^{+2} + 2 H_2O \rightarrow Ca^{+2} + 0.25$$
$$\text{Tremolite} + 0.667 \text{ Chrysotile} + 0.167 \text{ Talc} \qquad (1)$$
and,
$$2 Ca^{+2} + 1.66 \text{ Talc} + 1.33 H_2O + 1.33 SiO_{2(aq)} =$$
$$\text{Tremolite} + 4 H^+ \qquad (2)$$

Clearly, reaction (2) is not possible in ultramafic systems if olivine effectively titrates dissolved SiO_2 to levels predicted assuming full equilibrium. This is definitely not the case on experimental time scales as indicted by results reported by *Allen and Seyfried* [2003a]. Whether or not the sluggishness of olivine reaction kinetics can be overcome in natural systems is best indicated by the dissolved SiO_2 concentration in vent fluids issuing from ultramafic-hosted hydrothermal systems. As noted previously, the Rainbow hydrothermal system is hosted in just such a lithology. The dissolved SiO_2 concentrations of the Rainbow fluids, however, are significant [*Douville et al.*, 2002]. Indeed, the measured SiO_2 concentrations are in excess of values observed in the experimental simulations [*Allen and Seyfried*, 2003a] (Figure 4), the likely abundance of olivine in the hydrothermal reaction zone notwithstanding. As with the experiments, however, the Rainbow vent fluids appear to plot near the talc-tremolite join, with tremolite likely the pH-controlling mineral, since the aqueous fluid contains high dissolved Ca^{+2} [*Douville et al.*, 2002]. Although pyroxene dissolution at Rainbow and Logatchev likely contributes to the relatively high dissolved SiO_2 concentrations in the vent fluids flowing from these systems, it is also possible that gabbroic bodies associated with the ultramafic rocks at depth are involved in this as well. The coexistence of serpentinized peridotite and gabbros is well recognized in many ultramafic rocks dredged from slow spreading ridges [*Cannat*, 1993; *Cannat et al.*, 1992], and such may be the case for the Rainbow and Logatchev. *Douville et al.* [2002] have in fact supported this inference based on the pattern of REE mobility in vent fluids at Rainbow, which reveals a prominent Eu anomaly and LREE enrichment, analogous to REE patterns from basalt/gabbro hosted hydrother-

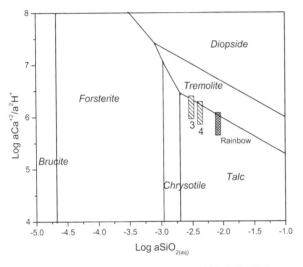

Figure 4. Phase relations in the CaO-MgO-SiO2-HCl-H2O system at 400°C, 500 bars depicting the stability fields of minerals likely in hydrothermally altered ultramafic-hosted hydrothermal systems. Symbols marked "3" and "4" refer to experiments performed by Allen and Seyfried [2003a], in which enstatite and diopside (50:50) and olivine, enstatite and diopside (76:17:7), respectively, were reacted with chloride bearing aqueous fluid. Dissolution of pyroxene minerals release dissolved SiO2(aq), which encourages precipitation of silica-rich secondary phases, such as talc and tremolite. It is likely that such a process accounts for the relatively high dissolved SiO2(aq) reported for Rainbow [Douville et al., 2002], as depicted here. Calculations were performed using the EQ3/6 software package [Wolery and Daveler, 1992] taking explicit account of recent upgrades in the SUPCRT92 database [Johnson et al., 1992]. Portions of this figure have been published elsewhere [Allen and Seyfried, 2003] and reproduced here with permission of Elsevier Press.

Table 1. Chemical composition and stoichiometry of minerals used in the text.

Mineral Name	Stoichiometry
Magnetite	Fe_3O_4
Pyrite	FeS_2
Pyrrhotite	FeS
Brucite	$Mg(OH)_2$
Forsterite	Mg_2SiO_4
Chrysotile	$Mg_2Si_3O_5(OH)_4$
Talc	$Mg_3Si_4O_{10}(OH)_2$
Enstatite	$MgSiO_3$
Tremolite	$CaMgSiO_3$
Millerite	NiS
Heazlewoodite	Ni_3S_2
Bornite	Cu_5FeS_4
Chalcopyrite	$CuFeS_2$
Chalcocite	Cu_2S

mal systems. *Allen and Seyfried* [2003b], however, have demonstrated that REE patterns such as these can be a function of constraints imposed by fluid chemistry (dissolved chloride, redox), and can not be taken, *a priori*, as evidence for interaction with plagioclase bearing substrates, all of which makes more complicated the role of magmatic heat sources associated with ultramafic-hosted hydrothermal systems. Regardless of the source of silica, however, it is still the slow rate of olivine reactivity at elevated temperatures that ultimately permits the relatively high $SiO_{2(aq)}$ concentrations to be achieved, with attendant effects on acid generation.

Fluid pH in subseafloor reaction zones in ultramafic-hosted hydrothermal systems can be calculated from mineral solubility constraints, assuming temperature, pressure and dissolved concentration of key aqueous species. For example, if we assume magnetite-fluid equilibria, which is consistent with the relatively high dissolved $H_{2(aq)}$ concentrations observed at Rainbow and Logatchev [*Charlou et al.*, 2002; *Donval et al.*, 1997], pH values of approximately 4.6 to 4.9 and 5.0 to 5.3 are predicted for temperatures of 350 to 400°C for Rainbow and Logatchev vent fluids, respectively, when explicit account is taken of dissolved Fe, chloride and $H_{2(aq)}$ concentrations together with distribution of aqueous species calculations (Figure 5). The range of pH values estimated for the reaction zone at Rainbow is distinctly more acidic than neutrality, which provides a sense of the effectiveness of H^+ generation by Ca-fixation reactions (tremolite formation) in the relatively $SiO_{2(aq)}$ rich fluids. The same is true for Logatchev, although pH is higher due to lower dissolved chloride and Fe concentrations. Moreover, the predicted range of pH values at both sites contrasts sharply with fluids involved in hydrothermal alteration of ultramafic rocks at lower temperatures where values in excess of 10 are possible, if not probable [*Kelley et al.*, 2001], due to the effects of active olivine hydrolysis. As might be expected by now, the relatively low pH estimated

for Rainbow and Logatchev using a partial equilibrium approach (magnetite-fluid equilibria) contrasts with that predicted for ultramafic hydrothermal systems at elevated temperatures and pressures assuming *full* equilibrium in the $MgO-H_2O-SiO_2$ system. Indeed, results based on the assumption of full equilibrium yield pH values on the order of 6.5 [*Wetzel and Shock*, 2000]. Thus, the lack of Fe and $SiO_{2(aq)}$ is not surprising, but very different from what is actually observed from the field or measured in the lab.

3. REDOX CONSTRAINTS IN HIGH-TEMPERATURE REACTION ZONES IN ULTRAMAFIC-HOSTED HYDROTHERMAL SYSTEMS

Vent fluids at Rainbow and Logatchev reveal some of the lowest dissolved $H_2S_{(aq)}$ concentrations yet reported for high temperature hydrothermal fluids at mid-ocean ridges [*Douville et al.*, 2002]. Considering the potential importance of $H_2S_{(aq)}$ on hydrothermal alteration processes and metal mobility, it

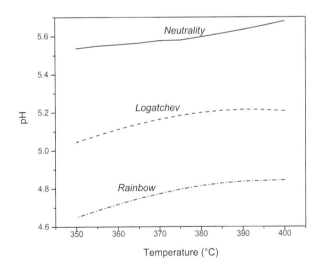

Figure 5. Predicted pH values in subseafloor reaction zones at Rainbow and Logatchev assuming magnetite-fluid equilibria at 350 to 400°C, 500 bars. In addition to temperature and pressure, determination of fluid pH requires knowledge of total dissolved Fe, chloride and $H_{2(aq)}$, which were fixed at values of 25 and 2.5 mmol/kg, 750 and 515 mmol/kg, and 12 and 16 mmol/kg for Rainbow and Logatchev, respectively [*Charlou et al.*, 2002; *Donval et al.*, 1997; *Douville et al.*, 2002]. The predicted range of pH values is distinctly acidic as indicated by the neutrality line over a similar temperature range. Although constraints imposed by magnetite solubility can be used to calculate the pH in reaction zone fluids, the cause of the acidity is fundamentally linked to the relatively high $SiO_{2(aq)}$ concentrations coexisting with tremolite (see text), which enhance Ca-fixation reactions. Calculations were performed using the EQ3/6 software package [*Wolery and Daveler*, 1992] taking explicit account of recent upgrades in the SUPCRT92 database [*Johnson et al.*, 1992].

is essential to more quantitatively investigate the cause and effect of the observed low $H_2S_{(aq)}$ concentrations.

Phase relations in the $FeO-Fe_2O_3-CuO-H_2S-H_2O-HCl$ system provide a useful frame of reference to assess the relative stability of Fe and Cu oxide and sulfide minerals (Table 1) at conditions likely for high temperature reaction zones at Rainbow and Logatchev (Figure 6). In general, these data indicate that for the high H_2 and low H_2S concentrations measured for the vent fluids, magnetite is stable, while pyrrhotite, and especially pyrite, can be ruled out as components of the subseafloor alteration assemblage. This contrasts with basaltic rocks where these phases can be shown to play prominent roles in hydrothermal alteration processes, especially during incipient stages of alteration [*Seyfried and Ding*, 1995].

Magnetite formation is linked to the oxidation of ferrous silicate components in the ultramafic rock. It is important to again emphasize, however, that the extent to which this occurs is very much a function of temperature. At the relatively high temperatures envisaged for reaction zones at Rainbow and Logatchev (>400°C), irreversible dissolution of olivine- the primary Fe-bearing silicate in the system, is unlikely. As a consequence of this, extremely reducing conditions charac-

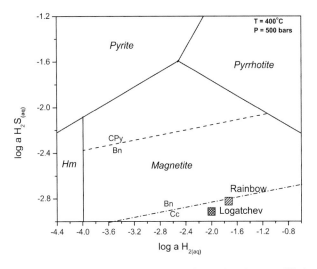

Figure 6. Activity-activity diagram depicting redox phase equilibria for the $FeO-Fe_2O_3-CuO-H_2S-H_2O-HCl$ system at 400°C, 500 bars. The relatively low $H_2S_{(aq)}$ and high $H_2_{(aq)}$ conditions that characterize Rainbow and Logatchev vent fluids strongly suggest magnetite-bornite-chalcocite-fluid equilibria (see text), as indicated by the position of the symbols. Clearly, the high $H_2_{(aq)}$ and low $H_2S_{(aq)}$ effectively rule out the presence of pyrite and pyrrhotite in high temperature subseafloor reaction zones in ultramafic systems. Mineral abbreviations are as follows: Hm, hematite; Bn, bornite; cc, chalcocite; Cpy, chalcopyrite. Calculations were performed using the EQ3/6 software package [*Wolery and Daveler*, 1992] taking explicit account of recent upgrades in the SUPCRT92 database [*Johnson et al.*, 1992].

teristic of low temperature serpentinization where olivine is often completely replaced by serpentine and magnetite [*Alt and Shanks*, 1998], are not to be expected. This does not imply the fluids are not reducing, clearly, the measured $H_2_{(aq)}$ concentrations provide adequate evidence of this, only that the level is moderated by the relative stability of olivine at reaction zone conditions. $H_2S_{(aq)}$ concentrations, which are invariably linked to $H_2_{(aq)}$, as in all geochemical systems, are low, apparently due to constraints imposed by sulfur availability in the rock. For example, taking explicit account of oxidation effects in serpentinites from the Hess Deep, *Alt and Shanks* [1998] estimated sulfur concentrations of 120 ± 70 ppm for the least altered rocks, which is slightly below that proposed for the fertile mantle [*Lorand*, 1991; *Hartmann and Wedepohl*, 1993], and in combination provides a key constraint on the base sulfur budget of the rock prior to alteration. Thus, the relatively low $H_2S_{(aq)}$ concentrations of Rainbow and Logatchev vent fluids are not surprising, in spite of the relatively high $H_2_{(aq)}$ and high temperatures in the subseafloor reaction zone. Together with magnetite, these conditions impose constraints on the composition of Cu-bearing phases, which likely include chalcocite and bornite (Figure 6 and Table 1), although uncertainties in temperature, pressure and supporting thermodynamic data preclude unambiguous interpretation. Assuming the coexistence of both phases, however, together with constraints imposed by redox, pH and dissolved chloride, dissolved concentrations of Cu and mFe/mCu (molar) ratio of the coexisting fluid can be calculated, as discussed below.

3.1. NiO-FeO-H_2O-H_2S-H_2 System

Owing to the relative abundance of Fe and Ni in peridotite [*Alt and Shanks*, 1998; *Alt and Shanks*, 2003; *MacLean*, 1977], it is apparent that these components in the form of compositionally variable Fe and Ni sulfides, oxides and possibly native metal alloys respond to redox constraints imposed by dissolved $H_2_{(aq)}$ and $H_2S_{(aq)}$. For example, *Frost* [1985] demonstrated from available thermodynamic data that irreversible dissolution of olivine, which occurs at temperatures below approximately 350°C, 500 bars, can result in $\log fO_{2(g)}$ values sufficiently low to stabilize a number of Ni-Fe alloys and sulfur-poor phases, such as heazelwoodite (Ni_3S_2) and millerite (NiS). In effect, olivine dissolution provides Fe for magnetite formation and H_2 generation, while Ni, also released during olivine hydrolysis, is available for Ni-Fe alloy formation, when H_2 becomes sufficiently elevated to stabilize native metal alloys.

At any temperature, phase relations in the $Fe-Ni-H_2S-H_2O$ system are exceedingly complex and poorly constrained thermodynamically, which often prevents full compositional variability from being expressed. Moreover, the dearth of data for activity-composition relations for native metals and sulfide phases underscores the uncertainties attending geochemical modeling of this compositional system, especially at elevated temperatures and pressures. Thus, applications must be restricted compositionally and results viewed with caution, as emphasized by *Frost* [1985]. Owing to these constraints, it is difficult to construct completely and unambiguously a diagram illustrating phase relations in the Ni-bearing system at conditions appropriate for reaction zones at Rainbow and Logatchev. Data that are available, however, suggest a broad field of stability for heazelwoodite. In the presence of this phase, the native metal field is restricted in H_2-H_2S space to highly reducing, low H_2S conditions (Figure 7). In fact, the presence of H_2S at concentrations nearly a 1000 times lower than observed for vent fluids at Rainbow or Logatchev is still sufficient to preclude native Ni, assuming measured H_2 concentrations. It is likely that even lower H_2S and higher H_2 conditions would be necessary if explicit account of solid solution effects involving Fe-Ni alloys were considered [*Frost*, 1985]. Native Ni-Fe alloys have been reported for oceanic serpentinites from the Hess Deep [*Alt and Shanks*, 1998],

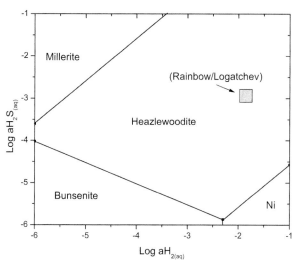

Figure 7. Activity-activity diagram depicting phase relations in the NiO-H_2S-H_2O-HCl system at 400°C, 500 bars. Heazelwoodite displays a broad range of stability in $H_{2(aq)}$-$H_2S_{(aq)}$ space and likely controls Ni solubility in subseafloor reaction zones in ultramafic-hosted hydrothermal systems. The stability of this phase effectively precludes the existence of Ni-Fe alloys at temperatures likely for the reaction zone at Rainbow and Logatchev, although this may not be the case at lower temperatures and pressures where dissolved H_2 and H_2S concentrations can be expected to be higher and lower, respectively. Data for $H_{2(aq)}$ and $H_2S_{(aq)}$ for Logatchev and Rainbow are from *Douville et al.* [2002] and *Charlou et al.* [2002]. Calculations were performed using the EQ3/6 software package taking explicit account of recent upgrades in the SUPCRT92 database (see text).

which to us implies temperatures of alteration sufficiently low to maximize the redox potential imposed by the irreversible dissolution of olivine.

3.2. Fe/Cu Ratio

Previously, we have shown that the low pH and high $H_{2(aq)}$ concentrations in Rainbow and Logatchev vent fluids contribute to the observed high Fe concentrations. It is the low H_2S concentrations, however, that are as distinctive as the high Fe [*Douville et al.*, 2002]. For example, recent experimental data [*Foustoukos and Seyfried*, 2004] have shown that dissolved $H_2S_{(aq)}$ tends to correlate inversely with dissolved Cu for relatively oxidizing systems and the same is likely also true for reducing systems, such as at Rainbow and Logatchev. Clearly, Cu concentrations of vent fluids issuing from these ultramafic-hosted hydrothermal systems [*Douville et al.*, 2002] are in excess of values typical of most vent fluids from basaltic hydrothermal systems where $H_2S_{(aq)}$ concentrations are relatively high [*Seyfried and Ding*, 1993; *Von Damm*, 1995].

To better illustrate the effect of redox relations on mFe/mCu (molar) ratios in fluids coexisting with minerals in the FeO-Fe_2O_3-CuO-H_2O-H_2S-HCl system (Table 1), we constructed an ion ratio activity diagram at 400°C, 500 bars depicting the extent to which the aFe^{+2}/aCu$^+$ ratio changes with aH$_2$S$_{(aq)}$ (Figure 8). For $H_2S_{(aq)}$ concentrations reported for Rainbow or Logatchev, aFe^{+2}/aCu$^+$ ratio is compatible with phase equilibria involving magnetite-bornite-chalcocite (Figs. 7 and 9). In effect, dissolved $H_{2(aq)}$ and $H_2S_{(aq)}$ buffer aFe^{+2}/aCu$^+$ at a low value, which can be expected to enhance the solubility of Cu relative to Fe—all else being equal. Moreover, aFe^{+2}/aCu$^+$ versus aH$_2$S$_{(aq)}$ data (Figure 8) clearly show how far removed the Rainbow or Logatchev vent fluids are from pyrite and chalcopyrite stability fields. Thus, from these data, one can infer very different Fe-Cu solubility patterns for basaltic and ultramafic-hosted hydrothermal systems.

Taking explicit account of distribution of aqueous species at 400°C, 500 bars, while assuming magnetite-bornite-chalcocite-fluid equilibria at a pH of 5 and 0.75 mol/kg chloride, we can calculate dissolved Fe and Cu concentrations, which can then be compared with measured data for vent fluids at Rainbow [*Douville et al.*, 2002]. Model results indicate Fe and Cu concentrations of approximately 16 and 1 mmol/kg, respectively (Figure 9a). Higher pH values and/or lower temperatures result in a dramatic decrease in both species, although Fe is more sensitive to pH, while Cu is more sensitive to temperature (Figure 9a and 9b). That Rainbow vent fluids reveal Fe concentrations actually higher than illustrated here strongly suggests lower pH values or higher temperatures in the hydrothermal reaction zone from which the fluids are derived [*Allen and Seyfried*, 2003a]. The predicted Cu concentration

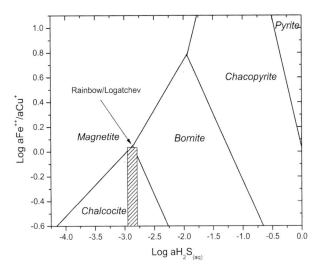

Figure 8. Activity-activity diagram depicting phase relations in the FeO-Fe$_2$O$_3$-CuO-H$_2$O-H$_2$S-HCl system at 400°C, 500 bars. The coexistence of magnetite-bornite-chalcocite-fluid equilibria at dissolved H$_{2(aq)}$ reported for the Rainbow system (~15 mmol/kg) constrains log aFe^{+2}/aCu$^+$ and H$_2$S$_{(aq)}$. The relatively low H$_2$S$_{(aq)}$ predicted and measured [*Douville et al.*, 2002]enhances the likelihood for Cu mobility in ultramafic-hosted hydrothermal systems, in good agreement with vent fluid data from the Rainbow and Logatchev systems (see text). Calculations were performed with SUPCRT92 [*Johnson et al.*, 1992].

at 400°C and pH 4.9, however, is in excess of that observed by approximately a factor 8. The most likely interpretation of this is precipitation of Cu during fluid ascent to the seafloor. It is of interest to note, that the Cu concentration predicted for a temperature of approximately 360–370°C, is virtually identical to that actually measured (~140 um/kg), as is the mFe/mCu (molal) ratio, which is predicted to increase with decreasing temperature due largely to the temperature dependent changes in dissolved Cu (Figure 9b). This clearly implies that a large proportion of the dissolved inventory of Cu initially mobilized at depth at Rainbow and/or Logatchev accumulates in chimney and stock-work deposits at or beneath the seafloor. *Fouquet et al.* [2003] have reported the widespread distribution of Cu-rich chimney material associated with the Rainbow vent field, confirming the localized accumulation very near the seafloor. It is noteworthy as well that the Cu-rich deposits also reveal enrichment of Ni, in the form of Ni-sulfides [*Fouquet, personal communication*]. We hypothesize that the low H$_2$S$_{(aq)}$ moderately acidic and reducing conditions that are so important to the transport of Cu in the ultramafic-hosted hydrothermal systems, enhance Ni transport as well. We have already documented the likely existence of heazelwoodite in the subseafloor reaction zone from phase equilibria constraints, and it is likely that the solubility of this phase at elevated temperatures and pressures controls Ni trans-

port at Rainbow and Logatchev. In addition to Ni-mineralization, vent fluids at Rainbow are significantly enriched in Ni relative to fluids at Logatchev and numerous other hydrothermal systems [*Douville et al.*, 2002]. The lower temperature and lower dissolved chloride of the Logatchev vent fluids may account for the lower dissolved Ni concentrations, while the relative lack of Ni in basalt/gabbro may constrain Ni solubility in these systems.

Additional evidence for the transport of significant Cu and Fe in fluid venting from the ultramafic-hosted Rainbow hydrothermal system can be discerned from the study of metaliferous sediments in the near and far field around the Rainbow hydrothermal system. For example, *Cave et al.* [2002] conducted a detailed geochemical investigation of sediment cores collected directly beneath the hydrothermal plume at variable distances from the Rainbow vent site. The cores nearest the vent reveal anomalous enrichments in Cu and Fe, and relatively low mFe/mCu ratio, while the mFe/mCu ratio increases with increasing distance from the plume source. Thus, once again we see evidence of the precipitation efficiency of Cu. What is particularly noteworthy about these data is that the mFe/mCu ratio of the sediments has remained relatively constant for the past approximately 8 to 12 kyr, although low sedimentation rates and bioturbation effects do not permit distinction of whether or not venting has been continuous or episodic over that time. [*Cave et al.*, 2002]. In any event, that metals such as Fe and Cu, the solubility of which is so sensitive to temperature, pH, redox and chloride, would accumulate in such abundance over so many years provides important clues to the nature and mechanism of heat and mass transfer in ultramafic-hosted hydrothermal systems at mid-ocean ridges.

4. CONTROLS ON DISSOLVED CHLORIDE AND TEMPERATURE

Hydrothermal vent fluids at Rainbow and Logatchev reveal a range of dissolved chloride concentrations that help to constrain the chemical and physical conditions that these fluids have experienced along their respective flow paths. Although vent fluids at Rainbow and Logatchev have similar temperatures, 353°C and 364°C, respectively, [*Charlou et al.*, 2002; *Douville et al.*, 2002], dissolved chloride concentrations are significantly different. For example, dissolved chloride concentrations of Rainbow vent fluids (750 mmol/kg) are significantly greater than seawater, while vent fluids at Logatchev indicate lower than seawater chloride concentrations (515 mmol/kg) [*Douville et al.*, 2002]. As with most hydrothermal systems at mid-ocean ridges, however, chloride variability is likely related to phase separation in high temperature reaction zones, which in turn

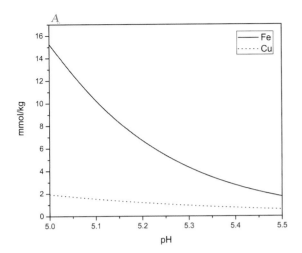 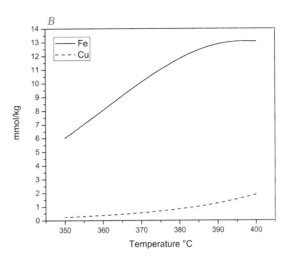

Figure 9. Predicted concentrations of dissolved Fe and Cu as a function of pH at 400°C (a) and temperature at pH 5 (b), assuming magnetite-bornite-chalcocite-fluid equilibria. The effect of pH on Fe solubility and temperature on Cu solubility is particularly noteworthy (see text). Calculations were performed using the EQ3/6 software package [*Wolery and Daveler*, 1992] taking explicit account of recent upgrades in the SUPCRT92 database [*Johnson et al.*, 1992].

suggests the existence of a robust heat source, although to our knowledge no geophysical expression of a magma lens has been imaged or inferred at sites on the MAR near the Logatchev or Rainbow hydrothermal systems.

At temperatures below the critical point of seawater (~408°C), the process of phase separation is equivalent to boiling. Thus, a low-density vapor separates from a higher-density liquid. Usually such a process is to be expected at relatively shallow ridge sections, and most often characterized by extremely low chloride vent fluids, especially when the corresponding fluid temperatures are high, as is often the case following diking or volcanic events [*Von Damm et al.*, 1997; *Von Damm et al.*, 1995]. If phase separation is assumed for both the Logatchev and Rainbow sites, the relatively high dissolved chloride concentrations of the vent fluids, especially at Rainbow, suggest supercritical conditions. Thus, a small amount of brine separates from the liquid, causing the liquid to become more vapor like [*Bischoff*, 1991; *Bischoff and Pitzer*, 1989]. The actual separation mechanism is referred to as brine condensation [*Bischoff and Rosenbauer*, 1987; *Fournier*, 1986]. The lack of pressure and/or temperature constraint at the site of phase separation for Logatchev or Rainbow, makes it difficult to model unambiguously the origin of the dissolved chloride other than to infer conditions greater than approximately 300 bars and 408°C, respectively (the critical point of seawater). It is clear, however, that if the zone of phase separation were greater than 2 km or so below the seafloor (~450 bars), relatively high temperatures would be needed to account for the dissolved chloride concentration of the vapor phase [*Bischoff and Pitzer*, 1989]. This has important implications for the nature of the heat

source, since a non-magmatic source would likely require relatively great depths of circulation beneath the vent sites at the seafloor [*Bach et al.*, 2002].

Although the relatively low chloride vent fluid at Logatchev can be accounted for by direct ascent of a supercritical vapor directly derived from a seawater source fluid, the same cannot be the case at Rainbow. At Rainbow, mixing of brine with variable amounts of evolved seawater or a supercritical Cl-bearing vapor is likely. Alternatively, the relatively chloride rich vent fluids at Rainbow could also be accounted for by separation of a Cl-bearing vapor from a brine enriched in chloride from previous phase separation events. Mixing reactions of this type involving brines and vapors generated by phase separation processes have often been proposed to account for chloride variability in hydrothermal vent fluids issuing from basaltic systems [*Butterfield et al.*, 1994; *Von Damm and Bischoff*, 1987], and perhaps the same is true for ultramafic-hosted hydrothermal systems. Possible existence of deep-seated concentrated brine at Rainbow not only would suggest temperatures considerably greater than observed at the present site of venting, but would also require an earlier stage of venting characterized by relatively low chloride fluids, perhaps more analogous to the fluids presently venting at Logatchev.

Assuming supercritical phase separation to account for the chloride variability observed for vent fluids at Rainbow and Logatchev, significant heat loss must then occur during fluid circulation and upon ascent to the seafloor. Part of this may be conductive, although mixing reactions are more likely, especially considering the chloride variability at Rainbow. As noted earlier, if significant heat loss were to occur this would affect the dissolved concentration of transition metals, especially Cu, although other species, such as dissolved $H_2S_{(aq)}$ might remain relatively unchanged provided fluid-mineral equilibria can be maintained (Figure 10). This is important because of the significance of $H_2S_{(aq)}$ as a primary redox indicator of hydrothermal alteration processes.

The high temperatures (>365°C) and fluid flow rates (500 kg/sec) at Rainbow [*Douville et al.*, 2002; *Thurnherr and Richards*, 2001], suggest the existence of a significant source of heat, which is consistent with phase separation processes controlling the chloride variability in vent fluids, as noted above. The magnitude of heat required to fuel the Rainbow system, however, is considerable indeed. When vent temperature and fluid flow rate are considered with constraints imposed by metal accumulation data for the Rainbow system, which suggest vigorous high-temperature venting for at least 8 kyr [*Cave et al.*, 2002], a time integrated heat flux of approximately 3×10^{20} Joules is indicated. This amount of heat is approximately an order of magnitude greater than estimated for the TAG hydrothermal system [*Humphris and Cann*, 2000]. The magnitude of the heat flux estimated for Rainbow clearly rules out a steady state magmatic heat source for any crustal segment and spreading rate reasonable for the MAR, suggesting an alternative or additional heat source. One such heat

source that may be unique to slow spreading ridges, however, involves heat mined from the lithospheric mantle [*Bach et al.*, 2002; *German and Lin*, 2003]. Calculations by *German and Lin* [this volume] suggest that to satisfy the thermal constraints imposed by uninterrupted venting at Rainbow for 10,000 years would require extraction of heat from a crustal segment (~6–8 km) emplaced within this time (~260 m) along ~100 km of the MAR. Although the tectonically active slow spreading MAR may provide effective conduits for seawater to access these deep (hot) lithospheric units where temperatures are sufficiently high for supercritical phase separation in the NaCl-H_2O system, the precise mechanism by which this occurs, is unclear. There can be no question as to the need for further investigation of heat and mass transport along tectonically active, slow spreading ocean ridges.

4.1. Serpentinization as a Possible Heat Source at Rainbow or Logatchev

An alternate or supplementary heat source that may be applicable to the Rainbow and Logatchev systems involves heat released during serpentinization reactions [*Lowell and Rona*, 2002; *Macdonald and Fyfe*, 1985]. *Lowell and Rona* [2002], for example, showed from heat balance calculations that the heat generated by serpentinization increases with decreasing fluid flow rate and increasing rates of serpentinization for a given background heat flux. In effect, the lower the fluid/rock mass ratio for serpentinization, the greater the potential heat generation relative to an imposed geotherm. With increasing temperature, however, the thermodynamic drive for serpentinization decreases, which is consistent with the observed decrease in rate of serpentinization recognized from experimental studies [*Allen and Seyfried*, 2003a; *Martin and Fyfe*, 1970]. We can illustrate these effects by showing the predicted change in temperature (ΔT) for one or ten kilograms of fluid involved in the conversion of one kilogram of olivine to serpentine ±brucite at a range of initial temperatures (50–400°C) (Figure 11). As expected, the lower the fluid/rock mass ratio the greater the potential temperature increase. Furthermore, temperature maxima are predicted for fluids having an initial temperature of approximately 250°C, due to a combination of factors involving the enthalpy of reaction and heat capacity effects. Even under the most favorable conditions, however, the temperature gain is unlikely to be greater than 50°C [*Allen and Seyfried*, 2003c]. Moreover, in real or open systems, complete conversion of olivine to serpentine is unlikely, especially at relatively low and high temperatures, as suggested by the rate data of *Martin and Fyfe* [1970]. If we assume that these data limit serpentinization, then significantly lower temperature gains are indicated, but once again greatest for fluids with an initial temperature

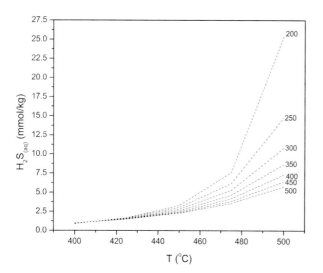

Figure 10. Predicted dissolved $H_2S_{(aq)}$ concentrations assuming coexistence of magnetite-bornite-chalcocite-fluid equilibria as a function of temperature and pressure. $H_{2(aq)}$ is assumed to be 16 mmol/kg in keeping with Rainbow vent fluids [*Charlou et al.*, 2002]. In contrast with other species (e.g., Cu and Fe), data indicate that $H_2S_{(aq)}$ concentrations would not be expected to change greatly during fluid ascent to the seafloor from ultramafic reaction zone provided equilibrium is maintained (see text). This is important in terms of using $H_2S_{(aq)}$ to constrain subseafloor hydrothermal alteration processes.

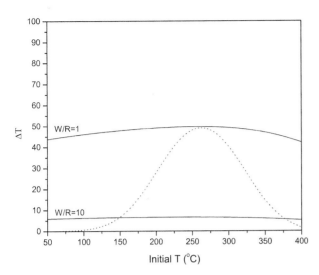

Figure 11. Predicted change in temperature due to heat of serpentinization as a function of initial reaction temperature using the equation, 2 Forsterite + 3 Water = Chrysotile + Brucite, and a simple heat balance model modified after that developed by Lowell and Rona [2002]. Thus, where H is the enthalpy of the reaction in Jkg^{-1}, Cw and Cr ($Jkg^{-1}°C^{-1}$) are the heat capacity of water and rock (olivine) respectively, W/R is the fluid/rock mass ratio, while λ is an empirical conversion factor that varies from 1 (100% serpentinization) to 0 (0% serpentinization) (see text). The solid lines are drawn for a W/R of 1 and 10 and are constructed assuming complete serpentinization (λ=1) at all temperatures. The dashed curve, however, is constructed taking explicit account of the temperature dependence of reaction rate data for olivine to serpentine + brucite as determined by Martin and Fyfe [1970]. The temperature change maxima depicted by the curve corresponds to an initial temperature of approximately 250°C, where it intersects the solid line indicating complete serpentinization at a W/R = unity.

$$\Delta T = \frac{H}{\left(\dfrac{W}{R\lambda} C_w + C_r \right)}$$

of approximately 250°C (Figure 11). Although the conversion rate data are based on lab experiments, and thus, not strictly applicable to subseafloor sites of serpentinization, because of uncertainties in assigning time integrated mineral surface areas to natural systems, the data still illustrate the highly non-linear feedback between serpentinization rates and temperature that almost certainly do exist and would need to be accounted for in fully coupled reaction path models. The same is true of conductive cooling, which may further limit the effectiveness of exothermic mineralization reactions as a source of heat in ultramafic-hosted hydrothermal systems [*Lowell and Rona*, 2002]. Thus, it is highly unlikely that serpentinization reactions can significantly affect the temperature of vent fluids in the Rainbow or Logatchev hydrothermal systems. The temperatures and fluid flow rates for these systems are simply too high to allow any significant heat pro-

duction linked directly to the exothermic nature of the olivine hydrolysis reaction. As inferred earlier, heat sources related to deep lithospheric cooling likely dominate the origin and evolution of these hydrothermal systems.

Although the high temperatures and high flow rates of vent fluids at Rainbow and Logatchev argue against exothermic heat sources from serpentinization reactions, it has been proposed recently that it is precisely these sorts of reactions that may fuel hydrothermal circulation at the relatively low temperature Lost City hydrothermal system, a vent site approximately 15km west of the MAR at 30°N [*Kelley et al.*, 2001]. These vent fluids, which do not exceed approximately 75°C, while issuing from chimney structures composed of carbonate minerals and brucite, are more in keeping with exothermic heating effects, provided alteration takes place at relatively low fluid/rock mass ratios. *Allen and Seyfried* [2003c], however, have shown from constraints imposed by geochemical modeling calculations and dissolved Cl, K and Na concentrations in Lost City vent fluids that little hydration has occurred. Apparently, the fluid/rock mass ratio associated with serpentinite formation at Lost City is sufficiently high to effectively compensate for hydration effects. Since exothermic heating necessarily involves conversion of olivine to serpentine ± brucite, the relative lack of hydration effectively precludes this at Lost City, at least based on the chemistry of vent fluids sampled to date. Moreover, the geochemical modeling calculations performed by Allen and Seyfried [2003c] actually suggest temperatures as high as approximately 200°C in the source region from which the Lost City fluids are derived. Thus, in spite of the relatively low vent temperatures at Lost City, and off axis location, a more distal heat source is suggested. Once again, the tectonically active nature of the slow spreading MAR may provide fluid flow paths facilitating heat and mass transfer over relatively long distances.

5. CARBON REDUCTION REACTIONS AND MINERAL CATALYSIS IN ULTRAMAFIC-HOSTED HYDROTHERMAL SYSTEMS

The conspicuous abundance of reduced carbon species in vent fluids issuing from ultramafic-hosted hydrothermal systems is an important feature that distinguishes these vent systems from their basaltic counterparts (Figure 1). Although some fraction of these species may be derived directly from leaching of mantle hydrocarbons [*Sugisaki and Mimura*, 1994], it is also possible, if not probable, that the abundance of methane and other dissolved hydrocarbons in the Rainbow vent fluids, for example, are the result of Fischer Tropsch Type (FTT) synthesis [*Charlou et al.*, 1998; *Holm and Charlou*, 2001], in which, oxidized forms of dissolved carbon are reduced to hydrocarbon species. Thus, in recent years, there has been a significant effort to better understand the mechanism by which carbon reduction reactions occur at conditions likely in hydrothermal systems hosted in ultramafic lithologies. Clearly, the high $H_{2(aq)}$ concentrations that are typically associated with these systems can be expected to enhance carbon reduction, as follows:

$$CO_{2(aq)} + 4H_{2(aq)} = CH_{4(aq)} + 2H_2O \qquad (3)$$

Indeed, *Janecky and Seyfried* [1986] reported methane formation during a series of peridotite alteration experiments at 300°C. The methane generated, however, was reported only for the final sample from the more than 2000-hour experiment, precluding unambiguous interpretation of reaction mechanism. *Berndt et al.* [1996] also investigated hydrocarbon generation during serpentinization and reported the production of modest amounts of methane and trace levels of other alkanes. The distribution of the hydrocarbon products suggested a mineral catalyzed reaction scheme, likely involving magnetite. Magnetite has often been proposed to play a role in carbon reduction and methane generation in both subareal and submarine settings [*Kelley*, 1996; *Salvi and Williams-Jones*, 1996; *Sherwood Lollar et al.*, 2002], especially when the mineral surface achieves a cation-excess condition (non-stoichiometric abundance of FeO), which is thought to facilitate electron transfer [*Madon and Taylor*, 1981; *Schultz*, 1999; *Tamaura and Tabata*, 1990]. Even under extremely reducing conditions, where cation-excess magnetite formation can be expected, the rate of methane generation is slow on experimental time scales. For example, *McCollom and Seewald* [2001], repeating the earlier study of Berndt et al. [1996] using isotopic doping techniques to avoid possible ambiguities introduced by carbon contamination effects, reported bicarbonate to methane conversion rates of only ~0.05% for a magnetite-bearing experiment at 300°C. The magnetite in this experiment, however, was derived during the normal course of serpentinization of olivine, in a manner consistent with procedures used by *Berndt et al.* [1996]. Thus, the coexisting aqueous fluid contained elevated dissolved $H_{2(aq)}$ concentrations (~60 mmol/kg), which increased with reaction progress. More recently, Foustoukos and Seyfried [2002] also performed experiments, which involved magnetite and an aqueous CO_2-bearing fluid. These experiments were performed at 390°C, 400 bars, while the dissolved $CO_{2(aq)}$ was isotopically labeled with anomalous ^{13}C, and buffered with H_2 concentrations as high as 150 mmol. The relatively high temperature at which the experiments were conducted was motivated by reaction zone conditions inferred for Rainbow, although the experiments bear on carbon reduction processes in general, in a wide range of chemical systems. Results of the experiments indicate that the

combination of increasing temperature and $H_{2(aq)}$ increases the rates of carbon reduction. After approximately 3000 hours of reaction, $C^{13}O_{2(aq)}$ conversion to $C^{13}H_{4(aq)}$ (methane) exceeded ~0.95%, although the rate at which this occurred clearly depended on dissolved $H_{2(aq)}$ (Figure 12), as might be expected.

The role of mineral catalysts has been investigated in a number of studies. *Horita and Berndt* [1999] investigated the role of Ni-Fe alloy on abiotic methane formation at 200–400°C, 500 bars. Results showed very significant methane generation, especially at the intermediate temperature of 300°C where conversion rates may have been limited only by $CO_{2(aq)}$ availability. Rates of methane generation, however, were directly linked to the abundance of the Ni-Fe alloy, confirming the role of the catalyst in the methane generation process. Moreover, these experiments showed that catalyst-induced carbon isotopic fractionation resulted in $^{13}\delta C$ values of the synthesized methane that are as low as typically associated with biologically mediated methane production, suggesting abiotic methane formation may be more wide spread than

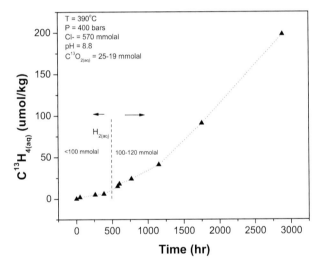

Figure 12. Mineral catalyzed carbon reduction at 390°C, 400 bars as a function of reaction [*Foustoukos and Seyfried*, 2002]. Dissolved chloride and pH of the aqueous fluids were fixed at values of approximately 570 mmol/kg and 8.8, respectively. The $CO_{2(aq)}$ ranged from 25 to 19 mmol/kg. The decrease in $CO_{2(aq)}$ was not the result of formation of formate species as observed in similar experiments at temperatures less than 350°C [*McCollom and Seewald*, 2001], but more likely reveals formation of more complex solid and aqueous hydrocarbons, yet to be identified. Following an early stage of reaction where dissolved $H_{2(aq)}$ was controlled at concentrations less than 100 mmol/kg, dissolved $H_{2(aq)}$ was increased to greater values, which resulted in an increase in the rate of methane formation. To distinguish better between abiotic methane formation and that which might have resulted from the thermogenic breakdown of hydrocarbon contaminates, aqueous $CO_{2(aq)}$ was isotopically labeled with ^{13}C.

previously envisaged. The exclusive formation of methane relative to other hydrocarbon species was another characteristic of the Ni-Fe alloy bearing experiments. In keeping with the constraints imposed by the stability of the alloy phase, however, dissolved H_2 concentrations were in the vicinity of 200–300 mmol/kg [*Horita and Berndt*, 1999].

A key element to the extent of conversion of $CO_{2(aq)}$ to methane and other hydrocarbons in experimental and natural systems likely involves the existence of reaction intermediates, which are often controlled by the composition of the catalyst, as well as temperature and pressure [*Madon and Taylor*, 1981]. For example, *McCollom and Seewald* [2001] showed that in their magnetite and serpentine-bearing experiments, $CO_{2(aq)}$ was rapidly converted to formate, which may have served as the intermediary phase resulting in subsequent methane formation. Formate was not specifically determined by Horita and Berndt in their Ni-Fe alloy experiments, although, it, too, likely served as an intermediate carbon compound, as suggested by these investigators. It is unlikely, however, that formate serves in this capacity during carbon reduction in the ultramafic-hosted hydrothermal systems at Rainbow and Logatchev due to the lack of stability of this species at temperatures and pressures applicable to these systems. *Foustoukos and Seyfried* [2002] and *Fu and Seyfried* [2001] have confirmed this in a series of experiments at 390–400°C 400, 500 bars. These studies also confirmed the loss of $CO_{2(aq)}$ in amounts significantly greater than can be accounted for by simple alkanes, which suggests formation of other hydrocarbon species in solution or attached to the surfaces of coexisting magnetite. XPS analysis of reactant magnetite indeed revealed the existence of an unidentified hydrocarbon phase [*Fu et al.*, 2002]. The recently reported existence of complex hydrocarbons in Rainbow vent fluids [*Holm and Charlou*, 2001], as well as the possible role that these species could play as intermediates for the formation of other carbon-bearing phases, enhances the significance of these results.

There can be no question that hydrocarbon formation during mineral fluid interaction at elevated temperatures and pressures is exceedingly complex. At this point in time, the abundances and distribution of hydrocarbon species observed in the Rainbow and Logatchev vent fluids have not been experimentally simulated. Part of this undoubtedly results from the restricted composition of the mineral catalysts investigated, while uncertain reaction mechanisms, which affect the existence of intermediary carbon compounds possibly needed for the formation of specific hydrocarbon species, could contribute further to this. The relative lack of synthesis of $C_2 - C_3$ hydrocarbons in hydrothermal experiments performed by *Horita and Berndt* [1999] and *McCollom and Seewald* [2001] likely relates more to the nature of the mineral catalysts, especially in the former study, than to other factors.

5.1. Phase Equilibria and Kinetic Effects in the CO_2-H_2O-H_2 System

The Rainbow and Logatchev vent fluids reveal dissolved methane concentrations in good agreement with coexisting concentrations of $CO_{2(aq)}$ and $H_{2(aq)}$ assuming full equilibrium in the CO_2-H_2O-H_2 system (Figure 13). Owing to the well-known difficulties involving reversible equilibrium for components in this system, even at elevated temperatures and pressures, it is more likely that the distribution of these species represent concentrations governed more by kinetic (catalytic) processes and source term effects than thermodynamic equilibrium. We hypothesize that it is the similar bulk chemistry and subseafloor temperatures and pressures of these ultramafic-hosted hydrothermal systems that influence catalytic tendencies of carbon reduction reactions and account most for the relatively high methane concentrations, as well the concentrations of longer chain hydrocarbons, as emphasized earlier. Although significant methane was generated in the mineral catalyzed experiments [*Foustoukos and Seyfried*, 2002], it is still below that theoretically predicted considering

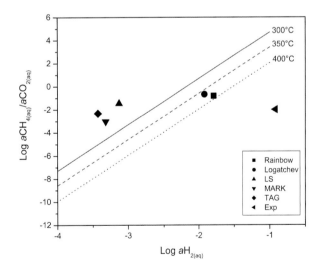

Figure 13. Predicted $CH_{4(aq)}/CO_{2(aq)}$ ratios versus dissolved $H_{2(aq)}$ at 300–400°C, 500 bars in comparison with measured concentrations of $CO_{2(aq)}$ and methane from ultramafic hosted hydrothermal systems (Rainbow and Logatchev), as well as from basaltic-hosted systems along the Mid-Atlantic Ridge [*Charlou et al.*, 2002, and references therein]. Although the measured carbon species from Rainbow vent fluids are consistent with constraints imposed by phase equilibria, the well known kinetic limitations in reversing methane-$CO_{2(aq)}$ equilibria make this unlikely. Catalytically induced excess methane formation by Fischer Tropsch type synthesis may affect all hydrothermal systems, although this may occur to a greater degree in ultramafic systems owing to compositional constraints unique to these systems (see text).

the availability of dissolved $CO_{2(aq)}$ and $H_{2(aq)}$ concentrations (Figure 13). Interestingly, dissolved methane concentrations reported for other high-temperature MAR vent fluids, reveal methane in excess of that predicted from equilibrium phase relations, which provides additional insight on the effectiveness in some systems of mineral catalyzed carbon reduction, while at the same time adding to the already considerable body of evidence for disequilibria between $CO_{2(aq)}$ and methane and H_2-bearing aqueous fluids at elevated temperatures and pressures (Figure 13).

In contrast, experimental data have shown that $CO_{2(aq)}$-$CO_{(aq)}$ equilibria proceeds rapidly in H_2-bearing aqueous fluids, especially at temperatures in excess of 350°C, where formic acid decomposition occurs virtually instantaneously [*Fu et al.*, 2002; *Yu and Savage*, 1998] (Figure 14). With this in mind, it is surprising that the high-temperature vent fluids at Rainbow [*Charlou et al.*, 2002] indicate a relatively low $CO_{(aq)}/CO_{2(aq)}$ (molal) ratio in comparison with that which is predicted from the measured dissolved $H_{2(aq)}$, assuming temperature and pressures of 400°C and 400 bars, respectively (Figure 14). One interpretation of this may involve requilibration at lower temperatures, possibly during sampling, with a block-in temperature of approximately 325°C, below which $CO_{(aq)}$-$CO_{2(aq)}$ fail to equilibrate. Unfortunately data for $CO_{(aq)}$ are not available for Logatchev, but considering the lower $H_{2(aq)}$ and $CO_{2(aq)}$ for vent fluids at this site [*Charlou et al.*, 2002], lower $CO_{(aq)}$ can be expected.

6. CONCLUSIONS

The recently discovered ultramafic-hosted hydrothermal systems at Rainbow and Logatchev at 36°N and 14°N respectively, on the Mid-Atlantic Ridge, represent exciting opportunities to assess the role of ultramafic lithologies on the chemistry and chemical evolution of hot spring vent fluids. Theoretical calculations assuming full equilibrium in the MgO-SiO_2-H_2O-HCl system indicate extremely low dissolved SiO_2 concentrations for fluids coexisting with olivine and serpentine at temperatures and pressures applicable to the Rainbow and Logatchev systems. The vent fluids, however, reveal moderately high dissolved SiO_2 concentrations (~7 mmol/kg), which can best be accounted for by partial disequilibria permitted by sluggish reaction kinetics involving olivine, in keeping with results of recent experiments. Thus, $SiO_{2(aq)}$ from pyroxene dissolution or dissolution of gabbroic components in the ultramafic rocks, may result in the formation of relatively SiO_2-rich secondary phases, including talc and tremolite, the overwhelming abundance of olivine notwithstanding. These phases coexisting with Ca-bearing fluids can buffer pH at relatively low values, which permit relatively high dissolved transition metal concentrations. Indeed, dissolved Fe con-

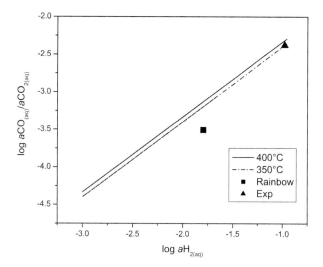

Figure 14. Predicted $CO_{(aq)}/CO_{2(aq)}$ ratios versus dissolved $H_{2(aq)}$ at 350–400°C, 500 bars in comparison with measured concentrations of these species from the Rainbow hydrothermal system [*Charlou et al.*, 2002, and references therein], and from hydrothermal experiments at 400°C, 500 bars [*Fu et al.*, 2002]. Experimental results indicate that $CO_{(aq)}$-$CO_{2(aq)}$ equilibrates rapidly at elevated temperatures and pressures. Thus the relatively low $CO_{(aq)}$ reported for Rainbow is surprising. This either indicates requilibration at lower temperatures during sampling, or possibly analytical uncertainties. The latter possibility is consistent with the approximate nature of the reported $CO_{(aq)}$ concentration [*Charlou et al.*, 2002]. This point notwithstanding, experimental and theoretical data indicate that $CO_{(aq)}$ in vent fluids is only possible under relatively reducing conditions.

centrations reported for the Rainbow system are unparalleled in vent fluids elsewhere—regardless of petrologic affinity, and provide unequivocal evidence of the very significant level of acidity that characterizes the aqueous fluid in the high-temperature region in the crust from which these fluids are derived. The relatively low pH underscores further the ineffectiveness of olivine dissolution to "titrate" fluid pH in either of these ultramafic-hosted hydrothermal systems.

Redox reactions in ultramafic systems are capable of creating very reducing environments, but again this is moderated at Rainbow and Logatchev as indicted by dissolved H_2 concentrations. The H_2 that is available, however, together with constraints imposed by the limited availability of rock derived $H_2S_{(aq)}$, define a very specific redox condition under which Cu solubility is relatively high, especially when considered with constraints imposed by temperature and acidity. Experimental and theoretical data indicate the existence of magnetite-bornite-chalcocite in the root zone of the Rainbow and Logatchev systems, which account well for the observed high dissolved Cu concentration, while being consistent with low H_2S concentrations. Whereas dissolved Cu is found to be exceedingly sensitive to temperature and $H_2S_{(aq)}$, Fe solubility is affected most by pH and dissolved chloride. Clearly, the high dissolved chloride of the Rainbow vent fluids contributes significantly to the high Fe concentrations, which accounts in part for the high mFe/mCu (molar) ratio in spite of the high Cu concentrations that are permissible by the low $H_2S_{(aq)}$ concentrations. Moreover, it is the combination of dissolved chloride and pH that accounts for the conspicuous difference in dissolved Fe between Rainbow and Logatchev vent fluids. Theoretical data also implicate the Ni-sulfide mineral, heazelwoodite, as a constraint on Ni solubility in ultramafic-hosted hydrothermal systems at mid-ocean ridges. In effect, the relative stability of this phase renders native Fe-Ni alloys unstable at all reasonable temperatures, pressures and redox conditions applicable to subseafloor reaction zones at Rainbow and Logatchev.

In addition to the high transition metals—in particular, Fe and Cu, the Rainbow and Logatchev systems reveal a number of other interesting characteristics. Relative to seawater, these systems have both higher (Rainbow) and lower (Logatchev) dissolved chloride concentrations. As in basaltic systems, these data are best interpreted to indicate phase separation in the $NaCl-H_2O$ system. More importantly, these data suggest temperatures higher than presently measured for the venting fluids. The diverging patterns in chloride variability relative to seawater for the two ultramafic-hosted hydrothermal systems, however, require variable mixing between vapor, brines and evolved seawater. Precisely how and where this occurs is uncertain. What is certain is the fact that temperatures sufficient to induce supercritical phase separation of seawater require a significant source of heat. Considering the limited magma supply at slow spreading ridges [*Cannat et al.,* 1992], it is likely that the heat source that fuels these systems is mined from the deep lithosphere [*Bach et al.,* 2002; *German and Lin,* this volume].

Ultramafic-hosted hydrothermal systems reveal an unusual capacity to generate significant methane concentrations and concentrations of other hydrocarbons and carbon compounds (e.g., $CO_{(aq)}$), as indicated by field and laboratory studies. A key element of this likely involves the relatively high $H_{2(aq)}$ concentrations in hydrothermal reaction zones, but also the existence of favorable catalytic pathways, which facilitate electron transfer, required for carbon reduction, especially in the case of hydrocarbon formation. That vent fluids at Rainbow and Logatchev reveal dissolved methane and $CO_2(aq)$ concentrations that are consistent with coexisting H_2 at temperatures and pressures reasonable for these systems is almost certainly not a reflection of full equilibrium in the $CO_2-H_2O-H_2$ system, but more likely a function of the existence of temperature dependent kinetic mechanisms involving minerals and fluids that have yet to be fully realized from results of laboratory experiments. Understanding these pathways is a critical challenge for future investigation.

Acknowledgments: This manuscript has benefited greatly from thoughtful reviews by Jeff Alt, Jean-Luc Charlou, Marv Lilley and Chris German, which significantly improved the final version. The paper also benefited from many discussions with attendees at the InterRidge Theoretical Institute, Pavia, Italy. The authors would also like to thank Ms. Sharon Kressler, Department of Geology and Geophysics, University of Minnesota, for assisting with preparation of the final layout and format of the paper. The senior author is appreciative of funding from IRTI. Funding from the U.S. National Science Foundation through research grants OCE-0117117, OCE-9911471 and OCE-9818908 played key roles in the development of concepts expressed in the paper.

REFERENCES

Agrinier, P., and M. Cannat, Oxygen-isotope constraints on serpentinization processes in ultramafic rocks from the Mid-Atlantic Ridge (23° N), in *Proceedings of the Ocean Drilling Program: Scientific Results,* pp. 381–388, 1997.

Allen, D. E., and W. E. Seyfried, Jr., Alteration and mass transfer in the $MgO-CaO-FeO-Fe_2O_3-SiO_2-Na_2O-H_2O-HCl$ system at 400°C and 500 bars: Implications for pH and compositional controls on vent fluids from ultramafic-hosted hydrothermal systems at mid-ocean ridges, *Geochimica et Cosmochimica Acta,* 67, 1531–1542, 2003a.

Allen, D. E., and W. E. Seyfried, Jr., REE controls in MOR hydrothermal systems: An experimental study at elevated temperature and pressure, *Geochimica et Cosmochimica Acta,* 2003b (submitted).

Allen, D. E., and W. E. Seyfried, Jr., Serpentinization and Heat Generation: Constraints from Lost City and Rainbow Hydrothermal Systems, *Geochimica et Cosmochimica Acta,* (in-press), 2003c.

Alt, J. C., and W. C. Shanks, III, Sulfur in serpentinized oceanic peridotites; serpentinization processes and microbial sulfate reduction, *Journal of Geophysical Research, B, Solid Earth and Planets,* 103 (5), 9917–9929, 1998.

Alt, J. C., and W. C. Shanks, III, Serpentinization of abyssal peridotites from the MARK area, Mid-Atlantic Ridge: Sulfur geochemistry and reaction modeling, *Geochimica et Cosmochimica Acta,* 67, 641–653, 2003.

Bach, W., N. R. Banerjee, H. J. B. Dick, and E. T. Baker, Discovery of ancient and active hydrothermal systems along the ultra-slow spreading Southwest Indian Ridge 10 degrees–16 degrees E, *Geochemistry Geophysics Geosystems,* 3, 31, 2002.

Berndt, M. E., D. E. Allen, and W. E. Seyfried, Jr., Reduction of CO_2 during serpentinization of olivine at 300°C and 500 bar, *Geology,* 24 (4), 351–354, 1996.

Bischoff, J. L., Densities of liquids and vapors in boiling $NaCl-H_2O$ solutions; a PVTX summary from 300° to 500°C, *American Journal of Science,* 291 (4), 309–338, 1991.

Bischoff, J. L., and K. S. Pitzer, Liquid-vapor relations for the system $NaCl-H_2O$: Summary of the P-T-X surface from 300°C to 500°C, *American Journal of Science,* 289, 217–248, 1989.

Bischoff, J. L., and R. J. Rosenbauer, Phase separation in seafloor geothermal systems: an experimental study on the effects of metal transport, *American Journal of Science*, *287*, 953–978, 1987.

Bougault, H., M. Aballea, J. Radford-Knoery, J. L. Charlou, P. J. Baptiste, P. Appriou, H. D. Needham, C. German, and M. Miranda, FAMOUS and AMAR segments on the Mid-Atlantic Ridge: ubiquitous hydrothermal Mn, CH_4, ^3He signals along the rift valley walls and rift offsets, *Earth and Planetary Science Letters*, *161* (1–4), 1–17, 1998.

Bougault, H., J. L. Charlou, Y. Fouquet, H. D. Needham, N. Vaslet, P. Appriou, P. J. Baptiste, P. A. Rona, L. Dmitriev, and S. Silantiev, Fast and slow spreading ridges: structure and hydrothermal activity, ultramafic topographic highs, and methane output, *Journal of Geophysical Research*, *98* (B6), 9643–51, 1993.

Bowers, T. S., A. C. Campbell, C. I. Measures, A. J. Spivack, M. Khadem, and J. M. Edmond, Chemical controls on the composition of vent fluids at 13°–11°N and 21° N, East Pacific Rise, *Journal of Geophysical Research*, *93* (5), 4522–4536, 1988.

Butterfield, D. A., and G. J. Massoth, Geochemistry of North Cleft Segment vent fluids—Temporal changes in chlorinity and their possible relation to recent volcanism, *Journal of Geophysical Research*, *99* (B3), 4951–4968, 1994.

Butterfield, D. A., R. E. McDuff, M. J. Mottl, M. D. Lilley, J. E. Lupton, and G. J. Massoth, Gradients in the composition of hydrothermal fluids from the Endeavor segment vent field: phase separation and brine loss, *Journal of Geophysical Research*, *99*, 9561–9583, 1994.

Campbell, A. C., T. S. Bowers, and J. M. Edmond, A time-series of vent fluid compositions from 21°N, EPR (1979, 1981, and 1985) and the Guaymas Basin, Gulf of California (1982, 1985), *Journal of Geophysical Research*, *93*, 4537–4549, 1988a.

Campbell, A. C., M. R. Palmer, G. P. Klinkhammer, T. S. Bowers, J.M. Edmond, J. R. Lawrence, J. F. Casey, G. Thompson, S. Humphris, P. Rona, and J. A. Karson, Chemistry of hot springs on the Mid-Atlantic Ridge, *Nature*, *335*, 514–519, 1988b.

Cannat, M., Emplacement of mantle rocks in the seafloor at mid-ocean ridges, *Journal of Geophysical Research*, *98*, 4163–4172, 1993.

Cannat, M., D. Bideau, and H. Bougault, Serpentinized peridotites and gabbros in the Mid-Atlantic Ridge axial valley at 15°37'N and 16°52'N, *Earth and Planetary Science Letters*, *109* (1–2), 87–106, 1992.

Cannat, M., and M. Seyler, Transform tectonics, metamorphic plagioclase and amphibolitization in ultramafic rocks of the Vema transform fault (Atlantic Ocean), *Earth and Planetary Science Letters*, *133* (3–4), 283–98, 1995.

Cave, R. R., C. German, J. Thompson, and R. W. Nesbitt, Fluxes to sediments underlying the Rainbow hydrothermal plume at 36°14'N on the Mid-Atlantic Ridge, *Geochimica et Cosmochimica Acta*, *66*, 1905–1923, 2002.

Charlou, J. L., L. Dmitriev, H. Bougault, and H. D. Needham, Hydrothermal methane between 12°N and 15°N over the Mid-Atlantic Ridge, *Deep-Sea Research*, *35* (1A), 121–31, 1988.

Charlou, J. L., J. P. Donval, E. Douville, J. Knoery, Y. Fouquet, H. Bougault, P. Jean-Baptiste, M. Stievenard, C. R. German, and

Anonymous, High methane flux between 15° N and the Azores triple junction, Mid-Atlantic Ridge; hydrothermal and serpentinization processes., *Eos, Transactions, American Geophysical Union*, *78* (46, Suppl.), 831, 1997.

Charlou, J. L., J. P. Donval, Y. Fouquet, P. Jean-Baptiste, and N. Holm, Geochemistry of high H2 and CH4 vent fluids issuing from ultramafic rocks at the Rainbow hydrothermal field (36° 14 ' N, MAR), *Chemical Geology*, *191* (4), 345–359, 2002.

Charlou, J. L., J. P. Donval, P. Jean-Baptiste, R. Mills, P. Rona, D. Von Herzen, and Anonymous, Methane, nitrogen, carbon dioxide and helium isotopes in vent fluids from TAG hydrothermal field, 26° N-MAR, *Eos, Transactions, American Geophysical Union*, *74* (43, Suppl.), 99, 1993.

Charlou, J. L., Y. Fouquet, H. Bougault, J. P. Donval, J. Etoubleau, P. Jean-Baptiste, A. Dapoigny, P. Appriou, and P. A. Rona, Intense CH_4 plumes generated by serpentinization of ultramafic rocks at the intersection of the 15° 20' N fracture zone and the Mid-Atlantic Ridge, *Geochimica et Cosmochimica Acta*, *62* (13), 2323–2333, 1998.

Charlou, J. L., Y. Fouquet, J.-P. Donval, J.-M. Auzende, P. Jean-Baptiste, M. Stievenard, and S. Michel, Mineral and gas chemistry of hydrothermal fluids on an ultrafast spreading ridge; East Pacific Rise, 17° to 19° S (Naudur cruise, 1993) phase separation processes controlled by volcanic and tectonic activity, *Journal of Geophysical Research*, *101* (7), 15,899–15,919, 1996.

Ding, K., and W. E. Seyfried, Jr., Determination of Fe-Cl complexing in the low pressure supercritical region (NaCl fluid) - Iron solubility constraints on pH of subseafloor hydrothermal fluids, *Geochimica et Cosmochimica Acta*, *56* (10), 3681–3692, 1992.

Donval, J. P., J. L. Charlou, E. Douville, J. Knoery, Y. Fouquet, E. Ponsevera, P. Jean-Baptiste, M. Stievenard, C. R. German, and Anonymous, High H_2 and CH_4 content in hydrothermal fluids from Rainbow site newly sampled at 36°14' N on the AMAR segment, Mid-Atlantic Ridge (diving FLORES cruise, July 1997); comparison with other MAR sites, *Eos, Transactions, American Geophysical Union*, *78* (46, Suppl.), 832, 1997.

Douville, E., J. L. Charlou, E. H. Oelkers, P. Bienvenu, C. F. Jove Colon, J. P. Donval, Y. Fouquet, D. Prieur, and P. Appriou, The Rainbow vent fluids (36° 14'N, MAR): the influence of ultramafic rocks and phase separation on trace metal content in Mid-Atlantic Ridge hydrothermal fluids, *Chemical Geology*, *184*, 37–48, 2002.

Evans, W. C., L. D. White, and J. B. Rapp, Geochemistry of some gases in hydrothermal fluids from the southern Juan de Fuca Ridge, *Journal of Geophysical Research*, *93*, 15,305–15,313, 1988.

Fouquet, Y., K. Henry, G. Bayon, P. Cambon, F. J. A. S. Barriga, I. Costa, H. Ondreas, L.M. Parson, A. Ribeiro, and J. M. R. S. Relvas, The Rainbow hydrothermal field: geologic setting, mineralogical and chemical composition of sulfide deposits, *Earth and Planetary Science Letters*, in-press, 2003.

Fournier, R. O., Conceptual models for brine evolution in magmatic-hydrothermal systems, edited by R. W. Decker, T. L. Wright, and P.H. Stauffer, pp. 45, U.S. Geological Survey, 1986.

Foustoukos, D., and W. E. Seyfried, Jr. (2002), Abiotic methane production in subseafloor hydrothermal systems: pH and compositional effects on rates of carbon reduction at elevated temperatures and pressures, in *Thermal Regime of Ocean Ridges and Dynamics of Hydrothermal Circulation,* InterRidge Theoretical Institute, Pavia, Italy.

Foustoukos, D., and W. E. Seyfried, Jr. (2004), Redox and pH constraints in the TAG hydrothermal system: 26°N, Mid-Atlantic Ridge, *Earth Planet Sci. Lett.* (submitted).

Frost, B. R., On the stability of sulfides, oxides and native metals in serpentinites, *Journal of Petrology, 26,* 31–63, 1985.

Fu, Q., J. Horita, and W. E. Seyfried, Jr., Isotopic fractionation in magnetite-catalyzed hydrothermal carbon dioxide reduction processes, *Eos, Transactions of the American Geophysical Union, 83,* P71C-0473, 2002.

Fu, Q., and W. E. Seyfreid, Jr., Hydrothermal reduction of carbon dioxide using magnetite, *Eleventh Annual V. M. Goldschmidt Conference, LPI Contribution No. 1088, Lunar and Planetary Institute, Houston.,* 2001.

German, C., and J. Lin, The thermal structure of the oceanic crust, ridge spreading and hydrothermal circulation: How do we understand their inter-connections, in *The Thermal Structure of the Oceanic Crust and The Dynamics of Hydrothermal Circulation,* edited by J. L. C. R. German, and L. M. Parson, American Geophysical Union, 2003 *(this volume).*

Hartmann, G., and K. H. Wedepohl, The compostion of peridotite tectonites from the Ivrea Complex, northern Italy: Residues from melt extraction, *Geochimica et Cosmochimica Acta, 57,* 1761–1782, 1993.

Holm, N. G., and J. L. Charlou, Initial indications of abiotic formation of hydrocarbons in the Rainbow ultramafic hydrothermal system, Mid-Atlantic Ridge, *Earth and Planetary Science Letters, 191,* 1–8, 2001.

Horita, J., and M. E. Berndt, Abiogenic methane formation and isotopic fractionation under hydrothermal conditions, *Science, 285* (5430), 1055–1057, 1999.

Humphris, S., and J. R. Cann, Constraints on the energy and chemical balances of the modern TAG and ancient Cyprus seafloor sulfide deposits, *Journal of Geophysical Research, 105,* 28,477–28,488, 2000.

Janecky, D. R., and W. E. Seyfried, Jr., Hydrothermal serpentinization of peridotite within the oceanic crust: Experimental investigations of mineralogy and major element chemistry, *Geochimica et Cosmochimica Acta, 50,* 1357–1378, 1986.

Johnson, J. W., E. H. Oelkers, and H. C. Helgeson, SUPCRT92 - A software package for calculating the standard molal thermodynamic properties of minerals, gases, aqueous species, and reactions from 1-bar to 5000-bar and 0°C to 1000°C, *Computers and Geosciences, 18* (7), 899–947, 1992.

Kelley, D. S., Methane-rich fluids in the oceanic crust, *Journal of Geophyical Research, 101* (B2), 2943–2962, 1996.

Kelley, D. S., J. R. Delaney, M. D. Lilley, D. A. Butterfield, and Anonymous, Unusual sulfide structures and venting style in the newly discovered Mothra hydrothermal field, Endeavour Segment of the Juan de Fuca Ridge, *Eos, Transactions, American Geophysical Union, 78* (46, Suppl.), 773, 1997.

Kelley, D. S., J. A. Karson, D. K. Blackman, G. L. Fruh-Green, D. A. Butterfield, M. D. Lilley, E. J. Olson, M. O. Schrenk, K. K. Roe, G. T. Lebon, and P. Rivizzigno, An off-axis hydrothermal vent field near the Mid-Atlantic Ridge at 30° N, *Nature, 412* (6843), 145–149, 2001.

Lilley, M. D., D. A. Butterfield, E. J. Olson, J. E. Lupton, S. A. Macko, and R. E. McDuff, Anomolous CH_4 and NH_4 concentrations at an unsedimented mid-ocean ridge hydrothermal system, *Nature, 364,* 45–47, 1993.

Lilley, M. D., M. A. De Angelis, and L. I. Gordon, CH4, H2, CO and N2O in submarine hydrothermal vent waters, *Nature, 300,* 48–50, 1982.

Lilley, M. D., J. R. Delaney, R. E. McDuff, E. J. Olson, E. A. McLaughlin, J. E. Lupton, D. A. Butterfield, and Anonymous, Volatile chemistry at the Endeavour Segment, Juan de Fuca Ridge, *Eos, Transactions, American Geophysical Union, 73* (43, Suppl.), 253, 1992.

Lorand J. P. (1991) Sulphide petrology and suphur geochemistry of orogenic lherzolites: A comparative study of the Pyrenean bodies (France) and the Lanzo Massif (Italy). In *Orogenic Lherzolites and Mantle Processes* (eds. M. A. Menzies, et al.), pp. 77–95. Oxford University Press, London.

Lowell, R. P., and P. Rona, Seafloor hydrothermal systems driven by the serpentinization of peridotite, *Geophysical Research Letters, 29* (11), 26–1–26–3, 2002.

Macdonald, A. H., and W. S. Fyfe, Rates of serpentinization, *Tectonophysics, 116,* 123–135, 1985.

MacLean, W. H., Sulfides in Leg 37 drill core from the Mid-Atlantic Ridge, *Canadian Journal of Earth Sciences, 14,* 674–683, 1977.

Madon, R. J., and W. F. Taylor, Fischer-Tropsch Synthesis on a precipitated iron catalyst, *Journal of Catalysis, 69,* 32–43, 1981.

Martin, B., and W. S. Fyfe, Some experimental and theoretical observations on the kinetics of hydration reactions with particular reference to serpentinization, *Chemical Geology, 6,* 185–202, 1970.

McCollom, T. M., and J. S. Seewald, A reassessment of the potential for reduction of dissolved CO_2 to hydrocarbons during serpentinization of olivine, *Geochimica et Cosmochimica Acta, 65,* 3769–3778, 2001.

Merlivat, L., F. Pineau, and M. Javoy, Hydrothermal vent waters at 13°N on the East Pacific Rise; isotopic composition and gas concentration, *Earth and Planetary Science Letters, 84* (1), 100–108, 1987.

Michard, G., Behavior of major elements and some trace elements (Li, Rb, Cs, Sr, Fe, Mn, W, F) in deep hot water from grantitic areas, *Chemical Geology, 89,* 117–134, 1990.

Mottl, M. J., Metabasalts, axial hot springs, and the structure of hydrothermal systems at mid-ocean ridges, *Geological Society of America Bulletin, 94* (2), 161–80, 1983.

Oosting, S. E., and K. L. Von Damm, Bromide/chloride fractionation in seafloor hydrothermal fluids from 9–10° N East Pacific Rise, *Earth and Planetary Science Letters, 144* (1–2), 133–145, 1996.

Palmer, M. R., Controls over the chloride concentration of submarine hydrothermal vent fluids: evidence from Sr/Ca and Sr^{87}/Sr^{86} ratios, *Earth and Planetary Science Letters, 109,* 37–47, 1992.

Rona, P. A., H. Bougault, J.L. Charlou, P. Appriou, T. A. Nelsen, J. H. Trefry, G. L. Eberhart, A. Barone, and H. D. Needham, Hydrothermal circulation, serpentinization, and degassing at a rift valley fracture zone intersection - Mid-Atlantic Ridge near 15° N, 45° W, *Geology, 20* (9), 783–786, 1992.

Salvi, S., and A. E. Williams-Jones, The role of hydrothermal processes in concentrating high-field strength elements in the Strange Lake peralkaline complex, northeastern Canada, *Geochimica et Cosmochimica Acta, 60,* 1917–1932, 1996.

Schultz, H., Short history and present trends of Fischer Tropsch synthesis, *Applied Catalysis, 186,* 3–12, 1999.

Seyfried, W. E., Jr., and K. Ding, The effect of redox on the relative solubilities of copper and iron in Cl⁻ bearing aqueous fluids at elevated temperatures and pressures: An experimental study with application to subseafloor hydrothermal systems, *Geochimica et Cosmochimica Acta, 57,* 1905–1917, 1993.

Seyfried, W. E., Jr., and K. Ding, Phase equilibria in subseafloor hydrothermal systems: A review of the role of redox, temperature, pH and dissolved Cl on the chemistry of hot spring fluids at mid-ocean ridges, in *Seafloor Hydrothermal Systems: Physical, Chemical, Biologic and Geological Interactions,* edited by S.E. Humphris, R.A. Zierenberg, L.S. Mullineaux, and R.E. Thompson, pp. 248–273, American Geophysical Union, 1995.

Sherwood Lollar, B., T. D. Westgate, J. A. Ward, G. F. Slater, and G. Lacrampe-Couloume, Abiogenic formation of alkanes in the Earth's crust as a minor source for global hydrocarbon reservoirs, *Nature, 416,* 522–524, 2002.

Sugisaki, R., and K. Mimura, Mantle Hydrocarbons—Abiotic or Biotic?, *Geochim. Cosmochim. Acta, 58* (11), 2527–2542, 1994.

Tamaura, Y., and M. Tabata, Complete reduction of carbon dioxide to carbon using cation excess magnetite, *Nature, 346,* 255–346, 1990.

Thurnherr, A. M., and K. J. Richards, Hydrography and high-temperature heat flux of the Rainbow hydrothermal site (36 degrees 14' N, Mid-Atlantic Ridge), *Journal of Geophysical Research Oceans, 106* (C5), 9411–9426, 2001.

Trefry, J. H., D. B. Butterfield, S. Metz, G. J. Massoth, R. P. Trocine, and R.A. Feely, Trace metals in hydrothermal solutions from Cleft Segment on the Southern Juan de Fuca Ridge, *Journal of Geophysical Research, 99* (B3), 4925–4935, 1994.

Von Damm, K. L., A comparison of Guaymas Basin hydrothermal solutions with other sedimented systems and experimental results, in *AAPG Memoir,* edited by J. P. Dauphin, and B. R. T. Simoneit, pp. 743–751, 1991.

Von Damm, K. L., Controls on the chemistry and temporal variability of seafloor hydrothermal fluids, in *Seafloor Hydrothermal Systems: Physical, Chemical, Biologic and Geologic Interactions,*
edited by S. E. Humphris, R. A. Zierenberg, L. S. Mullineaux, and R.E. Thompson, pp. 222–248, American Geophysical Union, 1995.

Von Damm, K. L., Chemistry of hydrothermal vent fluids from 9°–10° N, East Pacific Rise; "time zero," the immediate posteruptive period, *Journal of Geophysical Research, 105* (5), 11203–11222, 2000.

Von Damm, K. L., and J. L. Bischoff, Chemistry of hydrothermal solutions from the Southern Juan de Fuca Ridge, *Journal of Geophysical Research, 92,* 11334–11346, 1987.

Von Damm, K. L., L. G. Buttermore, S. E. Oosting, A. M. Bray, D. J. Fornari, M. D. Lilley, and W. C. Shanks, III, Direct observation of the evolution of a seafloor "black smoker" from vapor to brine, *Earth and Planetary Science Letters, 149,* 101–111, 1997.

Von Damm, K. L., S. E. Oosting, R. Kozlowski, L. G. Buttermore, D. Colodner, H. N. Edmonds, J. M. Edmond, and J. M. Grebmeir, Evolution of East Pacific Rise hydrothermal vent fluids following a volcanic eruption, *Nature, 375,* 47–50, 1995.

Welhan, J. A., and H. Craig, Methane and hydrogen in East Pacific Rise hydrothermal fluids, *American Geophysical Union; 1979 fall annual meeting, 60* (46), 863, 1979.

Welhan, J. A., and H. Craig, Methane, hydrogen and helium in hydrothermal fluids, in *Hydrothermal Processes at Seafloor Spreading Centers,* edited by P. A. Rona, K. Bostrom, L. Laubier, and K. L. Smith, Jr., pp. 391–411, Plenum Press, New York, 1983.

Welhan, J. A., and J. E. Lupton, Light hydrocarbon gases in Guaymas Basin hydrothermal fluids: thermogenic versus abiogenic origin, *American Association of Petroleum Geologists Bulletin, 71,* 215–223, 1987.

Welhan, J. A., and M. Schoell, Origins of methane in hydrothermal systems, *Geological Society of America Abstracts with Program, 71* (1–3), 183–198, 1988.

Wetzel, L. R., and E. L. Shock, Distinguishing ultramafic- from basalt-hosted submarine hydrothermal systems by comparing calculated vent fluid compositions, *Journal of Geophysical Research, 105* (4), 8319–8340, 2000.

Wolery, T. J., and S. A. Daveler, EQ6, A computer program for reaction path modeling of aqueous geochemical systems: Theoretical manual, users guide, and related documentation (version 7.0), *UCRL-MA-110662 PT IV,* 1–337, 1992.

Yu, J. H., and P. E. Savage, Decomposition of formic acid under hydrothermal conditions, *Industrial & Engineering Chemistry Research, 37,* 2–10, 1998.

D. E. Allen, D. I. Fousoutkos, and W. E. Seyfried, Jr., Department of Geology and Geophysics, University of Minnesota, 310 Pillsbury Drive SE, Minneapolis, Minnesota 55455

Evolution of the Hydrothermal System at East Pacific Rise 9°50′N: Geochemical Evidence for Changes in the Upper Oceanic Crust

Karen L. Von Damm

Complex Systems Research Center, Institute for the Study of Earth, Oceans and Space, University of New Hampshire, Durham, New Hampshire

The hydrothermal vent fluids at 9°50′N on the East Pacific Rise have undergone unprecedented chemical changes from 1991–2002. The Cl contents of the sampled fluids have varied by more than an order of magnitude during this time period, and are accompanied by profound changes in the concentrations of other chemical species as well as temperature. In this study the Cl concentrations, in conjunction with Si and temperature data are used to infer the temperature and pressure conditions of phase separation and reaction within the oceanic crust. The model proposed suggests changes in the depth to the heat source driving the hydrothermal system at this site during these 12 years. My interpretation is that it has shoaled significantly since ~2000. Although this interpretation may not be unique, it is the simplest one, containing the least number of assumptions that is consistent with all of the chemical and temperature data. If this model is correct, its results provide finer resolution on the depth and time scales on which hydrothermal and possibly magmatic processes behave in the oceanic crust at fast spreading ridges than is currently available from other data sets. The fluid chemistry data are combined with other available records to infer a sequence of events within the shallow oceanic crust at this site from 1991–2002. For most of the time period studied, only fluids with chloride concentrations less than the seawater value were collected, creating a mass balance problem for this element. The proposed model should be tested in the future by collection of contemporaneous fluid and seismic data.

1. INTRODUCTION

One of the ultimate goals of research on mid-ocean ridge (MOR) hydrothermal systems, particularly vent fluids, is to evaluate their net flux of heat and mass to the ocean. In order to assess this flux, the controls on the chemistry and heat content of the fluids must be understood, as well as how and why

these parameters vary over time. For the first decade that MOR hydrothermal systems were studied, the stability in the chemical composition of fluids exiting from a given vent was striking [e.g., *Bowers et al.,* 1988; *Campbell et al.,* 1988]. Just as striking was how each vent had a chemically unique fluid composition, distinct from others that had been sampled and analyzed [e.g., *Bowers et al.,* 1988; *Campbell et al.,* 1988]. One of the great advances in our understanding of seafloor hydrothermal systems during the 1990s was the observation of pronounced temporal variability in the chemistry and temperature of the fluids exiting from individual vents on the mid-ocean ridge system [*Butterfield et al.*, 1994, 1997;

Mid-Ocean Ridges: Hydrothermal Interactions Between the Lithosphere and Oceans
Geophysical Monograph Series 148
Copyright 2004 by the American Geophysical Union
10.1029/148GM12

Von Damm et al., 1995; 1997]. Most of this temporal variability was tied to sites where volcanic events, whether lava flows on the seafloor and/or dike intrusions were known to have occurred [*Haymon et al.*, 1993; *Charlou et al.*, 1996; *Perfit and Chadwick*, 1998 and references therein]. This magmatic activity was often evidenced by the presence of "megaplumes," now more commonly referred to as "event plumes" [*Baker et al.*, 1987; *Baker et al.*, 1989]. This temporal variability was observed by repeat sampling from submersibles because at the start of the 1990s we had no *in situ* sensors that we could leave at these sites for time periods of more than a few days. By the mid-1990s we could at least deploy temperature sensors to provide some information between the repeat sampling cruises [*Fornari et al.*, 1998; *Fornari et al.*, this volume].

One of the most dynamic and well-studied places on the global mid-ocean ridge system is the East Pacific Rise from 9–10°N latitude [e.g., see bibliography at http://www. ridge2000.bio.psu.edu/IPs/EPR/EPRBibliography.htm]. At 9°50′N on the East Pacific Rise we have had an unprecedented opportunity to track the evolution of a hydrothermal system for over a decade, since a volcanic eruption in 1991. Based on work with the ARGO-I system in 1989 [*Haymon et al.*, 1991] this area was known to have abundant hydrothermal activity. During the *DSV Alvin* dive series here in 1991, the objectives of which were to sample the hydrothermal systems prior to drilling by the Ocean Drilling Program, we found dramatic changes from the earlier ARGO-I work that were indicative of a volcanic eruption and dike intrusion, and the initiation of new hydrothermal vents and biological communities [*Haymon et al.*, 1993; *Shank et al.*, 1998; *Von Damm*, 2000]. Since 1991 the changes in the hydrothermal systems on the section of the ridge affected by volcanic eruptions in 1991/2, approximately 9°45–51′N, have been dramatic [e.g., *Von Damm et al.*, 1995; *Oosting and Von Damm*, 1996; *Von Damm and Lilley*, 2004]. These changes have continued into at least 2002, more than a decade after the last eruption in this area. Some of the most dramatic changes in the hydrothermal systems have been focused right at 9°50′N, an area sometimes referred to as the "Hole-to-Hell" (Plate 1). This area lies within the BIOGEOTRANSECT (or "Transect") [*Shank et al.*, 1998], a 1.46 km length of ridge delineated by 210 sequentially numbered Biomarkers (BM) deployed in 1992. The "Transect" has been frequently visited and sampled. The high temperature vents "Bio9," "Bio9Prime" (Bio9′) and "P" are three high temperature vents right at 9°50′N. While all the vents in the eruptive area, 9°45–51′N, have continued to evolve chemically from 1991–2002 [e.g., *Von Damm et al.*, 1995; *Oosting and Von Damm*, 1996; *Von Damm and Lilley*, 2004], it is precisely at 9°50′N that the temporal sampling has been most dense (Table 1), and accompanied by large

temporal variations in both fluid chemistry and temperature (Figures 1–6). For some of the parameters measured (e.g., Cl, temperature), the observed trends have even undergone reversals in sign during this time period (Figure 1).

The chemistry and temperature of vent fluids reflect and integrate the processes that have occurred along the hydrothermal flow path through the oceanic crust. The significant temporal variation on short time scales provides the opportunity to link changes in one part of the system with responses in other parts of the system. When natural systems are stable it can be difficult to understand the controlling factors, but with significant natural perturbations, such as volcanic eruptions and/or dike intrusions, one can begin to quantitatively address "cause and effect." The changes in composition and temperature in hydrothermal vent fluids, therefore, potentially provide sensitive indicators of changes occurring within the upper oceanic crust. Changes in the fluid compositions and temperatures have also been linked to changes in the biological communities at this site [*Shank et al.*, 1998]. It is ultimately the heat source that drives these hydrothermal systems. This paper presents chemical and temperature data for the high temperature hydrothermal vents occurring at 9°50′N on the East Pacific Rise, and interprets these data to elucidate and constrain crustal processes, including possible changes in the nature and location of the heat source underlying this section of the mid-ocean ridge.

2. GEOLOGIC SETTING AND SAMPLE SITES

The East Pacific Rise at 9°50′N is a fast spreading ridge with total opening rate of 11cm/yr [*Carbotte and Macdonald*, 1992]. The water depth is 2500±10 m. A multi-channel seismic study in the area in 1985 [*Detrick et al.*, 1987] placed the depth of the low velocity zone (melt lens) at ~1.5 km below the seafloor. All of the high temperature vents at 9°50′N are located within the axial summit collapse trough (ASCT), close to what is considered the eruptive fissure for the 1991/2 eruptions. Within this area there is also abundant diffuse flow with its associated animal communities [*Shank et al.*, 1998; *Von Damm and Lilley*, 2004]. The Bio9 area is about 60 m north of P vent (Plate 1b). Prior to 1994, Bio9 was the only high temperature vent sampled in the Bio9 area. By 1994, the Bio9′ vent was a distinct individual black smoker, and we also began to collect samples of its fluid and measure its temperature (Table 1). By 1999, the number of smokers around Bio9 had begun to increase, replacing some of the areas that had previously been characterized by diffuse flow. For some cruises it is unclear which vent was actually sampled and in this case the samples are simply referred to as coming from the "Bio9 Complex." In 2002, a third distinct smoker was also sampled in the Bio9

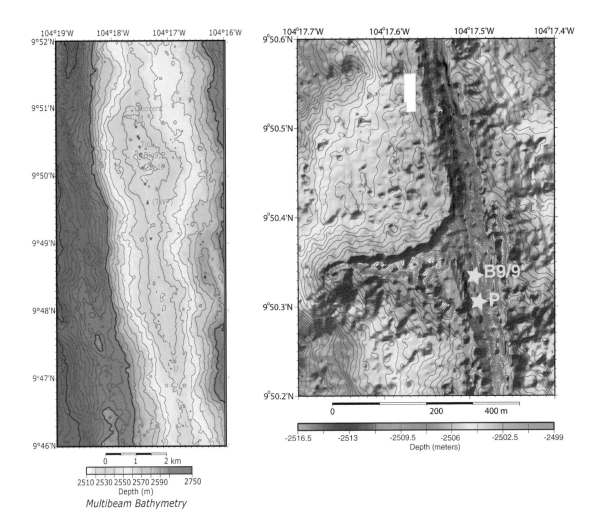

Plate 1. Map showing the distribution of vents in the 9°50′N area on the East Pacific Rise. (a) Vents from 9°46–51′N with high temperature vents shown in red (TWP=Tube Worm Pillar which as of 2003 was no longer active) and sites of diffuse flow venting are shown in blue. The multiple stars indicate multiple vents, as in the area of M, Bio9 and P vents. (b) Bio9 and P vents areas plotted on ABE microbathymetry. Figures edited from D. J. Fornari.

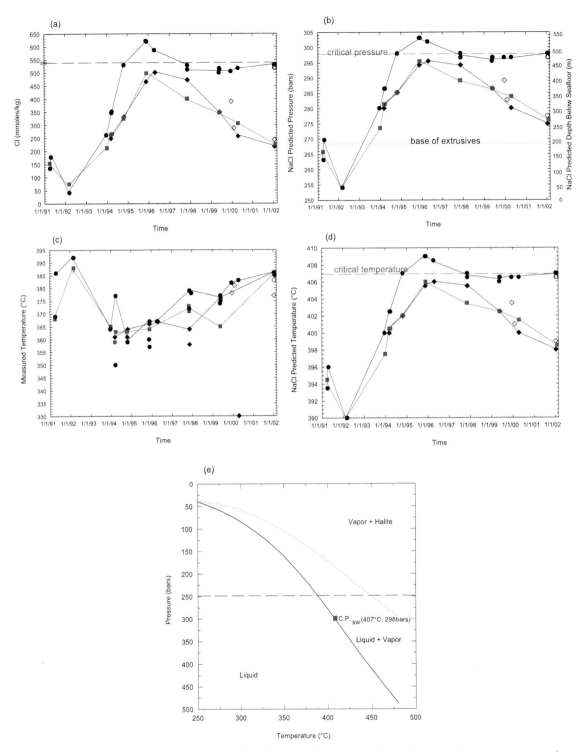

Figure 1. (a) End member Cl versus time, (b) Predicted pressure (depth) versus time, (c) measured temperature versus time, (d) predicted temperature versus time, (e) phase relationships for seawater based on *Bischoff* [1991] and references therein. Data from Bio9 vent are plotted as squares, Bio9′ as diamonds, P vent as circles, other P-area vents are plotted as unfilled circles, and other Bio9-area vents are plotted as unfilled diamonds. SW designates the seawater concentration, which is shown as a dashed line when appropriate.

Table 1. Sampling Information

Vent	Date	Days	Dive	T °C	End Member Cl mmol/kg	Minimum Measured Mg Majors mmol/kg	GT mmol/kg	Number of Samples Majors	GT
B9'.1	3/11/94	1075	2735	361	249±1	7.82	2.10	4	2
B9'.2	10/20/94	1298	2842	361	330±2	7.77	n.m.	2	0
B9'.3	10/30/94	1308	2854	364	332±2	1.98	0.42	6	3
B9'.4	11/21/95	1695	3026	366	466±2	4.40	2.10	4	3
B9'.5	4/25/96	1851	3074	367	501±3	1.36	n.m.	4	0
B9'.6	11/14/97	2419	3167	358	473±2	1.40	13.6	4	2
B9'.7	11/16/97	2421	3169	364	473±2	n.m.	26.5	0	1
B9'.8	5/28/99	2980	3413	376	349±2	2.57	n.m.	2	0
B9'.9	4/18/00	3306	3544	330	257±1	8.34	9.76	4	2
B9'.10	1/18/02	3945	3754	386	217±3	4.27	33.8	4	1
B9complex.1	12/20/99	3186	3514	378	390±2	n.m.	2.78	0	2
B9complex.2	02/06/00	3234	3531	381	288±1	n.m.	3.33	0	1
B9".1	1/16/02	3943	3752	377	243±1	2.24	10.1	7	2
B9.1	4/1/91	1	2351	368	154±1	4.03	25.2	2	1
B9.2	3/6/92	340	2498	>388	75±1	6.78	5.73	4	2
B9.3	12/28/93	1002	2693	365	212±2	5.00	2.48	6	3
B9.4	3/10/94	1074	2734	359	263±7	1.05	2.07	4	2
B9.5	3/29/94	1093	2754	363	267±1	2.13	1.62	4	2
B9.6	10/19/94	1297	2841	363	325±2	4.69	n.m.	2	0
B9.7	10/24/94	1302	2846	359	330±2	2.05	1.10	4	2
B9.8	11/25/95	1699	3030	364	498±2	1.80	2.10	3	3
B9.9	11/4/97	2409	3157	373	400±3	14.0	2.31	4	2
B9.10	11/10/97	2415	3163	371	401±2	1.67	17.2	4	1
B9.11	5/21/99	2973	3405	365	348±1	12.5	21.3	2	1
B9.12	4/20/00	3308	3546	379?	306±1	4.07	8.20	3	2
B9.13	2/4/02	3962	3769	386	226±2	4.38	18.5	6	2
P.1	4/7/91	7	2357	369	135±1	9.11	40.5	4	1
P.2	4/23/91	23	2372	386	178±6	42.5	43.8	2	1
P.3	3/4/92	337	2496	392	n.m.	n.m.	3.22	0	1
P.4	3/9/92	342	2501	392	41.6±0.43	20.4	34.6	3	1
P.5	12/18/93	992	2685	364	262±1	13.5	38.4	4	1
P.6	3/18/94	1082	2743	350	347±2	3.51	1.00	7	3
P.7	3/27/94	1091	2752	377	352±2	2.28	11.6	8	3
P.8	10/25/94	1303	2849	359	530±3	2.58	1.36	6	3
P.9	11/16/95	1690	3021	360	622±3	7.60	3.30	7	4
P.10	11/24/95	1698	3029	357	566±3	n.m.	46.6	0	2
P.11	11/28/95	1702	3033	367	620±3	20.2	n.m.	2	0
P.12	4/11/96	1837	3067	367	587±3	3.25	n.m.	4	0
P.13	4/22/96	1848	3071	368	584±3	3.80	n.m.	4	0
P.14	11/11/97	2416	3164	372	529±3	3.02	13.0	4	2
P.15	11/14/97	2419	3167	379	512±3	n.m.	3.77	0	1
P.16	12/16/97	2451	3190	378	n.m.	12.3	n.m.	3	0
P.17	12/19/97	2454	3193	378	n.m.	2.70	n.m.	4	0
P.18	05/16/99	2968	3400	374	500±3	n.m.	2.34	0	2
P.19	5/23/99	2975	3407	377	517±3	36.3	n.m.	1	0
P.20	5/28/99	2980	3413	375	512±3	5.63	n.m.	2	0
P.21	12/12/99	3178	3506	382	505±3	n.m.	2.64	0	2
P.22	4/17/00	3305	3543	383	517±3	1.85	2.26	4	2
P.23	1/16/02	3943	3752	386	530±3	2.53	2.36	2	2
P.24	1/30/02	3957	3764	385	531±3	1.96	2.24	2	1
Pmiddle.1	1/21/02	3948	3758	383	518±3	1.65	2.07	4	2
Pmiddle.2	1/30/02	3957	3764	385	519±3	4.90	2.55	2	1
Seawater				2	540	52.2	n.m.		

n.m. = not measured

GT = gas tight bottle

area, Bio9". The P vent site also has more than one vent, although samples were always collected from the same chimney. In 2002, a second vent at P was also sampled, referred to as "P-middle." These vents are all located in an area where the ASCT is somewhat wider than is typical for ±1 km along strike (Plate 1b). Within this ~2 km of ridge crest, there are approximately three additional black smoker vents to the north, and three to the south of the Hole-to-Hell area (Plate 1a). (An exact number of smokers cannot be given as it has changed over time, with several new ones forming and two becoming inactive.) While fluids from all of these vents have undergone considerable chemical changes during the 1991–2002 time period, the chemical evolution of each vent appears to be largely decoupled from other vents. This was also observed in the relationship between geographically separated high- and low-temperature vent fluids throughout the Transect [*Von Damm and Lilley,* 2004]. Because this area is so dynamic, and because there appears to be minimal coupling between vents further along the axis, this paper focuses solely on the high temperature vents precisely at 9°50′N (the "Hole-to-Hell").

The hydrothermal vents at 9°50′N EPR, have been sampled frequently between 1991 and 2002 (Table 1), and significant chemical and temperature changes have occurred during this time period (Table 2 and Figures 1–6). Not every vent was sampled on every cruise, hence not every time point has samples from each of these vents. Still, this remains one of the most densely sampled time series of vent fluid chemistries available on a decadal time scale.

3. SAMPLING AND ANALYTICAL METHODS

All fluid samples were collected in either the titanium majors or gas tight bottles, using the *DSV Alvin.* At most sampling times, these bottles were used either individually (gas tight bottles) or configured as pairs (majors bottles). In a few cases, both types of bottles may have been mounted on the NOAA manifold sampler [*Massoth et al.*, 1988]. Temperatures were measured with either the *Alvin* high temperature probe, the temperature probe on the manifold sampler, or the inductively coupled link (ICL) temperature probes which can be mounted directly on the bottle snorkels. In all cases the temperature probes were calibrated to an NIST traceable standard. The low measured Mg contents of the sampled fluids (Table 1) demonstrate that significant mixing of seawater into the vent fluids did not occur prior to or during sampling at most sampling time points. Data reported in Table 2 and Figures 1–6 are the calculated end member values for pure hydrothermal fluid, assumed to contain no Mg. Analytical methods and the calculation of the end member data are as reported in *Von Damm* [2000], except that after 1992 H_2S was determined by iodi-

metric titration. As the colorimetric method for H_2S also relies on this titration for standardization, the analytical precision is the same as previously reported.

4. RESULTS AND DISCUSSION

Hydrothermal circulation is driven by a heat source within the oceanic crust, the nature of which can vary, including a dike intrusion, melt lens, or simply hot rock. Seismic data can indicate low velocity zones, which are interpreted to contain a certain percentage of melt [e.g., *Detrick et al.,* 1987]. Brittle fracture can only occur in rocks that are cool enough to no longer be plastic. Similarly, hydrothermal circulation can only occur in rocks that are cool and brittle enough to maintain fractures, and hence to sustain fluid flow through fractures. Therefore, the depth indicated by the seismic data to the low velocity zone may also indicate the maximum depth to which hydrothermal fluids can circulate. The modeling results of *Lowell and Germanovich* [this volume] suggest that the thickness of the "lid" separating the fluid circulation from the area where melt is located may only be ~2–20 m thick. The seismic data collected for the 9°50′N area are for a few brief time periods in 1991 [*Hildebrand et al.,* 1991, 1992] and 1994–5 [*Sohn et al.,* 1998], while fluid sampling occurred at numerous intervals during these 12 years (Table 1). The Cl content, Si content, and measured temperature of the vent fluids at the seafloor provide three independent lines of evidence that may constrain the depth to the heat source at any given sampling time. Hence the fluid data potentially provide a much more detailed record of the temporal variability in the heat source depth than is currently available or accessible by other methods or from other data sets.

Phase separation is one of the fundamental processes controlling mid-ocean ridge vent fluid chemistry [e.g., *Von Damm,* 1995]. If one assumes the starting fluid has the composition of seawater which has lost Mg and SO_4 due to water-rock reaction, one can then use the measured Cl content of vent fluids to infer the pressure and temperature (PT) conditions at which phase separation occurred within the oceanic crust. The Cl content of the fluids in turn controls the total cation content of the fluids, as the fluids must maintain electroneutrality. The other major anions in seawater have been lost by precipitation and/or reduction (SO_4^{2-}), or titration (alkalinity or HCO_3^-/CO_3^{2-}), leaving Cl as by far the predominant anion; Br is of next greatest abundance. The changes in the Cl content of vent fluids, which are a result of the changing PT conditions of phase separation [*Bischoff,* 1991 and references therein] may, therefore, provide constraints on the depth to the low velocity zone. If it is assumed that the fluids were in equilibrium with quartz when they left the reaction zone, then the silica content of the fluids may also provide constraints on

Table 2. End Member Compositions

Vent	Date	T, °C	Cl, mmol[1]	Si, mmol	H2S-LR, mmol	H2S-Max, mmol	Na-Meas, mmol	Na-Calc, mmol	Na-CB, mmol	K, mmol	Li, µmol[1]	Ca, mmol	Sr, µmol	Fe, µmol	Mn, µmol	SO4, mmol	Br, µmol	pH (25°C)	Alk[1], meq
B9'.1	3/11/94	361	249±1	12.4±0.1	5.95	8.5	216±4	222±4	217	8.67±0.17	213±7	9.81±0.10	35.8±0.7	1270±10	382±4	0.035±0.007	374±11	3.66	-0.155
B9'.2	10/20/94	361	330±2	14.0±0.1	3.62	5.7	286±6	285±6	287	9.06±0.09	269±7	14.7±0.1	57.1±1.1	2620±30	621±6	1.28±0.03	n.m.	3.17	-0.304
B9'.3	10/30/94	364	332±2	13.3±0.1	5.71	6.2	285±6	287±6	287	9.87±0.10	285±3	15.5±0.2	57.1±1.1	2730±30	636±6	1.55±0.17	n.m.	3.15	-0.510
B9'.4	11/21/95	366	466±2	14.5±0.1	5.28	6.7	357±7	391±8	391	14.1±0.1	400±4	23.4±0.2	87.4±1.7	5840±60	1150±10	0.647±0.092	824±25	3.10	-0.561
B9'.5	4/25/96	367	501±3	15.1±0.2	5.68	7.2	425±9	416±8	417	15.1±0.1	438±4	26.1±0.3	91.6±1.8	6700±70	1260±10	0.005±0.053	732±22	3.09	-0.924
B9'.6	11/14/97	358	473±2	13.4±0.1	5.31	5.8	399±8	398±8	400	15.4±0.2	437±4	21.3±0.2	91.6±1.8	6640±70	1090±10	0.499±0.087	720±22	n.m.	-0.682
B9'.7	11/16/97	364	473±2	14.9±0.1	n.m.[2]	n.m.	423±8	406±8	408	15.1±0.2	415±4	22.6±0.2	92.0±1.8	5380±50	1050±10	4.17±0.04	581±17	n.m.	n.m.
B9'.8	5/28/99	376	349±2	11.7±0.1	8.21	8.3	327±7	297±6	296	10.8±0.1	922±9	15.7±0.2	63.1±0.3	3750±170	691±7	-0.691±0.020	397±12	n.m.	n.m.
B9'.9	4/18/00	330	257±1	11.8±0.1	7.91	8.2	219±4	220±4	221	8.58±0.09	237±2	9.40±0.09	38.6±0.8	2430±30	499±5	-1.24±0.12	336±10	3.52	-0.320
B9'.10	1/18/02	386	217±3	7.98±0.10	n.m.	n.m.	181±4	188±4	189	6.77±0.07	182±2	8.32±0.27	34.8±1.3	2980±30	380±4	1.33±0.31	n.m.	3.22	-0.555
B9complex	12/20/99	378	390±2	12.5±0.1	n.m.	n.m.	342±7	337±7	337	12.4±0.1	375±4	17.3±0.2	72.0±1.4	5220±50	845±8	2.58±0.17	658±20	n.m.	n.m.
B9complex	02/06/00	381	288±1	n.m.	n.m.	n.m.	243±5	248±5	250	9.60±0.10	292±3	12.7±0.1	54.6±1.1	3820±40	604±6	2.61±0.03	464±14	n.m.	n.m.
B9".1	1/16/02	377	243±1	8.55±0.09	8.79	12.5	209±4	208±4	209	7.70±0.08	227±2	9.40±0.09	40.1±0.8	2940±30	482±5	-0.083±0.083	390±12	3.17	-0.451
B9.1	4/1/91	368	154±1	9.90±0.10	21.0	23.2	128±3	137±3	139	3.69±0.06	97.9±1.0	2.07±0.07	7.47±0.15	2190±20	285±3	-0.203±0.095	225±7	2.57	-1.78
B9.2	3/6/92	≥388	75.5±1.0	6.98±0.09	15.2	26	65.5±1.3	63.8±1.3	64.7	1.56±0.04	27.4±0.3	3.95±0.05	36.7±6.2	1670±30	172±2	1.42±0.19	100±3	3.29	-0.481
B9.3	12/28/93	365	212±2	11.3±0.1	5.27	7.3	192±4	186±4	188	5.92±0.09	183±5	7.81±0.10	32.1±2.3	1060±10	280±3	0.175±0.502	n.m.	3.58	-0.315
B9.4	3/10/94	359	263±7	12.1±0.3	4.12	4.5	236±5	239±6	232	8.52±0.09	223±2	9.97±0.10	33.8±0.7	1280±10	367±4	0.354±0.116	417±13	3.23	-0.531
B9.5	3/29/94	363	267±1	12.6±0.1	5.06	5.5	236±5	234±5	235	8.46±0.08	240±2	10.6±0.1	37.3±0.7	1430±10	432±4	0.643±0.010	419±13	3.45	-0.313
B9.6	10/19/94	363	325±2	13.9±0.1	4.39	4.4	271±5	279±6	281	9.91±0.25	188±5	14.7±0.1	54.9±1.1	2310±20	590±6	1.03±0.14	n.m.	3.66	-0.280
B9.7	10/24/94	359	330±2	14.1±0.1	4.29	4.9	287±6	286±6	287	8.90±0.09	294±3	15.2±0.2	55.4±1.1	2430±30	618±6	1.53±0.11	n.m.	3.19	-0.470
B9.8	11/25/95	364	498±2	14.8±0.1	4.63	5.1	418±8	420±8	420	14.7±0.1	438±5	24.6±0.4	89.5±1.8	6030±60	1190±10	0.715±0.028	622±19	3.00	-0.547
B9.9	11/4/97	373	400±3	13.2±0.1	7.34	8.6	346±7	342±7	337	12.5±0.1	394±4	18.7±0.2	77.2±1.5	4950±50	929±11	-0.731±0.043	649±19	≤ 4.30	≤ -0.253
B9.10	11/10/97	371	401±2	12.8±0.1	5.08	5.3	347±7	339±7	338	12.9±0.1	390±4	19.3±0.5	75.9±1.5	5100±50	917±9	0.385±0.054	564±17	3.20	-0.485
B9.11	5/21/99	365	348±2	11.7±0.1	n.m.	n.m.	299±6	298±6	300	10.1±0.1	336±4	13.3±0.1	57.2±1.1	3100±50	685±7	-1.62±0.12	486±15	3.31	-0.735
B9.12	4/20/00	379?	306±2	11.0±0.1	6.79	7.5	261±5	261±5	262	10.2±0.1	302±3	12.6±0.2	54.2±1.2	3610±40	622±6	-0.137±0.102	486±15	3.29	-0.735
B9.13	2/4/02	386	226±2	8.15±0.08	7.77	7.8	195±4	197±4	197	7.20±0.09	209±2	8.68±0.09	37.1±0.7	2680±30	441±4	1.48±0.07	367±11	3.29	-0.359
P.1	4/7/91	369	135±1	8.70±0.09	24.9	27.4	94.7±2.9	110±2	114	2.56±0.05	22.6±0.2	1.17±0.07	-2.3±0.4	4420±50	175±2	-2.58±0.14	185±6	2.58	-1.92
P.2	4/23/91	386	178±6	4.06±0.04	47.1	47.1	52.8±1.1	166±3	174	-0.88±0.07	4.0±0.3	6.93±0.13	20.2±0.4	5870±60	142±1	10.70±0.20	n.m.	< 4.21	≤ 0.128
P.3	3/4/92	392	n.m.	3.66±0.04	n.m.	n.m.	33.5±0.7	n.m.	n.m.	0.93±0.01	22.8±0.2	4.35±0.04	25.0±0.5	63.9±0.6	70.9±0.7	3.90±0.04	n.m.	n.m.	-0.383
P.4	3/9/92	392	41.6±0.2	3.91±0.04	24.8	26.2	69.9±12.8	37.1±0.7	38.2	0.63±0.04	23.5±0.8	1.59±0.14	76.3±2.6	687±7	70.1±0.7	1.21±0.15	54±3	≤ 3.35	-0.373
P.5	12/18/93	364	262±1	12.3±0.1	7.22	8.1	230±5	227±5	228	7.20±0.07	204±4	11.3±0.1	43.6±1.3	1700±20	436±4	0.328±0.313	400±12	≤ 3.60	-0.654
P.6	3/18/94	350	347±2	14.2±0.1	5.37	8.4	289±6	302±6	303	10.3±0.1	295±3	14.3±0.2	50.9±1.0	2890±30	732±7	1.71±0.08	n.m.	3.14	-0.654
P.7	3/27/94	377	352±2	14.3±0.1	4.57	6.1	299±6	308±6	303	10.5±0.1	298±10	16.9±0.2	57.4±1.1	3040±30	735±14	1.76±0.03	534±16	3.15	-0.712
P.8	10/25/94	359	530±3	16.2±0.2	3.06	3.3	447±9	447±9	449	16.8±0.2	501±5	26.3±0.3	87.9±1.8	5390±50	1140±10	1.27±0.08	n.m.	3.23	-0.554
P.9	11/16/95	360	622±3	15.8±0.2	6.96	11.2	487±10	522±10	525	19.4±0.2	540±7	30.0±0.3	107±2	7100±70	1460±20	0.802±0.068	254±8	3.20	-1.10
P.10[3]	11/24/95	357	566±3	16.2±0.2	n.m.	n.m.	466±9	440±9	455	16.6±0.2	687±7	106±1	387±8	8280±80	1490±20	69.4±0.7	n.m.	n.m.	n.m.
P.11	11/28/95	367	620±3	16.4±0.2	5.79	5.9	490±10	518±10	520	16.6±0.2	565±6	28.9±0.3	109±2	7490±80	1470±20	-2.09±0.02	947±28	< 3.10	-0.556
P.12	4/11/96	367	587±3	16.6±0.2	5.12	8.2	491±10	491±10	492	18.9±0.2	583±6	29.4±0.5	102±2	6820±70	1430±10	-0.049±0.006	947±28	3.25	-0.877
P.13	4/22/96	368	584±3	16.0±0.2	5.70	6.3	496±10	489±10	492	18.7±0.2	591±6	28.5±0.3	105±2	6810±70	1410±10	-0.090±0.025	822±25	3.16	≤ -0.391
P.14	11/11/97	372	529±3	15.3±0.2	6.01	8.6	445±9	448±9	449	18.6±0.2	521±5	23.9±0.2	93.9±1.9	6020±60	1250±10	0.231±0.015	838±25	3.18	-0.641
P.15	11/14/97	379	512±3	15.6±0.2	n.m.	n.m.	460±9	450±9	450	17.5±0.2	537±5	24.3±0.2	97.3±1.9	6030±60	1280±10	9.58±0.10	811±24	< 3.19	n.m.
P.16	12/16/97	378	n.m.	15.1±0.2	n.m.	n.m.	436±9	n.m.	n.m.	17.6±0.2	501±5	24.2±0.3	105±2	5370±50	1200±10	3.73±0.04	834±25	3.02	n.m.
P.17	12/19/97	378	15.4±0.2	15.4±0.2	n.m.	n.m.	448±9	432±9	433	16.9±0.2	519±5	21.3±0.2	96.6±1.9	6000±190	1240±10	0.166±0.044	823±25	n.m.	n.m.
P.18	05/16/99	374	500±2	14.2±0.1	n.m.	n.m.	438±9	438±9	442	17.0±0.2	501±5	23.7±0.2	90.1±1.8	6530±70	1070±10	3.17±0.48	860±26	n.m.	n.m.
P.19	5/23/99	377	517±3	15.7±0.2	n.m.	n.m.	421±8	436±9	436	15.9±0.2	501±5	22.1±0.2	103±2	6210±60	1230±10	1.87±0.02	843±25	n.m.	n.m.
P.20	5/28/99	375	512±3	14.9±0.1	n.m.	n.m.	433±9	434±9	434	16.1±0.2	507±5	21.4±0.2	90.9±1.8	6130±60	1100±10	-0.702±0.013	838±25	n.m.	n.m.
P.21	12/12/99	382	505±3	14.7±0.1	6.00	6	424±8	438±9	440	16.5±0.2	507±5	21.7±0.2	93.8±1.9	6570±70	1250±10	2.64±0.41	803±24	n.m.	n.m.
P.22	4/17/00	383	517±3	14.4±0.1	5.64	6	432±9	453±9	451	16.2±0.2	500±5	21.5±0.2	92.6±1.9	6820±70	1260±10	-0.670±0.068	851±26	n.m.	n.m.
P.23	1/16/02	386	530±3	13.0±0.1	6.40	6.7	450±9	464±9	453	16.5±0.2	512±5	21.7±0.2	96.7±1.9	7250±70	1290±10	-0.134±0.033	842±25	3.10	-0.805
P.24	1/30/02	385	531±3	12.8±0.1	n.m.	n.m.	450±9	n.m.	n.m.	17.7±0.2	507±5	21.5±0.2	96.4±1.9	7400±70	1280±10	0.312±0.008	n.m.	3.21	-0.460
P.middle	1/21/02	383	518±3	12.8±0.1	6.48	6.7	440±9	440±9	441	16.9±0.2	499±5	21.1±0.2	92.9±1.9	7260±70	1250±10	0.055±0.022	844±25	3.07	-0.628
P.middle	1/30/02	385	519±3	12.6±0.1	6.77	8.1	437±9	444±9	445	16.7±0.2	492±5	20.8±0.2	93.9±1.9	7030±70	1250±10	0.641±0.056	822±25	3.23	-0.361
Seawater		2	540	0.2	0	0	464	464	464	10.10	26	9.95	87.0	0	0	28.2	840	7.8	2.4

[1] all are per kilogram
[2] n.m. = not measured
[3] highly uncertain value, see Table 1

the temperature and depth of circulation/reaction for the fluids. Third, our observations have shown that fluids cool conductively as they rise through the oceanic crust [e.g., Von Damm et al., 1995, 2000], hence very high measured temperatures at the seafloor (>375°C) also argue for a very shallow heat source (on the order of 100s of meters rather than a kilometer or more below the seafloor).

The parameters Cl, Si and measured temperature are discussed first because of the constraints they may provide on physical conditions within the oceanic crust. While each of these three indicators (Cl, Si, temperature) provides independent bounds on the PT conditions within the oceanic crust where the fluids acquired their properties, in some cases, more than one PT region may be able to generate the observed value for a given indicator. However, all three indicators must be internally consistent to provide a robust interpretation of conditions within the crust. In many cases the use of all three will provide a unique set of bounds for the PT conditions, even if an individual indicator considered alone does not.

The Cl concentration set by phase separation will in turn greatly affect water-rock interaction, hence elements controlled primarily by this process are discussed second. As the two major processes that control vent fluid compositions, phase separation and water-rock interaction, are integrally linked, it is important to note that the separation of the two here is purely to facilitate discussion and does not imply a true separation of the two processes. As all fluids discussed here have temperatures >120°C, biologically mediated changes are not of concern. There is also evidence for magmatic degassing on this section of the ridge [*Lilley et al.,* 2002], and this is discussed last, after water-rock reaction.

The first part of the discussion is organized by chemical element (species). The data are then presented in the context of a time line.

4.1. Chloride

Ambient seawater has a Cl content of 540 mmoles/kg at the 9°N site. Because vent fluids lose Mg and SO_4, to a first approximation they are equivalent to a 3.2 wt. % NaCl solution, as has been demonstrated experimentally [*Bischoff and Rosenbauer,* 1988; *Bischoff,* 1991 and references therein]. With no known mineralogic source or sink of any significance in basalts, the only mechanism that can cause substantial changes (≥10%) in the Cl content of vent fluids compared to seawater is phase separation. (Halite is known to form at certain conditions, but appears to be rapidly re-dissolved [*Oosting and Von Damm,* 1996; *Von Damm,* 2000].) All of the high temperature vent fluids sampled at 9°N have Cl contents significantly different from seawater (Figure 1a), therefore all of the vent fluids issuing from the seafloor have

undergone phase separation within the oceanic crust. For example, between 1991 and 2002, the chloride content in P vent fluids varied from 42–620 mmoles/kg, with the minimum occurring in 1992, and the maximum in 1995 (Figure 1a). During this same time period the Bio9 vent fluids varied from 75–498 mmoles/kg, with the minimum occurring in 1992, and the maximum in 1995. The Bio9′ vent fluids were first sampled in March 1994, and their range is 217–501 mmoles/kg, with the maximum occurring in 1996 and the minimum in 2002. While the three vents follow the same general trends, there are differences in their specific trajectories, and individual compositions. In the case of P vent, the fluid compositions vary by over an order of magnitude within the 12 year time period. Because Cl is the major anion, the total cations, dominated by Na, will vary in concert with Cl. Of the three vents, only P has emitted a fluid with a chlorinity greater than the ambient seawater composition (540 mmoles/kg), and then only in 1995–6. Therefore, over the1991–2002 time period, vapor phase fluids have been the dominant fluid emitted. With the exception of a single site at a single time point in 1995, all of the diffuse flow fluids sampled in the 9°50′N area have also contained a lower chlorinity than seawater, and have generally tracked the high temperature vent fluid compositions [*Von Damm and Lilley,* 2004].

4.1.1. Significance of the Cl Data. None of the fluids sampled from Bio9, Bio9′ and P vents from 1991–2002 have chlorinities equal to the seawater value. Therefore, they must have undergone phase separation. As the Cl contents of the vapor and liquid (brine) phases formed as a result of phase separation vary as a function of the pressure and temperature conditions at which phase separation occurs, the changing Cl-contents of the hydrothermal fluids from 9°50′N point to significant changes in the hydrothermal circulation cell within the oceanic crust at this site. Since a hydrothermal system is ultimately driven by its heat source, these changes also imply significant changes in the local heat source at this site during the last 12 years. The phase relations in the system $NaCl$-H_2O are well known due to extensive experimental studies [*Bischoff and Rosenbauer,* 1988; *Bischoff,* 1991 and references therein]. Therefore if certain assumptions are made, the measured Cl content of the vent fluids can be used to calculate the pressure and temperature (PT) conditions at which phase separation occurred. The following assumptions are made in the subsequent discussion.

1) Phase separation is the only process that may cause a significant (≥10%) change in the Cl content of the fluids.

• While some hydration of the basaltic substrate undoubtedly occurs, the amount of water in the basalts is at most a few percent, and to effect the changes in Cl observed here would require much more water than this to be either lost from the fluids, or gained from the rock.

- No major permanent mineral sink for Cl has been observed in oceanic rocks. Based on Br/Cl and other element/Cl ratios, halite dissolution has been inferred at the 9°50′N site in 1991–2 [*Oosting and Von Damm*, 1996; *Von Damm*, 2000]. Assuming Br is conservative allows the amount of Cl gained from halite dissolution to be calculated (see discussion in *Von Damm* [2000]) and these changes in Cl-content are at most 36 mmoles/kg, which does not quantitatively change the interpretations presented below.

2) The starting fluid for phase separation has the NaCl content of seawater. Once phase separation occurs, the vapor or brine rises directly to the seafloor without undergoing additional phase separation.

- While it is possible that multiple episodes of phase separation have occurred within the oceanic crust, we have no way to constrain if this is indeed the case. Hence I have invoked the simplest assumption.

- *Ravizza et al.* [2001] demonstrated, based primarily on Sr-isotope systematics, that ~3–10% of the fluids exiting the high temperature Biovent, P and Bio9′ vents in 1996 may have been due to mixing in of a partially reacted "crustal groundwater," most similar in composition to a <150°C bore hole fluid reported by *Magenheim et al.* [1995]. Such a fluid would have a Mg-content less than seawater, but a Cl-content identical to seawater. The data presented by *Ravizza et al.* [2001] suggest that this entrainment likely occurs preferentially in fluids that have very low densities, i.e., high temperatures and low Cl contents. Mixing in of a seawater-like fluid component would tend to raise the Cl-contents of the vent fluids but the maximum identified 10% input would also not change the model results presented below.

Invoking these assumptions I have used the measured Cl contents of the vent fluids to calculate both the pressure (i.e., depth below the seafloor) and temperature when phase separation occurred (Figure 1). Figures 1b and 1d show the inferred PT conditions for phase separation in fluids from the vents from 1991–2002. The critical point for seawater is 407°C and 298 bars [*Bischoff and Rosenbauer*, 1988] (Figure 1e), therefore fluids crossing the two phase curve for seawater below this temperature and pressure will phase separate sub-critically (boil) and a low Cl vapor phase will be generated. For tem-

peratures greater than this, the fluid will phase separate supercritically and a small amount of brine will be condensed. As the Cl contents in the Bio9 vent fluids are always less than the seawater value, the conditions of phase separation for these fluids must always be subcritical, i.e., less than 407°C. For P vent, in 1995 and 1996 the Cl contents are greater than local ambient seawater, hence phase separation is occurring supercritically, i.e., at temperatures greater than 407°C. The subsurface temperatures at which phase separation occurred in both P and Bio9 vents (Figure 1d) were at a minimum in 1992 (~390°C), were slightly higher than this in 1991 (394–396°C), and reached a maximum in 1995–6, 409°C in P and 406°C in Bio9 and Bio9′, before decreasing again to ~407°C in P and ~399°C in Bio9 and Bio9′ vents. Note that these temperature estimates do not rely at all on the actual measured exit temperatures (Figure 1c), but only the Cl content of the fluids and the assumptions stated above.

The pressures predicted in this same manner (Figure 1b) reached a minimum in 1992 of 250–255 bars, essentially at, or within 50 m of, the seafloor. This is in good agreement with other observations at this time, including a second seafloor eruption [*Rubin et al.*, 1994]. In 1991, the Cl data suggest phase separation occurred at pressures between 263 and 270 bars, or 130–200 m below the seafloor. After the minimum in 1992 the pressure started to rise, reaching a maximum in 1995–6 of ~304 bars or 540 m below the seafloor for P, and ~295 bars or 450 m for Bio9 and Bio9′. Since early 1996 the pressure at P vent has decreased and in 2002 was ~298 bars or ~480 m below the seafloor, while for Bio9 and Bio9′ vents it decreased to ~277 bars or ~270 m below the seafloor, the shoalest phase separation has occurred since ~1993, one year after the 1992 eruption.

In general, the Bio9 and P vent fluids cannot be simply related to each other for most of the time series, even though they are within ~60 m of each other. Both vents usually issued low chlorinity fluids and hence could not be the conjugate pairs to each other that result from the phase separation of seawater. It was only in 1995 that vent fluids from P were high chlorinity and those from Bio9 were low chlorinity. Both are relatively close to the seawater Cl content at that time, as is expected when the inferred conditions of phase separation are very close to the critical point for seawater.

4.2. Temperature

Temperature is a record that is less reliable than Cl, because if the various temperature probes are not inserted completely into the vent, temperatures that are too low may be recorded. All of the temperature data reported here are point measurements made at the time the water samples were collected. The high minimum Mg concentrations for B9′.1, B9.11, P.1, P.2 and

P.5 (Table 1) may imply mixing with seawater prior to the fluids exiting the seafloor, and this could result in lowered temperature readings for these time points. In 1991 it was especially difficult to insulate the probe from the influence of local seawater when measuring temperature, because there were no chimneys. Subsurface mixing with a "crustal groundwater" with temperatures <150°C could also lower the temperature of the exit fluids. Artificially low temperatures would also be recorded on recording temperature probes if the sulfide structure grew such that the temperature probe was isolated from the fluid flow path [e.g., *Fornari et al.*, 1998]. Measured temperature values can only be too low, however, not too high, which is why I report the maximum temperature measured for each given time point. The measured temperature record (Figure 1c) is, therefore, subject to more potential artifacts than is the Cl record (Figure 1a).

At the seafloor depth in this area (~2500 m), temperatures of ≥389°C are required to phase separate seawater. As pressure increases, as with depth within the hydrothermal circulation cell, higher temperatures are required for phase separation to occur. The maximum measured temperature for the P vent fluids, 392°C, was measured in 1992. The temperature probe unplugged on the dive to Bio9 vent in 1992, hence no temperature measurement exists but based on the extreme chemical composition, we inferred the temperature must be >388°C [e.g., *Oosting and Von Damm*, 1996; *Von Damm*, 2000]. After 1992, the temperatures in both Bio9 and P vent fluids decreased and measured values were mostly in the 350s and 360s until 1997, when values >370°C were again measured at both vents. Temperatures >370°C were first measured at Bio9′ in 1999. P vent fluids were measured at >380°C in December 1999. By January–February 2002, all three vents had measured fluid temperatures of 386°C, which is unusually high compared to the global data set. The measured temperatures at the seafloor therefore are anti-correlated with the measured Cl-contents, as expected for the simple phase separation model invoked here.

It has also been observed that the longer the distance on the hydrothermal upflow zone, the greater the amount of conductive cooling that occurs, and hence the lower the measured exit temperatures of the vent fluids. To measure temperatures of >380°C at the seafloor requires a heat source within a few hundred meters of the seafloor. Bio9 and P vent fluid temperatures therefore suggest a significant shoaling of the heat source after early 1996. In 2002, all of the vent fluids from 9°50′N had very high temperatures, ~386°C (Figure 1c). It is unusual to sample such hot fluids at the seafloor and the only times it occurs are when the heat source is relatively close to the seafloor (few hundred meters or less) as has been observed at the times of the 1991/2 eruptions, or Brandon vent on the SEPR which is close to the critical point

for seawater at its sampled conditions [*Von Damm et al.*, 2003]. Hence the fact that the **measured** seafloor temperatures are also very high is in agreement with the predictions based on the Cl content alone, that the heat source was very shallow in 2002. Due to adiabatic expansion as the fluids rise they would be expected to cool ~5°C, hence the measured temperatures at the seafloor are expected to be less than their *in situ* values within the oceanic crust. Since ~1999 the number of black smokers at Bio9 has increased significantly, also suggesting that more heat is being supplied in this area than in the ~6 immediately preceding years.

4.3. Silica

The silica content of the fluids (Figure 2a) can be used as an independent indicator of the depth of fluid circulation (1) if it is assumed that the fluids are in equilibrium with quartz, and (2) if it is assumed that insignificant Si is lost from the fluids during their rise to the seafloor. Quartz solubility is not only dependent on the pressure and temperature conditions (Figure 2b), but also on the salt content of the fluids, but this effect is much smaller [*Von Damm et al.*, 1991]. Increasing salt content increases the solubility of quartz. Quartz solubility decreases above ~375°C at pressures relevant to these seafloor systems, hence this retrograde solubility must also be considered in any interpretation of the data. Using the semi-empirical equation of *Von Damm et al.* [1991] which takes into account the salt content in predicting quartz solubility, one can (1) predict the P conditions based on the measured temperature and dissolved silica content, and (2) calculate what the silica content of the fluids in equilibrium should be, based on the PT conditions of phase separation calculated based on the Cl content. For the Bio9 and P vent fluids, because of the PT region we are in, small decreases in T will lead to large increases in dissolved Si (Figure 2b). Therefore the silica concentrations do not provide definitive conditions but only bounds, which are internally consistent with, and within the range predicted by, the inferred PT conditions at depth based on Cl and the measured seafloor conditions. It should be noted that if a supercritical model was invoked to generate the observed Cl contents, this would require Si concentrations that are significantly *less* than what was actually measured in these fluids, in addition to *higher* temperatures. While it is possible to lose Si to precipitation during upflow, it is difficult to explain significant increases in the silica contents of the fluids at the observed temperatures. Hence the silica data support the earlier choice of a model invoking subcritical, rather than supercritical, phase separation to produce the observed Cl contents in the fluids.

Several other observations and records provide potentially complementary but independent information on changing

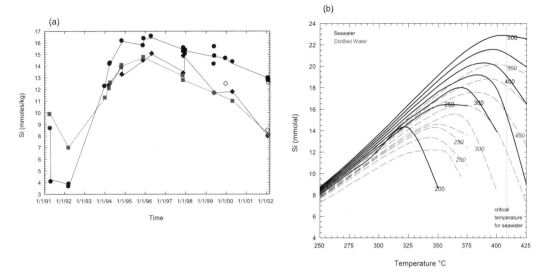

Figure 2. (a) End member silica versus time. (b) Solubility of quartz along isobars in pure water (black solid lines) and in seawater (red dashed lines) calculated with the equation of *Von Damm et al.* [1991]. The critical point (c.p.) for seawater is noted.

conditions within the oceanic crust. Shortly after the eruption was discovered in April 1991 and the vent fluids sampled, four ocean bottom seismometers (OBSs) were deployed [*Hildebrand et al.,* 1991, 1992]. The microearthquakes observed in May 1991 had hypocenter depths from 3–208 m, with a mean depth of 116±40 m [*Hildebrand et al.,* 1992], in excellent agreement with our calculated depth based on the Cl-content of the vent fluids sampled the month before of 130–200m (Figure 1b). The seismic swarm noted by *Sohn et al.* [1998] in March 1995 suggests deeper depths than either the *Hildebrand et al.* data from 1991 or our Cl data for 1994–5 (~400–500 mbsf). The *Sohn et al.* [1998] data are more consistent with the axial magma chamber depth of 1.1 km measured in 1987 [*Kent et al.,* 1993], but have a several hundred meter range in depth.

A temperature spike [*Fornari et al.,* 1998] in the Bio9 fluids was recorded early in 1995, beginning 11 days after the seismic swarm. These events in early 1995 occur approximately mid-way between the 1994 and 1995 point samplings of the fluids. The depth of phase separation had begun to deepen by 1993 (Figure 1b), and the seismic swarm does not change this downward trend. In fact, the fluid compositions do not reverse their trajectories until about a year after the 1995 seismic/temperature event, between the December 1995 and limited April 1996 samplings. The temperature data recorded on Hobo probes from Bio9 and Bio9´ vents, appear to begin a slow upward trend in late 1995 or early 1996, but no specific large changes accompany this reversal of fluid chemistries [*D. Fornari and T. Shank,* pers. comm.]. As the East Pacific Rise acoustic hydrophone array (AHA) did not

begin monitoring in this area until May 1996, we have no other records that provide signals when the fluid chemistries began to reverse.

Another important event that occurred subsurface in this area is a postulated replenishment with fresh lava of the magma chamber, based on the unusually high CO_2 and He that first appeared in the vent fluid samples collected in December 1993 and March 1994 [*Lilley et al.,* 2002]. These high gas concentrations were first noted in M vent (Biovent was not sampled until late 1994), furthest to the north, and then migrate southward to Tube Worm Pillar (TWP) (Plate 1). These high CO_2 contents were maintained into at least 2002, and no large changes were noted through the late 1995–1997 time period, when the reversal in chlorinity occurred [*M. Lilley,* pers. comm.].

4.4. Ions Controlled Principally by Water-Rock Interaction

Phase separation has modified the chemistry of the fluids by changing the Cl content substantially. Therefore, in order to truly evaluate what has been gained or lost as a result of water-rock interaction, not only must the absolute concentrations of elements be examined, but also their ratios-to-Cl compared to ambient seawater. The validity of this approach has been demonstrated both experimentally [*Berndt and Seyfried,* 1990] and in the field [*Von Damm et al.,* 2003], i.e., that most elements behave conservatively with respect to Cl during phase separation, the main exception being boron [*Bray and Von Damm,* 2004].

A potential complicating factor is that halite (NaCl) may form and subsequently re-dissolve within the oceanic crust. The best indicators of this are changes in the Br/Cl ratios in the flu-ids, as Br is excluded from the halite structure. Hence Br/Cl ratios that are elevated with respect to seawater are indica-tive of halite formation, while ratios that are lower than sea-water are indicative of halite dissolution. An independent check on this interpretation is that the dissolution of halite with its 1:1 molar Na:Cl ratio will raise the Na/Cl ratio of the vent fluid above the 0.86 ratio observed in seawater. Con-versely, the precipitation of halite will lower the Na/Cl ratio of a vent fluid. Because albitization will also lower the Na/Cl ratio in the fluids, a Na/Cl ratio lower than the seawater value must be observed to be internally consistent with the inter-pretation that halite is forming. However, the observation of a Na/Cl ratio less than the seawater value does not alone pro-vide definitive indication that halite is precipitating. The amounts of halite calculated to be dissolving (≤ 1.5g) on a /kg of solution basis [*Von Damm*, 2000] would have an insignif-icant impact on the pressure and temperature calculations as discussed previously in section 4.1.

4.4.1. Bromide. The absolute concentrations of Br (Fig-ure 3a) vary from less than the seawater value (early and late in the time series) to greater than the seawater value. The Br concentrations therefore generally track the Cl, as expected because Br behaves conservatively with respect to Cl during phase separation [*Berndt and Seyfried*, 1990; *Von Damm et al.*, 2003]. Br/Cl ratios (Figure 3b) signifi-cantly less than the seawater value, when accompanied by Na/Cl ratios (Figure 3d) significantly greater than the sea-water value (but less than 1) suggest dissolution of halite in 1991–2 [*Oosting and Von Damm*, 1996; *Von Damm*, 2000]. The elevated Br/Cl ratios which occur later in the time series would need to be accompanied by Na/Cl ratios significantly less than the seawater value to support a model invoking halite precipitation. Such variations are not clearly observed (Figures 3b,d).

4.4.2. The Alkali and Alkaline Earth Metals. Sodium is the dominant cation in seawater and vent fluids, hence its concentration tracks that of Cl (Figure 3b,c). It is well known that in some cases Na is removed from hydrothermal fluids and added to the rock as a result of albitization, which also leads to an increase in the Ca content of the fluids. All of the alkali- (Li, Na, K) and alkaline earth- metals (Ca, Sr) (Fig-ures 3 and 4) had concentrations in the vent fluids in 1991–2 that were lower than their concentrations in seawater. This was also true for K, Ca and Sr in some of the fluids at the very end of the time series. However, if the ratios-to-Cl are exam-ined in order to correct for the effects of Cl- complexing and phase separation, it can be observed that after 1992–3, except in the case of Na, that these metals were added to

the fluids from the rock. The ratios-to-Cl that were less than seawater in 1991–2 can be attributed to addition of small amounts of Cl from the dissolution of halite (NaCl) consis-tent with my interpretation (discussed above) of the low Br/Cl ratios at this site. What is striking is that for several of these elements (Li, K, Sr) their ratios to Cl were quite con-stant after ~1994 (Figure 4), even though their absolute con-centrations continued to change. This provides strong evidence that the fluid compositions for these elements were controlled by fluid-mineral equilibria, or at least steady-state. Our interpretation is that water-rock interaction was rel-atively minimal prior to 1994, but that the fluids and the rocks reached equilibrium (or steady-state) in ~1994. The pattern for Na and Ca is somewhat different (Figure 3). The Na/Cl ratio reached a minimum in 1996–7, and the Ca/Cl reached a maximum at this time. These results suggest that albitization may have been occurring at this time. *Seyfried and Shanks* [2004 and references therein] have noted that many seafloor hydrothermal fluids are suggestive of equili-bration with rocks between An60 and An82, and have sug-gested this is location dependent. The maximum in Ca/Cl observed here in ~1996, when subseafloor temperatures are inferred to have been at a maximum, may suggest that the amount of albitization at this site is varying temporally as a function of changes in the PT conditions, i.e., at higher PT conditions a more albite rich (lower An number) plagioclase is in equilibrium with the fluids. Unfortunately detailed data on plagioclase equilibrium concentrations are only avail-able at 400°C and 500 bars, and not at a variety of PT con-ditions around this window.

4.4.3. Manganese. Although Mn is a transition metal it rarely forms sulfide minerals, and therefore behaves more similarly to the alkaline earth metals with its predominately +2 charge. Manganese was extremely elevated in vent fluids com-pared to its concentration in seawater at all points in the time series (Figure 5). It too reached a maximum in ~1996, and then decreased after that time, especially in the Bio9 vent area fluids, where the Cl concentration decreased more than in the P vent fluids. In 2002 the Mn/Cl values were distinctly higher in P vent compared to Bio9, as was also true for Sr/Cl. This was presumably related to the different PT conditions within the reaction zones for these two vents.

4.4.4. pH and Alkalinity. The pH (measured shipboard at 25°C and 1 atmosphere) was acid, and the (total) alkalinity was negative (Figure 5) throughout this time period. These are both typical observations for black smoker fluids exiting from bare basalt. The 1991 values were unusually low, but aside from this there were no observable temporal trends in either parameter.

4.5. Sulfur and Iron

Sulfur, whose abundance and form in the vent fluids are controlled by both phase separation and water-rock reaction, and Fe, are discussed together as the abundance of both are ultimately controlled by the solubility of Fe-S minerals. The sulfate concentrations were close to zero in most cases (Figure 5). Sulfate is subject to artifacts, especially in samples collected in the gas tight bottles. If any anhydrite was incorporated into the sample, the measured sulfate concentrations were too high. Although gross contamination can be identified by coincident spikes in Ca and/or Sr, small anomalies in sulfate are more difficult to recognize. All of the analyses reported here are consistent with quantitative removal of sulfate. Negative sulfate end members can be the result of substantial mixing with seawater prior to sampling [McDuff and Edmond, 1982]. While mixing may have occurred in P vent fluids in 1991 when no chimneys were present, it does not appear to have been a major process in these vents throughout this time series. All of the sulfur was present in the reduced form of H_2S (Figure 6). As H_2S is a gas, it will preferentially partition into the vapor phase. The H_2S concentrations were at a maximum at the start of the time series, reached minima in ~1994–5, and then increased once again.

For Fe (Figure 6), water-rock reaction clearly is a major source as Fe concentrations in the fluids generally exceeded 1 mmole/kg, and reached a maximum in 1995–6 when Cl was also at a maximum. Iron, like the other cations, was present primarily as a chlorocomplex in the vent fluids. The maximum observed Fe/Cl ratio occurred in 1991–2, at the time of the eruptions, and my preferred interpretation is that this is related to critical phenomena [Von Damm et al., 2003]. Further evidence of the preferential mobilization of Fe early in the time series is also demonstrated by the high Fe/Mn ratios observed at this time (Figure 6d).

The iron and sulfide contents of the vent fluids, as reduced species, provide potential energy sources for chemosynthetic organisms. The concentrations of these species are tightly linked to each other, as both are usually limited by the temperature-dependent solubility of Fe-S mineral phases. With the Fe present mostly as Cl-complexes and the H_2S preferentially partitioned into the vapor phase, the relative concentrations of these two species, in conjunction with Cl-data, for example, can also provide constraints on proximity to the critical point [Von Damm et al., 2003] and other information about subsurface conditions, such as the redox state [Seyfried and Ding, 1993]. The rise in the absolute concentration of Fe (Figure 6) correlated with the inferred increase in reaction depth and temperature (Figure 1), as well as with the decrease in the Fe/H_2S ratio in the diffuse flow fluids which correlated with the demise of the BM9 Riftia community [Shank et al.,

1998; Von Damm and Lilley, 2004]. The decrease in the Fe content of the high temperature Bio9 vent fluids in 2002 may be related to the appearance of small (young) Riftia observed on the smoker in 2002.

While I infer that chlorocomplexing and critical phenomena were most responsible for the high Fe/Cl early in the time series, a secondary affect may also have been the redox state. Early in the time series the H_2, as well as H_2S, concentrations in the vent fluids were extremely high [Von Damm and Lilley, 2004]. The high H_2 in particular is an indicator of low $f(O_2)$.

5. TIME LINE

It is important to correlate the changes observed in the vent fluids with other specific records of events in this area. In 1991 there was an eruption in this area, and $^{210}Po/^{210}Pb$ dating of the basalts [Rubin et al., 1994] confirmed a second eruption in 1992 surrounding these particular vents. Starting in December 1993 the CO_2 levels in the fluids became extremely elevated, although based on sample quality it is difficult to tell if the maximum was reached in late 1993 or March 1994 [Lilley et al., 2002 and pers. comm]. This result suggests that fresh magma moved into deeper parts of the system and was degassing. It is important to note that this occurred after both the 1991–2 eruptions and also that the major and minor element chemistry of the vent fluids showed no evidence for an eruption or shallow emplacement of magma at that time. By late 1993 the chloride content of these vent fluids began to increase and all of the chemical evidence was consistent with a deepening of the reaction zone, which could also be viewed as an increasing path length for the vent fluids within the hydrothermal system. By early 1994, although most of the chemical species continued to change in absolute concentration, at about the 1000 day point many of the elemental ratios-to-Cl became stable (e.g., Li, K). The increase in Cl and Si suggests a deepening/lengthening of the fluid flow paths that continued gradually into late 1994. In early 1995, a seismic swarm was noted [Sohn et al., 1998] as well as a temperature increase in the continuously recorded temperature record from the Bio9 vent [Fornari et al., 1998]. Presumably the increase in temperature was related to a pulse of warmer water that entered the hydrothermal system related to the seismic event. By the time we returned to the area with Alvin in fall 1995, there was obviously much more iron-oxide present on the tube worm colonies in this area [Shank et al., 1998], and the Fe/H_2S ratio in the diffuse flow fluids where the animals lived had also increased [Von Damm and Lilley, 2004]. All of these observations are consistent with a deepening of the reaction zone, with the higher pressure and temperature conditions of reaction resulting in changes in the Cl, Si and increases in the Fe content of the vent fluids. However, it must

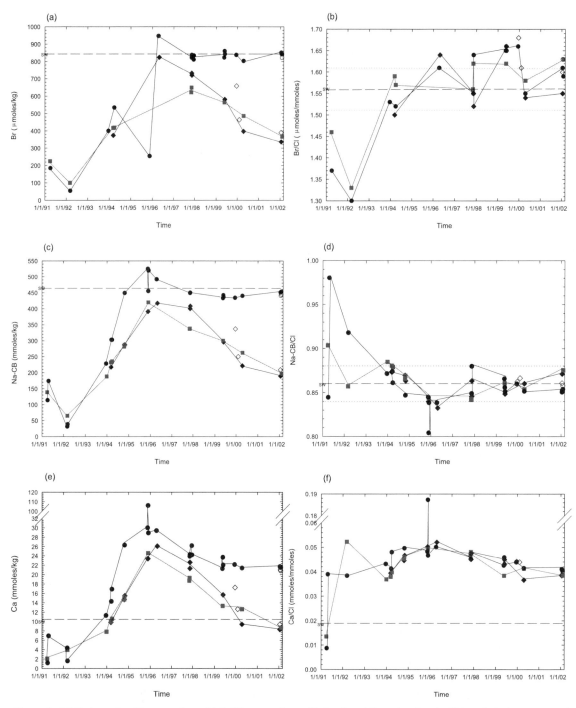

Figure 3. (a) End member Br versus time, (b) Br/Cl versus time with the dotted lines showing the 3% analytical error around the seawater value, (c) End member Na-charge balance (CB) versus time, (d) Na-charge balance/Cl versus time with the dotted lines showing the 2% analytical error around the seawater value, (e) End member Ca versus time, (f) Ca/Cl versus time. The data that scatter high in figures (e) and (f) are likely from gas tight samples (Table 1) that have entrained anhydrite, and similar spikes will be seen in the Sr and SO$_4$ data in Figures 4 and 5. Symbols as in Figure 1.

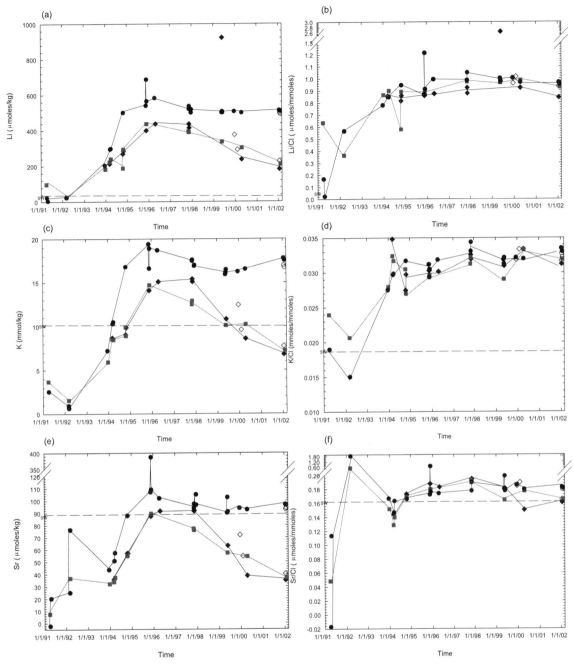

Figure 4. (a) End member Li versus time, (b) Li/Cl versus time, (c) End member K versus time, (d) K/Cl versus time, (e) End member Sr versus time, (f) Sr/Cl versus time. Symbols as in Figure 1.

be noted that the fluid compositions for the high temperature vents continued the trajectory they were on *prior* to these event(s), with increasing Cl and Si, and relatively stable element-to-Cl ratios, i.e. there were no pronounced changes in the chemical evolution of the vent fluids that correlated specifically with the seismic swarm, and no reversal in direction for any of the chemical trends. One model to explain what

occurred is that due to hydrothermal cooling, the reaction zone suddenly deepened, as in a cracking front model. However an equally valid but perhaps more speculative model is that the cracking was a result of magma injection from below, perhaps as a time lag associated with the magma resupply first noted in late 1993, and continued degassing. The seismic data are not sufficient to determine if this was a cracking

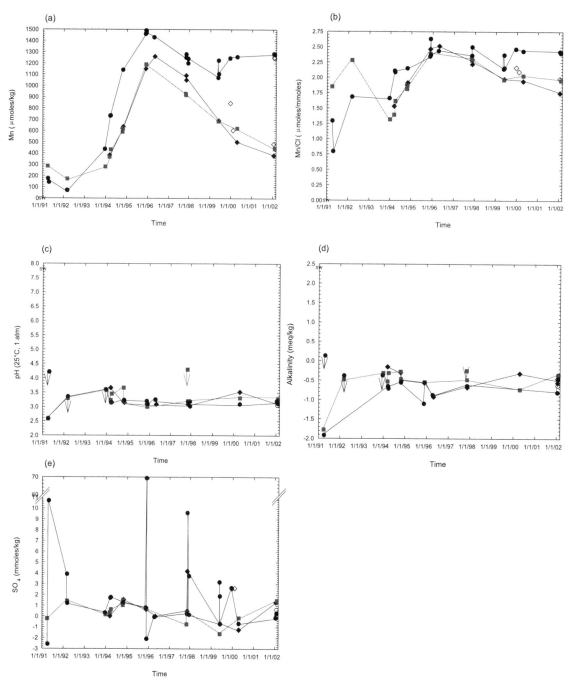

Figure 5. (a) End member Mn versus time, (b) Mn/Cl versus time, (c) End member pH (25°C and 1 atm) versus time, (d) End member total alkalinity versus time, (e) End member SO_4 versus time. For pH and alkalinity if end members could not be calculated due to poor sample quality, maximum values are indicated by arrows (see *Von Damm* [2000] for additional discussion). Symbols as in Figure 1.

event "up" or "down." What is clear, however, is that the hydrothermal fluids were reacting at deeper and hotter PT conditions than they were the previous year. Perhaps it is the seismic events that provided the cracks that facilitated, about

a year later in January 1996, the small spikes of increasing temperature that were noted in both the high and low temperature vents as recorded on the Hobo and vemco temperature probes [*Fornari et al.,* this volume]. By the time of the sampling of

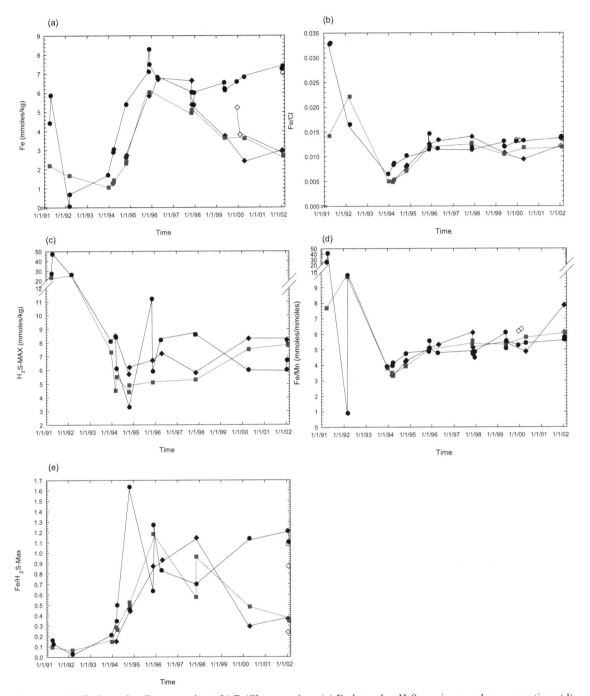

Figure 6. (a) End member Fe versus time, (b) Fe/Cl versus time, (c) End member H_2S-maximum values versus time, (d) Fe/Mn versus time, (e) Fe/H_2S-maximum versus time. For discussion of H_2S-maximum calculation see *Von Damm* [2000]. Symbols as in Figure 1.

fluids from P vent in April 1996 the Cl and Si contents of the fluids had begun to decrease, and the temperature to increase. This decrease in Cl and Si and increase in temperature was also observed in late 1997, when fluids from the Bio9 vents and P were next collected. These changes continued into 2002, the

last data set reported here. Therefore it appears that something occurred in early 1996 which I would interpret as the beginning of melt rising in the crust. Unfortunately there are no seismic data from this time, and as the acoustic hydrophone array (AHA) did not begin recording in this area until May

1996, thus there are no records of T-phase signals for this location from this time. By approximately 1999 there were also more black smokers in the Bio9 area, which also suggests an increase in the amount of available heat at this site. The very high temperatures in 2002, which were the highest measured since a year after the last volcanic eruption at this site in 1992, as well as the chemistry of the Bio9 vents which I interpret as demonstrating that the heat source is very shallow, suggest to me that this area is poised for a new eruption. *Macdonald and Fox* [1988] predicted that the recurrence interval for volcanic eruptions on this inflated area of ridge was ~10 years. Therefore geophysical predictions are in agreement with our predictions based on detailed measurements of the hydrothermal systems.

These results are, however, in contrast to some models previously proposed to describe the evolution of a hydrothermal system such as the "Lupton Curves" [*Baker,* 1995; *Butterfield et al.,* 1997]. The vent fluids at Bio9 never become the "brine" or liquid phase formed as a result of phase separation. The chlorinity in this vent began to drop again before it ever reached the levels in seawater. I would therefore argue that this model does not apply to a fast spreading ridge system where the magmatic time scale is too rapid to allow the brine phase to be expelled from the seafloor. This, however raises two additional points. The first is that there is a Cl mass balance problem, and the second is: where is the missing Cl being stored. The higher Cl fluids must either be stored within the oceanic crust or, as this is the high point of the ridge, perhaps they are traveling downhill subsurface away from the bathymetric high and venting elsewhere. Higher Cl-containing fluids are found venting >10 km to the south [*Von Damm et al.,* 1995; *Oosting and Von Damm,* 1996; *Von Damm,* 2000]. When the first fluids were collected from this site in 1991, it was clear that they were not brines that had been stored previously in the crust based on their elemental composition [*Von Damm,* 2000]. It is, of course, possible that the very first outward pulse of fluids preceded our sample collection and did contain these fluids. If we are able to track the 9°50′N site up to the next eruptive event, we may be able to address whether the brines are stored locally in the crust in this area. Such storage has been observed further south on this ridge at 9°16.8′N [*Von Damm et al.,* 1998] where a vent expelling vapor was found to be expelling the conjugate liquid phase three years later, without renewed volcanic activity.

6. CONCLUSIONS

The hydrothermal system in the 9°50′N area is very dynamic, and continues to experience pronounced temporal variability >10 years after the last known volcanic eruption at this site. The changes in the composition of the vent fluids

from the 9°50′N area on the East Pacific Rise have been profound (Figures 1–6). The temporal variability at this site is unmatched by any other site that has been studied on the global ridge system. Furthermore this temporal evolution does not follow previously proposed models and I argue that these models may not be applicable to a magmatically robust ridge crest, where the timing of replenishment of the magma supply is relatively short (decadal). Several types of records exist demonstrating temporal variability, but our ability to tie them together to understand the system as a whole has been compromised by their limited overlap in time. My preferred interpretation is that the fluid compositions provide evidence of magma migration within the crust on time scales that have not previously been discernable in the oceanic crust. My hope is that future studies will help to calibrate and test this proposed model, in order that previously collected data on fluid compositions may be used to better understand crustal processes.

Acknowledgments. We thank the chief scientists and science parties on AdVenture cruises 1–9, and Legacy for indulging our sample collection needs. Numerous people at UNH contributed to the data presented here including A. Bray, M. Brockington, L. Buttermore, R. Gallant, R. Kozlowski, J. Loveless, K. O'Grady, S. Oosting and C. Parker. We also express our gratitude to our colleagues at other institutions who contributed through discussion, bottle cleaning, sharing of data, etc. who are too numerous to name here. We also thank the officers and crew of the *R/V Atlantis II* and *R/V Atlantis* and the *Alvin* group. This research was funded by the National Science Foundation through grants OCE-9101440, OCE-9296158, OCE-9300508, OCE-9303678, OCE-9419156 and OCE-0002458 to K.L. Von Damm.

REFERENCES

Baker, E. T., Characteristics of hydrothermal discharge following a magmatic intrusion, *Geol. Soc. Spec. Publ., 87,* 65–76, 1995.

Baker, E. T., G. J. Massoth, and R. A. Feely, Cataclysmic hydrothermal venting on the Juan de Fuca Ridge, *Nature, 329,* 149–151, 1987.

Baker, E. T., J. W. Lavelle, R. A. Feely, G. J,. Massoth, S. L. Walker, and J. E. Lupton, Episodic venting of hydrothermal fluids from the Juan de Fuca Ridge, *J. Geophys. Res., 94,* 9237–9250, 1989.

Berndt, M. E. and W. E. Seyfried Jr., Boron, bromine, and other trace elements as clues to the fate of chlorine in mid-ocean ridge vent fluids, *Geochim. Cosmochim. Acta., 54,* 2235–2245, 1990.

Bischoff, J. L., Densities of liquids and vapors in boiling NaCl-H_2O solutions: A PVTX summary from 300° to 500°C, *Am. J. Sci., 291,* 309–338, 1991.

Bischoff, J. L. and R. J. Rosenbauer, Liquid-vapor relations in the critical region of the system NaCl-H_2O from 380° to 415°C: A refined determination of the critical point and two-phase boundary of seawater, *Geochim. Cosmochim. Acta, 52,* 2121–212, 1988.

Bowers, T. S., A. C. Campbell, C. I. Measures, A. Spivack, M. Khadem, and J. M. Edmond, Chemical controls on the composition of vent fluids at 13°–11°N and 21°N, East Pacific Rise, *J. Geophys. Res., 93,* 4522–4536, 1988.

Bray, A. M. and K. L. Von Damm, The role of phase separation and water-rock reactions in controlling the boron content of mid-ocean ridge hydrothermal vent fluids, *Geochim. Cosmochim. Acta*, in revision, 2004.

Butterfield, D. A., and G. J. Massoth, Geochemistry of North Cleft segment vent fluids: temporal changes in chlorinity and their possible relation to recent volcanism, *J. Geophys. Res., 99,* 4951–4968, 1994.

Butterfield, D. A., I. R. Jonasson, G. J. Massoth, R. A. Feely, K. K. Roe, R. E. Embley, J. F. Holden, R. E. McDuff, M. D. Lilley and J. R. Delaney, Seafloor eruptions and evolution of hydrothermal fluid chemistry, *Phil. Trans. R. Soc. Lond. A, 355,* 369–387, 1997.

Campbell, A. C., T. S. Bowers, C. I. Measures, K. K. Falkner, M. Khadem, and J. M. Edmond, A time series of vent fluid compositions from 21°N, East Pacific Rise (1979, 1981, 1985), and the Guaymas Basin, Gulf of California (1982, 1985), *J. Geophys Res., 93,* 4537–4549, 1988.

Carbotte, S. and K. C. Macdonald, East Pacific Rise 8–11°30′N: evolution of ridge segments and discontinuities from SeaMarc II and three-dimensional magnetic studies, *J. Geophys. Res., 97,* 6959–6982, 1992.

Charlou, J.-L., Y. Fouquet, J. P. Donval, J. M. Auzende, P. Jean Baptiste, M. Stievenard, and S. Michel, Mineral and gas chemistry of hydrothermal fluids on an ultrafast spreading ridge: East Pacific Rise, 17° to 19°S (NAUDUR cruise, 1993) phase separation processes controlled by volcanic and tectonic activity, *J. Geophys. Res., 101,* 15,899–15,919, 1996.

Detrick, R. S., P. Buhl, E. Vera, J. Mutter, J. Orcutt, J. Madsen and T. Brocher, Multi-channel seismic imaging of a crustal magma chamber along the East Pacific Rise, *Nature, 326,* 35–41, 1987

Fornari, D. J., T. Shank, K. L. Von Damm, T. K. P. Gregg, R. M. Haymon, M. Lilley, G. Levai, A. Bray, M. R. Perfit, and R. Lutz, A dike intrusion or crustal fracturing event inferred from time-series temperature measurements at high-temperature hydrothermal vents on the East Pacific Rise 9°49–51′N, *Earth & Planet. Sci. Lett., 160,* 419–431, 1998.

Fornari, D. J., M. Tivey, H. Schouten, M. Perfit, K. Von Damm, D. Yoerger, A. Bradley, M. Edwards, R. Haymon, T. Shank, D. Scheirer and P. Johnson, Submarine lava flow emplacement processes at the East Pacific Rise 5°50′N: implications for hydrothermal fluid circulation in the upper oceanic crust, *The Thermal Structure of the Oceanic Crust and the Dynamics of Hydrothermal Circulation*, C. German, J. Lin and L. Parson., eds., *AGU Monograph,* submitted, 2004 [this volume].

Haymon, R. M., D. J. Fornari, M. H. Edwards, S. Carbotte, D. Wright, and K. C. Macdonald, Hydrothermal vent distribution along the East Pacific Rise crest (9°09′–54′N) and its relationship to magmatic and tectonic processes on fast-spreading mid-ocean ridges, *Earth Planet. Sci. Lett., 104,* 513–534, 1991.

Haymon, R. M., et al., Volcanic eruption of the mid-ocean ridge along the East Pacific Rise crest at 9°45–52′N: Direct submersible observations of seafloor phenomena associated with an eruption in April, 1991, *Earth Planet. Sci. Lett., 119,* 85–101, 1993.

Hildebrand, J. A, S. C. Webb, and L. M. Dorman, Monitoring Ridge

Crest Activity with Ocean-Bottom Microseismicity, *RIDGE Events, 2,* 6–8, 1991.

Hildebrand, J. A, S. C. Webb, L. M. Dorman, A. E. Schreiner, M. A. McDonald, and W. C. Crawford, Microseismicity of a Mid-Ocean Ridge Volcanic Eruption: The East Pacific Rise at 9°50′N, *Eos, Transactions, American Geophysical Union, 73,* 530, 1992.

Hooft, E. E. and R. S. Detrick, The role of density in the accumulation of basaltic melts at mid-ocean ridges, *Geophys. Res. Letts., 20,* 423–426, 1993.

Kent, G. M., A. J. Harding and J. A. Orcutt, Distribution of magma beneath the East Pacific Rise between the Clipperton transform and the 9°17′N Deval from forward modeling of common depth point data, *J. Geophys. Res., 98,* 13,945–13,969, 1993.

Lilley, M. D., J. E. Lupton and E. A. Olson, Using CO_2 and He in vent fluids to constrain along axis magma dimension at 9°N, EPR, *Eos Trans. AGU, 83, Fall Meeting Suppl.,* F1386, 2002.

Lilley, M. D. D. A. Butterfield, J. E. Lupton and E. J. Olson, Magmatic events can produce rapid changes in hydrothermal vent chemistry, *Nature, 422,* 878–881, 2003.

Lowell, R. P. and L. N. Germanovich, Hydrothermal processes at mid-ocean ridges: results from scale analysis and single-pass models, *The Thermal Structure of the Oceanic Crust and the Dynamics of Hydrothermal Circulation*, C. German, J. Lin and L. Parson., eds., *AGU Monograph,* submitted, 2004 [this volume].

Macdonald, K. C. and P. J. Fox, The axial summit graben and cross-sectional shape of the East Pacific Rise as indicators of axial magma chambers and recent volcanic eruptions, *Earth Planet. Sci. Lett., 88,* 119–131, 1988.

Magenheim, A. J., A. J. Spivack, J. C. Alt, G. Bayhurst, L.-H. Chan, E. Zuleger and J. M. Gieskes, 13. Borehole fluid chemistry in Hole 504B, Leg 137: Formation water or in situ reaction, *Proc. of the Ocean Drilling Program, Sci. Results*, J. Erzinger, K. Becker, H. J. B. Dick and L. B. Stokking, eds., 137/140, 141–152, 1995.

McDuff, R. E., and J. M. Edmond, On the fate of sulfate during hydrothermal circulation at mid-ocean ridges, *Earth Planet. Sci. Lett., 57*, 117–132, 1982.

Oosting, S. E. and K. L. Von Damm, Bromide/chloride fractionation in seafloor hydrothermal fluids from 9–10°N East Pacific Rise, *Earth & Planet. Sci. Lett., 144,* 133–145, 1996.

Perfit, M. R. and W. W. Chadwick, Jr., Magmatism at mid-ocean ridges: constraints from volcanological and geochemical investigations, *Faulting and Magmatism at Mid-Ocean Ridges*, W. R. Buck et al., eds. *AGU Monograph 106 ,* 59–115, 1998.

Ravizza, G., J. Blusztajn, K. L. Von Damm, A. M. Bray, W. Bach and S. R. Hart, Sr isotope variations in vent fluids from 9°46–9°54′N East Pacific Rise: Evidence of a non-zero-Mg fluid component, *Geochim. Cosmochim. Acta*, 65, 729–739, 2001.

Rubin, K. H., J. D. MacDougall, and M. R. Perfit, $^{210}Po/^{210}Pb$ dating of recent volcanic eruptions on the sea floor, *Nature, 468*, 841–844, 1994.

Seyfried, W. E. Jr., and K. Ding, The effect of redox on the relative solubilities of copper and iron in Cl-bearing aqueous fluids at elevated temperatures and pressures: an experimental study with application to subseafloor hydrothermal systems, *Geochim. Cosmochim. Acta, 57,* 1905–1917, 1993.

Seyfried, W. E. Jr., and W. C. Shanks, III, Alteration and mass transport in high-temperature hydrothermal systems at mid-ocean ridge: controls on the chemical and isotopic evolution of axial vent fluids, *Hydrology of the Oceanic Crust,* H. Elderfield, ed., in press, 2004.

Shank, T. M., D. J. Fornari, K. L. Von Damm, M. D. Lilley, R. M. Haymon, and R. A. Lutz, Temporal and spatial patterns of biological community development at nascent deep-sea hydrothermal vents (9°50′N, East Pacific Rise), *Deep-Sea Res. II, 45,* 465–515, 1998.

Sohn, R. A., D. J. Fornari, K. L. Von Damm, J. A. Hildebrand, and S.C. Webb, Seismic and hydrothermal evidence for a cracking event on the East Pacific Rise at 9°50′N, *Nature, 396,* 159–16,1998.

Von Damm, K. L., Chemistry of hydrothermal vent fluids from 9–10°N, East Pacific Rise: "Time zero" the immediate post-eruptive period, *J. Geophys. Res., 105,* 11203–11222, 2000.

Von Damm, K. L. and M. D. Lilley, 2004. Diffuse flow hydrothermal fluids from 9°50′N East Pacific Rise: Origin, evolution and biogeochemical controls, *The Subsurface Biosphere at Mid-Ocean Ridges, AGU Monograph 144,* W. S. D. Wilcock, E. F. DeLong, D.S. Kelley, J. A. Baross and S. C. Cary, eds., 243–266.

Von Damm, K. L., J. L. Bischoff and R. J. Rosenbauer, Quartz solubility in hydrothermal seawater: an experimental study and equation describing quartz solubility for up to 0.5 M NaCl solutions, *Am. J. Sci., 291,* 977–1007, 1991.

Von Damm, K. L., S. E. Oosting, R. Kozlowski, L. G. Buttermore, D. C. Colodner, H. N. Edmonds, J. M. Edmond and J. M. Grebmeier, 1995. Evolution of East Pacific Rise hydrothermal vent fluids following a volcanic eruption, *Nature, 375,* 47–50, 1995.

Von Damm, K. L., L. G. Buttermore, S. E. Oosting, A. M. Bray, D. J. Fornari, M. D. Lilley, and W. C. Shanks, III, Direct observation of the evolution of a seafloor "black smoker" from vapor to brine, *Earth & Planet. Sci. Lett., 149,* 101–112, 1997.

Von Damm, K. L., M. D. Lilley, W. C. Shanks III, M. Brockington, A. M. Bray, K. M. O'Grady, E. Olson, A. Graham, G. Proskurowski and the SouEPR Science Party, Extraordinary phase separation and segregation in vent fluids from the southern East Pacific Rise, *Earth & Planet. Sci. Lett., 206,* 365–378, 2003.

Karen L. Von Damm, University of New Hampshire, Complex Systems Research Center, Institute for the Study of Earth, Oceans, and Space, 39 College Road, Durham, NH 03824–3525.

Vigorous Venting and Biology at Pito Seamount, Easter Microplate

D. F. Naar[a], R. Hekinian[b], M. Segonzac[b], J. Francheteau[c], and the Pito Dive Team*

A *Nautile* submersible investigation of Pito Seamount documents vigorous hydrothermal venting at 23° 19.65′S, 111° 38.41′W and at a depth of 2270 m. The data indicate the volcano is young and recently active, as predicted from analyses of SeaMARC II side-scan and swath bathymetry, and geophysical data. Pito Seamount lies near Pito Deep (5980 m), which marks the tip of the northwestward propagating East rift of the Easter microplate. Bathymetry surrounding Pito Seamount consists of a series of ridges and valleys with relief up to ~ 4 km. The 4-km submersible-transect to the summit of Pito Seamount crossed areas of very glassy basalt with little or no sediment cover, suggesting the lava flows are very young. Most of the lava samples from Pito Seamount are depleted normal MORB (mid-ocean ridge basalt). Lava samples associated with active and dead hydrothermal vents consist of phyric and aphyric transitional and enriched MORB. Sulfides consist primarily of sphalerite and pyrite, with traces of chalcopyrite. The active hydrothermal chimney on Pito Seamount has a small, undiversified biological community similar to northern East Pacific Rise vent sites (alvinellid worms, bythograeid crabs and bythitid fishes) and western Pacific back-arc basin sites (alvinocaridid shrimps). No vestimentiferan worms were observed. Previous geophysical data, and new geochemical data and visual observations, suggest that the vigorous black smoker is a result of deep, extensive crosscutting faults formed by extensive tectonic thinning of Pito Deep, and a very robust magmatic supply being supplied from upwelling asthenosphere. Although no biological or vent fluid samples were obtained, geological and biological observations, such as the large number of inactive chimneys, old hydrothermal deposits, and starfish, as well as the occurrence of dead mollusks (gastropod and mussels), suggest a recent waning of hydrothermal activity near the summit. The speculative interpretation that Pito Seamount is acting as a focal point for the formation of a new seafloor spreading axis trending northwest (310°) from the seamount summit towards Pito

---

* (R. Armijo[d], J.-P. Cogne[d], M. Constantin[b], J. Girardeau[e], R. N. Hey[f], and R. C. Searle[g])

[a] College of Marine Science, University of South Florida, St Petersburg, Florida, USA

Mid-Ocean Ridges: Hydrothermal Interactions Between the Lithosphere and Oceans
Geophysical Monograph Series 148
Copyright 2004 by the American Geophysical Union
10.1029/148GM13

[b] IFREMER, Centre de Brest, Plouzané, France
[c] Institut Universitaire Européen de la Mer, Université de Bretagne Occidentale, Plouzané, France
[d] Institut de Physique du Globe (IPG), Paris, France
[e] Département des Sciences de la Terre, Université de Nantes, Nantes, France
[f] SOEST, University of Hawaii, Honolulu, Hawaii, USA
[g] Department of Geological Sciences, University of Durham, Durham, UK

Deep is supported by the new data and observations reported here. These include the similar geochemistry of young lava samples obtained from Pito Seamount and from a small volcanic mound within Pito Deep and the strong SeaMARC II side-scan backscatter amplitudes along most of the ~ 50 km rift zone connecting the summit of Pito Seamount to Pito Deep.

1. INTRODUCTION

In November 1993, the French Pito expedition used the *Nautile* submersible to investigate the tectonic boundaries of the extensively-studied Easter microplate (Rusby and Searle, 1993, 1995; Rusby, 1992; Bird and Naar, 1994; Searle et al., 1993; Naar and Hey, 1986, 1989, 1991; Martinez et al., 1991; Naar et al., 1991; Francheteau et al., 1988; Zukin and Francheteau, 1990; Searle et al., 1989; Hey et al., 1985; Engeln and Stein, 1984; Handschumacher et al., 1981; Anderson et al., 1974; Herron, 1972a,b). Details regarding the twenty dives (which are located in Figure 1) are reported in Hekinian et al. (1996). One of these dives was at the Pito Seamount summit (23°19.65′S, 111°38.41′W, 2270 m) located ~50 km southeast from Pito

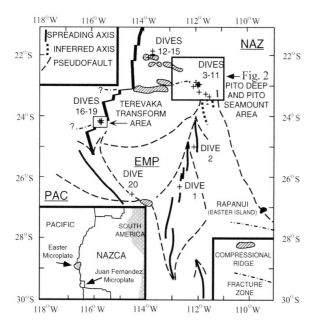

Figure 1. Location map of the tectonic boundaries of the Easter microplate. Legend insets are shown in the NW and SE corners of map. Large-scale inset map of the southeast Pacific Ocean is shown in SW corner of map. The locations of the twenty *Nautile* submersible dives are shown as plus (+) symbols. PAC, EMP, and NAZ represent the Pacific Plate, Easter Microplate, and Nazca plates, respectively. Pito Deep and Pito Seamount are within the large black box in the NE part of the microplate, which denotes the extent of Figure 2.

Deep (23°00′S, 111°56.5′W, 5980 m), which marks the northwestward extent of propagation of the East rift of the Easter Microplate (Francheteau et al., 1988; Naar et al., 1991; Martinez et al., 1991). Two other dives were made nearby: one within Pito Deep (dive 04) and one to the east of Pito Seamount (dive 08). We focus primarily on the dive to the summit (dive 07).

The plate tectonic history of the Easter Microplate (Naar and Hey, 1991; Rusby and Searle, 1995) appears to have played an important role in the formation of the Pito Deep area. The East rift of the microplate has propagated steadily northward and then northwestward since ~ 5 Ma (Figure 1). The change in propagation direction appears to have occurred at ~ 2 Ma when more rigid-like rotation of the microplate appears to replace suspected shear deformation of the microplate interior (Naar and Hey, 1991). The cause of this change could be due to edge-driven microplate rotation (Schouten et al., 1993) and/or other dynamic mechanisms (e.g., Neves et al., 2003). This rapid change in plate motion (Searle et al., 1993) resulted in the Easter-Nazca Euler vector moving closer to the propagating rift tip (Naar and Hey, 1989). The nearby Euler vector predicts a "slow" 30 mm/yr full spreading rate at Pito Seamount. The larger Pito rift area is surrounded by smooth seafloor formed by "superfast" Pacific-Nazca spreading rates of ~150 mm/yr (using Hey et al. (1995) updated Easter-Nazca rates from Naar and Hey (1989) and the Nuvel-1A Pacific-Nazca rates from DeMets et al. (1994)). This change in propagation direction and seafloor spreading rate within the Pito Rift area appears to have caused large block faulting with vertical relief reaching up to ~ 4 km (Martinez et al., 1991; Naar et al., 1991). This wedge of rough seafloor formed by slow seafloor spreading appears to be still propagating northwestward into the smoother "superfast" ~ 3 Ma seafloor. Continued extensional activity is indicated by recent nearby earthquake activity (Engeln and Stein, 1984; Naar et al., 1991), analysis of magnetic and gravity data (Martinez et al., 1991), and SeaMARC II 12 kHz side-scan and swath bathymetry data (Naar et al., 1991).

Naar et al. (1991) propose that Pito Seamount is young and serves as focal point for a new seafloor spreading axis to develop. An elongated volcanic rift zone trending 310° from the summit in a general direction towards Pito Deep (which is actually at an orientation of 320° from Pito Seamount—see dark stipple between PS and PD in Figure 2 and the area within the ellipse shown in Plate 1b). This rift zone appears to be young along most of its length, based on the seafloor fault pattern, volcanic character, and the strong homogenous amplitudes of the SeaMARC II side-scan data within the ellipse shown in Plate 1a. One of the main objectives of dive 07 was to test if this volcanic rift zone is an active seafloor spreading center. As will be shown, the submersible observations support the prediction. However, dive 07 only covered 4 km out of the 50 km between the summit and Pito Deep.

Thus, based on side-scan data and young similar rocks analyzed from a small volcano within Pito Deep obtained during dive 04 (Hekinian et al., 1996), we can only infer that the remainder of the axis is active between the Seamount and the Deep.

2. DIVE OBSERVATIONS

Locations of observations and samples from dive 07 are shown in plan view and in cross-section in Figures 3 and 4. Most lava flows are young to very young, fresh, glassy, and display sheet and lobate flow morphology. The glassy nature of the basalt is emphasized by the reflective nature of the rocks in the video and photographs (see background behind the corallimorphid sea-anemone in Plate 2a). Small pillow lava are occasionally observed during the dive, while at the shallowest, easternmost edge of the dive transect, large massive pillows composed of more viscous phyric rock types are observed. Shimmering water was also observed in a few places along the submersible track. Submersible sensors near stations 3 and 10 measured a 0.1°C and 0.2°C increase in the ambient water temperature, respectively (Figures 3 and 4).

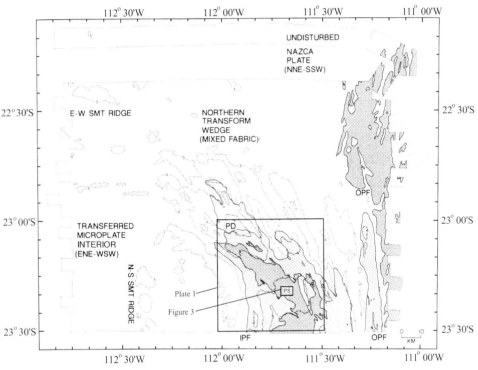

Figure 2. Volcanic interpretation map of the Pito Deep (PD) and Pito Seamount (PS) area (with one-km isobaths superimposed), based on the interpretation of magnetic, gravity, and SeaMARC II side-scan, and swath bathymetry data (Martinez et al., 1991; Naar et al., 1991). Dark stipple represents areas of suspected recent volcanic flows with little sediment cover. Intermediate stipple represents indeterminate areas (suspected to be pre-existing Nazca lithosphere extensively deformed by extension, although recent volcanic activity in certain areas cannot be ruled out). Light stipple near the PD represents regions of obscured side-scan data resulting from the deep rugged bathymetry. IPF and OPF stand for inner and outer pseudofaults of the east rift of the Easter Microplate, respectively. E–W SMT RIDGE and N–S SMT RIDGE refer to the orientation of small chain of volcanoes or compressional ridges (Rusby and Searle, 1993) thought to have formed from the rotation of the Easter microplate into the Nazca plate (see Figure 1 for further distribution of these ridges west of Pito Deep area). Thin line surrounding interpretation represents the extent of the SeaMARC II bathymetry and side-scan mosaic. Submersible dive 04 observed and sampled young basalt on a small volcano within Pito Deep (small closed hill between the "D" of the PD label and the top line denoting the extent of Plate 1). Dive 07 sampled young basalt under the PS label and within the rectangle denoting Figure 3. In addition, dive 08 (located southeast of PS label where the dark stipple crosses the eastern edge of the box denoting Plate 1) sampled young basalt corroborating the interpretation shown. The dark stipple trending northwest from PS to PD is the volcanic rift zone referred to throughout the text. Figure modified from Naar et al. (1991).

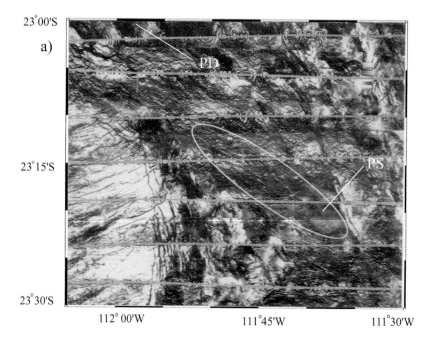

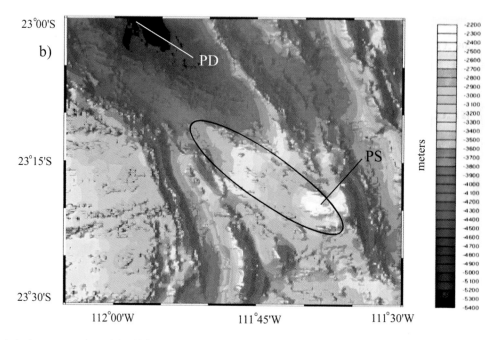

Plate 1. a) A close-up portion of the 12 kHz SeaMARC II side-scan mosaic displaying the volcanic flows between Pito Seamount (PS) and Pito Deep (PD). The dark areas represent strong backscatter amplitudes. Geometry of the seafloor with respect to the towed system and the amount of sediment cover are the primary controls on the backscatter intensity. The area proposed to have the most recent volcanic flows and to be forming a new seafloor spreading axis is enclosed in the white oval. b) Color-shaded bathymetric relief with illumination from the west. The ridge axis that is trending 310 degrees from the summit of Pito Seamount (area within oval) is interpreted to be a new seafloor spreading axis being formed in a direction towards Pito Deep, which lines up with the predicted relative pole of rotation for the Easter and Nazca plates (Naar and Hey, 1989). Where and how the volcanic rift zone may curve and connect with Pito Deep (PD) north of the oval is not clear. Figure modified from Naar et al. [1991].

Plate 2. Video stills (frame grabs) from dive 07: a) *Corallimorphus*-like Corallimorphidae sea anemone attached to a very young lava flow located between sampling stations 3 and 4 (on Figures 3 and 4, ~600 m along track). b) *Freyella*-like Brisingidae asterid. c) Numerous spires with small biological community. d) Vigorous "locomotive-like" venting of black smoker. e) Another view of a chimney spire. f) Black smoker with *Chorocaris*-like Alvinocarididae shrimp on the right wall of the chimney at shadow edge near center of image. g) *Chorocaris*-like Alvinocarididae shrimp, *Bythograea* Bythograeidae crab (see "C"), and *Alvinella pompejana* Alvinellidae polychaete worms (e.g., see "A"). h) Black smoker chimney and white starfish with short arms.

Sheet flows and ponded lava were observed near sampling sites 9A and 11 and in several places along the submersible transect, including near the active hydrothermal vent (shown as "flat flows" on either side of summit in Figures 3 and 4). Sulfide blocks and oxidized hydrothermal sediments were observed (near sampling sites 1, 3, 4, 7, 9, 10, and 11 in Figures 3 and 4). On the upper flank of Pito Seamount there are few faunas, few fish, one *Corallimorphus profundus* (?) corallimorphid sea-anemone (ca 15 cm high) attached on a young glassy basalt (Plate 2a), and some *Freyella*-like brisingid starfish. Also, numerous inactive chimneys standing up to 10 m high were observed near the summit. The main hydrothermal field at the summit of the Pito Seamount is composed of several distinct tall edifices overgrown by sulfide chimneys. These edifices support 70 to 80 nine-armed *Freyella*-like brisingid asterid starfish, especially near their tops (Plate 2b). At the base of this edifice the following biological community was observed: bacterial mats, a shell of a *Phymorhynchus*-like turrid gastropod, ca 5–6 cm long, and some dead *Bathymodiolus*-like mytilid mussels. The most vigorous vent is located at the top of an 8-m tall edifice within the active field, which consists of a few black smokers (Plate 2c). Videotape of this vigorous vent is reminiscent of a "steaming locomotive" (Plate 2d–e). Videotape also shows a thin transition zone of clear liquid exiting the orifice before metallic sulfide precipitates, but the still image of the videotape is not as clear as the playback of the original videos

(Plate 2f). These observations suggest a large flux rate of high temperature vent fluid. Unfortunately, the *Nautile* was not prepared to take temperature probe measurements, vent fluid samples, or biological samples. However, visual observations suggest similarities to other hot vents along the East Pacific Rise and Juan de Fuca. Many of these vents display evidence of phase separation in the vent fluids (e.g., Lilley et al., 1993; Seyfreid and Ding, 1995; Butterfield et al., 1994, 1997; Von Damm, 1995a,b; Von Damm et al., 1997, 2003—and references therein).

Around the vent, the associated fauna consist of about ten *Alvinella pompejana* alvinellid polychaetes (and as many empty tubes), about ten *Bythograea* sp. bythograeid crabs (Plate 2g), about ten *Chorocaris*-like alvinocaridid shrimps (Plate 2f), and a few galatheid crabs. About ten brisingid asteroid starfish have colonized the black-smoker walls, as well as one white starfish with short arms (Plate 2h). Two fish (one synaphobranchid and one *Bythites*-like bythidid) were observed swimming next to the edifice.

3. COMPOSITION OF SULFIDE DEPOSITS AND ROCK SAMPLES

The sulfide deposits were sampled from inactive and active hydrothermal vent sites on the northwest rift zone and near the summit of Pito Seamount (Figures 3 and 4). They are

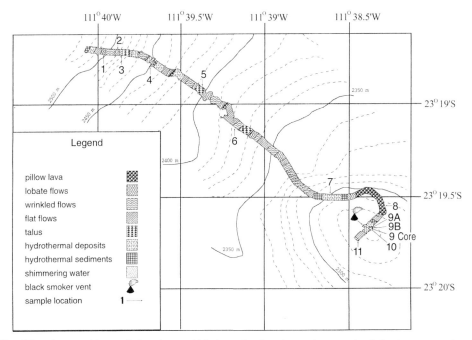

Figure 3. Dive 07 track map with sample locations and black smoker location at the summit of Pito Seamount (closed dashed contours). Contours are from SeaMARC II swath bathymetry data. Actual depth profile measured by submersible is shown in Figure 4. Southwestern edge of track represents actual transponder navigation; northeastern edge is a smoothed version for the purposes of pattern filling. Symbols explained in legend.

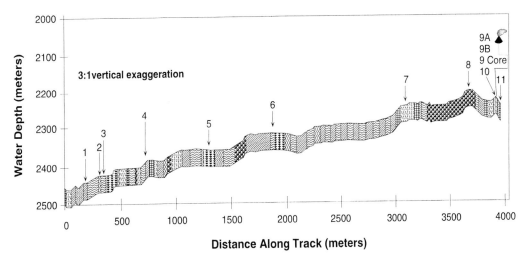

Figure 4. Dive 07 cross-section with locations of samples and blacksmoker. Symbols identified in Figure 3. Upper profile represents actual bathymetry measured by the submersible. Lower profile is shown for purposes of pattern filling.

associated with lobate sheet flows of normal MORB composition (Tables 1 and 2). Polished sections and microprobe analyses of the samples show that the chimneys consist mainly of Zn-Fe sulfides. Sphalerite and pyrite occur as major phases, while chalcopyrite forms a minor constituent (Table 3). A few marcasite crystals with radiating texture associated with pyrite and rare pyrrhotite were also observed in the samples from the active chimney (07–10, -10a). Chalcopyrite surrounding large agglomerations of pyrite grains was observed. Occasional magnetite crystals with exsolution lamellae of ilmenite were also observed. Si- and Fe-oxyhydroxide phases often form thin laminae around pyrite grains. Collomorph sphalerite and opaline material was also detected in sample 07–10. Opal and Fe oxyhydroxide precipitate late, when the temperature of hydrothermal fluid decreases and when the oxidation stage increases (Figure 5). The hydrothermal phases encountered on the Pito Seamount show comparable paragenesis to what has been observed elsewhere (Renard et al., 1985; Bäcker et al., 1985; Hekinian and Bideau, 1986; Duckworth et al., 1995; Verati et al., 1999). The mineral paragenesis indicates that the hydrothermal fluid precipitates pyrite as an early phase, to be followed by sphalerite and some chalcopyrite replacing pyrite and/or formed as a reaction rim around the pyrite (Figure 5).

The sulfides containing traces of As (700–3400 ppm) are associated with the Cu- and Fe- bearing phases (chalcopyrite and mainly pyrite). The zinc content in both the pyrite and chalcopyrite does not exceed 1.6 wt% (Table 3). Cu is pref-

Table 1. Sample distribution and rock types collected from the Pito Seamount in the Easter Microplate

Sample[1]	Lat(S)	Lon(W)	Depth (m)	Rock type	Geologic setting (Pito Seamount)
07-01	23° 18.73'	111° 39.97'	2456	aphyric basalt	*in situ,* glassy lobated flow
07-02	23° 18.74'	111° 39.88'	2457	aphyric basalt	*in situ,* glassy ropey flow
07-03	23° 18.74'	111° 39.86'	2457	aphyric basalt	*in situ,* glassy ropey flow
07-04	23° 18.79'	111° 39.69'	2392	aphyric basalt	*in situ,* glassy ropey flow
07-05	23° 18.94'	111° 39.40'	2364	aphyric basalt	*in situ,* glassy pillow, Mn coating
07-06	23° 19.13'	111° 39.18'	2328	aphyric basalt	*in situ,* glassy drape-like flow
07-07	23° 19.52'	111° 38.61'	2256	massive sulfide	*in situ,* dead vent
07-08	23° 19.55'	111° 38.32'	2243	HPPB[2]	*in situ,* «yam like» from pillow
07-09A	23° 19.65'	111° 38.41'	2254	aphyric basalt	*in situ,* lobated flow (ponded lava)
07-09B	23° 19.65'	111° 38.41'	2254	massive sulfide	active vent (black smoker)
07-09C	23° 19.65'	111° 38.41'	2255	sediment	*in situ,* hydrothermal core
07-10	23° 19.65'	111° 38.41'	2243	massive sulfide	active vent (black smoker)
07-11	23° 19.72'	111° 38.47'	2268	aphyric basalt	*in situ,* lobated flow (ponded lava)

[1] Sample positions are also shown in Figures 3 and 4
[2] HPPB = Highly phyric plagioclase basalt (megacrysts > 15%).

Table 2. Bulk rock analyses of basalts from Pito Seamount

Samples wt%	PI07-06 basalt	PI07-08 basalt	PI07-11 basalt	PI08-01 basalt	
SiO_2	50.68	48.86	49.98	48.43	
TiO_2	1.55	1.06	1.55	1.71	
Al_2O_3	15.18	20.09	15.03	17.32	
Fe_2O_3	1.08	1.15	1.31	1.97	
FeO	8.08	5.74	7.74	6.74	
MnO	0.16	0.15	0.16	0.14	
MgO	7.19	5.81	7.12	7.46	
CaO	11.63	12.86	11.71	11.17	
Na_2O	3.40	2.98	3.32	3.08	
K_2O	0.08	0.08	0.13	0.48	
P_2O_5	0.20	0.16	0.21	0.31	
LOI	-0.26	0.26	0.13	0.40	
TOTAL	98.97	99.20	98.39	99.21	
K/Ti	0.09	0.13	0.14	0.47	
Mg#	0.64	0.67	0.65	0.69	
Ba	11.2	8.12	10.6	46.4	Traces ppm
Sr	184	190	182	181	
Nb	3.2	1.85	3.1	10.1	
Zr	129	83.1	125	132	
Y	32.6	23.	31.8	31.4	
V	292	206	287	257	
Co	36.3	29.2	34.1	34.5	
Cr	266	267	256	267	
Ni	61.6	61.9	56.3	143	
As	0.25	0.81	0.27	0.97	
Cu	77.8	62.0	73.8	49.9	
Zn	77.1	56.1	74.9	73.2	
Cd	0.16	0.07	0.12	0.19	
Pb	2.77	3.76	1.12	1.10	
La	4.92	3.25	4.67	8.73	R.E.E. ppm
Ce	14.41	9.32	14.37	21.73	
Nd	12.58	8.28	12.39	14.03	
Sm	4.21	2.91	4.03	4.17	
Eu	1.44	1.09	1.45	1.57	
Gd	4.32	3.01	4.44	4.58	
Dy	5.18	3.89	5.22	5.24	
Er	3.32	2.38	3.25	3.17	
Tm	0.46	0.35	0.48	0.46	
Yb	3.03	2.21	2.98	3.21	
Lu	0.49	0.35	0.49	0.46	

LOI = lost in ignition (H_2O, CO_2, S). The analyses were done by Induced Coupled Plasma Mass Spectroscopy (IC-PMS) at the Centre de Recherche Pétrographique et Géochimique de Nancy (France). Methods and accuracy are found in Govindaraju (1989). PI = Pito cruise. Mg# = Mg^{2+} / ($Mg^{2+} + Fe^{2+}$); {Fe^{2+} / ($Fe^{2+} + Fe^{3+}$)} = 0.9

erentially concentrated (0.2–0.8 wt%) in the sphalerite (Table 3). The FeS mol% of the sphalerite varies between 7% and 23%. This range of variability for the hydrothermal samples from Pito Seamount is generally lower than that observed for the southern EPR near 17°26′S (FeS mol% = 18–45%) (Renard et al., 1985) and is more similar to that found in the 18–21ºS sulfide chimneys (Renard et al., 1985; Bäcker et al., 1985; Verati et al., 1999) (Figure 6). Hannington and Scott (1989)

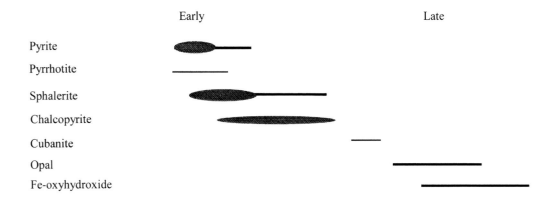

Paragenesis of hydrothermal chimney samples PI 07-10a, 07-10b and 07-07

Figure 5. Mineral paragenesis of zinc-iron rich sulfides from the active and inactive hydrothermal deposits on Pito Seamount.

have shown that the different FeS contents in sphalerite could reflect gold mineralization of the bulk sulfide deposit. Although we did not analyze the hydrothermal samples for gold, the relatively high arsenic content (700–3400 ppm) in the pyrite and chalcopyrite suggests a lower reducing condition of the hydrothermal fluid (Table 3).

The bulk rock samples from Pito Seamount consist of aphyric and highly phyric plagioclase basalts (HPPB). The HPPB (i.e., sample 07–08) consists of more than 15% plagioclase megacrysts forming isolated phases (> 1mm in length) (Table 1 and 2). Based on their chemical characteristics (K_2O and K/Ti ratios) they are classified as N-MORBs ($K_2O<0.1$, K/Ti <0.1, Ba = 8–12 ppm, and La < 5 ppm) and enriched E-MORBs (sample 8–01) ($K_2O>0.25$, and K/Ti = 0.47, Ba = 46 and La = 8.7) (Table 2). The E-MORBs consist of older flows erupted on the eastern flank of Pito Seamount away from the hydrothermal field (Table 1, Figure 3). The distribution of some transitional metals such as Cu, Zn, and Ni are compatible with a process of crystal fractionation. The liquid line of descent (LLD, Nielsen, 1989) is calculated for a melt having a composition close to that of the least evolved melt (N-MORB, 19–02) recovered from the median ridge of Terevaka transform fault (see Figure 1 for location) (Hekinian et al., 1996). The general compositional variability of the rocks follows the trend of the LLD (Figure 7a–c). The lavas from Pito Seamount have undergone a limited range of crystal liquid fractionation (<25%) when compared to the other basalts from the Easter microplate area (Hekinian et al., 1996) (Figure 7a–c). Both the Pito and other basalt samples from the Easter microplate show an increase

in Zn and a variable range of Cu contents as crystal-liquid fractionation increases (Figure 7b–c). Zn increases in the residual liquid as opposed to Ni, which is strongly dependent on the crystallization of olivine (Figure 7a–b). The behavior of Cu is more difficult to predict because of the scattering of the data away from the theoretical fractionation trend (Figure 7c). It is likely that Cu might have more affinities with sulfur and will therefore form magmatic sulfides such as those found in globules and vesicle-filling products during extended crystal-liquid fractionation (Ackermand et al., 1998).

4. DISCUSSION

The chemistry of the samples from dives 04 (small volcano within Pito Deep), 07 (Pito Seamount) and 08 (which sampled slightly coated elongated pillow lavas draping down the southeast side of Pito Seamount) are compatible with a slow seafloor spreading origin producing depleted MORBs at a shallower depth with less crystal fractionation (Hekinian et al., 1996). Furthermore, this lack of enriched basalts implies that some melts have not reached the surface due to rapid lithospheric cooling (or perhaps there was insufficient sampling).

Insight into the processes responsible for distribution of transitional metals in basalts will improve our understanding of the origin and precipitation of sulfides, especially with respect to their lithospheric environment. The chemistry and mineralogy of the rocks play an important role in determining the composition of the hydrothermal fluid (Von Damm, 1995a,b) and that of the hydrothermal precipitates. The distribution of the transitional elements such as Ni, Zn, and Cu

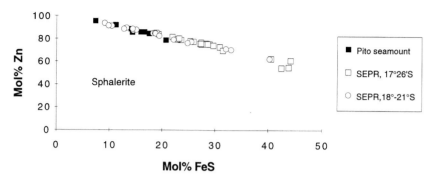

Figure 6. Zn-Fe (Mol %) variation diagram of sphalerite from the Pito Seamount sulfide deposits compared to other Southern East Pacific Rise (SEPR) samples near 17°26′S and 18°-21°S (Renard et al., 1985; Bäcker et al., 1985, and unpublished data).

between the southern EPR, the Easter microplate, and Pito Seamount volcanic samples does not show any substantial variation. The Ni and Zn variation in the basalts from the southern EPR and the microplate indicate that early crystallization of olivine and spinel has had an opposite effect on these two elements. The increase of Zn during crystal-liquid fractionation suggests that it is less compatible than Ni with respect to early crystallizing minerals and it is concentrated in the evolved melt. Cu shows more scattered values around a narrow (60–90 ppm) variation trend during MORB fractionation, and is more likely to reflect the content of the original magma source. Another alternative explanation is that Cu has segregated during the early accumulation of some ore forming bodies.

Similarities also exist between the sulfide deposits encountered along the southern EPR segments (from 17°S to 21°S, Renard et al. 1985; Bäcker et al., 1985; Auzende et al., 1994; Fouquet et al., 1994) and those from Pito Seamount. These similarities suggest that there are comparable ongoing hydrother-

mal processes of leaching and fluid circulation within the MORB-type of volcanic complexes in these regions. Furthermore, the presence of black smokers on top of the Pito seamount suggests high exit temperatures (200°–400°C) as observed elsewhere (Lilley et al., 1993). The sulfides from the Pito Seamount consisting essentially of pyrite, sphalerite and chalcopyrite are similar to those found elsewhere on the EPR (i.e., 18°–22°S, Renard et al., 1985; Bäcker et al., 1985) and on Juan de Fuca (Koski et al., 1981). However, the lead isotopic values of both the sulfides and the glassy basalts are more radiogenic than was found for those encountered on the southern EPR (20°–22°S, Verati et al., 1999). Also, the variability of lead isotopes for the Pito Seamount sulfides suggests heterogeneous crustal source components. This crustal heterogeneity could be related to the presence of more evolved and/or alkali enriched rocks than those encountered on the southern EPR (Verati et al., 1999).

As previously stated, vent biology, vent fluids, or *in situ* vent temperatures were not measured (only ambient tem-

Table 3. Average electron microprobe analyses of sulfide minerals from Pito Seamount

sample	07-10b	07-10b	07-10a	07-07	07-07	07-10b	07-10b	07-07	07-10b	07-10b	07-10b
mineral	py	py	py	py	py	cp	po	cp	sph	sph	sph
average	4	2	4	3	1	2	1	9	3	6	2
Fe wt%	45.37	46.06	46.54	46.29	46.16	30.23	61.38	29.28	13.49	9.01	5.71
Zn	0.49	0.77	-	0.56	-	1.24	-	1.53	50.67	56.26	60.93
S	54.12	53.34	53.08	52.69	52.37	34.74	32.61	34.63	33.40	33.31	33.51
Cu	-	-	-	-	-	32.95	0.10	34.00	0.83	0.34	0.18
As	0.34	-	0.08	0.26	-	-	-	0.07	-		
Total	100.32	100.17	99.70	99.80	98.53	99.15	94.09	99.51	98.39	98.91	100.33

py = pyrite, cp = chalcopyrite, po = pyrrhotite, sph = sphalerite
The analyses were performed by electron microprobe SX50 CAMEBAX (Microsonde de l'Ouest, IFREMER). The analytical conditions consist of 15 KV (accelerating voltage), 15 nA (beam current) and 1 μm beam diameter. The standards used are native copper and iron for Cu and Fe, Ga - As for As, FeS$_2$ (pyrite) for S and ZnS for Zn.

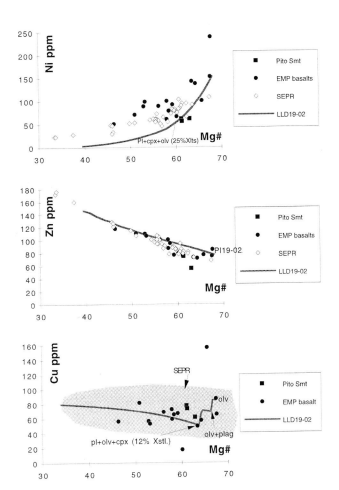

Figure 7. Mg# versus Ni (a), Zn (b), and Cu (c) variation diagram showing the bulk rock compositional variability of the basalts recovered from various geological settings of the Easter Microplate (EMP) ridge segments and Pito Seamount compared with Southeast Pacific Rise data (Bach et al., 1994; Sinton et al., 1991). Cpx = clinopyroxene, pl = plagioclase and olv = olivine. The trend of crystal fractionation (liquid line of descent = LLD) was calculated from the least evolved basalt (sample 19-02) collected from the EMP using the Nielsen (1989) algorithm.

peratures at the submersible were recorded). Thus, our fauna descriptions are based solely on video observations. The biological community of Pito Seamount is not very diversified (Segonzac et al., 1997). The fauna observed on Pito Seamount is reminiscent of the other EPR vent communities (e.g., Geistdoerfer et al., 1995). The closest one observed (to date) is at 18°S (Guinot and Segonzac, 1997), and there is another at 31°S (Guinot et al., 2002). The few *Chorocaris* alvinocaridid shrimp genus present at Pito Seamount, have also been found abundantly in western Pacific back-arc basin sites (Hessler and Lonsdale, 1991; Desbruyères et al., 1994) and rarely observed at the recently described southern EPR

sites at 18°S. The number of inactive vents, the degree of rock freshness, the low diversity and density of the fauna, and the presence of dead bivalves, suggest that the hydrothermal activity of Pito Seamount is presently waning. The present thermal/chemical conditions of the vent fluids at Pito Seamount may also be playing a role (K. L. Von Damm, pers. comm., 1996). Such factors and the possible biogeographic barrier due to the presence of the 300 km-long compressional transform between Pito Deep and the southern EPR (see Figure 1 for geometry), the 2–4 km relief surrounding Pito Seamount (Plate 1b), and the slow spreading rate (~30 mm/yr) might also help explain the absence of numerous vent taxa such as serpulid and vestimentiferan worms, vesicomyid bivalves, cirripeds and zoarcid fishes, usually present on southern EPR vent sites located a few hundred km to the northwest.

The volcanic rift zone emanating from Pito Seamount should continue to develop and propagate in a northwesterly direction due to the microplate rotation resulting from its interaction with the Pacific and Nazca plates (Schouten et al., 1993; Neves et al., 2003). However, because of the slow 30 mm/yr-spreading rate, the volcanic and tectonic cycles at Pito Seamount are much more similar to those of the slow spreading Mid-Atlantic Ridge than the superfast southern EPR. Thus, alternate periods of reduced volcanic and/or hydrothermal activity are expected.

Likely reasons for volcanic activity along the northwest rift zone away from Pito Seamount is as follows: mantle upwelling in this region (Martinez et al., 1991), strong backscatter amplitudes further down the rift zone, seafloor fault patterns and structures, earthquake epicenter pattern (Naar et al., 1991; Engeln and Stein, 1984), and young fresh normal MORB samples obtained during dive 04 (Hekinian et al., 1996) from a small volcano within the center of Pito Deep (22°59.7′S, 111°56.5′W and at depths of 5632 and 5614 m).

That there is a homogenous pattern of strong backscatter amplitudes from the high-quality SeaMARC II 12 kHz side-scan data along most of the northwest rift zone (see ellipse within Plate 1a and interpretation between PS and PD in Figure 2), that the orientation of the rift zone lines up with the Easter-Nazca Euler pole, and that young fresh volcanic samples with normal MORB characteristics were obtained from both ends of the 50-km long volcanic rift zone, all strongly suggest the volcanic rift zone is an active seafloor spreading center. The new data described herein, along with the previous data make this spreading center interpretation less speculative than before. A deeply-towed high-resolution side-scan sonar or photographic/video survey and rock sampling would be necessary to verify the amount of activity along this 50-km rift zone and if indeed this entire rift zone is an incipient seafloor spreading center.

The slow 30 mm/yr spreading rate at Pito Seamount and the surrounding ~3 Ma lithosphere that the East rift has propagated into, is most likely causing rapid lithospheric cooling of the Pito rift area. Furthermore, due to the changing directions of rift propagation during the tectonic evolution of the Easter microplate (Naar and Hey, 1991; Naar et al., 1991), crosscutting faults have been formed (see western portion of Plate 1a near 23°17′S, 111°57′W). This pervasively fractured and thinned lithosphere (Naar et al., 1991; Martinez et al., 1991) suggests that hydrothermal circulation is deep and very efficient. Not only would this increase lithospheric cooling, but it might also explain the very vigorous hydrothermal flux observed at Pito Seamount. Furthermore it suggests that additional hydrothermal activity may be found in this general area, especially along this proposed incipient seafloor spreading axis.

5. CONCLUSIONS

A vigorous hydrothermal black smoker vent exists at 23°19.65′S, 111°38.41′W, near the summit of Pito Seamount (2270 m) located at the northeast boundary of the Easter microplate. Observations during dives 07 and 08 corroborate the prediction based on SeaMARC II side-scan data that Pito Seamount is young and active. Although no *in situ* temperature measurements were made of the vent and no vent fluid samples or biological samples were obtained, we summarize our findings (which are based on visual and geochemical data analysis of rock, sediment, and hydrothermal samples).

The hydrothermal deposits consist essentially of Fe and Zn sulfides. The major crystallizing phases are pyrite and sphalerite followed by small amounts of chalcopyrite. The rock samples from Pito Seamount are mainly depleted N-MORBs. An older flow of enriched E-MORB occurs on the flank of the seamount. Taking into consideration their compatible element variation (Zn, Ni, and Cu), it is inferred that the Pito and the Easter microplate basalts have undergone a similar trend of shallow level crystal-liquid fractionation. Zn increases in the melt during fractionation and Cu shows a scattering of values with a general decrease for the Pito Seamount samples.

On the upper flank of Pito Seamount very few faunas were observed: very few fish, only one corallimorphid sea anemone and an asterid starfish. This type of fauna is usually observed in the Atlantic and Pacific abyssal plains. Nevertheless, the number of brisingid starfish is unusually high on the inactive edifices, and also on the black smoker walls. This could be explained by the local organic enrichment due to the vent-associated community. Alvinellid worms, alvinocaridid shrimps and bythitid fishes, seen surrounding the black smok-

ers are reminiscent of the biological communities from both northern EPR and western Pacific back-arc basin sites, but their number is very limited. The reason for the limited amount may be attributed to three parameters: (1) the isolated location of the active seamount (due to the ~ 2–4 km bathymetric relief and 300 km distance from the southern EPR to the north of the microplate), (2) the present thermal/chemical conditions of the vent fluid, and/or (3) the physical-chemical fluctuations related to volcanic and hydrothermal activity. In any event, these hypotheses can only be tested if biological samples are collected during future expeditions to this area.

We speculate that the black smoker is the result of deep extensive crosscutting normal faults formed by the large-scale rifting and a very robust magmatic supply being supplied from up-welling asthenosphere resulting from the extensive tectonic thinning of the Pito Deep area. We also speculate that the northwest rift zone of Pito Seamount trending 310° is the site of an incipient seafloor spreading axis because of the young fresh glassy normal MORB sampled at the summit of Pito Seamount and at a small volcanic mound within the base of Pito Deep. Further support comes from homogenous strong backscatter returns from the 12 kHz SeaMARC II side-scan data along the majority of this volcanic rift zone.

Acknowledgments. We thank the IFREMER personnel, crew, and Captain of the R/V Nadir for a successful and safe expedition, Ronan Apprioual for serving as rock curator at sea, and the people of Rapa Nui for their kind hospitality. We thank Craig Cary for discussions at an earlier stage of this manuscript. Martha G. Kuykendall, Zhengrong J. Liu, Chad Edmington, and especially Shihadah Saleem, helped generate and improve the figures. DFN acknowledges support from the National Science Foundation (OCE9302802) and the University of South Florida Research Council. We are grateful to Dan Fornari, an anonymous reviewer, and the associate editors for their helpful comments and patience. We are indebted to Marcel Bohn (Microsonde de L'Ouest) for helping with the microprobe analyses.

REFERENCES

Ackermand, D., R. Hekinian, and P. Stoffers, Magmatic sulfides and oxides from the Pitcairn hotspot (South Pacific), *Miner. Petrol.*, *64*, 149–162, 1998.

Anderson, R. N., D. W. Forsyth, P. Molnar, and J. Mammerickx, Fault plane solutions of earthquakes on the Nazca plate boundaries and the Easter plate, *Earth Planet. Sci. Lett.*, *24*, 188–202, 1974.

Auzende, J.-M., V. Ballu, R. Batiza, D. Bideau, M.-H. Cormier, Y. Fouquet, P. Geistdoerfer, Y. Lagabrielle, J. Sinton, and P. Spadea, Activité magmatique, tectonique et hydrothermale actuelle sur la Dorsale Est Pacifique entre 17° et 19°S (campagne Naudur), *C. R. Acad. Sci., Paris*, *319*(sér. II), 811–818, 1994.

Bach, W., E. Hegner, J. Erzinger, and M. Satir, Chemical and isotopic variations along the superfast spreading East Pacific Rise from 6° to 30°S, *Contr. Mineral. Petrol.*, *116*, 365–380, 1994.

Bäcker, H., J. Lange, and V. Marchig, Hydrothermal activity and sulfide formation in axial valleys of the East Pacific Rise crest between 18° and 22°S, *Earth Planet. Sci. Lett.*, *72*, 9–22, 1985.

Bird, R. T. and D. F. Naar, Intratransform origins of mid-ocean ridge microplates. *Geology*, *22*, 987–990, 1994.

Butterfield, D. A., R. E. McDuff, M. J. Mottl, M. D. Lilley, J. E. Lupton, and G. J. Massoth, Gradients in the composition of hydrothermal fluids from the Endeavour segment vent field: Phase separation and brine loss. *J. Geophys. Res.*, 99: 9561–9583, 1994.

Butterfield, D. A., I. R. Jonasson, G. J. Massoth, R. A. Feely, K. K. Roe, R. Embley, J. F. Holden, R. E. McDuff, M. D. Lilley, and J. R. Delaney, Seafloor eruptions and evolution of hydrothermal fluid chemistry, *Philosophical Transactions of the Royal Society of London, A355*, 369–386, 1997.

DeMets, C., R. G. Gordon, D. F. Argus, and S. Stein, Effect of recent revisions to the geomagnetic reversal time scale on estimates of current plate motions, *Geophys. Res. Lett.*, *21*, 2191–2194, 1994.

Desbruyères, D., A.-M. Alayse-Danet, S. Ohta, S. P. O. Biolau, and Starmer Cruises, Deep-sea hydrothermal communities in Southern Pacific back-arc basins (the North Fiji and Lau basins): Composition, microdistribution and food-web, *Marine Geology*, *116*, 227–242, 1994.

Duckworth, R. C., R. Knott, A. E. Fallick, D. Rickard, B. J. Murton, and C. Van Dover, Mineralogy and sulfur isotope geochemistry of Broken Spur sulphides, 29°N, Mid-Atlantic Ridge, in *Hydrothermal Vents and Processes*, edited by L. M. Parson, C. L. Walker, and D. R. Dixon, *J. Geol. Soc. London Spec. Publ. 87*, 175–189, 1995.

Engeln, J. F. and S. Stein, Tectonics of the Easter plate, *Earth Planet. Sci. Lett.*, *68*, 259–270, 1984.

Fouquet, Y., J.-M. Auzende, V. Ballu, R. Batiza, D. Bideau, M.-H. Cormier, P. Geistdoerfer, Y. Lagabrielle, J. Sinton, and P. Spadea, Hydrothermalisme et sulfures sur la dorsale du Pacifique est entre 17° et 19°S (campagne Naudur), *C. R. Acad. Sci., Paris, 319(sér. II)*, 1399–1406, 1994.

Francheteau, J., P. Patriat, J. Segoufin, R. Armijo, M. Doucoure, A. Yelles-Chaouche, J. Zukin, S. Calmant, D. F. Naar, and R. C. Searle, Pito and Orongo fracture zones: The northern and southern boundaries of the Easter Microplate (Southeast Pacific), *Earth Planet. Sci. Lett.*, *89*, 363–374, 1988.

Geistdoerfer, P., J.-M. Auzende, V. Ballu, R. Batiza, D. Bideau, M.-H. Cormier, Y. Fouquet, Y. Lagabrielle, J. Sinton, and P. Spadea, Hydrothermalisme et communautés animales associées sur la dorsale du Pacifique oriental entre 17° et 19°S (campagne Naudur, décembre 1993), *C. R. Acad. Sci., Paris, 320(sér. IIa)*, 47–54, 1995.

Govindaraju, J., Compilation of working values and sample description for 272 geostandards, *Geostand. Newsl. Spec. Issue, 13*, 1–113, 1989.

Guinot, D. and M. Segonzac, Description d'un crabe hydrothermal nouveau du genre *Bythograea* (Crustacea, Decapoda, Brachyura) et remarques sur les Bythograeidae de la dorsale du Pacifique oriental, *Zoosystema, 19(1)*, 117–145, 1997.

Guinot, D., L. A. Hurtado, and R. Vrijenoek, New genus and species of brachyuran crab from the southern East Pacific Rise (Crustacea Decapoda Brachyura Bythograeidae), *Comptes Rendus Biologies, 325(11)*, 1143–1152, 2002.

Handschumacher, D. W., R. H. Pilger, J. A. Foreman, and J. R. Campbell, Structure and evolution of the Easter plate, *Mem. Geol. Soc. Am.*, *154*, 63–76, 1981.

Hannington, M. D. and S. D. Scott, Sulfide equilibria as a guide to gold mineralization in volcanogenic massive sulfides: evidence from sulfide mineralogy and the composition of sphalerite, *Economic Geology, 84*, 1978–1995, 1989.

Hekinian, R. and D. Bideau, Volcanism and mineralization of the oceanic crust on the East Pacific Rise, in *Metallogeny of basic and ultrabasic rocks*, edited by M. J. Gallagerm, R. A. Ixer, C. R. Neary, and H. M. Prichard, published by Institution of Mining and Metallogeny, 44 Portland place, London, 3–20, 1986.

Hekinian, R., J. Francheteau, R. Armijo, J.-P Cogne, M. Constantin, J. Girardeau, R. Hey, D. F. Naar, and R. Searle, Petrology of the Easter Microplate region in the South Pacific, *J. of Volcanology and Geothermal Research*, *72*, 259–289, 1996.

Herron, E. M., Sea-floor spreading and the Cenozoic history of the East-Central Pacific, *Geol. Soc. Am. Bull.*, *83*, 1672–1692, 1972a.

Herron, E. M., Two small crustal plates in the South Pacific near Easter Island, *Nature, 240*, 35–37, 1972b.

Hessler, R. R. and Lonsdale, P. F., Biogeography of Mariana Trough hydrothermal vent communities, *Deep-Sea Research, 38*, 185–199, 1991.

Hey, R. N., D. F. Naar, M. C. Kleinrock, J. P. Morgan, E. Morales, and J.-G. Schilling, Microplate tectonics along a superfast seafloor spreading system near Easter Island, *Nature, 317*, 320–325, 1985.

Hey, R. N., P. D. Johnson, F. Martinez, J. Korenaga, M. L. Somers, Q. J. Huggett, T. P. LeBas, R. I. Rusby, and D. F. Naar, Plate boundary reorganization at a large-offset, rapidly propagating rift, *Nature, 378*, 167–170, 1995.

Koski, R. A., D. A. Clague, and E. Oudin, Mineralogy and chemistry of massive sulfide deposit from the Juan de Fuca Ridge, *Geol. Soc. Am. Bull.*, *95*, 930–945, 1981.

Lilley, M. D., D. A. Butterfield, E. J. Olson, J. E. Lupton, S. E. Mako, and R. E. McDuff, Sediment evolvement and phase separation in volcanic-hosted mid-ocean ridge hydrothermal system, *Nature, 364*, 45–47, 1993.

Martinez, F., D. F. Naar, T. B. Reed IV, and R. N. Hey, Three-dimensional SeaMARC II, gravity, and magnetics study of large-offset rift propagation at the Pito Rift, Easter microplate, *Mar. Geophys. Res.*, *13*, 255–285, 1991.

Naar, D. F. and R. N. Hey, Fast rift propagation along the East Pacific Rise near Easter Island, *J. Geophys. Res.*, *91*, 3425–3438, 1986.

Naar, D. F. and R. N. Hey, Pacific-Easter-Nazca plate motions, in *Evolution of Mid Ocean Ridges, IUGG Symposium 8*, edited by J. M. Sinton, *AGU Geophysical Monograph*, *57*, 9–30, 1989.

Naar, D. F. and R. N. Hey, Tectonic evolution of the Easter Microplate, *J. Geophys. Res.*, *96*, 7961–7993, 1991.

Naar, D. F., F. Martinez, R. N. Hey, T. B. Reed IV, and S. Stein, Pito Rift: How a large-offset rift propagates, *Mar. Geophys. Res.*, *13*, 287–309, 1991.

Nielsen, R. L., A model for the simulation of combined major and trace element liquid lines of descent, *Geochim. Cosmochim. Acta*, *52,* 27–38, 1989.

Neves, M. C., R. C. Searle, and M. H. P. Bott, Easter microplate dynamics, *J. Geophys. Res.*, *108*, 2213, doi:10.1029/2001JB000908, 2003.

Renard, V., R. Hekinian, J. Francheteau, R. D. Ballard, and H. Backer, Submersible observations at the axis of the ultra-fast-spreading East Pacific Rise (17°30′S to 21°30′S), *Earth Planet. Sci. Lett.*, *75*, 339–353, 1985.

Rusby, R. I., GLORIA and other geophysical studies of the tectonic pattern and history of the Easter Microplate, southeast Pacific, in *Ophiolites and their Modern Oceanic Analogues,* edited by L. M. Parson, B. J. Murton, and P. Browning, *Geol. Soc. London Spec. Publ., 60,* 81–106, 1992.

Rusby, R. I. and R. C. Searle, Intraplate thrusting near the Easter Microplate, *Geology, 21,* 311–314, 1993.

Rusby, R. I. and R. C. Searle, R. C., A history of the Easter microplate, 5.25 Ma to Present, *J. Geophys. Res., 100,* 12617–12640, 1995.

Schouten, H., K. D. Klitgord, and D. G. Gallo, Edge-driven microplate kinematics, *J. Geophys. Res., 98,* 6689–6701, 1993.

Searle, R. C., R. I. Rusby, J. Engeln, R. N. Hey, J. Zukin, P. M. Hunter, T. P. LeBas, H.-J. Hoffman, and R. Livermore, Comprehensive sonar imaging of the Easter Microplate, *Nature, 341,* 701–705, 1989.

Searle, R. C., R. T. Bird, R. I. Rusby, and D. F. Naar, The development of two oceanic microplates: Easter and Juan Fernandez Microplates, East Pacific Rise, *Geological Society of London Journal, 150,* 965–976, 1993.

Segonzac, M., R. Hekinian, J.-M. Auzende, and J. Francheteau, Recently Discovered Animal Communities on the Southeast Pacific Rise (17–19°S) and Easter Microplate Region, *Cahiers Biologie Marine, 38,* 140–141, 1997.

Seyfried, W. E. Jr. and K. Ding, Phase equilibria in subseafloor hydrothermal systems: A review of the role of redox, temperature, pH, and dissolved Cl on the chemistry of hot spring fluids at mid-ocean ridges, in *Physical, Chemical, Biological, and Geological Interactions within Seafloor Hydrothermal Systems*, edited by S. Humphris, J. Lupton, L. Mullineaux, and R. Zierenberg, *AGU Monograph, 91,* 248–272, 1995.

Sinton, J. M., S. M. Smaglik, J. J. Mahoney, and K. C. Macdonald, Magmatic processes at super fast spreading oceanic ridges: Glass variations along the East Pacific Rise, 13°S–23°S, *J. Geophys. Res., 96,* 6133–6155, 1991.

Verati, C. J., J. Lancelot, and R. Hekinian, Pb isotope study of black-smokers and basalts from Pito Seamount site (Easter Microplate), *Chem. Geol., 155,* 45–63, 1999.

Von Damm, K. L., Controls on the chemistry and temporal variability of seafloor hydrothermal fluids, In *Physical, Chemical, Biological, and Geological Interactions within Seafloor Hydrothermal Systems*, edited by S. Humphris, J. Lupton, L. Mullineaux, and R. Zierenberg, *AGU Monograph, 91,* 222–247, 1995a.

Von Damm, K. L., Temporal and compositional diversity in seafloor hydrothermal fluids, *Reviews of Geophysics, Supplement, U.S. National Report to IUGG 1991–1994,* 1297–1305, 1995b.

Von Damm, K. L., L. G. Buttermore, S. E. Oosting, A. M. Bray, D. J. Fornari, M. D. Lilley, and W. C. Shanks III, Direct observation of the evolution of a seafloor "black smoker" from vapor to brine, *Earth Planet. Sci. Lett., 149,* 101–112, 1997.

Von Damm, K. L., M. D. Lilley, W. C. Shanks III, M. Brockington, A. M. Bray, K. M. O'Grady, E. Olson, A. Graham, G. Proskurowski, and the SouEPR Science Party, Extraordinary phase separation and segregation in vent fluids from the southern East Pacific Rise, *Earth Planet. Sci. Lett., 206,* 365–378, 2003.

Zukin, J. and J. Francheteau, A tectonic test of instantaneous kinematics of the Easter Microplate, *Oceanologica Acta (special edition), 10,* 183–198, 1990.

David F. Naar, College of Marine Science, University of South Florida, 140 Seventh Avenue South, St. Petersburg, FL 33701-5016. (naar@usf.edu)